EXCLUSIVE REACTIONS

at High Momentum Transfer IV

EXCLUSIVE REACTIONS

at High Momentum Transfer IV

Proceedings of the 4th Workshop

Thomas Jefferson National Accelerator Facility, USA, 18–21 May 2010

Editor

Anatoly Radyushkin

Old Dominion University, USA & Jefferson Lab, USA

World Scientific

NEW JERSEY • LONDON • SINGAPORE • BEIJING • SHANGHAI • HONG KONG • TAIPEI • CHENNAI

Published by

World Scientific Publishing Co. Pte. Ltd.

5 Toh Tuck Link, Singapore 596224

USA office: 27 Warren Street, Suite 401-402, Hackensack, NJ 07601

UK office: 57 Shelton Street, Covent Garden, London WC2H 9HE

British Library Cataloguing-in-Publication Data
A catalogue record for this book is available from the British Library.

EXCLUSIVE REACTIONS AT HIGH MOMENTUM TRANSFER IV
Proceedings of the 4th Workshop

ISBN-13 978-981-4329-55-2
ISBN-10 981-4329-55-X

Printed in Singapore by World Scientific Printers.

ORGANIZING COMMITTEE

Harut Avakian (Jefferson Lab)
Volker Burkert (Jefferson Lab)
Carl Carlson (William & Mary)
Latifa Elouadrhiri (Jefferson Lab)
Rolf Ent (Jefferson Lab)
Tanja Horn (Catholic University of America)
Mark Jones (Jefferson Lab)
Victor Mokeev (Jefferson Lab)
Anatoly Radyushkin (Old Dominion & Jefferson Lab, Co-Chair)
Paul Stoler (Rensselaer, Co-Chair)
Christian Weiss (Jefferson Lab)
Bogdan Wojtsekhowski (Jefferson Lab)

PREFACE

These are the proceedings of the Workshop on Exclusive Reactions at High Momentum Transfer, which was held 18–21 May, 2010 at the Thomas Jefferson National Accelerator Facility (Jefferson Lab) in Newport News, Virginia, USA. The workshop was sponsored by Jefferson Lab.

The workshop was open to all interested scientists, and embraced both theorists and experimentalists. The Organizing Committee thanks the speakers and participants at the conference for the stimulating atmosphere that prevailed. We trust the written versions of the talks presented in the volume will reproduce some of that feeling.

Thanks are also due to the people who worked hard to organize the conference and make it a success. The Conference group at Jefferson Lab was indispensable. Allow us to give special thanks to Ruth Bizot, Cynthia Lockwood, Marty Hightower, Stephanie Vermeire, Mary Fox and Joanna Griffin for all their efforts before, during and since the conference.

Anatoly Radyushkin
Paul Stoler
October 2010

CONTENTS

HARD EXCLUSIVE PROCESSES : PERSPECTIVES

MARC VANDERHAEGHEN*

Institut für Kernphysik, University Mainz,
D-55099 Mainz, Germany,
**E-mail: marcvdh@kph.uni-mainz.de*

The hadron structure information contained in hard exclusive processes is discussed. I address the nucleon light-front densities as obtained from nucleon form factor data, its generalizations for particles of higher spin, and the correlation of quark transverse position with longitudinal momentum, as encoded in GDPs. The model-independent information on quark GPDs which has so far been extracted from DVCS data is also reviewed.

Keywords: Elastic scattering, Form Factors, Generalized Parton Distributions

1. Introduction

Hard exclusive processes using electroweak probes provide unique tools to investigate the internal quark-gluon structure of hadrons. By far, the best studied process is the elastic scattering of a lepton beam off a hadron, in which the matrix element of a well defined quark-gluon operator (in this case the vector operator $\bar{q}\gamma^\mu q$) is measured over the hadronic state. Experimentally, it was pioneered more than 50 years ago,[1] when the first form factor (FF) measurement of the proton allowed to get a first glimpse into the spatial distribution of electric charges inside the proton. Since then, electromagnetic FFs have been studied extensively, and became a high precision source of information on the nucleon structure; see Refs.[2–5] for recent reviews. Such lepton scattering experiments have also allowed to study e.g. transitions between nucleons (N) and its excited states (N^*). With the maturing of lattice QCD techniques, the non-perturbative matrix elements which are probed by these processes can be calculated ab initio, see e.g. Ref. [6] for a recent comprehensive review and references. Furthemore, the lattice calculations allow to go beyond a comparision with measured hadron structure observables, and predict such quantities as e.g. $\Delta(1232)$ resonance FFs,[7] which are so far not accessible in experiment.

Besides precision studies of electroweak FFs, it has been realized in more recent years that we are not limited in elucidating the hadron structure to probes such as photons or W and Z bosons. The phenomenon of asymptotic freedom of QCD, meaning that at short distances the interactions between quarks and gluons become weak, provides us with more sophisticated QCD operators to explore the structure of hadrons. Such operators can be accessed by selecting a small size configuration of quarks and gluons, provided by a hard reaction, such as deep inelastic scattering (DIS), or hard exclusive reactions such as deeply virtual Compton scattering (DVCS), $\gamma^*(q_h) + N(p) \rightarrow \gamma(q') + N(p')$, where the virtual photon momentum q_h is the hard scale, see Fig. 1. The common important feature of such hard reactions is the possibility to separate clearly the perturbative and non-perturbative stages of the interactions : this is the so-called factorization property, see Refs.[8–11] for pioneering works. The non-perturbative stage of such hard exclusive electroproduction processes is described by universal objects, so-called generalized parton distributions (GPDs), see Refs.[12–16] for recent reviews and references.

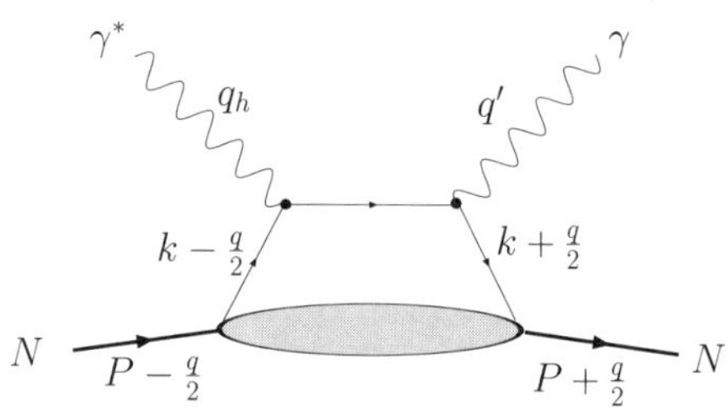

Fig. 1. 'Handbag' diagram for the nucleon DVCS process. Provided the virtuality of the initial photon is sufficiently large, the QCD factorization theorem allows to express the total amplitude as the convolution of a Compton process at the quark level and a non-perturbative amplitude parameterized in terms of GPDs (lower blob).

Among the new matrix elements which GPDs allow to study are e.g. FFs of the energy-momentum tensor of hadrons, allowing for an access to the angular momentum of quark (and gluons) in the nucleon[10] . In the forward direction, the GPDs reduce to forward parton distributions measuring the momentum densities of quarks in a hadron, whereas integration of GPDs over all parton momentum fractions gives us the elastic FFs. In this way GPDs provide a bridge connecting the (longitudinal) momentum information in parton distributions with (transverse) position information contained in elastic FFs, allowing for a quark imaging of the nucleon.[17] It has been realized that the full correlation between the quark three-momentum and its transverse position is contained in Wigner distributions.[18,19] Integrating the Wigner distributions over coordinate space allows to extract quark three-momentum distributions resulting in transverse momentum dependent distributions (TMDs), see e.g. Ref. [20].

On the experimental side, experimental programs measuring hard exclusive processes such as deeply virtual Compton scattering have been completed over the past decade at HERMES, HERA, and JLab@6GeV, indicating a dominance of the leading twist-2 mechanism. Such experiments allow for first, model independent, extractions of Compton form factors (CFFs) involving GPDs, as will be discussed below.

In Section 2, we review the empirical information on nucleon light-front densities as obtained from nucleon form factor data. This will be generalized in Section 3 for particles of spin higher than 1/2, allowing to quantify the shape of hadrons from its electromagnetic moments. In Section 4, the correlation of the transverse quark position with its longitudinal momentum, as encoded in GDPs, will be discussed. In Section 5, we briefly discuss the model-independent information on quark GPDs which has so far been extracted from DVCS data and discuss the dispersion relation formalism as a tool to constrain the analyses of forthcoming data.

2. Nucleon form factors and densities

Hadrons such as protons or neutrons are relativistic bound states, made up of almost massless quarks. The three valence quarks, which make up for the proton quantum numbers, constitute only a few percent of the total proton mass. In such a system, the wave function contains, besides the three valence quark Fock component $|qqq\rangle$, also components where additional $q\bar{q}$ pairs, so-called sea-quarks, or (transverse) gluons are excited, leading to an infinite tower of $|qqqq\bar{q}\rangle$, $|qqqg\rangle$, ... components. When probing such a system using electron scattering, the exchanged virtual photon will couple to any quark, both valence and sea. In addition, the virtual photon can also produce a $q\bar{q}$ pair, leading e.g. to a transition from a $3q$ state in the initial wave function to a $5q$ state in the final wave function. Such processes leading to non-diagonal overlaps between initial and final wave functions are not positive definite, and do not allow for a simple probability interpretation of the density anymore.

This relativistic dynamical effect of pair creation or annihilation fundamentally hampers the interpretation of density and any discussion of size and shape of a relativistic quantum system. An interpretation in terms of the concept of a density requires suppressing the contributions due to pair production. This is possible when viewing the hadron from a light front, allowing to describe the hadron state by an infinite tower of light-front wave functions.[21] Consider the electromagnetic (e.m.) transition from an initial hadron (with four-momentum p) to a final hadron (with four-momentum p') when viewed from a light front moving towards the hadron. Equivalently, this corresponds with a frame where the hadrons have a large momentum component along the z-axis chosen along the direction of the hadrons average momentum $P = (p + p')/2$. One then defines the light-front plus (+) component by $a^{\pm} \equiv a^0 \pm a^3$, which is always a positive quantity for the quark or anti-quark four-momenta in the hadron. When we now view the hadron in a so-called Drell-Yan frame,[22] where the virtual photon four-momentum q satisfies $q^+ = 0$, energy-momentum conservation will forbid processes where this virtual photon splits into a $q\bar{q}$ pair. Such a choice is possible for a space-like virtual photon, and its virtuality is then given by $q^2 = -\vec{q}_\perp^{\,2} \equiv -Q^2 < 0$, where $\vec{q}_\perp$is the transverse photon momentum (lying in the xy-pane). In such a frame, the virtual photon only couples to forward moving partons, and it allows to define a proper density operator through the + component of the four-current by $J^+ = J^0 + J^3$.[23] For quarks it is given by

$$J^+ = \bar{q}\gamma^+ q = 2q_+^\dagger q_+, \qquad \text{with} \qquad q_+ \equiv \frac{1}{4}\gamma^-\gamma^+ q, \tag{1}$$

where we introduced the q_+ fields through a field redefinition from the initial quark fields q, involving the $\pm$ components of the Dirac γ-matrices. The relativistic density operator J^+, as defined in Eq. (1), is a positive definite quantity. For systems consisting of e.g. light u and d quarks, multiplying this current with the quark charges yields a quark charge density operator given by $J^+(0) = +2/3\,\bar{u}(0)\gamma^+ u(0) - 1/3\,\bar{d}(0)\gamma^+(0)d(0)$. Using such quark charge density operator, one can then define quark (transverse) charge densities in a hadron as[17,24] :

$$\rho_\lambda(b) \equiv \int \frac{d^2\vec{q}_\perp}{(2\pi)^2}\, e^{-i\,\vec{q}_\perp\cdot\vec{b}}\, \frac{1}{2P^+}\, \langle P^+, \frac{\vec{q}_\perp}{2}, \lambda \,|\, J^+(0) \,|\, P^+, -\frac{\vec{q}_\perp}{2}, \lambda\rangle, \tag{2}$$

where λ denotes the hadron (light-front) helicity. In the two-dimensional Fourier transform of Eq. (2), the two-dimensional vector $\vec{b}$ denotes the quark position (in the xy-plane) from the transverse center-of-momentum (c.m.) of the hadron. It represents the position variable conjugate to the hadron relative transverse momentum, which equals the photon momentum $\vec{q}_\perp$.

The quantity $\rho_\lambda(b)$ has the interpretation of the two-dimensional (transverse) quark charge density at a distance $b = |\vec{b}|$ from the transverse c.m. of the hadron with helicity λ. In the light-front frame, it corresponds with the projection of the charge density in the hadron along the line-of-sight.

The quark charge densities in Eq. (2) do not fully describe the e.m. structure of the hadron, e.g. for a spin 1/2 hadron the densities with $\lambda = \pm 1/2$ yield the same information. We do know however that there are two independent e.m. FFs describing the structure of the nucleon. In general, a particle of spin S is described by $(2S + 1)$ c.m. moments. To fully describe the structure of a hadron one also needs to consider the charge densities in a transversely polarized hadron state, denoting the transverse polarization direction by $\vec{S}_\perp = \cos\phi_S\,\hat{e}_x + \sin\phi_S\,\hat{e}_y$. The transverse charge densities can be defined through matrix elements of the density operator J^+ in eigenstates of transverse spin as [25–27] :

$$\rho_{T\,s_\perp}(\vec{b}) \equiv \int \frac{d^2\vec{q}_\perp}{(2\pi)^2}\, e^{-i\,\vec{q}_\perp\cdot\vec{b}}\, \frac{1}{2P^+}\, \langle P^+, \frac{\vec{q}_\perp}{2}, s_\perp | J^+ | P^+, \frac{-\vec{q}_\perp}{2}, s_\perp \rangle, \quad (3)$$

where $s_\perp$ is the hadron spin projection along the direction of $\vec{S}_\perp$. Whereas the density ρ_λ for a hadron in a state of definite helicity is circularly symmetric for all spins, the density $\rho_{T\,s_\perp}$ depends also on the orientation of the position vector $\vec{b}$, relative to the transverse spin vector $\vec{S}_\perp$. Therefore it contains the information on the hadron shape, as projected on a plane perpendicular to the line-of-sight.

In Fig. 2, the transverse charge densities in a proton and neutron are extracted based on the empirical information on the nucleon e.m. FFs. The unpolarized proton density shows a broad positive central density, whereas the neutron charge density reveals the well known negative contribution at large distances, around 1.5 fm, due to the pion cloud, a positive contribution at intermediate b values, and a negative core at b values smaller than about 0.3 fm, as found in Ref. [24]. One can understand the negative value of the neutron $\rho_0(b = 0)$ from Eq. (2) and the observation that over the whole measured Q^2 range the neutron FF F_1 is negative. One notices furthermore from the lower panels in Fig. 2 that polarizing the proton along the x-axis leads to an induced electric dipole moment along the positive y-axis equivalent to the anomalous magnetic moment μ_N. This field pattern due to the induced electric dipole is a consequence of special relativity. The nucleon spin along the x-axis is the source of a magnetic dipole field $\vec{B}$. An observer moving towards the nucleon with velocity $\vec{v}$ will see an electric dipole field pattern with $\vec{E}' = -\gamma(\vec{v} \times \vec{B})$ giving rise to the observed asymmetry. For the neutron, one notices its charge density gets displaced significantly due to its large negative anomalous magnetic moment $\mu_N = -1.91$ yielding an induced electric dipole moment along the negative y-axis.

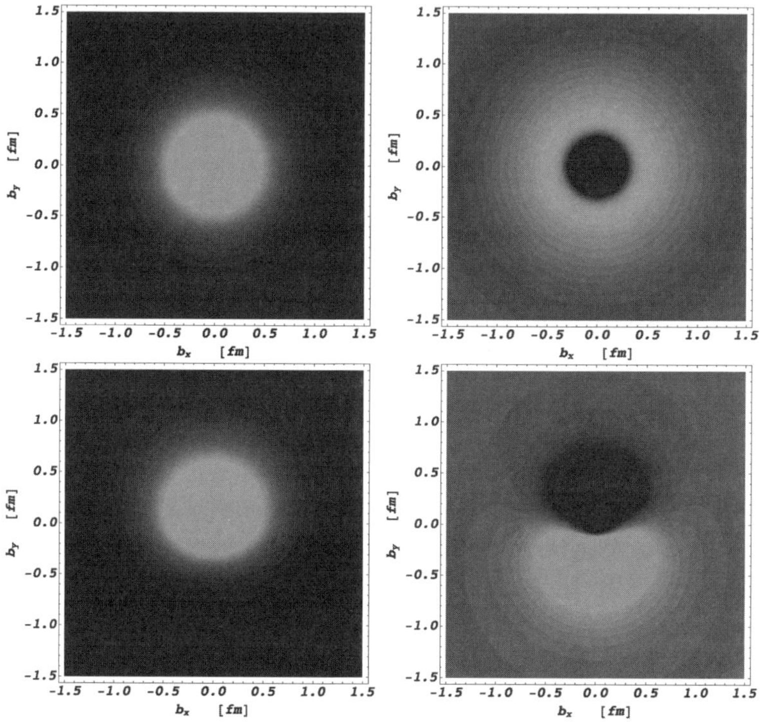

Fig. 2. Quark transverse charge densities in the *proton* (left panels) and *neutron* (right panels). The upper panels show the density in the transverse plane for a nucleon with definite helicity. The lower panels for a nucleon polarized along the x-axis. The light (dark) regions correspond with largest (smallest) values of the density. For the proton e.m. FFs, the empirical parametrization of Arrington *et al.*[28] is used. For the neutron e.m. FFs, the empirical parametrization of Bradford *et al.*[29] is used.

Combining the charge densities in proton and neutron, one can extract the charge densities for u- and d-quark in a proton, as shown in Fig. 3. If the down and up quarks would have the same spatial distribution in the proton, the ratio as displayed in Fig. 3 (left panel) would be one. We see however that in the center region of the proton, at distances smaller than about 0.5 fm, down quarks are less abundant than up quarks (when compensating for the different normalization of u and d quarks). The down quarks have a much wider distribution and are shifted to larger distances, dominating over up quarks between 0.5 to 1.5 fm. In the neutron, the roles of u and d-quarks are interchanged.

One can understand this behavior from empirical parametrizations of forward parton distributions and GPDs, which display a much faster fall-off for down quarks as for up quarks at large quark momentum fractions, see right panel of Fig. 3.

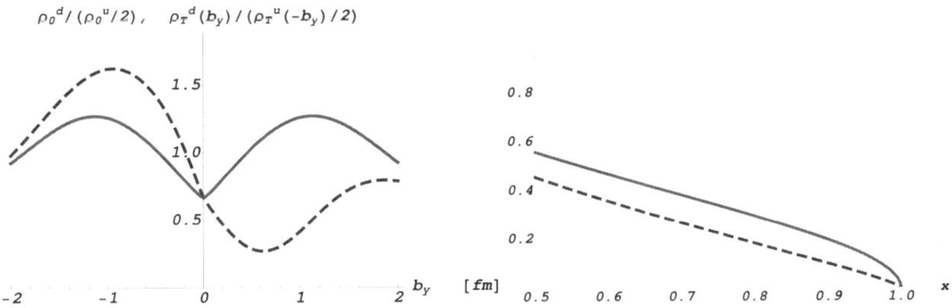

Fig. 3. Left panel : ratio of d/u quark charge density in a proton of definite helicity (solid red curve) and when polarized along the x-axis (dashed blue curve). The u-quark densities are divided by 2. The densities are based on a model[30] for GPDs H and E. Right panel : behavior of the ratio of parton distributions $d/(u/2)$ (solid red curve), and of the ratio of forward proton helicity flip GPDs $(e_d/\kappa_d)/(e_u/\kappa_u)$ (dashed blue curve), with $\kappa_u = 1.673$ and $\kappa_d = -2.033$.

As these light-front densities require us to develop some new intuition, we would like to close this section with some additional remarks. The light-front wave functions and light-front densities discussed above are defined at equal light-front time ($x^+ = 0$) of their constituents. When constituents move non-relativistically, it does not make a difference whether they are observed at equal time ($t = 0$) or equal light-front time ($x^+ = 0$), since the constituents can only move a negligible small distance during the small time interval that a light-ray needs to connect them. This is not the case, however, for bound systems of relativistic constituents such as hadrons, or even when considering bound systems of non-relativistic constituents but with the system in relativistic motion. Considering the latter, one may ask, for example, how the equal-time wave functions of bound state systems such as the hydrogen atom or positronium, which are non-relativistic bound states in QED at leading order in the fine structure constant α, transform when considering them in relativistic motion.[31,32] Positronium has been studied in Ref. [31], by using the exact Bethe-Salpeter equation. It was found that the equal-time wave function of its leading $|e^+e^-\rangle$ Fock state contracts as expected from classical relativity. However calculating the bound state in relativistic motion also necessitates the inclusion of an $|e^+e^-\gamma\rangle$ component in the wave function, whose probability is of order $O(\alpha)$. Unlike classical transformation laws, it was found that this photon amplitude, which can be considered as a quantum fluctuation, does not contract.[31] Therefore, the Fock components have a more complex shape change under relativistic boosts than a flattening of their distributions along the direction of the boost. Only very recently, relativistic bound states, such as $q\bar{q}$ states in QCD have also been studied at lowest order in $\hbar$.[32] It was found that the resulting equal-time wave functions have unique Lorentz transformation properties, ensuring the correct dependence of the bound state energy on the center-of-mass momentum.

3. Higher spin transverse densities

When investigating the structure of hadrons of spin higher than 1/2, one can obtain from the information on $\rho_{T\,s_\perp}$ of Eq. (3) e.m. moments quantifying its shape. As an example, for a hadron with spin $S > 1/2$ and transverse spin orientation $\vec{S}_\perp = \hat{e}_x$, the electric quadrupole moment is given by :

$$Q_{s_\perp} \equiv e \int d^2\vec{b}\,(b_x^2 - b_y^2)\,\rho_{T\,s_\perp}(\vec{b}). \tag{4}$$

It has been noticed[7,26] that from the requirement that for a particle without internal structure the e.m. moments of its transverse charge densities should vanish, the natural (non-zero) values of these e.m. moments can be extracted. The generalization to arbitrary spin has been demonstrated in Ref. [27]. For the case of a hadron as the $\Delta(1232)$ with an internal structure, it is thus interesting to observe from Eq. (4) that $Q^{\Delta}_{s_\perp}$ is only sensitive to the anomalous parts of the spin-3/2 magnetic dipole and electric quadrupole moments.

As an example of the predictive power of lattice QCD, we show in Fig. 4 the transverse charge density of Eq. (3) for a $\Delta^+(1232)$ that has a transverse spin projection $s_\perp = +3/2$. This charge density is obtained from the Δ e.m. FFs, which were calculated within lattice QCD.[7] It is seen from Fig. 4 that the quark charge density in a Δ^+ in a state of transverse spin projection $s_\perp = +3/2$ is elongated along the axis of the spin (prolate deformation) when observed from a light-front.

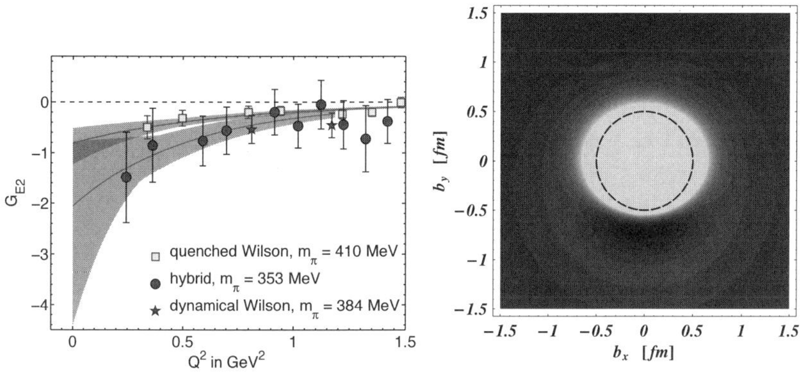

Fig. 4. Left panel : lattice QCD results for the $\Delta^+(1232)$ electric quadrupole FF G_{E2} in three lattice QCD simulations.[7] Right panel : results for the quark transverse charge density $\rho^{\Delta}_{T\,\frac{3}{2}}$ in a $\Delta^+(1232)$, polarized along the positive x-axis. The light (dark) regions correspond to the largest (smallest) values of the density. In order to see the deformation more clearly, a circle of radius 0.5 fm is drawn for comparison. The density is obtained from the quenched lattice QCD results at $m_\pi = 410$ MeV for the Δ e.m. FFs.

4. Coordinate and momentum space information contained in GPDs

The light-front densities discussed above provide an imaging of quark charges in the hadron. This spatial dependence is correlated with quark longitudinal momentum fractions in GPDs. It is the interplay between the x and t-dependence of the GPDs which contains new nucleon structure information beyond the information encoded in forward parton distributions depending only on x, or FFs depending only on t. A Fourier transform of the t-dependence of GPDs accesses the distributions of partons with a fixed momentum fraction x in the transverse plane.[17] For the vector GPD H^q (for skewedness value $\xi = 0$), this Fourier integral reads as:

$$H^q(x, \vec{b}_\perp) \equiv \int \frac{d^2\vec{q}_\perp}{(2\pi)^2}\, e^{-i\vec{q}_\perp\cdot\vec{b}_\perp}\, H^q(x, \xi = 0, t = -\vec{q}_\perp^{\,2}). \tag{5}$$

Such impact parameter GPDs have the physical meaning of measuring the probability to find a quark which carries longitudinal momentum fraction x at a transverse position $\vec{b}_\perp$ (relative to the transverse center-of-momentum) in a nucleon, see Refs.[17,33] Fig. 5 shows the GPD H^u , for an u-quark, within a GPD Regge based parameterization.[30] The distribution has a characteristic behavior, approaching a delta function behavior at large x, and becoming increasingly broad when going to smaller values of x. This behavior is reproduced by lattice QCD results as is shown in Fig. 6 for a recent state-of-the-art calculation[34] of three moments of the isovector GPD H^{u-d}. It is clearly seen that the higher moments, which become increasingly sensitive to the large-x region, show a much flatter t-dependence, resulting in a narrower spatial distribution (at large-x) upon Fourier transforming.

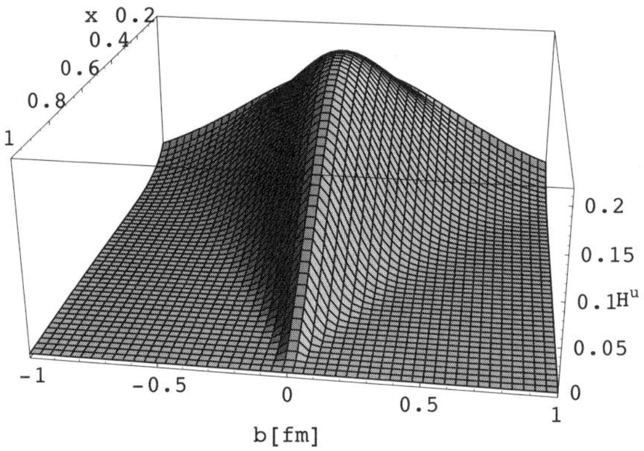

Fig. 5. The GPD H^u for a valence up-quark in the proton as function of the quark momentum fraction x and the quark position b in the transverse plane (perpendicular to the average direction of the fast moving nucleons). The calculation is based on a GPD Regge parametrization.[30]

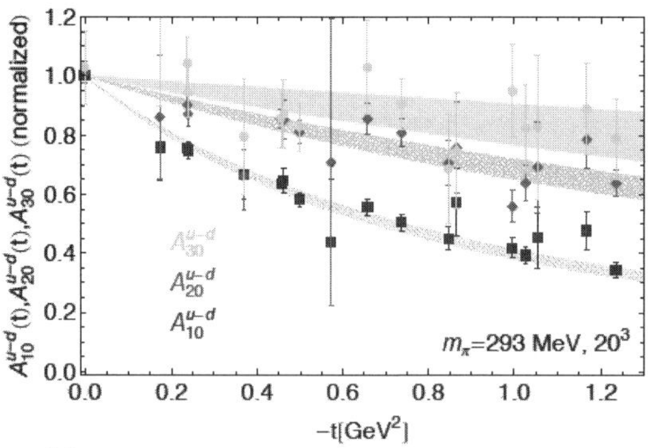

Fig. 6. Lattice QCD results for the x^n moments of the isovector GPD H^{u-d} at $\xi = 0$, for 3 values of n : $n = 0$ (blue symbols and band), $n = 1$ (red symbols and band), $n = 2$ (green symbols and band). The lattice results for 2+1 flavors of domain wall fermions are from Ref. [34] for $m_\pi = 293$ MeV.

5. DVCS observables and GPD fits

The nucleon structure information entering the nucleon DVCS process, can be parameterized at leading twist-2 level, in terms of four (quark chirality conserving) GPDs, depending on three variables: the quark longitudinal momentum fractions $(x+\xi)$ of intial quark, $(x-\xi)$ of final quark, and the momentum transfer t to the nucleon.

The contribution of the GPD H to the twist-2 DVCS amplitude, denoted by $A(\xi,t)$, is given by :

$$A(\xi,t) \equiv -\int_0^1 dx\, H^{(+)}(x,\xi,t)\left[\frac{1}{x-\xi+i\varepsilon}+\frac{1}{x+\xi-i\varepsilon}\right], \qquad (6)$$

which depends on the GPD singlet combination

$$H^{(+)}(x,\xi,t) \equiv H(x,\xi,t) - H(-x,\xi,t)$$

One directly notices that the imaginary part of the amplitude contains the GPD along $x=\xi$, i.e.

$$ImA(\xi,t) = \pi\, H^{(+)}(\xi,\xi,t). \qquad (7)$$

The real part of $A(\xi,t)$ can be written as a principal value (PV) convolution integral :

$$ReA(\xi,t) \equiv -PV\int_0^1 dx\, H^{(+)}(x,\xi,t)\left[\frac{1}{x-\xi}+\frac{1}{x+\xi}\right]. \qquad (8)$$

Eqs. (7) and (8) are the Compton form factors (CFFs), which together with the corresponding expressions for the three other GPDs E, $\tilde{H}$, and $\tilde{E}$, constitute the experimental information on GPDs which can be accessed from different DVCS observables, assuming twist-2 dominance.

In a series of recent works,[35–38] the available DVCS data from HERMES and JLab have been used in a model-independent fitting procedure of the CFFs. In these fits, the beam-polarized and unpolarized cross sections from JLab/Hall A[39] ; the beam-polarized[40–42] and longitudinally polarized target[43] spin asymmetries from JLab/CLAS; and a series of (seventeen) of HERMES beam-charge, beam-polarized, and transversely polarized target spin asymmetries[44–47] were used. Assuming twist-2 dominance, it was shown that these obserservables allow to extract, in a largely model-independent fashion, three CFF, notably the real and imaginary parts involving the GPD H, denoted by H_{Re}, and H_{im} respectively, as well as the imaginary part of the GPD $\tilde{H}$, denoted by $\tilde{H}_{im}$. The results for H_{Re} and H_{Im} are shown in Fig. 7. The comparison of HERMES and JLab results allows a few conclusions. As the energy increases (smaller x_B) the CFF H_{Im}

displays a steeper behavior in $-t$ indicating some shrinkage. This behavior of H_{im} is rather well reproduced in Regge based GPD model parameterizations. One also notices from Fig. 7 that the CFF H_{Re} has a very different t-dependence as H_{Im}, which is not so well understood based on present GPD models. The fit of the CFF $\tilde{H}_{im}$ is shown in Fig. 8, and shows a rather flat t-dependence at both energies.

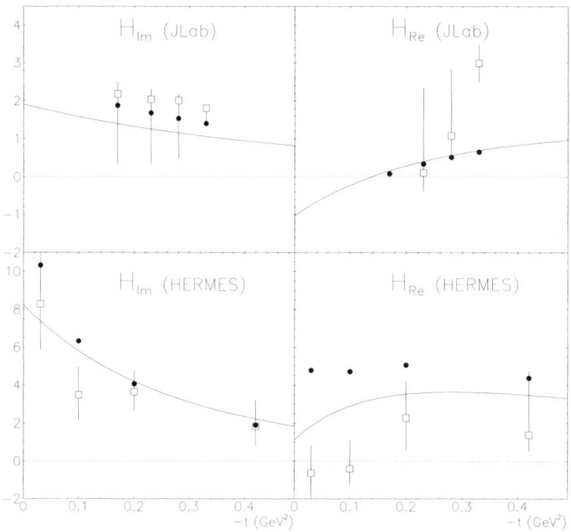

Fig. 7. t-dependence of the imaginary parts H_{Im} (left panels) and real parts H_{Re} (right panels) of the CFF, extracted from a fit of JLab data at $\langle x_B \rangle = 0.36$, $\langle Q^2 \rangle = 2.3$ GeV^2 (upper panels) and HERMES data at $\langle x_B \rangle = 0.09$, $\langle Q^2 \rangle = 2.5$ GeV^2 (lower panels) according to the analysis of Ref. [37]. For comparison, the VGG model predictions are given by the solid circles. The solid curve shows the model-dependent fit of the data performed in Ref. [49].

It is clear that such model independent analyses will be improved with forthcoming data, especially from the JLab@12GeV facility. In analyzing the new data, it will be worthwhile to include dispersion relation constraints on the DVCS amplitude. It has been shown in recent years[50–53] that Eq. (6) can be cast into the form of a *once-subtracted* dispersion relation (assuming one subtraction is enough to make it convergent) :

$$ReA(\xi, t) = \Delta(t) - PV \int_0^1 dx\, H^{(+)}(x, x, t) \left[\frac{1}{x - \xi} + \frac{1}{x + \xi} \right], \qquad (9)$$

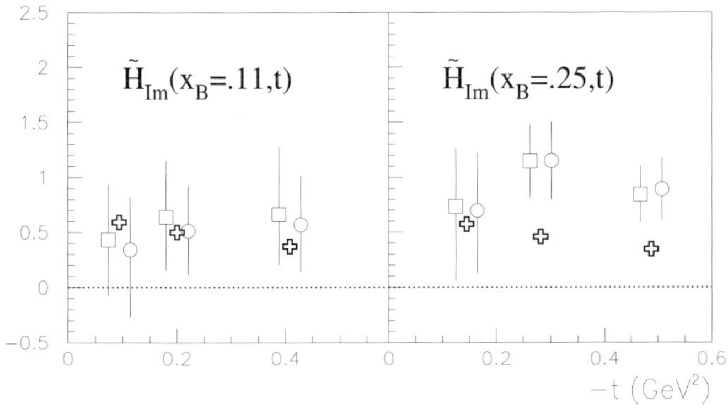

Fig. 8. t-dependence of the imaginary part of the CFF $\tilde{H}_{Im}$, extracted from a fit of HERMES data (left panel) and JLab/CLAS data (right panel) according to the analysis of Ref. [38]. The empty squares (circles) show the fit results when the boundary of the domain over which the CFFs are allowed to vary is 5 (3) times the VGG reference value.[48] For comparison, the VGG model predictions are given by the empty crosses.

where the subtraction constant (at zero energy) $\Delta(t)$ is directly proportional to the D-form factor (normalization of the D-term). One notices that in contrast to the convolution integral of Eq. (8), where the GPD enters for unequal values of its first and second argument, the integrand in the dispersion relation (spectral function) corresponds with the GPD where its first and second arguments are equal.

In Fig. 9, we show how such a dispersive analysis of future data may look like, based on two GPD model calculations which satisfy these dispersion relations. The imaginary part of the amplitude would be measured over some range from spin asymmetries. The real part of the amplitude at a fixed value of t, is then fixed up to a global normalization (subtraction constant) which is related with the D-form factor. A measurement of the real part at different values of ξ will allow to cross-check this procedure.

Acknowledgments

I would like to thank the organizers of the workshop on "Exclusive Reactions at High Momentum Transfer" for the invitation to speak.

References

1. Hofstadter R. and R. W. McAllister, Phys. Rev. **98**, 217 (1955).
2. C. E. Hyde-Wright and K. de Jager, Ann. Rev. Nucl. Part. Sci. **54**, 217 (2004).

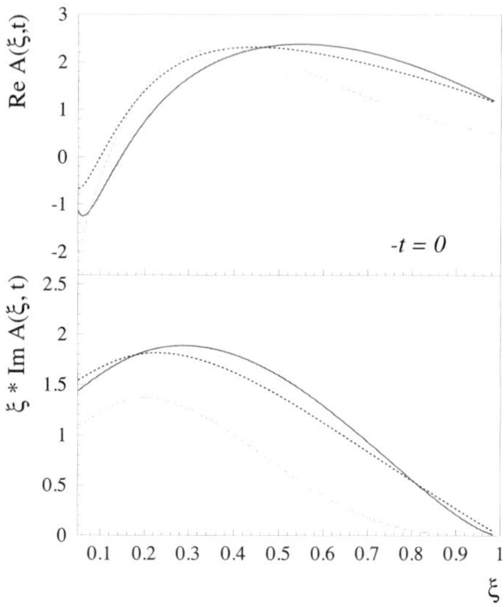

Fig. 9. Dispersion relation evaluation of the DVCS amplitude $A(\xi, t = 0)$ corresponding with the GPD H. The solid (red) curves are within the dual parameterization.[54] The dashed (blue) and dotted (green) curves are using the double distribution parameterization, see Ref. [12], for $b_v = b_s = 1$ (dashed curves) and $b_v = b_s = 20$ (dotted curves).

3. J. Arrington, C. D. Roberts and J. M. Zanotti, J. Phys. G **34**, S23 (2007).
4. C. F. Perdrisat, V. Punjabi and M. Vanderhaeghen, Prog. Part. Nucl. Phys. **59**, 694 (2007).
5. M. Vanderhaeghen and T. Walcher, arXiv:1008.4225 [hep-ph].
6. Ph. Hagler, Phys. Rept. **490**, 49 (2010).
7. C. Alexandrou *et al.*, Phys. Rev. D **79**, 014507 (2009); C. Alexandrou *et al.*, Nucl. Phys. A **825**, 115 (2009).
8. D. Mueller, D. Robaschik, B. Geyer, F. M. Dittes and J. Horejsi, Fortsch. Phys. **42**, 101 (1994).
9. A. V. Radyushkin, Phys. Lett. B **380**, 417 (1996).
10. X. D. Ji, Phys. Rev. Lett. **78**, 610 (1997); Phys. Rev. D **55**, 7114 (1997).
11. J. C. Collins, L. Frankfurt and M. Strikman, Phys. Rev. D **56**, 2982 (1997).
12. K. Goeke, M. V. Polyakov and M. Vanderhaeghen, Prog. Part. Nucl. Phys. **47**, 401 (2001).
13. M. Diehl, Phys. Rept. **388**, 41 (2003).
14. X. D. Ji, Ann. Rev. Nucl. Part. Sci. **54**, 413 (2004).
15. A. V. Belitsky and A. V. Radyushkin, Phys. Rept. **418**, 1 (2005).
16. S. Boffi and B. Pasquini, Riv. Nuovo Cim. **30**, 387 (2007).
17. M. Burkardt, Phys. Rev. D **62**, 071503 (2000); [Erratum-ibid. D **66**, 119903

(2002)]; Int. J. Mod. Phys. A **18**, 173 (2003).
18. X. d. Ji, Phys. Rev. Lett. **91**, 062001 (2003).
19. A. V. Belitsky, X. d. Ji and F. Yuan, Phys. Rev. D **69**, 074014 (2004).
20. S. Meissner, A. Metz and M. Schlegel, JHEP **0908**, 056 (2009).
21. S. J. Brodsky, H. C. Pauli and S. S. Pinsky, Phys. Rept. **301**, 299 (1998).
22. S. D. Drell and T. M. Yan, Phys. Rev. Lett. **24**, 181 (1970).
23. L. Susskind, Phys. Rev. **165**, 1547 (1968).
24. G. A. Miller, Phys. Rev. Lett. **99**, 112001 (2007).
25. C. E. Carlson and M. Vanderhaeghen, Phys. Rev. Lett. **100**, 032004 (2008).
26. C. E. Carlson and M. Vanderhaeghen, Eur. Phys. J. A **41**, 1 (2009).
27. C. Lorcé, Phys. Rev. D **79**, 113011 (2009).
28. J. Arrington, W. Melnitchouk and J. A. Tjon, Phys. Rev. C **76**, 035205 (2007).
29. R. Bradford, A. Bodek, H. Budd and J. Arrington, Nucl. Phys. Proc. Suppl. **159**, 127 (2006).
30. M. Guidal, M. V. Polyakov, A. V. Radyushkin and M. Vanderhaeghen, Phys. Rev. D **72**, 054013 (2005).
31. M. Järvinen, Phys. Rev. D **71**, 085006 (2005).
32. P. Hoyer, arXiv:0909.3045 [hep-ph].
33. M. Diehl and Ph. Hagler, Eur. Phys. J. C **44**, 87 (2005).
34. J. D. Bratt *et al.* [LHPC Coll.], arXiv:1001.3620 [hep-lat].
35. M. Guidal, Eur. Phys. J. A **37**, 319 (2008) [Erratum-ibid. A **40**, 119 (2009)].
36. H. Moutarde, Phys. Rev. D **79**, 094021 (2009).
37. M. Guidal and H. Moutarde, Eur. Phys. J. A **42**, 71 (2009).
38. M. Guidal, Phys. Lett. B **689**, 156 (2010); arXiv:1005.4922 [hep-ph].
39. C. Munoz Camacho *et al.* [Jefferson Lab Hall A Coll. and Hall A DVCS Coll.], Phys. Rev. Lett. **97**, 262002 (2006).
40. S. Stepanyan *et al.* [CLAS Coll.], Phys. Rev. Lett. **87**, 182002 (2001).
41. F. X. Girod *et al.* [CLAS Coll.], Phys. Rev. Lett. **100**, 162002 (2008).
42. G. Gavalian *et al.* [CLAS Coll.], Phys. Rev. C **80**, 035206 (2009).
43. S. Chen *et al.* [CLAS Coll.], Phys. Rev. Lett. **97**, 072002 (2006).
44. A. Airapetian *et al.* [HERMES Coll.], Phys. Rev. Lett. **87**, 182001 (2001).
45. A. Airapetian *et al.* [HERMES Coll.], JHEP **0806**, 066 (2008).
46. A. Airapetian *et al.* [HERMES Coll.], JHEP **0911**, 083 (2009).
47. A. Airapetian *et al.* [HERMES Coll.], JHEP **1006**, 019 (2010).
48. M. Vanderhaeghen, P. A. M. Guichon and M. Guidal, Phys. Rev. D **60**, 094017 (1999); Phys. Rev. Lett. **80**, 5064 (1998).
49. K. Kumericki and D. Mueller, arXiv:0904.0458 [hep-ph]; see also these proceedings (arXiv:1008.2762 [hep-ph]).
50. I. V. Anikin and O. V. Teryaev, Phys. Rev. D **76**, 056007 (2007).
51. K. Kumericki, D. Mueller and K. Passek-Kumericki, Nucl. Phys. B **794**, 244 (2008).
52. M. Diehl and D. Y. Ivanov, Eur. Phys. J. C **52**, 919 (2007).
53. M. V. Polyakov, Phys. Lett. B **659**, 542 (2008).
54. M. V. Polyakov and M. Vanderhaeghen, arXiv:0803.1271 [hep-ph].

FIXED ANGLE SCATTERING AND THE TRANSVERSE STRUCTURE OF HADRONS

GEORGE STERMAN

C.N. Yang Institute for Theoretical Physics, Stony Brook University
Stony Brook, New York, 11794-3840, USA
**E-mail: george.sterman@stonybrook.edu*

The perturbative treatment of high-energy fixed-angle hadron-hadron exclusive scattering is reviewed and related to the transverse structure of the proton and other hadrons.

Keywords: Elastic scattering, hadronic wave functions, factorization

1. Introduction

Hadron-hadron exclusive cross sections offer a sensitivity to transverse partonic substructure that is complementary to most photon-induced processes.[1] As we shall see, at high enough energies and momentum transfers, these cross sections probe hadronic Fock states with the minimal numbers of partons, the "valence" states. This talk reviews classic results on hadron-hadron exclusive reactions from this point of view[2–7] and recalls as well some related subsequent work.[8–10]

A full treatment of the transition to asymptotic behavior requires both perturbative resummation and nonperturbative input on the transverse structure of the hadronic valence states. It is not really known to what extent experiments have yet reached a truly asymptotic regime, and pictures involving higher Fock states may offer alternative descriptions. A challenge for the future is to find a unified, perhaps dual, description of exclusive reactions. This short presentation, however, will concentrate on the classic picture based on valence states.

The next section reviews the origin of the basic parton-model "quark counting"[2,3] predictions for elastic scattering, grounding them in a simple geometric picture, in which transverse structure plays only a passive role. Section 3 introduces the "Landshoff mechanism",[4] which reintroduces a dynamic role for transverse structure, and Sec. 4 shows how an analysis of radiative corrections links the two. In Sec. 5, some intriguing data on wide-angle particle-antiparticle scattering is briefly discussed.

2. Valence states, geometry and quark counting

We begin with the parton model applied to high-energy elastic scattering. In Refs. 2 and 3, elastic scattering is pictured as occurring via time-dilated Fock states with fixed numbers of partons. Suppose all the n_H partons in the valence state for hadron H have comparable momentum fractions x_i. A large coherent momentum transfer $t \sim -Q^2$ is necessary to redirect all these partons into another direction. Such a momentum transfer requires all of the n_H (anti-)quarks to be in a region of area $1/Q^2$ in the Lorentz-contracted wave functions of the colliding hadrons in the center-of-mass frame. This has to be the case for each colliding hadron, and also for the two hadrons that emerge from the scattering. The essential observation is that if the distribution of partons in the transverse direction is random, the likelihood for such a configuration is estimated by $\left(\frac{1}{Q^2} \times \frac{1}{\pi R_H^2}\right)^{n_H - 1}$ for each hadron. This geometric picture is illustrated in Figs. 1 and 2.

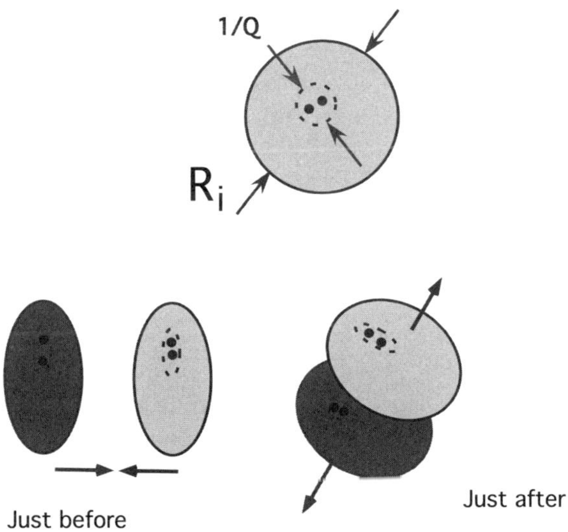

Fig. 1. Valence states before and after an elastic scattering.

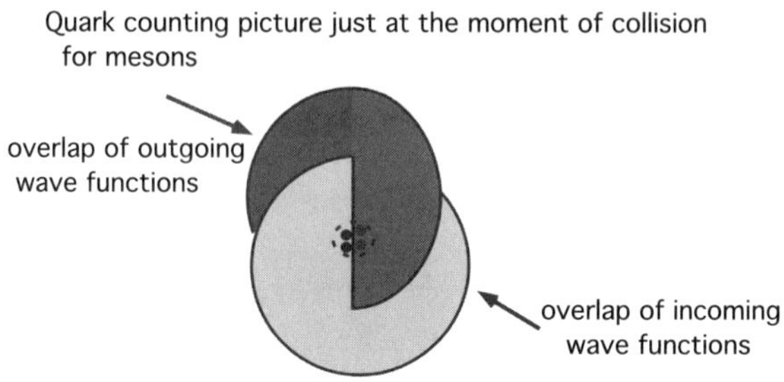

Fig. 2. Incoming and outgoing wave functions at the moment of collision.

If the geometric picture determines the overall energy and momentum dependence, then otherwise the amplitude is a function only of the scattering angle. At fixed s/t (that is, fixed c.m. scattering angle) we find,

$$\frac{d\sigma}{dt} = \frac{f(s/t)}{s^2} \left(\frac{m^2}{s}\right)^{\sum_{i=1}^{4}(n_i - 1)} . \tag{1}$$

This is the basic parton model result, the 'quark counting rule'. It clearly picks out the valence state: the cross sections associated with larger numbers of partons are power suppressed, at least as long as we assume that they do not have vanishing momentum fractions.

As the parton model took on a new life in the late 1970's as a limit of quantum chromodynamics, it was discovered in groundbreaking papers that the corresponding elastic amplitude could be written as[5,6]

$$\begin{aligned} \mathcal{M}(s,t;h_i) = \int \prod_{i=1}^{4} [dx_i]\, \phi(x_{m,i}, \lambda_{m,i}, h_i; \mu) \\ \times M_H \left(\frac{x_{n,i} x_{m,j} p_i \cdot p_j}{\mu^2}; \lambda_{n,i}, h_i \right) , \end{aligned} \tag{2}$$

in terms of a calculable hard-scattering amplitude M_H in convolution with factorized and evolved valence (light-cone) wave functions, $\phi(x_{m,i}, \lambda_{m,i}, h_i; \mu)$, with helicities h_i for hadrons and $\lambda_{n,i}$ for quarks. The mass μ is the factorization scale. The convolution is in terms of the partonic momentum fractions: for baryons,

$$[dx_i] = dx_{1,i} dx_{2,i} dx_{3,i}\, \delta\left(1 - \sum_{n=1}^{3} x_{n,i}\right) . \tag{3}$$

So far, the transverse dimensions of colliding hadrons enter only indirectly, in the geometric justification of power behavior.

In principle, all this is straightforward, but our knowledge of the wave functions is not complete, and in any case for nucleon-nucleon scattering there are too many diagrams even at low orders to make a direct calculation practical, at least up to this time.

3. Splitting the hard scattering

The geometric analysis for elastic scattering in the valence state is more flexible than it might at first seem, and there is an alternative picture of the rearrangement of parton momenta, shown in Fig. 3. This is the geometric interpretation of the process identified first by Landshoff.[4]

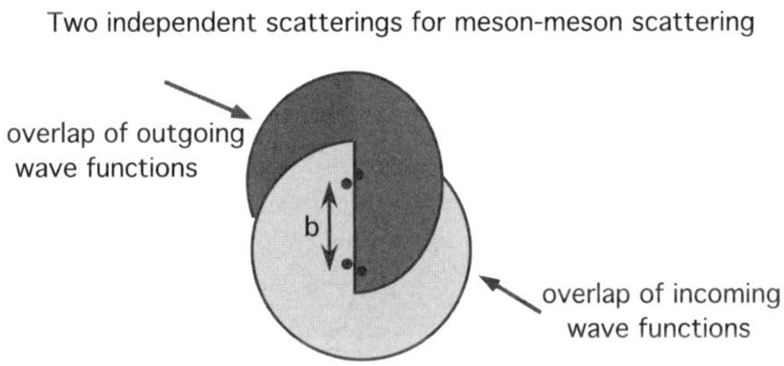

Fig. 3. Representation of independent scatterings.

Figure 3 shows that the single hard scattering of quark counting can in principle be split into as many hard scatterings as there are valence partons in each of the incoming hadrons. (The figure shows two, as for $\pi\pi$ scattering.) Each parton must overlap with an incoming and outgoing wave function. Relative to the single hard scattering picture, however, we no longer require all of the partons to be in the same region of radius $1/Q$. If there are, for example, two hard scatterings, we gain a geometric factor of the transverse size of the hadrons, $R_H Q$ relative to single scattering, with Q the momentum transfer. An enhancement $1/Q \to R_H$ in the amplitude gives a factor $1/Q^2 \to R_H^2$ in cross section. In this way, the Landshoff analysis provides an enhancement by a factor $Q^2 \sim s$ for fixed angle pion scattering cross sections and $Q^4 \sim s^2$ for proton-proton fixed angle elastic scattering. More specifically, this modified geometric configuration gives[4] for pp at fixed angle (*i.e.* fixed s/t)

$$\frac{d\sigma}{dt} = \frac{f(s/t)}{s^2} \left(\frac{1}{s\,\pi R_H^2} \right)^6 , \tag{4}$$

while for $s \gg -t \gg \Lambda_{\rm QCD}$ it gives

$$\frac{d\sigma}{dt} = \frac{F(s)}{t^2} \left(\frac{1}{t\,\pi R_H^2} \right)^6 . \tag{5}$$

Experimentally, the forward scattering proposal works well at a wide range of center-of-mass energies, but at fixed angles the data appear to follow the original quark counting rules. As I'll now argue, this distinction may be associated with the transverse substructure of hadrons.

4. The return of (approximate) quark counting at wide angles

The scattering of isolated color charges tends to produce radiation in the incoming and outgoing directions. Figure 4 illustrates this effect, superimposed on the parton model template of Fig. 3 for pion scattering. For finite impact parameter b, each of the two scattering processes along the vertical line of overlap involves colored partons. This should lead to the radiation of gluons of wavelength as small as order $1/Q$ and as large as b. Final states without such radiation are suppressed by virtual corrections, unless the impact parameters b is small. On the one hand the Landshoff mechanism is enhanced because there are more ways to produce two hard scatterings than one, but on the other hand it is suppressed because isolated scatterings of colored particles are only rarely elastic. The full amplitude is actually the result of a competition between geometric enhancement and radiative suppression.[7,11] The resulting balance, and its energy and momentum transfer dependence, was analyzed in Ref. 8.

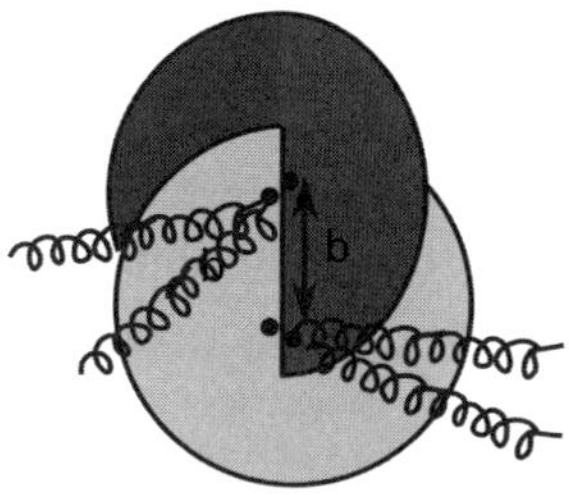

Fig. 4. Schematic representation of radiation that is absent in elastic scattering.

Impact parameter b is conjugate to $Q = \sqrt{-t}$. As $-t$ increases toward s, radiative corrections force the b's to $1/\sqrt{s}$ and the quark counting geometric

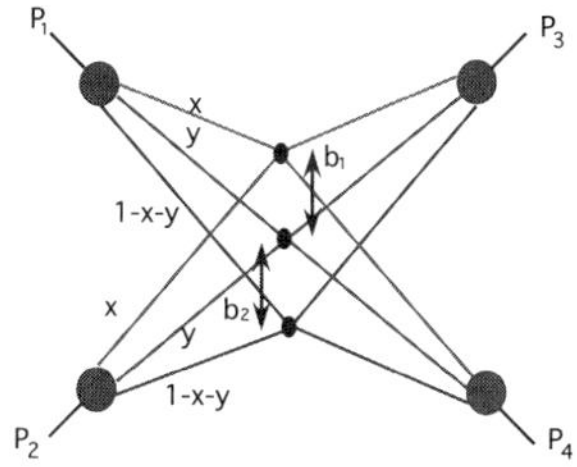

Fig. 5. The Landshoff mechanism in impact parameter space.

picture should be recovered approximately.[6,7,12] For color-singlet nucleons, we have as many as three independent scatterings, and this sort of analysis leads for baryons to the expression (see Fig. 5)[10]

$$\mathcal{M}(s,t) = \frac{N}{stu} \sum_f \int_0^1 \frac{dx dy}{x^2 y^2 (1-x-y)^2} \times \int db_1 db_2 \, \mathrm{Tr}_{\mathrm{color}} \left[U(b_i Q) M^1 M^2 M^3 \right] \times \prod_{m=1,2,3,4} \Psi_{H_m}(x,y,b_1,b_2)\,, \tag{6}$$

where N is a normalization factor. The Trace $\left[U(b_i Q) M^1 M^2 M^3\right]$ ties color together and includes ϵ_{abc} for colors of three quarks in each external hadron, with possible color exchange in each hard scattering $M^i(x_i p_j)$. The Ms are independent of the transverse separations. The matrix U, which depends on both transverse distances b_i, is necessary to incorporate coherent logarithms, left over after the factorization of the wave functions, Ψ.

The wave functions depend on the center-of-mass energy, rather than a factorization scale, as would be the case for parton distributions. This is because they represent the creation or absorption of the partons of the valence state at equal light cone time but fixed transverse separation. In this way, they reflect the amplitude for such a state to exist, without additional degrees of freedom, in a particular frame. They are not gauge invariant, but their product in Eq. (6) is invariant. In fact, it is convenient to construct the wave functions in a physical gauge[13] and to use their gauge dependence to determine their energy dependence. The pattern is illustrated by the simpler example of pion wave functions, for which one finds an exponentiated form,[8]

$$\Psi(x,b,Q) \sim \Psi_{NP}(x,b) \exp[-s(x,b,Q) - s(1-x,b,Q)]\,, \tag{7}$$

where the two partons of the valence state carry fractional momenta x and $1-x$, where $Q \equiv \sqrt{-t}$ and b is the distance between the hard scatterings in Fig. 4. The functions $s(\xi, b, Q)$ provide double-logarithmic suppression whenever the product Qb is large, being defined in terms of universal anomalous dimensions. These are $A(\alpha_s) = C_F(\alpha_s/\pi) + \ldots$, which organizes all leading logarithms, and $B(\alpha_s)$, which depends on the details of the factorization procedure,

$$s(x,b,Q) = \int_{C_1/b}^{C_2 x Q} \frac{d\mu}{\mu} \left[4 \ln \left(\frac{C_2 x Q}{\mu} \right) A(\alpha_s(\mu)) + B(\alpha_s(\mu)) \right] . \tag{8}$$

For fixed coupling and small b, one finds that the wave functions have double-logarithmic dependence on the impact parameters,

$$\Psi(x,b,Q) \to \phi_{asy}(x_j) \exp[-\text{const } \ln^2(1/Qb)] . \tag{9}$$

Assembling the pieces for baryon scattering, this gives the following asymptotic amplitude, with an example of "Sudakov resummation" when $b \gg 1/Q$, which drives the wave functions to their 'asymptotic' behaviors[6,8]

$$\mathcal{M}(s,t) = \frac{N}{stu} \sum_f \int_0^1 \frac{dx_1 dx_2}{x_1^2 x_2^2 (1-x-y)^2} \prod_{m=1,2,3,4} \phi_{m,asy}(x,y)$$
$$\times \int db_1 db_2 \, \mathrm{Tr}_{\mathrm{color}} \left[U(b_i Q) M^1 M^2 M^3 \right]$$
$$\times \, e^{-S_1(b_i Q) - S_2(b_i Q) - S_3(b_i Q)} . \tag{10}$$

At large Q for each scattering, radiation suppression also forces the hard scatterings back together. The S_i are sums of exponential functions like those in (8), one for each scattering. At moderate $(xQ)^2$, $(yQ)^2$, the amplitude is dominated by the "boundary conditions," $\Psi_{NP}(x, y, b_i)$ rather than asymptotic behavior. One of the attractive features of this expression is that the original eight integrals over momentum fractions $x_{m,i}$ in Eq. (2) are reduced to only two.

An important feature of Eq. (10) is that the scale of Sudakov suppression is set by the momentum transfer. Varying t at fixed s thus in principle modifies the role of radiative corrections, and as $|t|$ decreases, we anticipate that the cross section makes a transition from s^{-10} to t^{-8} behaviour, as seen in the data.[14] A study of the implications of Eq. (10) for the transverse size of the proton valence state was made in Ref. 10, where it was concluded that the data favor a quite small transverse extent. Fixed-angle scattering on nuclei offers another possible test of the transverse structure of hadrons, via nuclear transparency for small-size color-singlet states.[15–18]

5. Exchanging quarks

The formalism we've described so far is valid asymptotically, but it is not so clear how high the energy and momentum transfer have to be for corrections to these pictures to be negligible. For the multiple scattering picture, in particular, the invariant mass for the "hard scattering" described by the amplitude M_i in Eq. (10) is $x_i^2 s$, typically an order of magnitude smaller than the hadron-level invariants.[19] At the very least, it is clear that the wave functions must vanish sufficiently fast when any of the $x_i \to 0$ in the valence state. Considerations such as these suggest that at accessible energies, alternative descriptions of exclusive scattering, not necessarily limited to the valence state, should be examined. For the closely related case of form factors, descriptions based on QCD sum rules have shown success.[20]

With all this in mind, and because of the complexity of the calculations involved, further experimental comparisons of elastic amplitudes between different hadrons may lead to valuable insights. For example, a conceptual contrast was made between gluon and quark exchange processes early on, and has remained of interest.[21,22] Quark exchange should be highly sensitive to flavor content.

The perturbative factorization formalisms described above can be thought of as involving the exchange of quark degrees of freedom in addition to gluons, and generally these processes contribute to the amplitude in a complicated way. A case of interest, however, is the comparison of $p\bar{p}$ to pp scattering. For pp, there are 2^3 ways of connecting incoming and outgoing quarks compared to only one for $p\bar{p}$. If amplitudes are coherent sums of these processes, all with a similar weight, then we might expect a ratio of 1/64 for the elastic scattering cross sections of protons on antiprotons to that for protons on protons.

The relevance of this elegant observation was tested in experiments at Brookhaven National Laboratory's AGS,[23] and ratios of particle-antiparticle to particle-particle seemed roughly consistent with this counting, $R_N = \left.\frac{\frac{d\sigma_{N\bar{N}}}{dt}}{\frac{d\sigma_{NN}}{dt}}\right|_{90\ \mathrm{deg}} \sim \frac{1}{40}$. Any successful picture of these exclusive hadronic processes must explain this simple counting result.

Sotiropoulos[10] studied this issue in the Landshoff multiple scattering picture. At first sight, the situation seems promising: at each hard scattering in Fig. 5, quarks may be exchanged (or not) between the participating protons of pp scattering in all possible ways, but for $p\bar{p}$ scattering, the quarks and antiquarks have only one way to get from the initial to the final state. There is a caveat in any pQCD picture, however: we have to reform an antisymmetric color combination of quarks when they are exchanged, a process that requires them to exchange gluons. Sotiropoulos found that the pertubative picture sketched above works qualitatively only with a "color randomization" hypothesis in which the factor $\left[U(bQ)\prod_i M^i\right]$ is independent of the flavor flow. He found a ratio of cross sections of $R_p \sim 1/30$ at ninety degrees with color randomization, but only $\sim 1/3$ without it, with a qualitatively successful angular dependence.[10,24] Randomization is a plausible effect at the moderate BNL energies, $\sqrt{s} \sim 3.5$ GeV^2, and is easy to picture in the context of quark exchange. This leaves us with a description that is a mixture of perturbative and nonperturbative effects. On the other hand, the Landshoff/Sudakov model with or without randomization gives explicit predictions for angular and helicity dependence. Clearly, a decrease of R_{pp} with energy would be a compelling signal for an emerging role for color.

6. Conclusions

It has been some time since hadron-hadron elastic scattering has held center stage. It remains, however, an important probe of hadron structure, especially for information on pion and nucleon valence states. New facilities may push the frontier in s and t coverage, and clarify essential questions of the role of perturbative and nonperturbative scattering mechanisms. Insights flowing from the development of duality-based pictures of hadronic structure[25,26] and the increase in our capability to compute multi-parton scattering amplitudes[27] should provide new tools for these fundamental processes.

7. Acknowledgments

This work was supported in part by the National Science Foundation, grant PHY-0653342. I thank Mark Strikman for discussions, and the organizers of "Exclusive Reactions at High Momentum Transfer", especially Anatoly Radyushkin, Paul Stoler and Christian Weiss, for the invitation to speak at the workshop

References

1. X. D. Ji, J. Phys. G **24**, 1181 (1998) [arXiv:hep-ph/9807358]. A. V. Belitsky and A. V. Radyushkin, Phys. Rept. **418**, 1 (2005) [arXiv:hep-ph/0504030].

2. V. A. Matveev, R. M. Muradian and A. N. Tavkhelidze, *Lett. Nuovo Cim.* **7**, 719 (1973).
3. S. J. Brodsky and G. R. Farrar, *Phys. Rev.* D **11**, 1309 (1975).
4. P. V. Landshoff, *Phys. Rev. D* **10**, 1024 (1974).
5. A. V. Efremov and A. V. Radyushkin, *Phys. Lett.* B **94**, 245 (1980).
6. G. P. Lepage and S. J. Brodsky, *Phys. Rev.* D **22**, 2157 (1980).
7. A. H. Mueller, *Phys. Rept.* **73**, 237 (1981).
8. J. Botts and G. F. Sterman, Nucl. Phys. B **325**, 62 (1989).
9. M. G. Sotiropoulos and G. F. Sterman, *Nucl. Phys.* B **425**, 489 (1994) [arXiv:hep-ph/9401237].
10. M. G. Sotiropoulos, *Phys. Rev.* D **54**, 808 (1996) [arXiv:hep-ph/9512397].
11. B. Pire and J. P. Ralston, *Phys. Lett.* B **117**, 233 (1982).
12. P. V. Landshoff and D. J. Pritchard, *Z. Phys.* C **6**, 69 (1980).
13. J. C. Collins and D. E. Soper, Nucl. Phys. B **193**, 381 (1981) [Erratum-ibid. B **213**, 545 (1983)].
14. E. Nagy *et al.*, *Nucl. Phys.* B **150**, 221 (1979). W. Faissler *et al.*, *Phys. Rev.* D **23**, 33 (1981).
15. G. Bertsch, S. J. Brodsky, A. S. Goldhaber and J. F. Gunion, *Phys. Rev. Lett.* **47**, 297 (1981); S. J. Brodsky and A. H. Mueller, *Phys. Lett.* B **206**, 685 (1988).
16. P. Jain, B. Pire and J. P. Ralston, *Phys. Rept.* **271**, 67 (1996) [arXiv:hep-ph/9511333].
17. L. Frankfurt, G. A. Miller and M. Strikman, *Phys. Lett.* B **304**, 1 (1993) [arXiv:hep-ph/9305228];
18. A. S. Carroll *et al.*, *Phys. Rev. Lett.* **61**, 1698 (1988); E. M. Aitala *et al.* [E791 Collaboration], *Phys. Rev. Lett.* **86** (2001) 4773 [arXiv:hep-ex/0010044].
19. N. Isgur and C. H. Llewellyn Smith, *Nucl. Phys.* B **317**, 526 (1989).
20. V. A. Nesterenko and A. V. Radyushkin, *Phys. Lett.* B **115**, 410 (1982).
21. J. F. Gunion, S. J. Brodsky and R. Blankenbecler, *Phys. Rev.* D **8**, 287 (1973).
22. G. P. Ramsey and D. W. Sivers, *Phys. Rev.* D **45**, 79 (1992). *Phys. Rev.* D **52**, 116 (1995).
23. G. C. Blazey *et al.*, *Phys. Rev. Lett.* **55**, 1820 (1985). B. R. Baller *et al.*, *Phys. Rev. Lett.* **60**, 1118 (1988).
C. G. White *et al.*, *Phys. Rev.* D **49**, 58 (1994).
24. G. R. Farrar and C. C. Wu, *Nucl. Phys.* B **85**, 50 (1975).
25. S. J. Brodsky and G. F. de Teramond, Phys. Rev. Lett. **96**, 201601 (2006) [arXiv:hep-ph/0602252].
26. H. R. Grigoryan and A. V. Radyushkin, Phys. Rev. D **78**, 115008 (2008) [arXiv:0808.1243 [hep-ph]].
27. Z. Bern, L. J. Dixon and D. A. Kosower, Annals Phys. **322**, 1587 (2007) [arXiv:0704.2798 [hep-ph]].

TOWARDS A GLOBAL ANALYSIS OF GENERALIZED PARTON DISTRIBUTIONS

K. KUMERIČKI

Department of Physics, University of Zagreb,
Zagreb, HR-10002, Croatia
E-mail: kkumer@phy.hr

D. MÜLLER

Institut für Theoretische Physik II, Ruhr-University Bochum,
Bochum, D-44780, Germany
E-mail: dieter.mueller@tp2.rub.de

We discuss the complexity of GPD phenomenology, comment on the technological needs for a global analysis, and report on model and neural network fits to the photon electroproduction off unpolarized proton. We also point out that Radyushkin's double distribution ansatz is a 'holographic' GPD model.

Keywords: Generalized Parton Distributions; Exclusive Electroproduction.

1. Introduction

Generalized parton distributions (GPDs), introduced some time ago,[1–3] have received much attention from both the theoretical and the experimental side. This was triggered by the 'spin crisis', referring to the mismatch of quark spin, extracted from polarized deep inelastic scattering, and the constituent quark model. This is rather analogous to the 'momentum crisis', where almost 50% of the proton momentum is carried by gluons. Obviously, both 'crises' of the constituent quark model are 'solved' by switching to the fundamental degrees of freedom. Hence, we view the 'spin puzzle' first as a quest to quantify the partonic structure of the nucleon in terms of quark and gluon angular momenta and second as a wish for an appropriate matching of effective and partonic degrees of freedom. Furthermore, it has been realized that GPDs build up a whole framework for description of hadron structure,[4,5] including a three dimensional imaging.[6] Certainly, the 'spin puzzle' is one interesting and intricate aspect, which we will now discuss.

An appropriate decomposition of the nucleon spin in terms of energy-momentum tensor form factors has been suggested by X. Ji:[7]

$$\frac{1}{2} = J^Q + J^G, \quad J^Q = \sum_{q=u,d,\cdots} J^q, \quad J^i = A^i(Q^2) + B^i(Q^2), \qquad (1)$$

where the quark and gluon contributions are given by the first moments of parity even and target helicity (non-)conserved GPD H (E),

$$\left\{ \begin{matrix} A^i \\ B^i \end{matrix} \right\} (Q^2) = \lim_{t \to 0, \eta \to 0} \int_{-1}^{1} dx \, \frac{x}{2} \left\{ \begin{matrix} H^i \\ E^i \end{matrix} \right\} (x, \eta, t, Q^2) \, , \quad i \in \{u, d, \cdots, G\} \, , (2)$$

taken in the forward limit. Note that the momentum fractions A^i are obtained from global PDF fits and that momentum and angular momentum conservation imply that the total gravitomagnetic moment vanishes,

$$\sum_{q=u,d,\cdots} B^q(Q^2) + B^G(Q^2) = 0 \, .$$

Hence, the unknowns in the spin sum rule are the helicity flip form factors B^i. Any 'pure' quark model trivially predicts $A^G = 0$ and $B^G = 0$, whereas taking for A^G the phenomenological value $\simeq 0.45$ and relying on $B^G = 0$, one concludes that quarks and gluons carry roughly the same amount of angular momentum[a]. Indeed, lattice calculations indicate a rather small B^Q. Unfortunately, these simulations suffer from systematic uncertainties, in particular, so-called disconnected contributions are presently neglected. Thus, a phenomenological handle on the spin sum rule is highly desired.

Deeply virtual Compton scattering (DVCS) off nucleon is considered as the theoretically cleanest process offering access to GPDs. Its amplitude can be parameterized by twelve Compton form factors (CFFs),[9] which are given in terms of GPDs, e.g., at leading order (LO) in generic notation:

$$\left\{ \begin{matrix} \mathcal{H} \\ \mathcal{E} \end{matrix} \right\} (x_B, t, \mathcal{Q}^2) \stackrel{\text{LO}}{=} \int_{-1}^{1} dx \, \frac{2x}{\xi^2 - x^2 - i\epsilon} \left\{ \begin{matrix} H \\ E \end{matrix} \right\} (x, \eta = \xi, t, \mathcal{Q}^2) \, . \qquad (3)$$

The Bjorken variable x_B might be set equal to $2\xi/(1+\xi)$. Favorably, DVCS enters as a subprocess into the hard photon electroproduction where its interference with the Bethe-Heitler bremsstrahlung process provides variety of handles on CFFs. However, the target helicity conserved CFF $\mathcal{H}$ is often the dominant contribution, while $\mathcal{E}$ always appears with a kinematic suppression factor $t/4M_N^2$, induced by the helicity flip. In other words, one

[a]The orbital angular momentum of quarks is defined as $L^q = J^q - \Sigma^q/2$, where $\Sigma^q/2 \sim 0.15$ is the spin of quarks. For more detailed discussions see review.[8]

should look at observables for which $\mathcal{H}$ is suppressed, too, which requires a transversely polarized target. Also, in photon electroproduction off neutron $\mathcal{H}$ is suppressed in the interference term by the accompanying form factor F_1^n ($F_1^n(t=0)=0$). Unfortunately, one also has to worry about other non-dominant CFF contributions. Thus, the extraction of $\mathcal{E}$ requires a most complete measurement of all possible observables in dedicated experiments.

Supposing $\mathcal{E}$ is measured, the question arises: Can one deconvolute Eq. (3)? Apart from radiative and higher twist-contributions, one might view the GPD on the $\eta = x$ cross-over line as a "spectral function", which provides also the real part of the CFF via a dispersion relation:[10–13]

$$\Im\mathrm{m}\mathcal{F}(x_B,t,\mathcal{Q}^2) \stackrel{\mathrm{LO}}{=} \pi F(\xi,\xi,t,\mathcal{Q}^2)\,, \quad F=\{H,E,\widetilde{H},\widetilde{E}\}\,, \tag{4}$$

$$\Re\mathrm{e}\begin{Bmatrix}\mathcal{H}\\ \mathcal{E}\end{Bmatrix}(x_B,t,\mathcal{Q}^2) \stackrel{\mathrm{LO}}{=} \mathrm{PV}\int_0^1 dx\,\frac{2x}{\xi^2-x^2}\begin{Bmatrix}H\\ E\end{Bmatrix}(x,x,t,\mathcal{Q}^2) \pm \mathcal{D}(t,\mathcal{Q}^2). \tag{5}$$

The GPD support properties ensure that Eqs. (4,5) are in *one-to-one* correspondence to the perturbative formula (3), where the subtraction constant $\mathcal{D}$ can be calculated from either H or E. To pin down the GPD in the outer region $y \geq \eta = x$, one might employ evolution, e.g., in the non-singlet case,

$$\mu^2\frac{d}{d\mu^2}F(x,x,t,\mu^2) = \int_x^1 \frac{dy}{x}V(1,x/y,\alpha_s(\mu))F(y,x,t,\mu^2)\,. \tag{6}$$

A large enough $\mathcal{Q}^2$ range is not available in fixed target experiments. Hence, we must conclude that essentially only the GPD on the cross-over line (thanks to (5), also outside of the experimentally accessible part of this line,[13]) and the so-called D-term can be accessed. Moments, such as those entering the spin sum rule (1), can only be obtained from a GPD model, fitted to data, or more generally with help of some 'holographic' mapping:[13]

$$\{F(x,\eta=0,t,\mathcal{Q}^2), F(x,x,t,\mathcal{Q}^2)\} \quad\Longrightarrow\quad F(x,\eta,t,\mathcal{Q}^2)\,. \tag{7}$$

Here, $F^i(x,\eta=0,t,\mathcal{Q}^2)$ are constrained from form factor measurements and, additionally, GPDs $\widetilde{H}^i$ (H^i) by (un)polarized phenomenological PDFs.

2. GPD representations and modeling

Let us now turn to GPD modeling in different representations. First, GPDs might be defined as Radon transform of double distributions:[1,14] (DD)

$$F(x,\eta,t,\mu^2) = \int_0^1 dy\int_{-1+y}^{1-y} dz\,(1-x)^p\delta(x-y-z\eta)f(y,z,t,\mu^2)\,, \tag{8}$$

where $p\in\{0,1\}$. In this representation polynomiality, however, not positivity constraints are explicitly implemented. Moreover, with the right choice for p,[b] the polynomiality of x-moments can be completed to the required order in η. In the central, $-\eta \leq x \leq \eta$, and outer, $\eta \leq x \leq 1$, region the GPD can be interpreted as the probability amplitude of a t-channel meson-like and s-channel parton exchange, respectively. Mathematically, F is a twofold image of the DD f, where those in the central and the outer region can be mapped to each other.[11,13,15] The potential ambiguity, a term that lives only in the central region, is removed by requiring analyticity.[11,13]

Popular GPD models are based on Radyushkin's DD ansatz[14] for $t = 0$, where the DD factorizes into the PDF analogue $f(y)$ and a normalized profile function $\Pi(z)$. The GPD on the cross-over line is then given as

$$F(x,x) = \int_{-1}^{1} \frac{dz}{1-xz} f\left(\frac{x(1-z)}{1-xz}\right) \Pi(z)\,, \tag{9}$$

which is a linear integral equation of the first kind within the kernel $f(\frac{x(1-z)}{1-zx})/(1-xz)$. Knowing the GPD at $\eta = 0$, i.e., $f(y)$, and on the cross-over line, allows to determine the profile function and so to reconstruct the entire GPD[c], giving example of the 'holographic' mapping (7).

On the first glance a GPD in the outer region can be straightforwardly represented by an overlap of light-cone wave functions (LCWF),[16] which guarantees that positivity constraints are implemented. In simple models one might even reduce the number of non-perturbative functions, e.g., in a spectator diquark model, one only deals with one effective scalar LCWF for each struck quark species. This predicts for each of them four chiral even and four chiral odd GPDs. Moreover, one might use such a representation to evaluate also transverse momentum dependent parton distributions (TMDs). Unfortunately, there is a drawback. In the central region the GPD possesses an overlap representation in which the parton number is not conserved, and where the LCWFs are dynamically tied to those used in the outer region. A closer look reveals that Lorentz covariance already ties the momentum fraction and transverse momentum dependence of a LCWF.[17] Hence, a overlap representation is only usable if the LCWFs respect Lorentz symmetry, which would allow to restore the GPD in the central region.[17]

Strictly spoken, positivity constraints for GPDs are only valid at LO, since they can be violated by the factorization scheme ambiguity. Nevertheless, it would be desired to impose them on GPD models. One might

[b]Note that the factor $(1-x)$ might be replaced by a more general first order polynomial.
[c]An example is provided by $f(x) \propto x^{-\alpha}(1-x)^{\beta}$, which yields after some redefinitions the integral kernel $k = (1-xz)^{-\beta-1}$. The solution is then obtained in Mellin space.

follow the suggestion[18] and model GPDs as an integral transform of (triangle) Feynman diagrams, i.e., spectator quark models. A specific integral transformation, namely, a convolution with a spectator mass spectral function, can be used to include Regge behavior from the s-channel view. Such dynamical models provide also effective LCWFs or TMDs; however, simplicity is lost. In particular, PDF and form factor constraints cannot be implemented, i.e., one has to pin down such models within global fitting.

At present we neglect positivity constraints and model GPDs in the most convenient manner by means of a conformal SL(2,$\mathbb{R}$) partial wave expansion, which might be written as a Mellin-Barnes integral[15]

$$F(x, \eta, t, \mu^2) = \frac{i}{2} \int_{c-i\infty}^{c+i\infty} \frac{p_j(x, \eta)}{\sin(\pi j)} F_j(\eta, t, \mu^2) \,. \tag{10}$$

Here, $p_j(x, \eta)$ are the partial waves, given in terms of associated Legendre functions of the first and second kind, and the integral conformal GPD moments $F_j(\eta, t, \mu^2)$ are even polynomials in η of order j or $j + 1$. The advantages of this representation are: i. the conformal moments evolve autonomously at LO, ii. one can employ conformal symmetry to obtain NNLO corrections to the DVCS amplitude[11] and iii. PDF and form factor constraints can be straightforwardly implemented. Namely, $F_j(\eta = 0, t = 0, \mu^2)$ are the Mellin moments of PDFs, $F_{j=0}$ are partonic contributions to elastic form factors, $H_{j=1}$ and $E_{j=1}$ are the energy-momentum tensor form factors, and for general j one immediately makes contact to lattice measurements. One might expand the conformal moments in terms of t-channel SO(3) partial waves[19] $\hat{d}_j(\eta)$, expressed by Wigner rotation matrices and normalized to $\hat{d}_j(\eta = 0) = 1$. An effective GPD model at given input scale $\mathcal{Q}_0^2$ is provided by taking into account three partial waves,

$$F_j(\eta, t) = \hat{d}_j(\eta) f_j^{j+1}(t) + \eta^2 \hat{d}_{j-2}(\eta) f_j^{j-1}(t) + \eta^4 \hat{d}_{j-4}(\eta) f_j^{j-3}(t) \,, \tag{11}$$

valid for integral $j \geq 4$. Such a model allows us to control the size of the GPD on the cross-over line and its $\mathcal{Q}^2$-evolution, see right panel in Fig. 3.

3. Extracting CFFs and GPDs from DVCS measurements

Certainly, the access to GPDs from experimental data requires some software tools. The variety of both observables and models suggests setting up a flexible architecture, which would allow easy implementation of new models, processes, data sets and fitting strategies. We wrote software prototypes based on object-oriented programming paradigm (using Python) and, alternatively, on functional programming (using Mathematica). In both cases we group scattering processes, theoretical frameworks, models, and experimental data in classes, which also serve as databases. The cross sections are implemented as functions of electromagnetic form factors and CFFs, where models for them are set up separately. Experimental data is stored in ASCII files, similar to commonly used ones but more standardized. They contain all information needed for the evaluation of observables. After specifying data files, models, and conventions, our prototype software provides theory predictions for data points, depending on model parameters, which can be controlled by a fitting routine.

In a first global fit[20] to photon electroproduction off unpolarized proton we took sea quark and gluon GPD models with two SO(3) partial waves at small x, reparameterized the outcome from H1 and ZEUS DVCS fits at $\mathcal{Q}^2 = 2\,\mathrm{GeV}^2$, and employed it in fits of fixed target data within the scaling hypothesis. Thereby, we used the dispersion relation (4,5), where

$$H^{\mathrm{val}}(x,x,t) = \frac{1.35\,r}{1+x}\left(\frac{2x}{1+x}\right)^{-\alpha(t)}\left(\frac{1-x}{1+x}\right)^{b}\left(1-\frac{1-x}{1+x}\frac{t}{M^2}\right)^{-1} \tag{12}$$

specifies a valence-like GPD on the cross-over line. Here, the skewness ratio $r = \lim_{x\to 0} H(x,x)/H(x,0)$, $\alpha(t) = 0.43{+}0.85\,t/\mathrm{GeV}^2$, b controls the $x \to 1$ limit, and M the residual t-dependence, which we set to $M = 0.8\,\mathrm{GeV}$. The subtraction constant is normalized by d and M_d controls the t-dependence:

$$\mathcal{D}(t) = d\left(1-\frac{t}{M_d^2}\right)^{-2}. \tag{13}$$

We also included the parameter-free pion-pole model for the $\tilde{E}$ GPD[21] and parameterized the $\widetilde{H}$ GPD rather analogously to Eq. (12) with $b = 3/2$.

For the fixed target fits we chose two data sets. The first contains twist-two dominated (preliminary) beam spin asymmetry $A_{\mathrm{BS}}^{(1)}$ and beam charge asymmetry $A_{\mathrm{BC}}^{(i)}$ coefficients from HERMES[22,24] and 12 beam spin asymmetry coefficients $A_{\mathrm{BS}}^{(1)}$, which we obtained by Fourier transform of selected CLAS[25] data with small $-t$. The second data set includes also Hall A measurements[26] for four different t values. In light of the discussion[27] of Hall A data, we projected on the first harmonic of a *normalized* beam spin sum

$$\Sigma_{\mathrm{BS}}^{(1),w} = \int_0^{2\pi} dw\,\cos(\phi)\frac{d\sigma}{dx_{\mathrm{Bj}}dtd\mathcal{Q}^2d\phi}\Big/\int_0^{2\pi} dw\frac{d\sigma}{dx_{\mathrm{Bj}}dtd\mathcal{Q}^2d\phi}, \tag{14}$$

where $dw \propto \mathcal{P}_1(\phi)\mathcal{P}_2(\phi)d\phi$ includes the Bethe-Heitler propagators, and we also neglected then the helicity dependent cross sections (beam spin differences). We confirm that formed beam spin asymmetries are compatible with

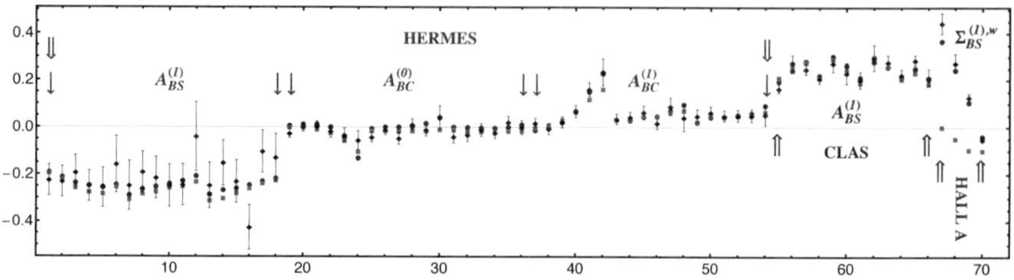

Fig. 1. Experimental vs. model results,[20] pinned down by fits without (squares) and with (circles) HALL A data, for fixed target kinematics: $A_{\mathrm{BS}}^{(1)}$ (1-18), $A_{\mathrm{BC}}^{(0)}$ (19-36), $A_{\mathrm{BC}}^{(1)}$ (37-54) from Ref. 23; $A_{\mathrm{BS}}^{(1)}$ (55-66) and $\Sigma_{\mathrm{BS}}^{(1),w}$ (67-70) are derived from Refs. 25 and 26.

CLAS ones[25] and we spell out that the second harmonics in HALL A data, i.e., effective twist-three contributions, are *tiny* or hard to separate from noise. Such contributions are small[d] in HERMES kinematics, too, where the constant $A_{\mathrm{BC}}^{(0)}$, appearing at twist-three level, is a twist-two dominated quantity that, as expected,[9] turns out to be correlated with $A_{\mathrm{BC}}^{(1)}$.

To relate the CFFs with the observables we employed the BKM formulas[9] within the 'hot-fix' convention.[28] First we excluded the Hall A data and set $\widetilde{H}$ to zero. A least squares fit (χ^2/d.o.f. ≈ 1) provides the parameters

$$r^{\mathrm{val}} = 0.95\,,\quad b^{\mathrm{val}} = 0.45\,,\quad d = -0.24\,,\quad M_d = 0.5\,\mathrm{GeV}\,, \tag{15}$$

from one local minimum, cf. description of recent data in Fig. 1 (squares). This should not be considered a unique solution; however, it is compatible with the expectation that the skewness effect at small x should be small, i.e., $r \sim 1$, and that, according to counting rules,[29] b should be smaller than the corresponding β value of a PDF. The smallness of b^{val} indicates a rather strong enhancement effect in the resonance region. In the second fit we included Hall A data, which remains challenging due to the steepness of data points ($-0.33\,\mathrm{GeV}^2 \leq t \leq -0.17\,\mathrm{GeV}^2$) — last four points in Fig. 1. Again we took some local minimum with χ^2/d.o.f. ≈ 1, giving

$$r^{\mathrm{val}} = 1.11\,,\quad b^{\mathrm{val}} = 2.4\,,\qquad d = -6.0\,,\quad M^{\mathrm{sub}} = 1.5\,\mathrm{GeV}\,. \tag{16}$$

One observes a slight increase of r^{val} and a larger value of b^{val}, i.e., the enhancement of the GPD H in the resonance region diminishes. In agreement with a chiral quark soliton model estimate,[30] the subtraction constant remains negative and is now sizable; however, at present a positive sign cannot be excluded. A closer look reveals that our first fit fails to describe Hall A beam spin sums and underestimates the beam spin differences by

[d]Except for 3×2 beam spin asymmetry data points at large $-t$, x_{B}, and $\mathcal{Q}^2$.

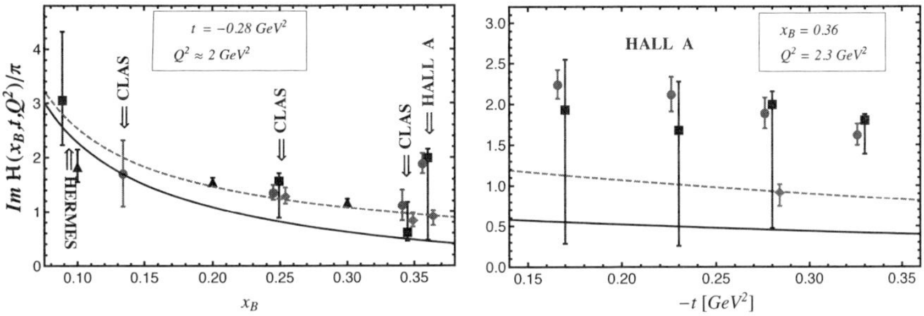

Fig. 2. $\Im\mathrm{m}\mathcal{H}/\pi$ obtained from different strategies: our model fits[20] (dashed (solid) curve excludes (includes) Hall A data), seven-fold CFF fit[32] with boundary conditions (squares), $\mathcal{H}$, $\widetilde{\mathcal{H}}$ CFF fit[31] (diamonds), smeared conformal partial wave model fit[33] within H GPD (circles). The triangles result from our neural network fit, cf. Fig. 3 (left).

about 50%, while our second one still underestimates all the cross sections by about 25%. For the latter we find a rather large remainder, effectively parameterized by $\widetilde{H}$, which is roughly five times bigger than expected. Longitudinally polarized target data provide a handle on $\widetilde{H}$,[9] where CFF fits[31] in JLAB kinematics provide at the means a two to three times bigger $\widetilde{H}$ contribution compared to our expectations ($r_{\widetilde{H}} \simeq 1, b_{\widetilde{H}} \simeq 2$). These findings are one to two standard deviations away from our big $\widetilde{H}$ ad hoc scenario.

So far we did not study model uncertainties or experimental error propagation, since both tasks might be rather intricate. To illuminate this, we compare in Fig. 2 our outcomes for $\Im\mathrm{m}\mathcal{H}(x_\mathrm{B}, t)/\pi$ versus x_B at $t = -0.28\,\mathrm{GeV}^2$ (left) and for Hall A kinematics x_B=0.36 versus $-t$ (right) with results that do provide error estimates. The squares arise from constrained least squares fits[32] at given kinematic means of HERMES and JLAB measurements on unpolarized proton, where the imaginary and real parts of twist-two CFFs are taken as parameters. Note that $\Im\mathrm{m}\widetilde{\mathcal{E}}$ and the other remaining eight CFFs are set to zero, however, all available observables, even those which are dominated by these CFFs, have been employed. This might increase 'statistics', however, yields also a growth of systematic uncertainties. The huge size of the errors mainly shows to which accuracy H can be extracted from unpolarized proton data alone.[9] A pure H GPD model fit[33] (circles) to JLAB data provides much smaller errors, arising from error propagation and some estimated model uncertainties. Both of our curves are compatible with the findings[32] and the H GPD model analysis[33] of CLAS data. However, for Hall A kinematics the deviation of the two predictions that are based on H dominance hypothesis, see dashed curve and circles in the right panel, are obvious and are explained by our underes-

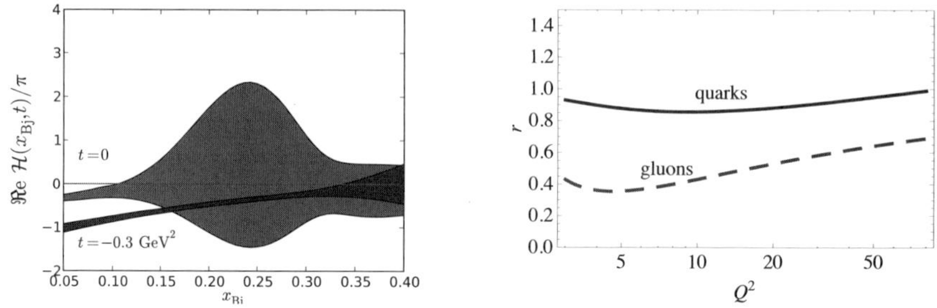

Fig. 3. Neural network extraction of $\Re\mathrm{e}\,\mathcal{H}(x_{\mathrm{Bj}},t)/\pi$ from BCA[23] and BSA[25] data (left). Ratio $r = H(x,x)/H(x,0)$ of LO sea quark and gluon GPDs for small x versus Q^2 (right).

timation of cross section normalization of about 50%. Moreover, the quality of fit[33] $\chi^2/\mathrm{d.o.f.} \sim 1.7$, might provide another indication that CLAS and Hall A data are not compatible within this hypothesis, see, e.g., the two rightmost circles in the left panel for CLAS ($x_B = 0.34$, $t = -0.3\mathrm{GeV}^2$, $Q^2 = 2.3\mathrm{GeV}^2$) and Hall A ($x_B = 0.36$, $t = -0.28\mathrm{GeV}^2$, $Q^2 = 2.3\mathrm{GeV}^2$). The pure $\mathcal{H}$ and $\widetilde{\mathcal{H}}$ CFF fit[31] (diamonds), including longitudinal polarized target data, is within error bars inconsistent with the H dominated scenario[33] (circles), however, (accidently) reproduces our dashed curve. All of these exemplifies that within (strong) assumptions and the present set of measurements the propagated experimental error cannot be taken as an estimate of GPD uncertainties. An error estimation in model fits might be based on twist-two sector projection technique,[9] boundaries for the superficial model degrees of freedom, and error propagation.

Let us give a short outlook. Neural networks may be an ideal tool to extract CFFs or GPDs. We present in Fig. 3 a first example in which, within H-dominance hypothesis, $\mathcal{H}$ is extracted by training 50 feed-forward nets with two hidden layers to HERMES BCA[23] and CLAS BSA[25] data. Hence, only the experimental errors were propagated, which in absence of a model hypothesis get large for the $t \to 0$ extrapolation. Furthermore, we note that our Fortran code, used so far for small x, is now combined with the dispersion relation model for valence quarks, which gives us the possibility to provide predictions for the photon electroproduction cross section over a wide kinematical range. Our new fitting results are compatible with the presented ones, where the GPDs are modeled with three effective SO(3) partial waves (11). Thus, to LO accuracy the skewness ratio for gluons is now $r \sim 0.5$ at small x and rather stable under evolution, cf. Fig. 3 and Fig. 4 of Ref. 20. Finally, we would like to add that the perturbative description of hard exclusive meson electroproduction data is under investigation.

In conclusion, a phenomenological access to the proton spin sum rule can be only reached within an understanding of GPD models, which can be also formulated in terms of an effective nucleon LCWF. A whole framework is available to reveal GPDs and to access the nucleon wave function, which should be considered as the primary task. To do so, reliable data, some mathematical understanding, and appropriate software tools are required.

References

1. D. Müller *et al.*, *Fortschr. Phys.* **42**, p. 101 (1994).
2. A. V. Radyushkin, *Phys. Lett.* **B380**, p. 417 (1996).
3. X. Ji, *Phys. Rev.* **D55**, p. 7114 (1997).
4. M. Diehl, *Phys. Rept.* **388**, p. 41 (2003).
5. A. V. Belitsky and A. V. Radyushkin, *Phys. Rept.* **418**, 1 (2005).
6. J. P. Ralston and B. Pire, *Phys. Rev.* **D66**, p. 111501 (2002).
7. X. Ji, *Phys. Rev. Lett.* **78**, p. 610 (1997).
8. M. Burkardt, A. Miller, W. D. Nowak, *Rept. Prog. Phys.* **73**, p. 016201 (2010).
9. A. V. Belitsky, D. Müller and A. Kirchner, *Nucl. Phys.* **B629**, p. 323 (2002).
10. O. V. Teryaev, hep-ph/0510031 (2005).
11. K. Kumerički, D. Müller, K. Passek-Kumerički, *Nucl.Phys.* **B794**, 244 (2008).
12. M. Diehl and D. Y. Ivanov, *Eur. Phys. J.* **C52**, 919 (2007).
13. K. Kumerički, D. Müller, K. Passek-Kumerički, *Eur.Phys.J.* **C58**, 193 (2008).
14. A. V. Radyushkin, *Phys. Rev.* **D56**, p. 5524 (1997).
15. D. Müller and A. Schäfer, *Nucl. Phys.* **B739**, 1 (2006).
16. M. Diehl *et al. Nucl. Phys.* **B596**, p. 33 (2001), Erratum-ibid. B605 (2001) 647; S. J. Brodsky, M. Diehl and D. S. Hwang, *Nucl. Phys.* **B596**, p. 99 (2001).
17. D. S. Hwang and D. Müller, *Phys. Lett.* **B660**, 350 (2008).
18. P. V. Pobylitsa, *Phys. Rev.* **D67**, p. 094012 (2003).
19. M. V. Polyakov, *Nucl. Phys.* **B555**, p. 231 (1999).
20. K. Kumeriĉki and D. Müller, *Nucl. Phys.* **B**, in print (2010), 0904.0458.
21. M. Penttinen, M. Polyakov and K. Goeke, *Phys. Rev.* **D62**, p. 014024 (2000).
22. F. Ellinghaus, *DVCS at HERMES: Recent Results*, arXive:0710.5768 (2007).
23. A. Airapetian *et al.*, *JHEP* **0911**, p. 083 (2009).
24. A. Airapetian *et al.*, *JHEP* **0806**, p. 066 (2008).
25. F. X. Girod *et al.*, *Phys. Rev. Lett.* **100**, p. 162002 (2008).
26. C. M. Camacho *et al.*, *Phys. Rev. Lett.* **97**, p. 262002 (2006).
27. M. V. Polyakov and M. Vanderhaeghen, arXiv:0803.1271 (2008).
28. A. V. Belitsky and D. Müller, *Phys. Rev.* **D79**, p. 014017 (2009).
29. F. Yuan, *Phys. Rev.* **D69**, p. 051501 (2004).
30. K. Goeke, M. Polyakov, and M. Vanderhaeghen, *Prog. Part. Nucl. Phys.* **47**, p. 401 (2001).
31. M. Guidal, *Phys. Lett.* **B689**, 156 (2010).
32. M. Guidal, *Eur. Phys. J.* **A37**, 319 (2008); M. Guidal and H. Moutarde, *Eur. Phys. J.* **A42**, 71 (2009).
33. H. Moutarde, *Phys. Rev.* **D79**, p. 094021 (2009).

MOMENTS OF GPDs AND THE NUCLEON SPIN STRUCTURE FROM LATTICE QCD

PH. HÄGLER

Institut für Theoretische Physik T39
Physik-Department der TU München
James-Franck-Strasse
D-85747 Garching, Germany
E-mail: phaegler@ph.tum.de

Lattice QCD calculations of hadron structure have seen impressive progress over the last years. In this contribution, we illustrate the recent achievements by presenting the latest lattice computation of moments of generalized parton distributions, and then focus on the impact of these results on the decomposition of the nucleon spin 1/2 in terms of quark spin and orbital angular momentum contributions. Potential discrepancies with expectations from theoretical analyses and model studies, as well as a comparison with previous lattice and phenomenological results, will be discussed in some detail.

Keywords: GPDs; Nucleon Spin Structure; Lattice QCD.

1. Introduction

Since the famous EMC measurement[1] of the spin structure function g_1^p more than two decades ago at the latest[a], the quest for a solid understanding of the origin of the nucleon spin 1/2 has been at the center of countless experimental, theoretical and phenomenological investigations. These studies provided remarkable insights into the quark and gluon structure of the nucleon, often on the basis of progress made with respect to elastic and deeply inelastic lepton-hadron- and related scattering processes.

Generalized parton distributions (GPDs)[b] turned out to provide a particularly successful framework for the understanding of hadron structure, that not only encompasses the well known hadron form factors and the

[a]that gave rise to what is known as the proton spin crisis
[b]for reviews, we refer to Refs. [2–4]

PDFs, but that also provides a solid basis for the decomposition of the nucleon spin in terms of spin and (orbital) angular momentum contributions of quarks and gluons. According to [5], the nucleon spin can be given in terms of x-moments of the GPDs $H(x,\xi,t)$ and $E(x,\xi,t)$,

$$\begin{aligned}\frac{1}{2} &= \sum_{q=u,d,s,\ldots;g} J_{q,g} = \sum_{q,g} \frac{1}{2} \int_{-1}^{1} dx x \{H^{q,g}(x,\xi,t) + E^{q,g}(x,\xi,t)\}_{t=0} \\ &= \sum_{q,g} \frac{1}{2} \big(A_{20}^{q,g}(t) + B_{20}^{q,g}(t)\big)_{t=0}\,, \qquad (1)\end{aligned}$$

where A_{20} and B_{20} are generalized form factors (GFFs) that will be discussed in greater detail below. It is important to keep in mind that the total angular momentum of quarks can be further decomposed in a manifestly gauge-invariant manner in terms of quark spin and orbital angular momentum contributions, $J_q = \Delta\Sigma/2 + L_q$ [5].

While the GPDs can be measured for example in deeply virtual Compton scattering employing QCD factorization, their extraction from experimental data turns out to be very challenging. This holds in particular for their dependence on x at fixed ξ, and therefore also for the x-moments in the decomposition in Eq. (1). As it turns out, this is different in the framework of lattice QCD, which represents at least in principle a straightforward (but in practice quite challenging) approach from first principles to the lowest x-moments of PDFs and GPDs. Lattice QCD calculations of hadron structure, which are in many cases complementary to phenomenological and experimental studies, have seen remarkable progress during the last 10 years. For reviews of lattice QCD studies of, e.g., form factors, moments of PDFs, GPDs and distributions amplitudes, we refer to Refs. [6–8].

Significant advances in particular with respect to moments of GPDs have been made since the pioneering calculations by the LHPC and QCDSF collaborations in 2003.[9,10] A comprehensive lattice study of GPDs by LHPC using a mixed action approach was presented in Ref. [11]. The discussions in the following sections will be based on a recent update of this study,[12] which includes an additional ensemble at a lower pion mass of $m_\pi \approx 300$ MeV, a factor of 8 increased statistics, and a strongly improved statistical analysis.

2. Moments of GPDs

The x-moments of generalized parton distributions can be written in terms of polynomials in powers of ξ with the GFFs as coefficients. For the $n=2$-

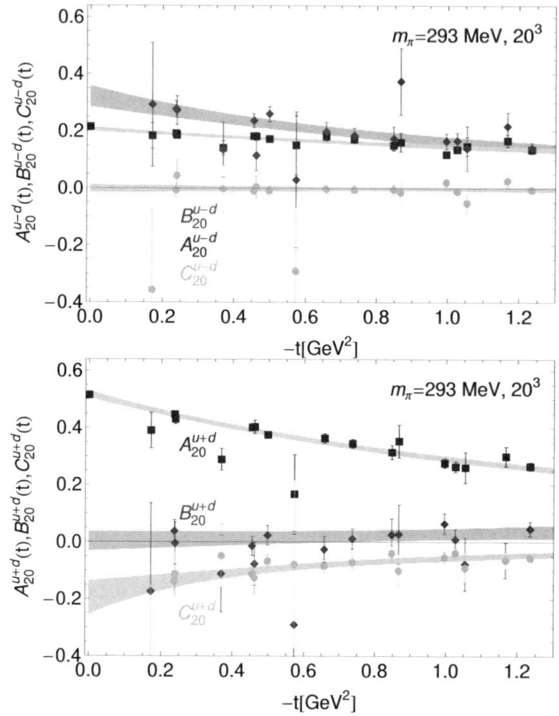

Fig. 1. Generalized form factors A, B, C as functions of the squared momentum transfer t, in the $\overline{\text{MS}}$ scheme at a scale of $\mu = 2$ GeV, for a pion mass of $m_\pi \approx 300$ MeV [12]. The t-dependence can be parametrized by, e.g., a dipole ansatz, as illustrated by the shaded error bands.

(x-)moments in the unpolarized case, the relevant GFFs A_{20}, B_{20}, C_{20} parametrize the off-forward matrix element of the (symmetric, traceless) energy momentum tensor, e.g. for quarks

$$\langle P'|\overline{q}\gamma^{\{\mu}D^{\nu\}}q|P\rangle = \langle P'|T_q^{\mu\nu}|P\rangle = \overline{U}(P')\Bigg\{\gamma^{\{\mu}\overline{P}^{\nu\}}A_{20}(t) - \frac{i\Delta_\rho\sigma^{\rho\{\mu}\overline{P}^{\nu\}}}{2m_N}B_{20}(t) + \frac{\Delta^\mu\Delta^\nu}{m_N}C_{20}(t)\Bigg\}U(P), \quad (2)$$

where $\overline{P} = (P' - P)/2$, $\Delta = P' - P$, and $t = \Delta^2$. Exemplary lattice results from LHPC[12] for the t-dependence of these GFFs in the isovector and isosinglet channels, for a comparatively low pion mass of ≈ 300 MeV, are displayed in Fig. 1. We note at least two characteristic features: First, while the B_{20}-GFF is dominant in the isovector case, it is very small and fully

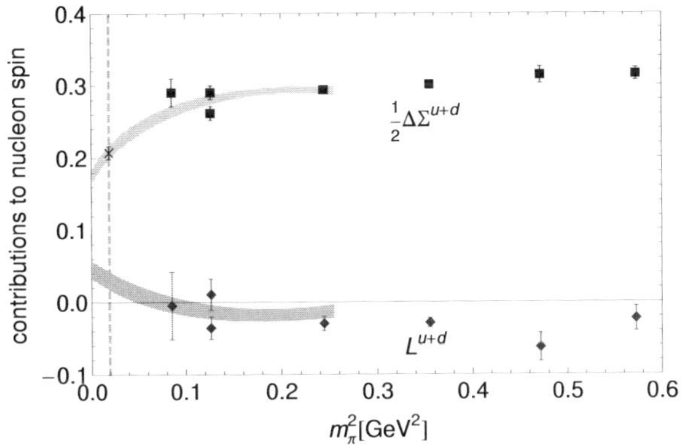

Fig. 2. Contributions to the nucleon spin in the $\overline{\text{MS}}$ scheme at a scale of $\mu = 2$ GeV [12].

compatible with zero within the uncertainties for all accessible values of t in the isosinglet channel. Conversely, the GFF C_{20} (which is directly related to the $n=2$-moment of the so-called D-term[13]) is zero within relatively small errors in the isovector channel, but clearly non-zero and negative for $u+d$-quarks. From the sum of A_{20} and B_{20} at zero momentum transfer, we can compute the quark angular momentum J_q, cf. Eq. (1). The above findings for B_{20} obviously have a significant impact on the decomposition of the nucleon spin in terms of quark and gluon degrees of freedom.

3. Decomposition of the nucleon spin

Instead of calculating matrix elements of the quark OAM operator directly on the lattice, we can employ the relation $L_q = J_q - \Delta\Sigma/2$, together with corresponding results for the quark spin fraction $\Delta\Sigma$, to compute the OAM carried by the quarks. The numerical results for $u+d$-quarks are shown in Fig. 2 as a function of the squared pion mass. For the extrapolation of $\Delta\Sigma_{u+d}$ to the physical pion mass, we employed the heavy baryon ChPT results of Ref. [14]. Although HBChPT at the given order is most probably not applicable at the accessible pion masses, the two-parameter fit to the lattice data points, indicated by the error band, shows an excellent (but likely accidental) agreement with the value from HERMES,[15] indicated by the cross, at m_π^{phys}. We used the results of a covariant baryon chiral perturbation theory calculation[16] for the simultaneous extrapolation of the GFFs A_{20}, B_{20}, C_{20} in m_π and in t, in particular to obtain $B_{20}(t=0, m_\pi^{\text{phys}})$. The results of the fits to the lattice data were subsequently used to compute L_q, which is shown in Fig. 2 for $q = u+d$ by the triangles and the error

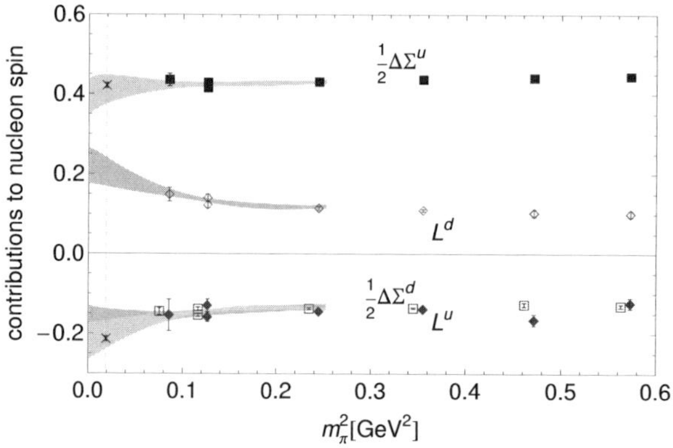

Fig. 3. Contributions from up and down quarks to the nucleon spin in the $\overline{\text{MS}}$ scheme at a scale of $\mu = 2$ GeV [12].

band. Most remarkably, following this procedure, we find a very small OAM contribution of only $L_{u+d} \approx (6 \pm 3)\%$ of 1/2 at the physical pion mass. Note that we obtain a slightly larger value, $L_{u+d} \approx (11 \pm 3)\%$, on the basis of a HBChPT extrapolation of J_{u+d} employing the results of Ref. [17]. These results may be surprising and at first sight even appear to be in clear contradiction to expectations from relativistic quark models, where the quark OAM contributes about $\approx 30 - 40\%$ to the total nucleon spin. Moreover, it has been frequently pointed out in the past that non-zero quark orbital angular momentum is strictly required with respect to certain non-vanishing single spin asymmetries related to, e.g., the Sivers effect, as well as for the Pauli form factor $F_2(Q^2)$ to be non-zero.[18–21]

Concerning the latter, it is important to be precise in the use of the term *quark orbital angular momentum*: The lattice results on the one hand correspond to the proton matrix element of the manifestly gauge invariant quark OAM operator as given in, e.g., Ref. [5], which is part of Jis nucleon spin sumrule, see Eq. (1) and the adjacent discussion. On the other hand, what is required with respect to a non-vanishing $F_2(Q^2)$ and the Sivers effect are light cone wave functions with non-zero orbital angular momentum in an overlap representation of the corresponding matrix elements. Also, the more heuristic approach to explain single spin asymmetries, e.g. the Sivers effect, proposed in Ref. [20], requires a generically non-zero *quark orbital motion*. The notions of OAM and orbital motion in these contexts are, however, not necessarily in one-to-one correspondence with the gauge invariant L_q discussed before.

Furthermore, we note that L_q is in general not positive definite. To fully appreciate the potential consequences of this basic observation, it is useful to study the individual up and down quark OAM contributions to the nucleon spin, which are shown in Fig. 3 as functions of m_π^2, together with $\Delta\Sigma_{u,d}/2$ and the corresponding chiral extrapolations (given by the error bands). It indeed turns out that the orbital angular momentum of up and of down quarks is in each case *large* and of similar magnitude, but opposite in sign, and therefore nearly cancels in the sum over a wide range of pion masses. From the chiral extrapolations, we find that the individual quark OAM contributions are also substantial at the physical pion mass, amounting to $|L_{u,d}| \approx 33\%$. For an understanding of the mechanism behind the observed approximate cancellation, one probably has to study more complex correlation functions on the lattice, or get inspiration from model calculations. We also note that cancellations of this sort in the isosinglet channel are not unusual. A well known example is the anomalous magnetic moment, which is large both for the proton and for the neutron, $\kappa_p = F_2^p(0) \sim 1.79$ and $\kappa_n = F_2^n(0) \sim -1.91$, respectively. Using isospin symmetry, however, one finds that the sizeable, individual contributions from up and down quarks to κ_p largely cancel out in the sum, $\kappa_p^{u+d} \approx -0.36$. Keeping in mind that $F_2(Q^2 = -t)$ correspond to the first, and the GFF $B_{20}(t)$ to the second x-moment of the GPD $E(x, \xi = 0, t)$, we also find this to be perfectly consistent with the cancellation observed above in Fig. 1, where in particular $B_{20}^{u+d}(t) \approx 0$, while $B_{20}^{u-d}(t)$ is large and positive.

In the following, we will briefly address the apparent discrepancy between the common expectations from relativistic quark model calculations that $L_{u+d} \approx 30 - 40\%$, and the lattice QCD result of a nearly vanishing total quark OAM, $L_{u+d} \approx 0$. As has been pointed already out some time ago,[22,23] the model calculations generically correspond to a low hadronic scale $\ll 1$ GeV, while a typical renormalization scale for which the lattice results are given is $\mu \sim 2$ GeV (e.g. in the $\overline{\text{MS}}$-scheme). Since OAM is a scale and scheme dependent quantity,[24] a naive, direct comparison of lattice and model values at the respective scales is therefore in general pointless. However, one might try to use QCD evolution equations to evolve the lattice values down to the lower model scale, where they may be compared to the model results. To this extent, we show in Fig. 4 the evolution of $\Delta\Sigma_{u+d}/2$ and L_{u+d} at NLO.[25] Although the evolution equations are quantitatively not applicable at very low scales, it is still interesting to see that the quark OAM indeed *increases* substantial at lower μ^2. The initial suspicion about a contradiction of lattice and model results hence appears to be premature.

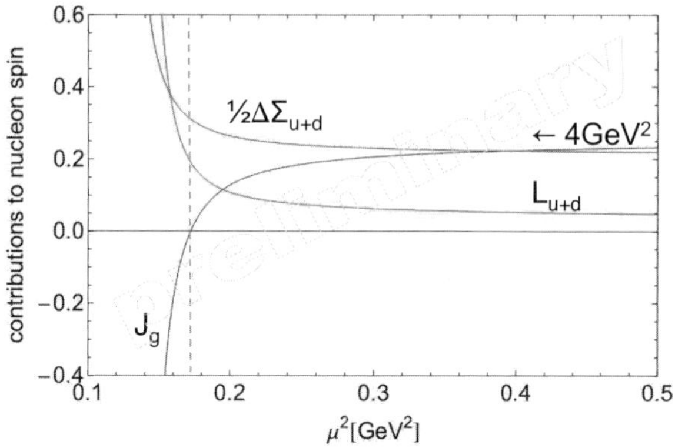

Fig. 4. Evolution of contributions to the nucleon spin [25].

Accidentally, one even finds a very good agreement with relativistic quark model results at the (model) scale where the gluon contribution vanishes identically[c], $J_g = 0$, and which is indicated by the vertical dashed line in Fig. 4.

Finally, we show a comparison of the results for the total quark angular momentum $J_q = (A^q_{20} + B^q_{20})_{t=0}/2$, as obtained from the chiral extrapolations of A_{20} and B_{20} mentioned above, with earlier lattice as well as model- and phenomenological calculations in Fig. 5 in the J_u, J_d-plane. It turns out that the numerical values,[12] $J_u = 0.236(6)$ and $J_d = 0.0018(37)$, are in surprisingly good agreement not only with the previous lattice computations, but also with most of the phenomenological and model results, at least when they are evolved to the common renormalization point of 4 GeV^2. We note in passing that in the lattice calculations, the smallness of J_d can be traced back to a cancellation of the separately sizeable spin- and OAM contributions of down quarks, as can be inferred directly from Fig. 3. It will be very interesting to see if the slight discrepancy to the cloudy bag model (CBM) study of Ref. [23] is due to systematic effects, missing contributions or similar on the side of the lattice[d] and the other approaches, or if it points towards the need for a further improvement of the CBM calculation.

[c] note that many basic quark models do not include any explicit gluon degrees of freedom

[d] most notable issues are: discretization and finite volume effects, missing disconnected contributions, and the so far neglected mixing of quark and gluon operators under renormalization and evolution in the singlet sector

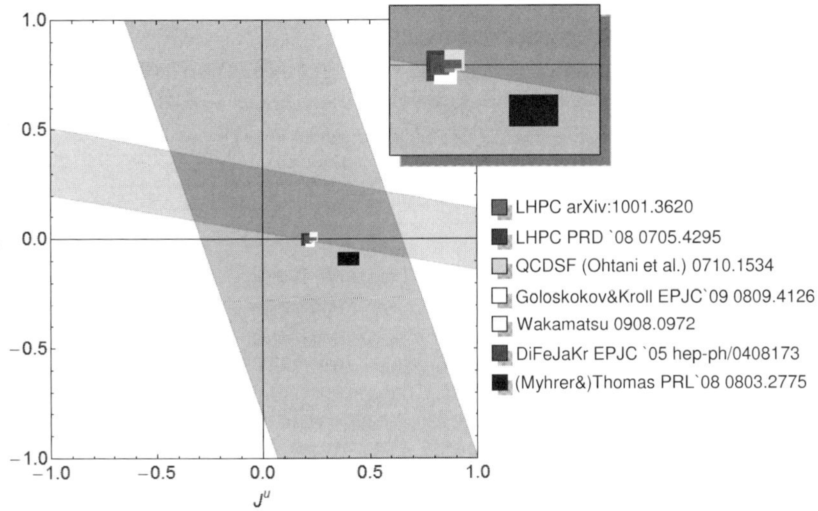

Fig. 5. Comparison of quark angular momentum contributions to the nucleon spin, obtained in different lattice (for uncertainties, see footnote [d]), model and phenomenological studies,[11,23,26–29] displayed in the J_u, J_d-plane in the style of Refs. [30,31], for the $\overline{\text{MS}}$ scheme at a scale of $\mu = 2$ GeV. Note that the shaded bands are mainly given for illustration purposes and only approximately represent the constraints obtained in Refs. [30,31], which are still strongly model dependent.

Acknowledgments

It is a pleasure to thank M. Altenbuchinger, B. Musch and W. Weise from the theory group T39 at the TU München, and the colleagues from LHPC. I gratefully acknowledge the support by the Emmy-Noether program and the cluster of excellence "Origin and Structure of the Universe" of the DFG.

References

1. J. Ashman *et al.*, *Nucl. Phys.* **B328**, p. 1 (1989).
2. K. Goeke, M. V. Polyakov and M. Vanderhaeghen, *Prog. Part. Nucl. Phys.* **47**, 401 (2001).
3. M. Diehl, *Phys. Rept.* **388**, 41 (2003).
4. A. V. Belitsky and A. V. Radyushkin, *Phys. Rept.* **418**, 1 (2005).
5. X.-D. Ji, *Phys. Rev. Lett.* **78**, 610 (1997).
6. J. M. Zanotti, *PoS* **LAT2008**, p. 007 (2008).
7. Ph. Hägler, *Phys. Rept.* **490**, 49 (2010).
8. D. B. Renner, *PoS* **LAT2009**, 018 (2009).
9. Ph. Hägler *et al.*, *Phys. Rev.* **D68**, p. 034505 (2003).
10. M. Göckeler *et al.*, *Phys. Rev. Lett.* **92**, p. 042002 (2004).
11. Ph. Hägler *et al.* [LHPC], *Phys. Rev.* **D77**, p. 094502 (2008).

12. J. D. Bratt *et al.* [LHPC], arXiv:1001.3620 [hep-lat], to appear in *Phys.Rev. D* (2010).
13. M. V. Polyakov and C. Weiss, *Phys. Rev.* **D60**, p. 114017 (1999).
14. M. Diehl, A. Manashov and A. Schäfer, *Eur. Phys. J.* **A31**, 335 (2007).
15. A. Airapetian *et al.*, *Phys. Rev.* **D75**, p. 012007 (2007).
16. M. Dorati, T. A. Gail and T. R. Hemmert, *Nucl. Phys.* **A798**, 96 (2008).
17. J.-W. Chen and X.-D. Ji, *Phys. Rev. Lett.* **88**, p. 052003 (2002).
18. S. J. Brodsky, D. S. Hwang, B.-Q. Ma and I. Schmidt, *Nucl. Phys.* **B593**, 311 (2001).
19. S. J. Brodsky, D. S. Hwang and I. Schmidt, *Phys. Lett.* **B530**, 99 (2002).
20. M. Burkardt, *Phys. Rev.* **D66**, p. 114005 (2002).
21. M. Burkardt and D. S. Hwang, *Phys. Rev.* **D69**, p. 074032 (2004).
22. M. Wakamatsu and Y. Nakakoji, *Phys. Rev.* **D77**, p. 074011 (2008).
23. A. W. Thomas, *Phys. Rev. Lett.* **101**, p. 102003 (2008).
24. X.-D. Ji, J. Tang and P. Hoodbhoy, *Phys. Rev. Lett.* **76**, 740 (1996).
25. M. Altenbuchinger, Ph. Hägler, and W. Weise, in preparation.
26. J. D. Bratt *et al.*, *PoS* **LATTICE2008**, p. 141 (2008).
27. D. Brömmel *et al.*, *PoS* **LAT2007**, p. 158 (2007).
28. M. Wakamatsu, *Eur. Phys. J.* **A44**, 297 (2010).
29. M. Diehl, T. Feldmann, R. Jakob and P. Kroll, *Eur. Phys. J.* **C39**, 1 (2005).
30. A. Airapetian *et al.*, *JHEP* **06**, p. 066 (2008).
31. M. Mazouz *et al.*, *Phys. Rev. Lett.* **99**, p. 242501 (2007).

GPD PROGRAM AT COMPASS

N. D'HOSE* for the COMPASS Collaboration

CEA, Centre de Saclay
Irfu/SPhN, 91191 Gif-sur-Yvette, France
**E-mail: nicole.dhose@cea.fr*

The study of exclusive reactions like Deeply Virtual Compton Scattering (DVCS) and Meson Production is one major part of the future COMPASS program[1] in order to investigate nucleon structure through Generalised Parton Distributions (GPD). The high energy of the muon beam allows to measure the x_B-dependence of the t-slope of the pure DVCS cross section and to study nucleon tomography. The use of positive and negative polarised muon beams allows to determine the Beam Charge and Spin Difference of the DVCS cross sections to access the real part of the Compton form factor related to the dominant GPD H.

Keywords: Nucleon Structure, Generalised Parton Distributions, Deeply Virtual Compton Scattering

1. Kinematic domain provided by COMPASS

The COMPASS apparatus is located at the unique CERN SPS M2 beam line that is able to deliver high-energy (50-280 GeV) and highly-polarised $\mu^{\pm}$ particles. It consists of a high-resolution forward spectrometer and an unpolarised, longitudinally or transversely polarised target. By installing a recoil proton detector around a liquid hydrogen target to ensure exclusivity of the Deeply Virtual Compton Scattering (DVCS) and Deeply Virtual Meson Production (DVMP) events, COMPASS would become a facility measuring exclusive reactions within a kinematic domain ranging from $x_B \sim 0.01$ to about 0.1, which cannot be explored by any other existing or planned facility in the near future (Fig. 1). COMPASS would thus explore the uncharted x_B domain between the HERA collider experiments H1 and ZEUS and the fixed-target experiments as HERMES and the planned 12 GeV extension of the JLab.

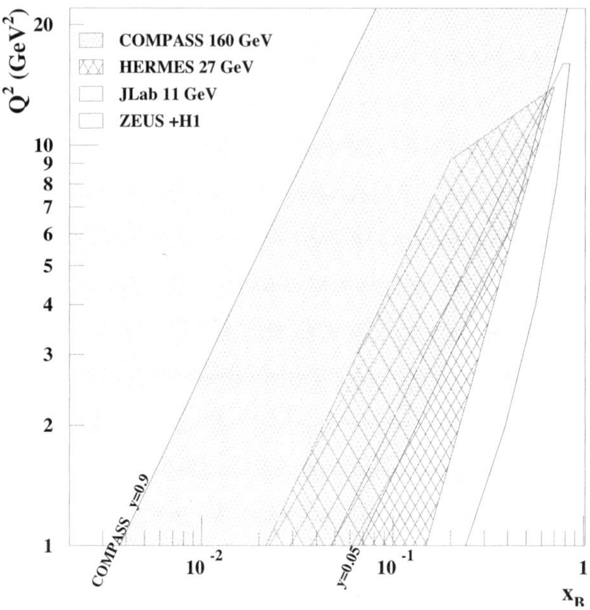

Figure 1. Kinematic domains for measurements of hard exclusive processes shown for the COMPASS (green area enclosed by the lines $y = 0.9$ and $y = 0.05$), HERMES and JLab fixed-target experiments and the HERA collider experiments H1 and ZEUS. COMPASS will explore the uncharted territory in between the collider region and that of the lower-energy fixed-target HERMES and JLab experiments.

2. Methodology for DVCS with high energy polarised μ^+ and μ^- beams

Deeply Virtual Compton Scattering has the same final state as the competing Bethe–Heitler (BH) process, which is elastic lepton–nucleon scattering with a hard photon emitted by either the incoming or outgoing lepton. The differential cross section for hard exclusive muoproduction of real photons off an unpolarised proton target can be written as[a]

$$\frac{\mathrm{d}^4\sigma(\mu p \to \mu p \gamma)}{\mathrm{d}x_B \mathrm{d}Q^2 \mathrm{d}|t| \mathrm{d}\phi} = \mathrm{d}\sigma^{BH} + \left(\mathrm{d}\sigma^{DVCS}_{unpol} + P_\mu\, \mathrm{d}\sigma^{DVCS}_{pol}\right) + e_\mu \left(\mathrm{Re}\, I + P_\mu\, \mathrm{Im}\, I\right), \tag{1}$$

[a]For simplicity $\mathrm{d}\sigma$ is used in the following instead of $\frac{\mathrm{d}^4\sigma(\mu p \to \mu p \gamma)}{\mathrm{d}x_B \mathrm{d}Q^2 \mathrm{d}|t| \mathrm{d}\phi}$.

where P_μ is the polarisation and e_μ the charge in units of the elementary charge, of the polarised muon beam. The interference term I arises as the DVCS and BH processes interfere on the level of amplitudes.

COMPASS offers the advantage to provide various kinematic domains where either BH or DVCS dominate (see Fig. 2). The Bethe-Heitler am-

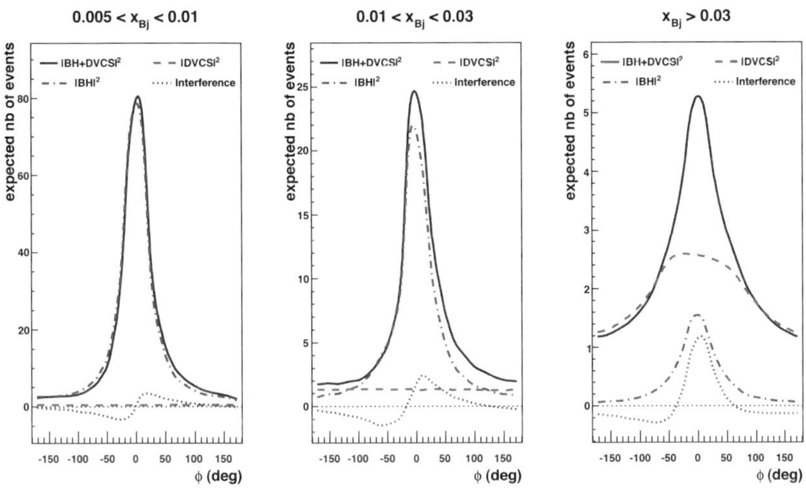

Figure 2. Monte Carlo simulation of the exclusive process $\mu^+ p \to \mu'^+ p\gamma$ for $Q^2 > 1\ \text{GeV}^2$, showing the ϕ angle distribution for three bins in x_B: $0.005 < x_B < 0.01$ (left), $0.01 < x_B < 0.03$ (middle) and $0.03 < x_B$ (right). The event yield shown is normalised to the integrated luminosity of the 2009 DVCS test run, as described in Sect. 5. It is based on the acceptance of the presently existing COMPASS set-up, *i.e.*, using the existing calorimeters ECAL1 and ECAL2.

plitude is well known (it relies only on the knowledge of elastic nucleon form factors). The collection of almost pure BH events at small x_B allows one to get an excellent reference yield and to control accurately the global efficiency of the apparatus. In contrast the collection of almost pure DVCS events at larger x_B will allow the measurement of the x_B-dependence of the t-slope of the cross section which is related to the tomographic partonic image of the nucleon. In the intermediate domain, the DVCS contribution will be boosted by the BH process through the interference term. COMPASS is presently the only facility to provide polarised leptons with either charge: polarised μ^+ and μ^- beams. As the BH is independent of charge and polarisation, this contribution can be removed by subtracting 2 separate measurements obtained for the two beam charges. Moreover the natural polarisation of the muon beam produced from pion decay changes sign when the beam charge is reversed and the different topologies of μ^+ and μ^-, polarised with opposite direction, allows one to select only the real part or the imaginary part of the complex amplitude of DVCS.

3. Beam Charge and Spin Difference of the DVCS cross sections to determine the real part of the Compton form factor $\mathcal{H}$

The Beam Charge (C) and Spin (S) Difference for Unpolarized (U) proton can be written as:

$$\mathcal{D}_{CS,U} \equiv d\sigma^{\overset{+}{\leftarrow}} - d\sigma^{\overset{-}{\rightarrow}} = 2[P_\mu d\sigma_{pol}^{DVCS} + e_\mu \text{Re}\, I] \quad (2)$$
$$\propto \left(\{s_1^{DVCS} \sin\phi\}\right) + \left(c_0^I + c_1^I \cos\phi + \{c_2^I \cos 2\phi + c_3^I \cos 3\phi\}\right)$$

in which the BH contribution *cancels out.* In this formula, the DVCS amplitude is expanded in $1/Q$ beyond leading twist-2 including all twist-3 contributions.[2] The coefficients c_n^{DVCS} and s_n^I are related to certain combinations of Compton Form Factors (CFFs). A CFF $\mathcal{F}$ is a sum over flavors f, of convolutions of the respective GPDs F^f with a perturbatively calculable kernel describing the hard $\gamma^* q$ interaction. Note that each contribution in Eq. 2 and Eq. 3 shown between a pair of braces corresponds to higher-twist or higher-order effects. The analysis of the ϕ-dependence of the beam charge and spin difference $\mathcal{D}_{CS,U}$ will provide via the term Re I the two leading twist-2 expansion coefficients c_0^I and c_1^I. In COMPASS kinematics they are mainly related to the *real* part of the Compton form factor $\mathcal{H}$ that is in LO given by a flavor sum of convolutions involving the GPDs H^f *i.e.* $\text{Re}\mathcal{H}^f = \mathcal{P}\int_{-1}^{+1} \frac{H^f(x,\xi,t)}{x-\xi} dx$. In fact the GPDs depend upon three kinematic variables: t the total four-momentum squared transferred between initial and final nucleon states, x and $\xi \sim x_B/(2-x_B)$ respectively average and half the difference between the initial and final longitudinal momentum fractions of the nucleon, carried by the partons throughout the process. In hard exclusive processes as DVCS, x is then an internal variable which is integrated over in a convolution.

Fig. 3 shows the projected statistical accuracy in a particular (x, Q^2) bin, for a measurement of the ϕ-dependence of the beam charge and spin difference $\mathcal{D}_{CS,U}$. Two of the curves are calculated using the "VGG" GPD model.[3] The model based on a "reggeized" parameterisation of the correlated x, t dependence of the GPDs uses a shrinkage parameter α' to describe the decrease in nucleon size with increasing x. As this model is meant to be applied mostly in the valence region, typically a large value $\alpha' = 0.8$ is used. For comparison, also the model for the "factorized" x, t dependence is shown, which corresponds to $\alpha' \approx 0.1$ in the 'reggeized' ansatz. A recent theoretical development exploits dispersion relations for Compton form factors. In this context, the additional curve is the result of a fitting procedure[4] including next-to-next-to leading order (NNLO) corrections which was developed and successfully applied to describe DVCS observables from very small values of x_B, for the HERA collider to large x_B for HERMES and JLab. It has to be noted that the real part of the Compton form factor $\mathcal{H}$ was found positive at H1 and ZEUS and negative at HERMES and JLab. The COMPASS kinematic domain is expected to provide the node of the real part of this amplitude, which is an essential input for any global fit analysis.

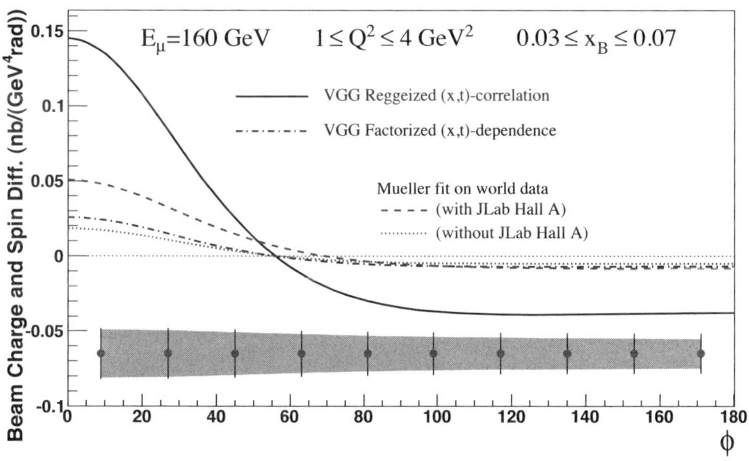

Figure 3. Projected statistical (error bars) and systematic (grey band) accuracy for a measurement of the ϕ dependence of the Beam Charge and Spin Difference at COMPASS for $0.03 \leq x \leq 0.07$ and $1 \leq Q^2 \leq 4$ GeV^2 and for 280 days of running time with a 2.5m LH_2 target, an intensity of $4.6 \times 10^8 \mu$ in a 48 s SPS spill period and an overall *global efficiency* $\epsilon_{global} = 0.1$. The solid and dash-dotted curves correspond to different variants of the VGG model[3] while the dashed and dotted curves show predictions based on first fits on world data.[4]

4. x_B-dependence of the t-slope of the pure DVCS cross section to study nucleon tomography

The Beam Charge and Spin Sum of cross sections can also be evaluated:

$$\begin{aligned} \mathcal{S}_{CS,U} \equiv d\sigma^{\overset{+}{\leftarrow}} + d\sigma^{\overset{-}{\rightarrow}} &= 2[d\sigma^{BH} + d\sigma^{DVCS}_{unpol} + e_\mu P_\mu \mathrm{Im}\, I] \\ &\propto 2[d\sigma^{BH}] + \left(c_0^{DVCS} + \left\{c_1^{DVCS}\cos\phi + c_2^{DVCS}\cos 2\phi\right\}\right) \\ &\quad + \left(s_1^I \sin\phi + \left\{s_2^I \sin 2\phi\right\}\right) \end{aligned} \tag{3}$$

in which the BH contribution *does not cancel out.*

(1) The analysis of the ϕ-dependence of the beam charge and spin sum $\mathcal{S}_{CS,U}$ will provide via the term Im I the leading twist-2 quantity s_1^I. Its dominant contribution is related to the *imaginary* part of the Compton form factor $\mathcal{H}$, *i.e.* $\mathrm{Im}\mathcal{H}^f \propto H(\xi, \xi, t)$.

(2) A parallel analysis can be performed subtracting the BH contribution when it is not too large, and integrating over ϕ to get rid of the complete interference term and of the ϕ-dependent terms of the DVCS contribution. Thus the DVCS leading twist-2 quantity c_0^{DVCS} can be isolated and its characteristic t-slope can be determined as a function of x_B, from which conclusions can be drawn on the evolution of transverse size of the nucleon over the x_B-range accessible to COMPASS. Such a mapping in x_B is referred to as "Nucleon Tomography".

Fig. 4 shows the projected statistical accuracy for a measurement at COMPASS of the x_B-dependence of the t-slope parameter $B(x_B)$ of the DVCS cross section. In the simple ansatz $\frac{d\sigma}{dt} \propto \exp(-B(x_B)\,|t|)$ and $B(x_B) = B_0 + 2\,\alpha' \log(\frac{x_0}{x})$. At small x_B, where amplitudes are predominantly imaginary, the overall transverse size of the nucleon $\langle r_\perp^2\,(x_B)\rangle \approx 2{\cdot}B(x_B)$. Data on B exist only for the HERA collider x_B-range from 10^{-4} to 0.01,[5,6] below the COMPASS range $0.01 < x_B < 0.1$. No evolution with x_B was observed and a first transverse proton radius has been determined $\langle r_\perp^2\rangle = 0.65 \pm 0.02$ fm using H1 data.[5] In the valence region, where no experimental determinations of B exist, some information comes from fits adjusted to form factor data which give $\alpha' \simeq 1$ GeV^2.[7,8] For the simulation two values $\alpha' = 0.125$ and $\alpha' = 0.26$ are shown which correspond to the half and the total of the value for Pomeron exchange in soft scattering processes. The largest value can be determined with an accuracy better than 2.5 sigma by using the two existing calorimeters ECAL1 and ECAL2 while the smallest value requires a new ECAL0 to get a better precision at large x_B.

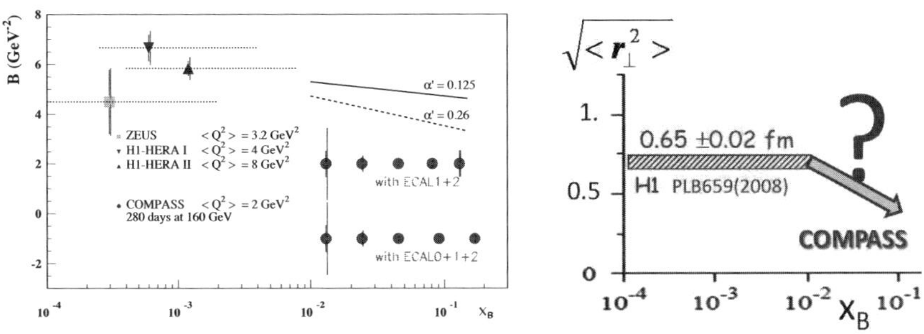

Figure 4. Left: Projections for measuring the x_B dependence of the t-slope parameter $B(x_B)$ of the DVCS cross section, calculated for $1 < Q^2 < 8$ GeV^2. For comparison some HERA results with similar $\langle Q^2 \rangle$ are shown.[5,6] The left vertical bar on each data point indicates the statistical error only while the right one includes also the systematic uncertainty, using only ECAL1 and ECAL2 (first row) and also ECAL0 (second row). Two different parameterisations are shown using $\alpha' = 0.125$ GeV^{-2} and 0.26 GeV^{-2}. **Right:** Transverse proton radius as a function of x_B. New and significant information will be obtained in the uncharted x_B region, further elucidating the issue of "nucleon tomography".

5. First Observation of BH and DVCS events in 2008 and 2009 tests

A preliminary study of the feasibility of DVCS measurements was performed in 2008 and 2009 with minimum disturbance of the ongoing hadron spectroscopy measurements. The experimental set-up makes use of the 40 cm long LH_2 target, surrounded by a Recoil Proton Detector (RPD). It includes all detectors presently availa ble including the ECAL1 and ECAL2 electromagnetic calorimeters for photon detection. Apart from the shorter target and the corresponding RPD lengths, this set-up has the principal features of the one foreseen for the proposed future GPD programme. DVCS test measurements were performed using both μ^+ and μ^- beams of 160 GeV energy, with an intensity of about 1/3 of the maximum in 2008 and close to the maximum intensity in 2009. The 2008 DVCS test data allowed one to clearly identify exclusive photon production and to obtain a first direct determination of the global overall efficiency. The latter, including also SPS beam and spectrometer availabilities, trigger and DAQ efficiencies, was found to be close to the value of 10% assumed for the projections. The 2009

DVCS tests data provided a first direct measurement of the relative contributions of the BH and DVCS in the x_B domain available at COMPASS (Fig. 5). Note that the predicted ϕ dependence of the DVCS cross section is not flat in ϕ due to acceptance effects in the 2009 set-up as there is no electromagnetic calorimeter at large photon angles. This will be improved by the addition of the new calorimeter ECAL0 already mentionned above.

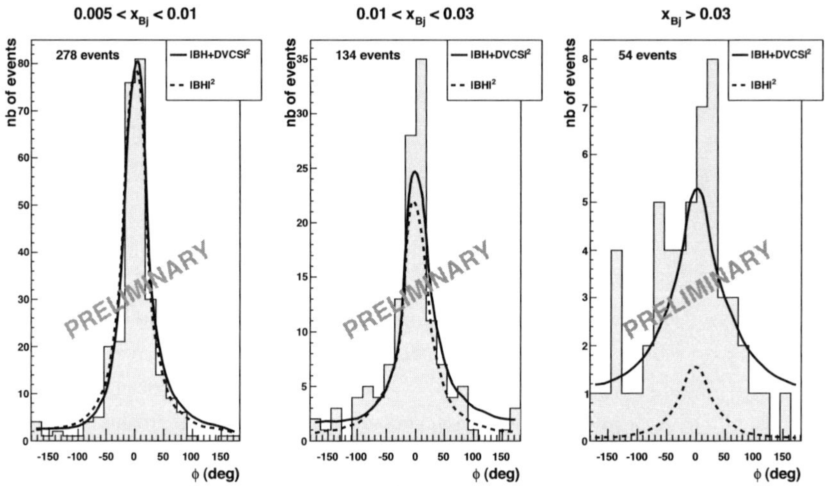

Figure 5. Distribution in the azimuthal angle ϕ for measured exclusive single-photon events, $\mu p \to \mu' p \gamma$ with $Q^2 > 1$ GeV^2, in the same three x_B bins as in Fig. 1. Shown here is a Monte Carlo simulation of only the BH process ($|\mathrm{BH}|^2$) and of both the BH and DVCS processes ($|\mathrm{BH} + \mathrm{DVCS}|^2$).

References

1. COMPASS-II proposal, SPSC-P-340, http://cdsweb.cern.ch/record/1265628/files/SPSC-P-340.pdf, May 17, 2010
2. A.V. Belitsky, D. Müller and A. Kirchner, Nucl. Phys. B **629** (2002) 323.
3. M. Vanderhaeghen, P.A.M. Guichon and M. Guidal, Phys. Rev. Lett. **80** (1998) 5064; Phys. Rev. D **60** (1999) 094017; K. Goeke, M.V. Polyakov and M. Vanderhaeghen, Prog. Part. Nucl. Phys. **47** (2001) 401.
4. K. Kumericki,D. Mueller and K. Passek-Kumericki, Nucl. Phys. **B 794** (2008) 244, K. Kumericki and D. Mueller, arXiv 0904.0458[hep-ph]
5. H1, A. Aktas *et al.*, Eur. Phys. J.C **44** (2005) 1, F.D. Aaron *et al.*, Phys. Lett. B **659** (2008) 796.
6. ZEUS, S. Chekanov *et al.*, DESY-08-178, arXiv:hep-exp:0812.2517v3.
7. M. Diehl, Th. Feldmann, R. Jakob and P. Kroll, Eur. Phys. J. C **39** (2005) 1.
8. M. Guidal, M.V. Polyakov, A.V. Radyushkin and M. Vanderhaeghen, Phys. Rev. D **72** (2005) 054013.

EXTRACTION OF THE COMPTON FORM FACTOR $\mathcal{H}$ FROM RECENT DVCS MEASUREMENTS AT JLAB

H. MOUTARDE* on behalf of the CLAS group at Saclay

CEA, Centre de Saclay, IRFU/Service de Physique Nucléaire,
F-91191 Gif-sur-Yvette, France
**E-mail: herve.moutarde@cea.fr*
http://irfu.cea.fr/Sphn/

Working at twist 2 accuracy and assuming the dominance of the Generalized Parton Distribution H we study the helicity-dependent and independent cross sections measured in Hall A and the beam spin asymmetries measured in Hall B at Jefferson Laboratory. We extract the real and imaginary parts of the Compton Form Factor $\mathcal{H}$, the latter being obtained with a 20–50 % uncertainty. We discuss our results and compare to other extractions as well as to the popular VGG model. We pay extra attention to the estimation of systematic errors on the extraction of $\mathcal{H}$.

Keywords: Generalized Parton Distributions, Compton Form Factors, Deeply Virtual Compton Scattering.

Introduction

The Deeply Virtual Compton Scattering (DVCS) process[1,2] was early recognized as the cleanest way to access Generalised Parton Distributions (GPD) and has been so far a very active field of research. Theoretical developments are reviewed in Refs. [3–6] and relevant experimental results are published in Refs. [7–19].

Even if getting a precise experimental knowledge of GPDs is a long-term program, we can already perform the first fits to get a flavour of the actual sensitivity of observables to GPDs and elaborate robust and efficient extracting methods. We address these issues by studying recent JLab data, namely beam spin asymmetries[14] (BSA) and helicity-dependent and independent cross sections.[12] These data offer the interesting features of a large kinematic coverage and a fine kinematic binning.

The first section of this paper describes the evaluation of the $ep \to ep\gamma$ cross sections and outlines our hypothesis. In the second part we explain our fitting strategy. In the third section we discuss the extracted values of $\mathcal{H}$ and compare them to other similar studies.

1. Preliminary analysis

1.1. *DVCS at leading twist*

Four GPDs H, E, $\tilde{H}$ and $\tilde{E}$ appear at twist 2, but the cross sections depend on the Compton Form Factors (CFF) $\mathcal{H}$, $\mathcal{E}$, $\tilde{\mathcal{H}}$ and $\tilde{\mathcal{E}}$. The convention of Ref. [20] is used to define the CFFs. The complex integration kernel yields a real and an imaginary part to a CFF. For example the CFF $\mathcal{H}$ satisfies :

$$Re\mathcal{H} = \mathcal{P}\int_{-1}^{+1} dx\, H(x,\xi,t)\left(\frac{1}{\xi - x} - \frac{1}{\xi + x}\right) \tag{1}$$

$$Im\mathcal{H} = \pi\Big(H(\xi,\xi,t) - H(-\xi,\xi,t)\Big) \tag{2}$$

where $\xi = x_B(1 + t/(2Q^2))/(2 - x_B + x_B t/Q^2))$ is the generalised Bjorken variable,[20,23] x_B the standard Bjorken variable, Q^2 the virtuality of the initial photon and t the square momentum transfer between initial and final protons. $\mathcal{P}$ denotes the principal value of the integral.

The use of dispersion relations relating real and imaginary parts of a CFF had been discussed by Anikin and Teryaev[24–26], Diehl and Ivanov,[27] and Goldstein and Liuti.[28–30] Dispersions relations are used in the model-dependent global fit of Kumericki and Müller.[31] However, the unknown subtraction (the D-term[32]) and the limited kinematic range of available data make this constraint rather weak in model-independent fitting. In this work we try to minimize the model-dependence of the extraction. Then we consider real and imaginary parts of CFFs as independent except when we use an explicit parametrisation of GPDs to compute CFFs.

1.2. $ep \to ep\gamma$ *observables at twist 2*

The harmonic analysis of $ep \to ep\gamma$ cross-sections has so far relied on the 2002 work of Belitsky, Müller and Kirchner[20] (BMK). In this formalism, the interference between the Bethe-Heitler (BH) and DVCS processes was treated with a leading order approximation of the BH part. This assumption was removed by Belitsky and Müller in the case of a spinless target[21] and for a (longitudinally polarised or unpolarised) spin 1/2 target[22] and by Guichon and Vanderhaegen in the case of a proton target.[23] In all the following, we will use the expressions from the latter. At last, the twist 2 approximation is motivated by the claim of early Q^2-scaling observed in Hall A.[12]

1.3. *H-dominance hypothesis*

Since we are considering experiments on unpolarised proton targets we can neglect E, $\tilde{H}$ and $\tilde{E}$ (H-dominance). The contribution of these GPDs to helicity-dependent cross-sections is indeed kinematically suppressed[33] and we can check (for instance thanks to the VGG model[34–37]) that $Im\mathcal{E}$ and $Im\tilde{\mathcal{H}}$ are usually smaller than $Im\mathcal{H}$. At last for small t and ξ we expect $\tilde{H}/H$ to be close to $\Delta q/q$ *i.e.* $\frac{1}{4}$.

Thus assuming H-dominance we may hope to extract information on H from BSAs or helicity-dependent cross sections with a systematic error of 20 % to 50 %[a], this approximation being better at small t. The advantage of this approach is the dramatic decrease of the number of degrees of freedom involved in fits. Guidal indeed showed[38] in the same kinematic region that it is not possible to extract sensible information about the real and imaginary parts of all CFFs by direct fits of helicity-dependent and independent cross sections.

2. Fitting strategies

The possibility to study GPDs in DVCS rests on factorisation theorems[39] which require a small value of $|t|/Q^2$. In the following, we restrict ourselves to kinematic configurations for which $|t|/Q^2 < 1/2$ and justify this choice *a posteriori*. We perform two kinds of fits :

Local fits $Re\mathcal{H}$ and $Im\mathcal{H}$ are the free parameters of the fits.

Global fit We use a "dual model"-like parametrization; it consists basically in a simultaneous expansion in Gegenbauer and Legendre polynomials. See Ref. [33] for details on the fitting Ansatz.

We estimate the systematic errors associated to our H-dominance hypothesis by first fitting data setting the subdominant GPDs to 0, then fitting the same data setting the subdominant GPDs to their VGG value, and computing the difference. We estimate the systematic uncertainty related to the global fitting Ansatz by varying the number of polynomials involved in the expansion.

A careful fitting strategy is elaborated to handle the high number of free parameters involved in the fits and make sure that the numeric implementation is under control. We adopt an iterative (5 steps) fitting procedure. See Ref. [33] for full details.

[a]A direct test of the H-dominance assumption with the VGG model gives an upper bound of 25 %. This is comparable to the typical statistical uncertainty on BSAs.

3. Results

3.1. *Extraction of $Im\mathcal{H}$ and $Re\mathcal{H}$*

The Fig. 1 displays our results for $Im\mathcal{H}$ and $Re\mathcal{H}$. Both local fits and global fit give results with comparable accuracy for $Im\mathcal{H}$, but as expected the results of the global fits are smoother. This is especially true concerning $Re\mathcal{H}$: in this case the local fits suffer from large fluctuations of $Re\mathcal{H}$ with values which fall outside the plot range. However, we could not reliably extract values of the CFFs for the larger values of ξ with the global fit.

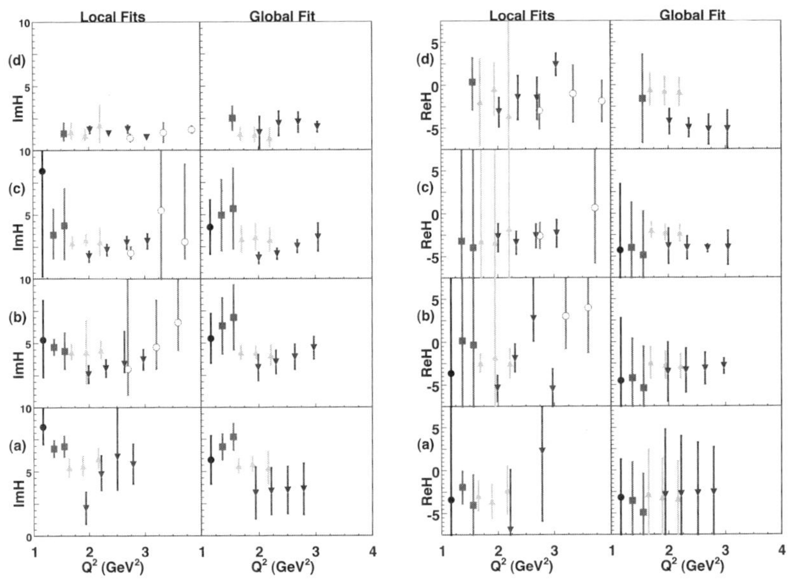

Fig. 1. Q^2-behaviour ($1 < Q^2 < 4$ GeV^2) of the extracted values of $Im\mathcal{H}$ (left) and $Re\mathcal{H}$ (right) of local fits and global fit on Hall B kinematics : $0.09 < -t < 0.2$ GeV^2 (a), $0.2 < -t < 0.4$ GeV^2 (b), $0.4 < -t < 0.6$ GeV^2 (c), and $0.6 < -t < 1.$ GeV^2 (d). The error bars include both statistics and systematics. $Im\mathcal{H}$ ranges between 0 and 10 and $Re\mathcal{H}$ ranges between -7.5 and +7.5. The black full circles correspond to x_B=0.125, red squares to x_B=0.175, green up triangles to x_B=0.250, blue down triangles to x_B=0.360 and magenta open circles to x_B=0.491.

The results for local and global fits are almost always compatible, which is a strong consistency check. Both rely on the assumptions of twist 2 accuracy and of H-dominance. On one hand, local fits suffer from numerical fluctuations (the 2-parameter local fits are not constrained enough on some bins) but are almost model-independent. On the other hand, global fits are smoother, but suffer from oscillations. That both methods give the same results indicates that fluctuations and oscillations are reasonably controlled in the bins for which results are displayed. Since, in both cases, the total error bars have the same size, we conclude that our estimation of systematic uncertainties on the global fit Ansatz is realistic.

All fits keep data satisfying $|t|/Q^2 < 1/2$. For local fits, changing the maximal value of $|t|/Q^2$ amounts to dropping points. For global fits, the whole results may be changed, but the good agreement between the results of both types of fits, and the slow Q^2-evolution of the extracted CFFs, indicate that this restricted kinematic region is suitable for an analysis in the GPD framework.

Our results are explicitly given in Ref. [33]. Error bars are dominated by systematic effects. Typically we obtain a relative accuracy of 20 to 50 % on $Im\mathcal{H}$, which is quite satisfactory under the assumption of H-dominance and given the statistical accuracy of JLab data. On the contrary, $Re\mathcal{H}$ is still largely undetermined, and is never extracted with a precision better than 50 %.

3.2. *Discussion*

The Fig. 2 compares our results on Hall A kinematics to twist 2 model-independent extractions[38,40] and two extractions with the BMK formalism.[12,31]. Firstly, the use of the GV expressions creates important deviations to the extraction of Ref. [12] Since the extracted combinations of GPDs are not the same, we will not make the argument more quantitative. The same discrepancy occur when comparing to the results of the global fit of the GPD H of Kumericki and Müller.[31] Secondly, we obtained results in good agreement with Ref. [38]. Note however the surprisingly big shift between the results of Ref. [38] and Ref. [40] induced by the inclusion of target spin asymmetries. These points will be elucidated in the near future.

For a given t, and at $Q^2 = 3$ GeV^2, the Fig. 3 displays the x_B-dependence of the CFFs we extracted, and compares them to the predictions of the VGG model and to the recent extraction of Guidal.[40] Since the errors are mostly due to systematic effects, the points in the coloured domain have the same probability.

The extracted values of $Im\mathcal{H}$ have an x_B-dependence similar to that predicted by the VGG model. We also observe that VGG tends to overestimate $Im\mathcal{H}$. The extracted CFFs are often but not always compatible with VGG. The errors on the extracted $Re\mathcal{H}$ and $Im\mathcal{H}$ in the hatched zone may be underestimated for $0.3 \leq x_B \leq 0.4$ (See Ref. [33]).

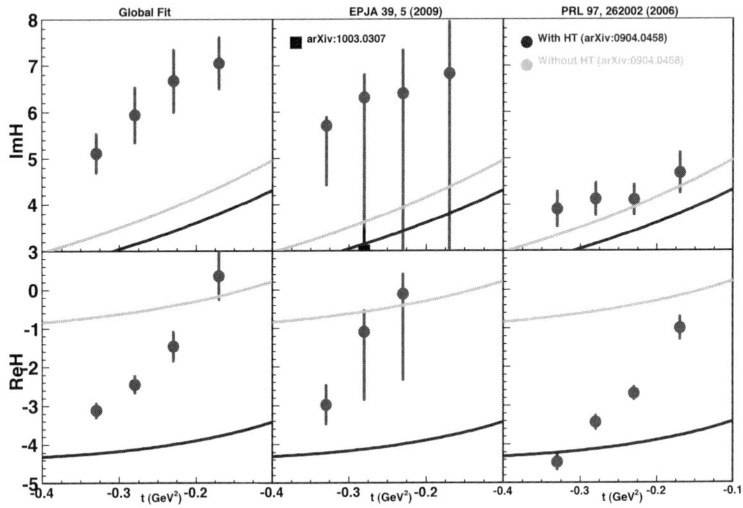

Fig. 2. $Im\mathcal{H}$ (up) and $Re\mathcal{H}$ (down) vs t ($0.1 < -t < 0.4$ GeV^2) on Hall A kinematics (x_B=0.36 and $Q^2 = 2.3$ GeV^2). $Im\mathcal{H}$ ranges between 3 and 8, and $Re\mathcal{H}$ between -5 and 1. We compare our results (left column) to those of Guidal[38] (middle column) and Muñoz-Camacho *et al.*[12] (right column). In the latter column $\mathcal{H}+\frac{x_B}{2-x_B}\left(1+\frac{F_2}{F_1}\tilde{\mathcal{H}}\right)-\frac{t}{4M^2}\frac{F_2}{F_1}\mathcal{E}$ is plotted, and not $\mathcal{H}$. The error bars include both statistical and systematic uncertainties. The green (resp. blue) curve is the result of the model-dependent global fit of Kumericki and Müller[31] with (resp. without) $\tilde{\mathcal{H}}$. The full black square is the result of Ref. [40].

Conclusions

Working at leading twist, and assuming H-dominance, we extracted $Im\mathcal{H}$ and $Re\mathcal{H}$ with 20–50 % accuracy on $Im\mathcal{H}$. The good agreement between the results of local and global fits is a strong consistency check. It demonstrates that the extraction of $Im\mathcal{H}$ and $Re\mathcal{H}$ and the estimation of the systematics related to this fitting procedure produce relatively model-independent results, at least for x_B not too large. This comparison and the weak Q^2-dependency of the results also validates *a posteriori* the restriction to kinematic configurations with $|t|/Q^2 < 1/2$.

Our results are dominated by systematic uncertainties. The systematic uncertainties related to H-dominance will decrease in the future using additional BSA measurements[41] on unpolarised and longitudinally polarised proton target which will put stronger constraints on the global fits.

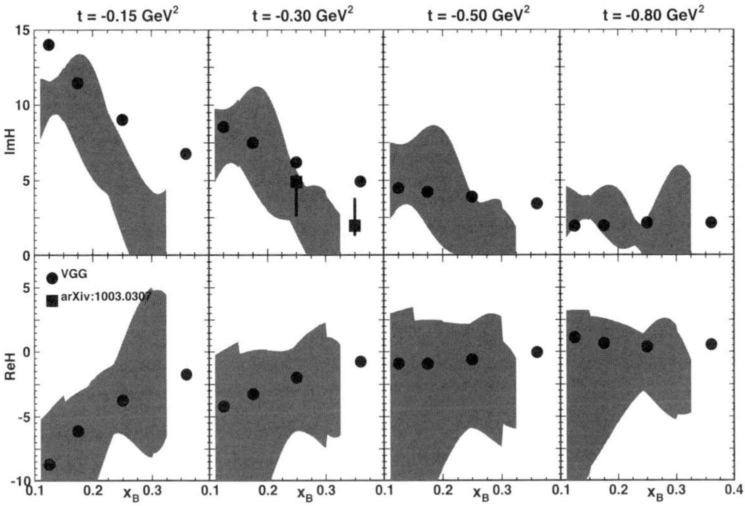

Fig. 3. Comparisons of the extracted $Im\mathcal{H}$ (up) and $Re\mathcal{H}$ (down) and the predictions of the VGG model (black circles) for $0.1 < x_B < 0.4$ and $0.09 < -t < 0.2$ GeV^2 (a), $0.2 < -t < 0.4$ GeV^2 (b), $0.4 < -t < 0.6$ GeV^2 (c), and $0.6 < -t < 1.$ GeV^2 (d). $Im\mathcal{H}$ ranges between 0 and 15, and $Re\mathcal{H}$ between -10 and +7.5. The full black square is the result of Ref. 40. Error bars include both statistical and systematic contributions. Since our errors are dominated by systematics, we can only say that the true value of $Im\mathcal{H}$ and $Re\mathcal{H}$ lie in the coloured bands. See Ref. 33 for more explanations about these bands.

Acknowledgments

The author would like to thank the CLAS group at Saclay, P. Guichon, M. Guidal and D. Müller for many fruitful and stimulating discussions. The author also thank the organizers of the Fourth Workshop on Exclusive Reactions at High Momentum transfer held in Jefferson Laboratory (May 18-21, 2010), Newport News, Virginia. This work was supported in part by the Commissariat à l'Energie Atomique and the GDR n° 3034 Physique du Nucleon.

References

1. X.-D. Ji., *Phys. Rev. Lett.* **78**, 610 (1997).
2. M. Diehl, Th. Gousset, B. Pire and J.P. Ralston, *Phys. Lett. B* **411**, 193 (1997).
3. K. Goeke, M.V. Polyakov and M. Vanderhaeghen, *Prog. Part. Nucl. Phys.* **47**, 401 (2001).
4. M. Diehl, *Phys. Rept.* **388**, 41 (2003).
5. A.V. Belitsky and A.V. Radyushkin, *Phys. Rept.* **418**, 1 (2005).
6. S. Boffi and B. Pasquini, *Riv. Nuovo Cim.* **30**, 387 (2007).

7. A. Airapetian *et al.*, *Phys. Rev. Lett.* **87**, 182001 (2001).
8. C. Adloff *et al.*, *Phys. Lett. B* **517**, 47 (2001).
9. S. Stepanyan *et al.*, *Phys. Rev. Lett.* **87**, 182002 (2001).
10. S. Chekanov *et al.*, *Phys. Lett. B* **573**, 46 (2003).
11. S. Chen *et al.*, *Phys. Rev. Lett.* **97**, 072002 (2006).
12. C. Muñoz Camacho *et al.*, *Phys. Rev. Lett.* **97**, 262002 (2006).
13. F.D. Aaron *et al.*, *Phys. Lett. B* **659**, 796 (2008).
14. F.X. Girod *et al.*, *Phys. Rev. Lett.* **100**, 162002 (2008).
15. A. Airapetian *et al.*, *JHEP* **06**, 066 (2008).
16. A. Airapetian *et al.*, *JHEP* **11**, 083 (2009).
17. A. Airapetian *et al.*, *JHEP* **06**, 019 (2010).
18. A. Airapetian *et al.*, *Nucl. Phys. B* **829**, 1 (2010).
19. A. Airapetian *et al.*, *Phys. Rev. C* **81**, 035202 (2010).
20. A.V. Belitsky, D. Mueller and A. Kirchner, *Nucl. Phys. B* **629**, 323 (2002).
21. A.V. Belitsky and D. Mueller, *Phys. Rev. D* **79**, 014017 (2009).
22. A.V. Belitsky and D. Mueller, arXiv:1005.5209 [hep-ph].
23. P.A.M. Guichon and M. Vanderhaeghen, *Analytic ee'γ cross section*, in *Atelier DVCS, Laboratoire de Physique Corpusculaire, Clermont-Ferrand, June 30 - July 01, 2008.*
24. O.V. Teryaev, arXiv:hep-ph/0510031.
25. I.V. Anikin and O.V. Teryaev, arXiv:0710.4211 [hep-ph].
26. I.V. Anikin and O.V. Teryaev, *Phys. Rev. D* **76**, 056007 (2007).
27. M. Diehl and D.Yu. Ivanov, *Eur. Phys. J. C* **52**, 919 (2007).
28. G.R. Goldstein and S. Liuti, *Phys. Rev. D* **80**, 071501 (2009).
29. G.R. Goldstein and S. Liuti, arXiv:0908.2215 [hep-ph].
30. G.R. Goldstein and S. Liuti, arXiv:1006.0213 [hep-ph].
31. K. Kumericki and D. Mueller, arXiv:0904.0458 [hep-ph].
32. M.V. Polyakov and C. Weiss, *Phys. Rev. D* **60**, 114017 (1999).
33. H. Moutarde, *Phys. Rev D* **79**, 094021 (2009).
34. P.A.M. Guichon and M. Vanderhaeghen, *Prog. Part. Nucl. Phys.* **41**, 125 (1998).
35. M. Vanderhaeghen, P.A.M. Guichon and M. Guidal, *Phys. Rev. Lett.* **80**, 5064 (1998).
36. M. Vanderhaeghen, P.A.M. Guichon and M. Guidal, *Phys. Rev. D* **60**, 094017 (1999).
37. M. Guidal, M.V. Polyakov, A.V. Radyushkin and M. Vanderhaeghen, *Phys. Rev. D* **72**, 054013 (2005).
38. M. Guidal, *Eur. Phys. J. A* **37**, 319 (2008).
39. J.C. Collins and A. Freund, *Phys. Rev. D* **59**, 074009 (1999).
40. M. Guidal, *Phys. Lett. B* **689**, 156 (2010).
41. V. Kubarovsky, *Deeply Virtual Exclusive Reactions with CLAS*, in *ICHEP Paris 2010.*

BEAM SPIN ASYMMETRY OF THE ELECTROPRODUCTION OF A Δ^+ RESONANCE AND A PHOTON ON THE PROTON WITH THE CLAS DETECTOR

B. MORENO* for the CLAS collaboration

CEA, Centre de Saclay, Irfu/Service de Physique Nucléaire,
Gif-sur-Yvette, 91191, France
**E-mail: brahim.moreno@cea.fr*
http://irfu.cea.fr/

Institut de Physique Nucléaire d'Orsay, Université Paris-Sud XI, CNRS-IN2P3,
Orsay, 91406, France

The Generalized Parton Distributions (GPDs) parameterize the non-pertubative nucleonic content and appear in the amplitude of hard exclusive processes such as DVCS on the nucleon ($ep \to ep\gamma$). They correlate the spatial and momentum structure of the nucleon in terms of partonic degrees of freedom. The GPD formalism has been extended to more general final states with the introduction of transition GPDs. These GPDs are used for instance to describe processes where the recoil nucleon is excited into a nucleon resonance. In this paper the focus is made on the N-Δ transition. The simplest process to access N-Δ transition GPDs is the ΔVCS ($ep \to e\Delta^+\gamma$). First experimental beam spin asymmetry results corresponding to this process and obtained from CLAS data are presented.

Keywords: N-Δ transition; N-Δ transition GPDs; ΔVCS; Jefferson Lab; CLAS detector; Beam Spin Asymmetry.

1. Phenomenology of the electroproduction of a Δ^+ resonance and a real photon on the proton

In the appropriate limit, the electroproduction of a Δ^+ resonance and a real photon on the proton is the simplest tool to acess N-Δ transition GPDs.[1] In this limit defined as,

$$Q^2 \to \infty, \quad x_B \text{ fixed}, \quad t_\gamma/Q^2 \to 0, \quad W^2/Q^2 \to 0, \tag{1}$$

the $ep \to e\Delta^+\gamma$ reaction, involves two mechanisms: ΔVCS described later in the text and the Bethe-Heitler process where the produced photon is emitted by the incoming or outgoing electrons. Associated diagrams are shown on figure 1. The quantity x_B is the Bjorken variable. The variables Q^2, W and t_γ are defined as:

$$Q^2 = -(k - k')^2 \tag{2}$$
$$2M\nu = Q^2 + W^2 - M^2 \tag{3}$$
$$t_\gamma = (p_\gamma - q)^2 \tag{4}$$

where k, k', p_γ and q are the four-momenta of the incoming electron, outgoing electron, produced photon and virtual photon respectively. The physical quantity ν is the virtual photon energy.

1.1. *The* Δ*VCS process*

The ΔVCS can be factorized[1] in the limit given above. This process is almost the same as DVCS, the main difference being that the recoil nucleon is excited into a Δ^+ state. Thus the non-perturbative part of the process is described through N-Δ transition GPDs,[1–3] these functions being an extension of nucleon GPDs to final states where the recoil particle is a Δ. This is the principal motivation for studying this process: it may provide a first partonic description of the N-Δ transition and on top of that a 3-dimensional picture of it.

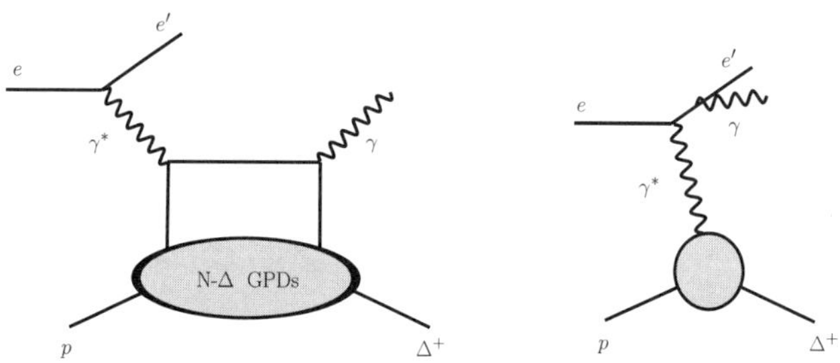

Fig. 1. Diagrams corresponding to the $ep \to e\Delta^+\gamma$ reaction. On the left-hand side: ΔVCS handbag diagram. On the right-hand side: Bethe-Heitler diagram.

Also in the *large* N_c limit, N-Δ transition GPDs can be linked to nucleon GPDs with an estimated 30% accuracy[1] providing a complimentary access to these functions. The description of the ΔVCS process done in Ref. 1 only takes into account GPDs having a significant contribution to the transition. The remaining transition GPDs entering the model are expressed in terms of nucleon GPDs through the following relations,

$$H_M = \frac{2}{\sqrt{3}}[E^u - E^d] \tag{5}$$

$$C_1 = \sqrt{3}[\widetilde{H}^u - \widetilde{H}^d] \tag{6}$$

$$C_2 = \frac{\sqrt{3}}{4}[\widetilde{E}^u - \widetilde{E}^d] \tag{7}$$

1.2. *Beam spin asymmetry of the electroproduction of a Δ^+ resonance and a real photon on the proton*

The ΔVCS and Bethe-Heitler processes interfere at the amplitude level. When using a polarized lepton beam this results in a cross section asymmetry with respect to the incoming lepton helicity. This asymmetry, shown in figure 2, has been calculated in Ref.1 based on the *large* N_c approximation as described above. These calculations take into account the contribution from non-resonant channels which are indiscernible from ΔVCS and Bethe-Heitler from an experimental point of view.

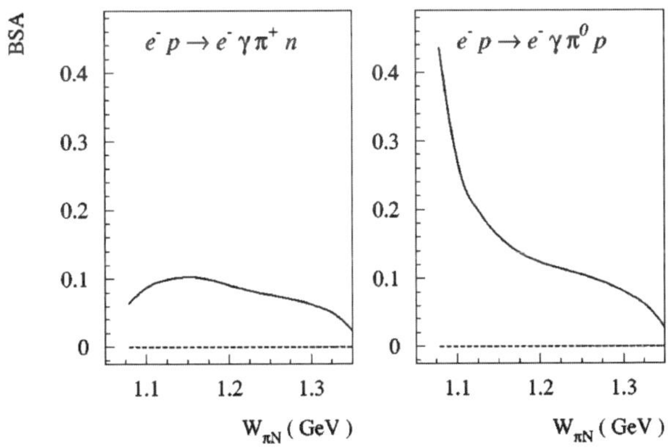

Fig. 2. Predicted beam spin asymmetry for the $ep \to eN\pi\gamma$ reactions.[1] The calculations are performed in the Δ^+ region. Asymmetries are plotted as a function of the invariant mass of the nucleon-pion system, $W_{\pi N}$, for typical Jlab kinematic: $E_e = 6$ GeV, $Q^2 = 2.5$ GeV^2, $x_B = 0.3$, $t = -0.5$ GeV^2, $\Phi = 90^o$ (Φ being the angle between hadronic and leptonic planes).

Studying the experimental ΔVCS beam spin asymmetry (BSA) means extracting experimentally the BSA of $ep \to eN\pi\gamma$ channels in the Δ^+ region, the Δ^+ decaying into either a $p\pi^0$ or a $n\pi^+$ pair. Corresponding experimental work is described in the next section.

2. Experiment and data analyses

2.1. *Experiment*

It took place in 2005 at Jefferson Lab as the first part of the E01-113 experiment[4] now referred to as E06-003,[5] the first experiment dedicated to DVCS performed in hall B. A 5.776 GeV polarized electron beam impeging a LH_2 target was used. Data were collected using the CEBAF Large Acceptance Spectrometer.[6] The photon acceptance of the CLAS detector (Fig. 3, l.h.s) was increased by the addition of a new calorimeter (Fig. 3, r.h.s) aiming at detecting photons at very forward angles where the majority of DVCS photon is expected.

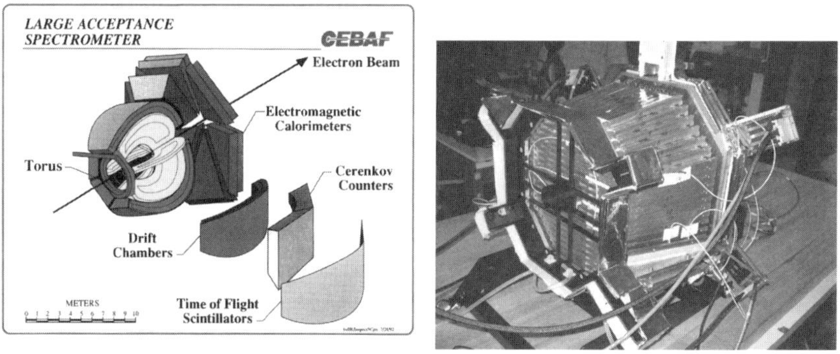

Fig. 3. CLAS apparatus. On the left-hand side is an exploded view of the standard CLAS detector. On the right-hand side is an image of the new inner calorimeter.

2.2. *Data analyses: $ep \to eN\pi\gamma$ in the Δ^+ region*

Both the $ep \to ep\pi^0\gamma$ and $ep \to en\pi^+\gamma$ reactions in the Δ^+ region were analysed and the corresponding BSAs extracted.[7] The first analysis step, particles identification, was performed according to standard CLAS procedure as described in Ref. 7.

2.2.1. *Event identification*

The main conclusion of a precedent analysis[8] of the $ep \to en\pi^+\gamma$ channel was that the non-detection of the final state photon makes it difficult to disantangle the signal, $en\pi^+\gamma$ events, from background contributions. That is why for the hereby described analyses all final state particles are required to have been detected. Then selection cuts are applied to several quantities, such as missing masses. For the $ep \to ep\pi^0\gamma$ analysis, the sideband method (a detailed example of which can be found in Ref. 9) is used as a complimentary tool to subtract the combinatorial background under the π^0 mass peak and identify the π^0s. The invariant mass of the nucleon-pion system (Fig. 4, r.h.s), $W_{\pi N}$, is then used to define the Δ^+ region by selecting events in the range:

$$1.08 \leq W_{\pi N}(\text{GeV}) \leq 1.32 \tag{8}$$

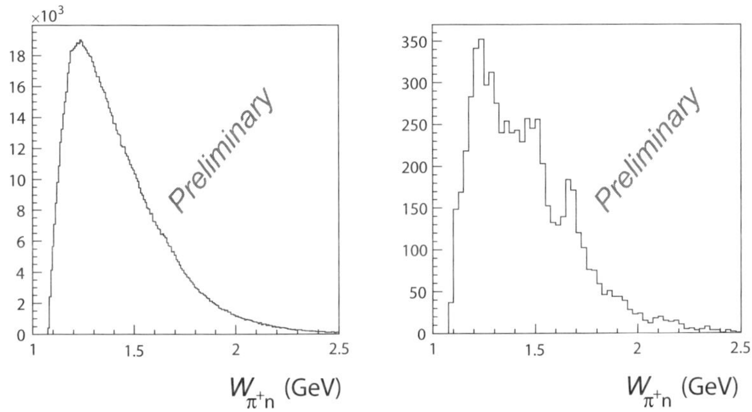

Fig. 4. Invariant of the nucleon-pion system for the $ep \to en\pi^+\gamma$ channel. On the left-hand side is the invariant mass W_{π^+n} before applying the selection cuts, on the right-hand side after. After event selection, the Δ^+ peak around 1.2 GeV is clearly visible. The second and third peaks around 1.45 and 1.7 GeV are associated with the second and third resonance regions respectively.

A background arising from double pions electroproduction is remaining after selecting $ep \to eN\pi\gamma$ events. It is coming from $ep \to ep\pi^0\pi^0$ events for the $ep \to ep\pi^0\gamma$ channel and from $ep \to en\rho^+ \to en\pi^+\pi^0$ and $ep \to en\pi^+\pi^0$ for the $ep \to en\pi^+\gamma$ reaction. The subtraction of these background contributions is the last step of the event identification. The background is estimated and then subtracted by using both Monte-Carlo simulations and exclusive double pions electroproduction data. It amounts to an average of 30% for the $ep \to ep\pi^0\gamma$ analysis and of 45% for the $ep \to en\pi^+\gamma$ study and varies with the kinematics. Further details concerning the analyses may be found in Ref.7.

2.2.2. *Beam spin asymmetry results*

The BSA of the $ep \to eN\pi\gamma$ channels in the Δ^+ region is presented on figure 5. It is the first time such asymmetries are extracted. Despite the size of the statistical uncertainties a variation of the asymmetries is clearly visible: it goes from positive to negative for the $ep \to ep\pi^0\gamma$ channel and varies from negative to positive for the $ep \to en\pi^+\gamma$ reaction. The observed behaviour may be interpreted as the first experimental observation of the ΔVCS process through its interference with Bethe-Heitler.

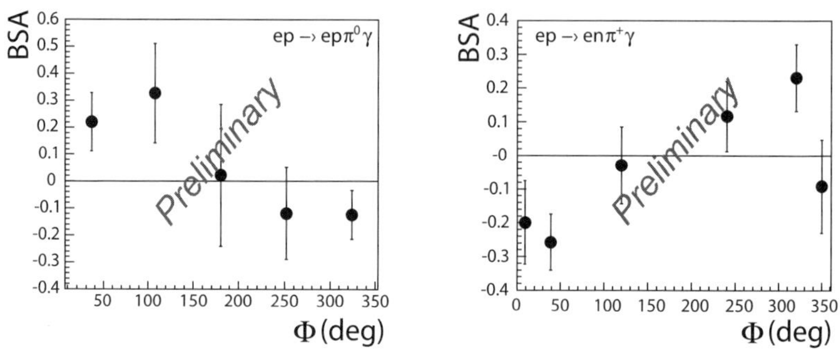

Fig. 5. Beam spin asymmetry of $ep \to eN\pi\gamma$ channels. On the left-hand side is the BSA of the $ep \to ep\pi^0\gamma$ reaction, on the left-hand side of the $ep \to en\pi^+\gamma$ channel. Asymmetries are plotted as a function of Φ and shown in arbitrary unit. The error bars are statistical.

Given the relatively low amount of statistics which is the major limit of these analyses and the subsequent fact that data are integrated over most of the accessible phase space, it would be unwise to push the interpretation further.

3. Conclusion

A pioneering investigation of ΔVCS performed with CLAS data was shown. Beam spin asymmetry results of $ep \to eN\pi\gamma$ reactions in the Δ^+ region exhibit a clear Φ-dependency that may be interpreted as the first experimental observation of ΔVCS. The major limit of the analyses: the low amount of statistics calls for future detailed analysis of this process to be performed on a dedicated experiment optimized for these particular channels. Contextually these results may be considered as the equivalent of the first observation of DVCS.[10–12] If they were to be confirmed experimentally in an independent way, it would renew the way the Δ^+ is studied by giving access to a first partonic description of the N-Δ transition and on top of that to a 3-dimensional picture of it.

References

1. P.A.M Guichon, L. Mossé, M. Vanderhaeghen, *PRD* **68**, 034018 (2003).
2. A.V. Belitsky, A.V. Radyushkin, *Phys. Rept.* **418**, 1-387 (2005).
3. K. Goeke, M.V. Polyakov, M. Vanderhaeghen, *Prog. Part. Nucl. Phys.* **47**, 401 (2001).
4. V. Burkert, L. Elouadrhiri, M. Garçon, S. Stepanyan (spokespersons), `http://www.jlab.org/exp\_prog/proposals/01/PR01-113.pdf`.
5. V. Burkert, L. Elouadrhiri, M. Garçon, S. Stepanyan (spokespersons), `http://www.jlab.org/exp_prog/proposals/06/PR06-003.pdf`.
6. B.A. Mecking *et al.*, *Nucl. Instr. and Meth. A* **503**, 513-553 (2003).
7. B. Moreno, Etude de l'asymétrie de spin du faisceau à 6 GeV pour l'électroproduction d'un état résonnant Δ^+ et d'un photon sur le proton ($ep \to e\Delta^+\gamma$) avec le détecteur CLAS, Université Paris-Sud XI, (FRANCE, 2009), pp. 61-74.
8. S. Bouchigny, Développement d'une cible polarisée de pur HD: Analyse et distillation du HD, Diffusion Compton Virtuelle Résonante sur le Nucléon à TJNAF, Université Paris-Sud XI, (FRANCE, 2004), pp. 87-198.
9. N. Nekipelov *et al.*, *J. Phys. G* **34**, 627 (2007).
10. S. Stepanyan *et al.*, *Phys. Rev. Lett.* **87**, 182002 (2001).
11. A. Airapetian *et al.*, *Phys. Rev. Lett.* **87**, 182001 (2001).
12. C. Adloff *et al.*, *Phys. Lett. B* **517**, 47 (2001).

SELF-ORGANIZING MAPS AND PARTON DISTRIBUTION FUNCTIONS

K. HOLCOMB

University of Virginia - UVACSE
112, Albert H. Small Building, Charlottesville, Virginia 22904 - USA
E-mail: kholcomb@virginia.edu

S. LIUTI[1], D.Z. PERRY[2]

University of Virginia - Physics Department
382, McCormick Rd., Charlottesville, Virginia 22904 - USA
[1] *E-mail: sl4y@virginia.edu*
[2] *E-mail: dzp3h@virginia.edu*

We present a new method to extract parton distribution functions from high energy experimental data based on a specific type of neural networks, the Self-Organizing Maps. We illustrate the features of our new procedure that are particularly useful for an anaysis directed at extracting generalized parton distributions from data. We show quantitative results of our initial analysis of the parton distribution functions from inclusive deep inelastic scattering.

1. Introduction

High energy processes can be described within Quantum Chromodynamics (QCD) through its factorization properties. These allow us to write the cross sections by separating the hard part, which is the well defined interaction of a quark with *e.g.* a virtual photon, from the soft part, described in terms of Parton Distributions Functions (PDFs). The latter cannot be calculated from first principles. Information about the PDFs can only be obtained directly from experiment. Several collaborative groups in the past decades have sought to use fitting techniques to study the PDFs behavior by devising models/analytic forms for the various PDFs, $f_i(x, Q^2)$, $i = u, d, s., c, b, t, g$, at light cone momentum fraction x, and scale Q^2, whose parameters are constrained by experiment.

More recently, new types of partonic distributions that improve considerably our ability to study the nucleon structure and its partons dynamics became accessible experimentally. The new distributions extend, in different ways, the concept of PDFs to observables that can measure the spin, spatial and momentum correlations in hadrons. These are the Transverse Momentum Distributions, TMDs Ref.4, and the Generalized Parton Distributions, GPDs (see review in Ref.5). For GPDs and TMDs one increases the range of kinematical variables that need to be measured by introducing both transverse momenta, and the Fourier conjugates of the partonic spatial degrees of freedom.

While for the PDFs one faces the problem of considering different results/extractions that vary significantly from group to group, lacking a well defined theory to constrain them (for a review of recent results see *e.g.* Refs.1,2), for TMDs and more crucially for GPDs, standard global analyses are doubtful due to the increased number of variables compared to the scarcity of the experimental data (see *e.g.* Ref.3 for a state of the art analysis).

A relatively new approach to PDFs fitting is the one proposed by the NNPDF collaboration Ref.6, who have replaced the standard analytic forms of PDFs with a more complex Neural Network (NN) solution. The estimated uncertainties for NNPDF fits are larger than those of global fits, possibly indicating that the global fit uncertainties might be underestimated. In Ref.7 a criticism was put forward about relying on purely automated fitting procedures such as the ones used by NNPDF. A new specific type of neural network, the Self-Organizing Map (SOM), was proposed. The main point is that since for NNPDFs the effect of modifying individual NN parameters is unknown, the result might not be under control in the extrapolation region, or in between the data points if the data are sparse. This issue is even more important when extending the fitting procedures to a wider set of semi-inclusive and exclusive observables. We therefore pushed forward with the SOM method, and we improved the preliminary work in Ref.7 by restructuring the original code in such a way that on one side, a fully quantitative error analysis can be implemented, and on the other we now have sufficient flexibility to allow for analyses of different observables, including the matrix elements for deeply virtual exclusive and semi-inclusive processes. Our first quantitative results for the unpolarized case using Next-to-Leading-Order (NLO) perturbative QCD in the MS scheme were presented in Ref.8. In this workshop we also discuss a possible procedure to extend SOMs to the extraction of GPDs.

2. Self-Organizing Maps

Recent years have seen an outstanding growth in the usage of neural networks as paradigms for computational methods. SOMs are a type of neural network developed by T. Kohonen in the '80s Ref.9 based on a topological mapping of the external environment onto the brain's internal neural connections.

In SOMs the nodes/neurons – map cells – are tuned to a set of input signals/data/samples according to a form of adaptation. The various nodes form a topologically ordered map during the learning process. This feature constitutes perhaps the main strength of SOMs in that it allows one to obtain a simplified two-dimensional representation of complex data otherwise depending on a "hard to control" set of parameters (see Figure 1). In this respect SOMs can be considered as a non linear form of principal component analysis which is often used to analyze highly dimensional data.

Another aspect that sets SOMs apart from other NNs is their learning process. SOMs learn via *unsupervised learning* whereas the learning of generic artificial NNs is *supervised*. In supervised learning a set of examples is given, and the goal is to force the data to match the examples as closely as possible. A cost function is defined that measures the importance to miss or detect the correct result. During the learning process the cost function is minimized. In unsupervised learning the cost function is minimized without introducing a definite set of examples, but just by similarity relations, or by finding how the data cluster or self-organize. Theoretical studies of the mechanism for map evolutions are in progress. In phenomenology, many new uses, one of which described below, might derive from this property.

SOMs are built as two dimensional arrays whose cells get sensitized/tuned to a specific set of input signals according to a given order. To illustrate this in Fig.1 we show the simple example of a "color map", where the input signals are represented by the three colors to be associated with each one of the map cells.

The SOM algorithm consists of three stages: *i)* Initialization; *ii)* Training; *iii)* Mapping.

During the initialization procedure weight vectors of dimension n are associated to each cell i:

$$V_i = [v_i^{(1)}, ..., v_i^{(n)}]$$

V_i are given spatial coordinates, *i.e.* one defines the geometry/topology of a 2D map that gets populated randomly with V_i. For the training, a set of

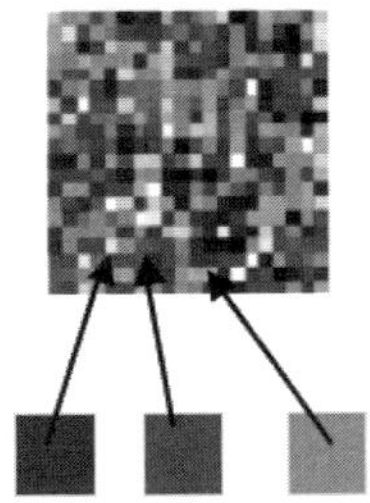

Fig. 1. A simple illustration of a 20×20 square map, and its input signals represented by colors. The arrows indicate the signal to cell matching procedure. In our more realistic situation, the map and signals will constitute of either PDFs or DIS structure functions.

input data

$$\xi = [\xi_i^{(1)}, ..., \xi_i^{(n)}],$$

(isomorphic to V_i) is then presented to V_i, or compared via a "similarity metric" that we choose to be

$$L_2(x, y) = \sum_{i=1,2} \sqrt{x_i^2 - y_i^2}$$

The most similar weight vector is the Best Matching Unit (BMU).

SOMs are based on unsupervised and "competitive" learning. This means that the cells that are closest to the BMU activate each other in order to "learn" from ξ. Practically, they adjust their values according to

$$V_i(n+1) = V_i(n) + h_{ci}(n)[\xi(n) - V_i(n)], \tag{1}$$

where n is the iteration number, and $h_{ci(n)}$ is the "neighborhood function" defining a radius on the map which decreases with both n, and the distance between the BMU and node i. In our case we use square maps of size L_{MAP}, and

$$h_{ci}(n) = 1.5 \left(\frac{n_{train} - n}{n_{train}} \right) L_{MAP} \tag{2}$$

where n_{train} is the number of iterations. At the end of a properly trained SOM, cells that are topologically close to each other will contain data which are similar to each other. In the final phase the actual data are distributed on the map and clusters emerge. (Note that the specific location of the clusters on the map is not relevant and will not necessarily be the same from one run to another; only the clustering properties are important.)

Since each map vector now represent a class of similar objects, the SOM is an ideal tool to visualize high-dimensional data, by projecting it onto a low-dimensional map clustered according to some desired similarity feature.

3. SOMPDF Parametrization

In Ref.Ref.7 a SOM algorithm was constructed which, together with a Genetic Algorithm (GA) was applied to PDF fitting. In this initial work it was proven that the SOM method works well as a minimization technique, in that it was shown that the χ^2/d.o.f. decreased towards unity with each GA iteration. However problems connected with the smoothing of the functions remained, related to the stochastic procedure used to generate them and to initialize and train the map. These prevented a fully quantitative analysis of the data, including uncertainty evaluations.

A solution to this problem was obtained in the present work Ref.8,10 by developing a new version of the SOMPDF code where several important improvements were introduced. By writing it in a single compiled language – fortran 95 – the code was made more flexible and faster since a parallel (MPI) version was easily implemented. As explained in more detail below, the increased flexibility and speed allow us to perform now random variations on the parameters of the various input PDFs, instead of on the PDF values themselves, thus providing continuous solutions which are amenable to standard error analyses.

Our ultimate goal is to provide a procedure to extract the GPDs, and their related observables, including *e.g.* the spatial partonic d.o.f. from experimental data. We therefore added flexibility in the code's inputs so as to accommodate the additional kinematical variables and parameters.

We now describe the main fitting procedure, originally applied to PDFs. This first phase of our work involves only marginally the clustering and visualization properties of SOMs. For PDFs this is, in fact, barely needed since the inclusive scattering databases provide a rather large kinematical coverage, and only two kinematical variables are necessary to describe the data. SOMs are used mostly to realize random variations of the PDF curves.

Our fitting procedure is based on the initialization, training, and mapping steps described for a general case in Section 2. In a nutshell, a set of database/input PDFs is formed by selecting at random from a range of existing PDF functions and varying their parameters. Baryon number and momentum sum rules are imposed at every step. These input PDFs, evolved to NLO to the desired Q^2, are used to initialize the map. A subset of input PDFs is then used to train the map. Notice that this phase of the procedure implies defining specific criteria for selecting the PDFs that will define each subset. Two possible criteria are discussed in Ref.7.

The similarity is tested by comparing the PDFs, according to Eqs.(1) and (2) at given x and Q^2 values. The new map PDFs are obtained by averaging the neighboring PDFs with the BMU PDFs. Once the map is trained a GA is implemented. The χ^2 per input PDFs is calculated with respect to the experimental data. We then take a subset of these functions with the best χ^2, and use them as seeds to form a new set of input PDFs. We train the map with the new set of input PDFs and repeat the process. The χ^2 was found to decrease monotonically towards $\chi^2 = 1$ with every GA iteration (Fig.2). Our stopping criterion is established when the χ^2 stops varying – its curve flattens.

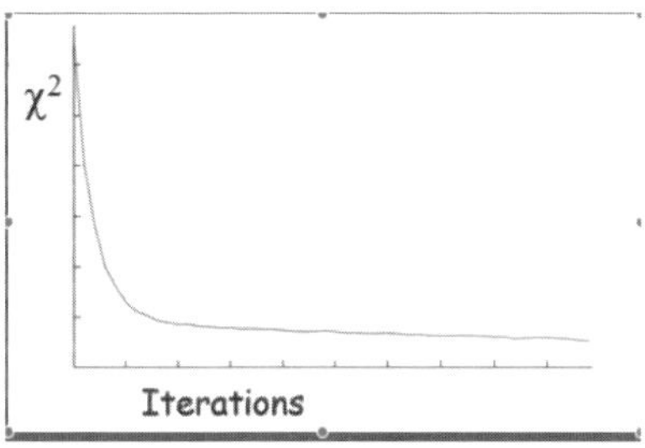

Fig. 2. Illustration of the behavior of the χ^2/d.o.f for one of our SOMPDF runs.

Our first runs, presented in Figure 3, used a "test" set of data from DIS consistent with the sets used in Refs.6,11,12. The data sets chosen were from BCDMS, H1, NMC, SLAC and ZEUS (see *e.g.* references in Ref.6). The shaded areas in the figure represent our error bands. In our preliminary runs we defined a statistical error on an ensemble of SOMPDF runs.

Finally, with a fully working procedure in hand, we comment on our capability to extend the PDF analysis to the GPD case. Here one has a total of 8 observables per parton component, given by the real and imaginary part of the Compton Form Factors (CFFs) containing the corresponding GPDs Ref.5. The number of observables that has been defined *e.g.* in Ref.3 is 17 of which 6 appear at Leading Order. An open question is therefore to establish which experiments, observables, and with what precision they can determine the various GPD components. Our analysis will provide a step in this direction in that by applying the clustering properties of the SOMs we will be able to quantitatively discern the various CFFs, that will become "dimensions" in our analysis, and to establish the sensitivity of the

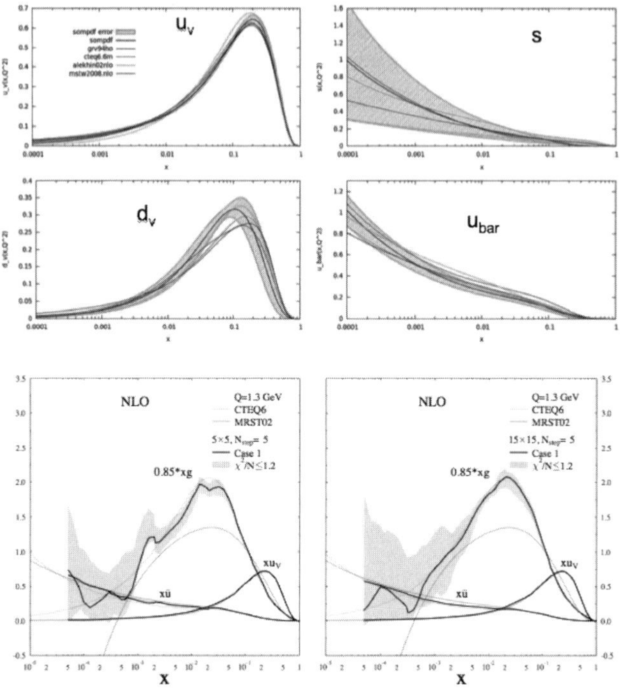

Fig. 3. Upper panel: Test results using a 5×5 map for a set of 43 runs. The panels represent (clockwise from upper left)the u_v, s, d_v, $\bar{u}$ distributions, respectively, at $Q^2 = 7.5$ GeV^2. The shaded areas are our results including the error analysis outlined in the text. For comparison we show also results from several other parametrizations. Lower panel: u_v, $\bar{u}$ and gluon distributions at $Q^2 = 1.3$ GeV^2, along with their variations obtained for $\chi^2/$ d.o.f. < 1.2, from our previous SOMPDF analysis. Left: 5×5 map, right: 15×15 map. These figures are shown to illustrate the improvement in the curves smoothing attained in our new procedure (adapted from Ref.7).

different experiments to each one of them. An illustration of how clustering of GPDs on the SOM might be represented is given in Fig.4. More work is in progress Ref.10.

4. Conclusions

In this work we described a new computational method based on Self-Organizing Maps for parametrizing nucleon PDFs. Future developments will also be directed at exploiting the full potential of SOMs that offer the capability of going beyond a fully automated procedure, by enabling one to control the fitting procedure at each step. The selection of the best

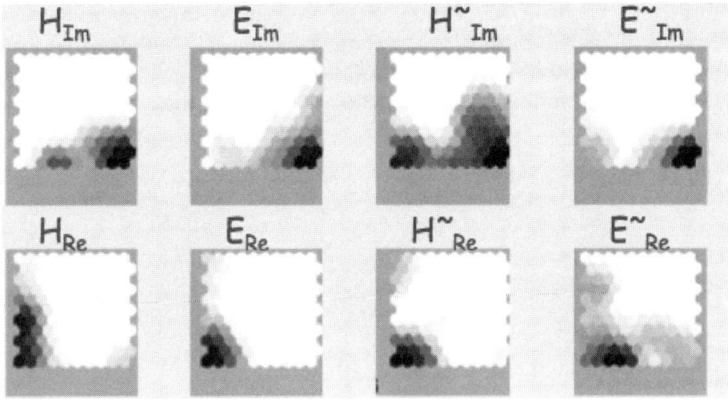

Fig. 4. Extension of our analysis to GPDs.

PDF candidates for the subsequent iteration could then be made based on the user's preferences instead of solely based on the χ^2. Our program can be extended to multivariable cases such as the Generalized Parton Distributions where the data are too sparse for stochastically generated, parameter-free, PDFs.

We thank the University of Virginia Alliance for Computational Science and Engineering for computer time, and the HPC group at Jefferson Lab, in particular David Richards and Chip Watson, for allotting us space on their clusters. This work is partially supported by the U.S. Department of Energy grant DE-FG02-01ER4120 (S.L and D.Z.P.).

References

1. A. De Roeck *et al.*, Eur. Phys. J. C **66**, 525 (2010)
2. R. Placakyte, S. Alekhin, D. Colferai and J.W. Huston, summary talk at "XVIII International Workshop on Deep-Inelastic Scattering and Related Subjects April 19 -23, 2010, Firenze, Italy"
3. M. Guidal and H. Moutarde, Eur. Phys. J. A **42**, 71 (2009)
4. P. J. Mulders and R. D. Tangerman, Nucl. Phys. B **461**, 197 (1996) [Erratum-ibid. B **484**, 538 (1997)]
5. A. V. Belitsky and A. V. Radyushkin, Phys. Rept. **418**, 1 (2005)
6. R. D. Ball, L. Del Debbio, S. Forte, A. Guffanti, J. I. Latorre, J. Rojo and M. Ubiali, Nucl. Phys. B **838**, 136 (2010)
7. H. Honkanen, S. Liuti, J. Carnahan, Y. Loitiere, P. R. Reynolds, Phys. Rev. D 79, 034022 (2009)
8. D. Z. Perry, K. Holcomb and S. Liuti, arXiv:1008.2137 [hep-ph].
9. T. Kohonen, Self-organizing Maps (Springer, New York, 2001), 3rd. ed.

10. K. Holcomb, S.Liuti, D.Z. Perry, and S.K. Taneja, *in preparation.*
11. J. Pumplin, D. R. Stump, J. Huston, H. L. Lai, P. Nadolsky, W. K. Tung (CTEQ6), Phys. Rev. D **65**, 014013 (2001)
12. S. Alekhin, K. Melnikov, F. Petriello (AMP06), Phys. Rev. D 74, 054033 (2006)

DEEPLY VIRTUAL PSEUDOSCALAR MESON PRODUCTION WITH CLAS

V. KUBAROVSKY[1,†], I. BEDLINSKY[2] and P. STOLER[3].

[1] *Jefferson Lab, Newport News, USA,*
[2] *Institute for Theoretical and Experimental Physics, Moscow, Russia*
[3] *Rensselaer Polytechnic Institute, Troy, USA,*
[†] *Email: vpk@jlab.org*

Introduction

Deeply virtual exclusive reactions offer a unique opportunity to study the structure of the nucleon at the parton level as one varies both the size of the probe, i.e. the photon virtuality Q^2, and the momentum transfer to the nucleon t. Such processes can reveal much more information about the structure of the nucleon than either inclusive electroproduction (Q^2 only), which gives us a one-dimensional snapshot of the overall longitudinal momentum distribution of the partons, or elastic form factors ($t = -Q^2$), which give us an overall view of the transverse extend of the parton disstributions. The mapping of the nucleon's structure in terms of the generalized parton distributions (GPD) is one of the major objectives of the Jefferson Lab. In this article we focus on deeply virtual pseudoscalar meson production (DVMP), specifically the reaction $\gamma^* p \to p(\pi^0, \eta)$. Schematic diagrams for DVCS and DVMP in the GPD, or handbag, framework are illustrated in Fig. 1.

Although DVCS is the cleanest way of accessing vector-type GPDs, H and E, DVMP amplitudes involves the axial vector-type GPDs, $\tilde{H}$ and $\tilde{E}$, as well as vector GPDs. These GPDs are closely related to the distribution of quark spin in the proton, and $\tilde{H}$ reduces to the polarized quark/antiquark densities in the limit of momentum transfer $t = 0$. Also the variety of DVMP channels allows one to separate flavor.

One of the primary goals of the CLAS12 program is to double the Q^2 range of the available data into a region where approached with lower twist corrections become more reliable. Since the extraction of GPDs from elec-

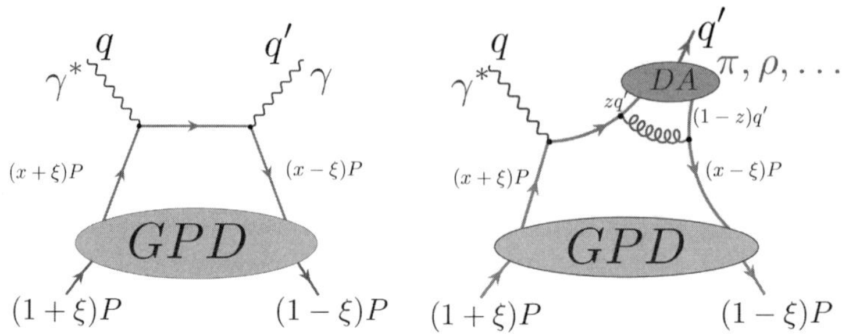

Fig. 1. Schematic illustration of the GPD approach to DVCS (left) and DVMP (right).

troproduction data can be difficult, a detailed understanding of the reaction mechanism is essential before one can compare with theoretical calculations. It is not yet clear at what values of Q^2 the application of GPDs to meson electroproduction becomes valid.[1–5] However, detailed measurements of observables may test model-independent features of the reaction mechanism.

Measurements of π^0 and η electroproduction.

The reactions $ep \to e'p'\pi^0$ and $ep \to e'p'\eta$ were measured with the CLAS spectrometer at a beam energy of about 6 GeV. The pions and etas are identified through their 2γ decay channels. This has been made possible by constructing a high quality electromagnetic calorimeter consisting of 424 lead-tungsten crystals covering an angular range from 4.5° to 15°, which was positioned into the existing CLAS large acceptance detector.

Cross sections measurements

In order to obtain quantitatively accurate cross sections for a large acceptance spectrometer such as CLAS which contains many detector elements in a complicated geometry a very large effort was undertaken to produce a GEANT based simulation which accurately describes known processes.

A major necessary aspect for all electron scattering experiments involves radiative corrections. The corrections procedure used the program EXCLURAD[6] which was developed to account for radiative corrections in single meson production. This program requires the input of realistic electroproduction structure functions, which were also obtained by self consistent successive fits to the observed distributions. Radiative corrections were found to be on the order of 10 to 15 percent depending on kinematics.

Finally, absolute cross sections were obtained over a large kinematic range and systematic uncertainty estimated at about 10 percent.

The virtual photon cross section can be written as

$$\frac{d\sigma}{d\Omega_\pi} = \sigma_T + \epsilon\sigma_L + \epsilon\sigma_{TT} cos2\phi + \sqrt{2\epsilon(1+\epsilon)/2}\sigma_{LT} cos\phi_\pi + h\sqrt{\epsilon(\epsilon-1)}\sigma'_{LT} sin\phi \tag{1}$$

where ϕ denotes the azimuthal angle between the hadronic and leptonic scattering planes and h is the electron beam polarization.

The large acceptance of CLAS enabled the data to be grouped into about 2000 intervals in Q^2, t x_B and ϕ for π^0 and about 1000 bins for η production. For unpolarized electrons ($h = 0$) the separation of the ϕ dependence in moments of a *constant*, $cos\phi$, and $cos2\phi$ allows us to obtain $\sigma_T + \epsilon\sigma_L$, σ_{TT} and σ_{LT}.

ϕ distributions

Examples of the ϕ distributions for $t = 0.35$ GeV^2, $x_B = 0.13$ and Q^2 =1.2 GeV^2 and $t = 0.35$ GeV^2, $x_B = 0.34$ and Q^2 =2.7 GeV^2 are shown in Fig. 2. Note the strong modulation as a function with ϕ which are indicative of strong interferences between transverse and longitudinal structure, and thus indicative of large non-factorizeable contributions to the reaction, so that the simple handbag approach cannot be applied in this low range of Q^2.

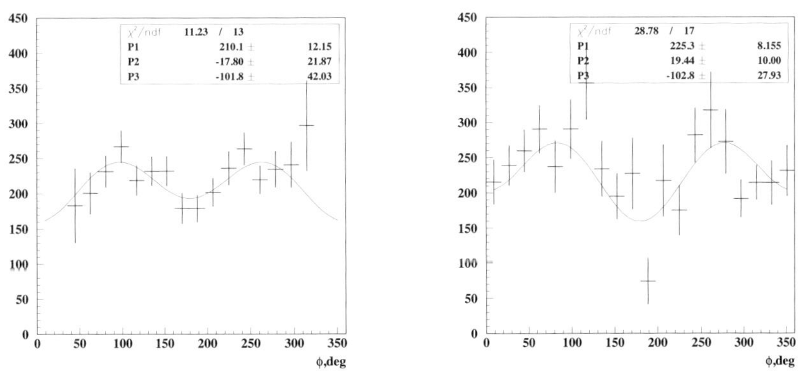

Fig. 2. Examples of the ϕ distributions. Left: $t = 0.35$ GeV^2, $x_B = 0.13$ and Q^2 =1.2 GeV^2. Right: $t = 0.35$ GeV^2, $x_B = 0.34$ and Q^2 =2.7 GeV^2.

Separated structure functions

Separated structure $\sigma_T + \epsilon\sigma_L$, σ_{TT}, and σ_{LT} as functions of $-t$ were obtained from fits to the differential cross section data for 15 intervals in Q^2 and x_B, two of which are shown in Fig 3. Since the structure functions at the present kinematics cannot be described by the leading order handbag approach, we also display the results of a Regge type analysis.[7] It appears that the Regge approach can account for the main features of the structure functions, both in their magnitude and their relationships to each other.

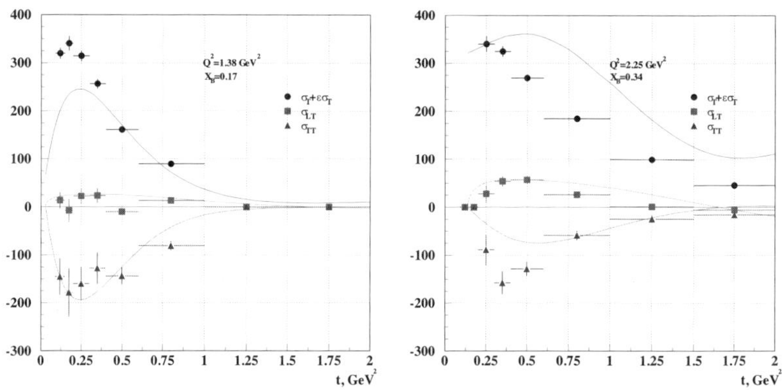

(a) (b)

Fig. 3. The separated structure functions $\sigma_T + \epsilon\sigma_L$ (circles), σ_{TT} (triangles) and σ_{LT} (squares) as a function of $-t$ at $Q^2 = 1.38$ and 2.25 GeV^2 obtained with the CLAS spectrometer. The curves are the results of a Regge based calculation of Ref.[7].

t-slopes and transverse spatial structure

In the Regge framework the cross sections can be expressed according to the following form: $d\sigma/dt \propto exp\,(B(x_B)t)$ with slope $B(x_B) = 2\alpha ln(1/x_B)$, where the parameter $\alpha \sim 1.1$. The t slope $B(x_B, Q^2)$ is plotted as a function of x_B for various fixed values of Q^2 in Fig. 4. Remarkably, the shape of the slope is seen to be almost completely independent of Q^2. Equally remarkably, it follows the Regge predicted shape with $\alpha \sim 1.1$ and no further adjustable parameters. The fact that the t-slope goes to zero for large x_B may be purely kinematical. However, even taking into account this factor, we note that $B_\perp$ falls with x_B in the region x_B from 0.1 to 0.5 where we have experimental data. This implies that the impact parameter distribution is

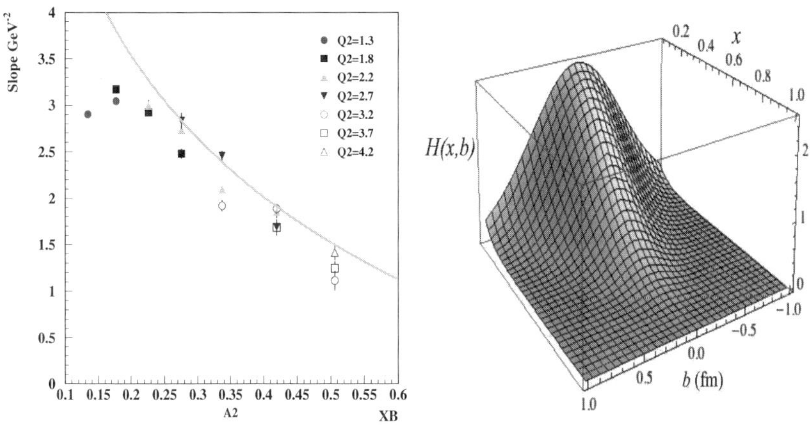

Fig. 4. Left: the experimental slope parameters B (preliminary) obtained from fits to the data for various values of Q^2 and x_B. The solid curve is the Regge inspired parametrization $B(x_B) = 2\alpha ln(1/x_B)$ with $\alpha = 1.1$. Right: The Fourier transform of the solid curve on the left vs. x_B and impact parameter b.

broadest at lowest x_B and becomes narrower at increasing x_B.

One may get information about the spacial distribution of the impact parameter b by obtaining the Fourier transform of the t dependence.[8] Using the above Regge-like parametrization of $d\sigma/dt$ one obtains the impact parameter distribution shown in Fig. 4. For the interpretation in terms of the impact parameter, the $\Delta_\perp^2$ slope is relevant, where $\Delta_\perp^2$ is the transverse component of the momentum transfer ($\Delta^2 = t$).

The ratio of cross sections for η and π^0

Even though current experiments are limited in Q^2 and t, it has been argued[2] that *precocious factorization* ratios of cross sections as a function of x_B could be valid at relatively lower Q^2 than for cross sections themselves. The ratio of cross sections for π^0 and η electroproduction from a proton is presented in Fig. 5 for the different values of t and Q^2 as a function of x_B. Note that this ratio is almost independent of x_B and varies from 0.3 to 0.4 with increasing t. This is in contrast with the prediction of Ref.[2], where this ratio is equal to ~1. However, we cannot compare directly with Ref. [2], since σ_L and σ_T were not separated.

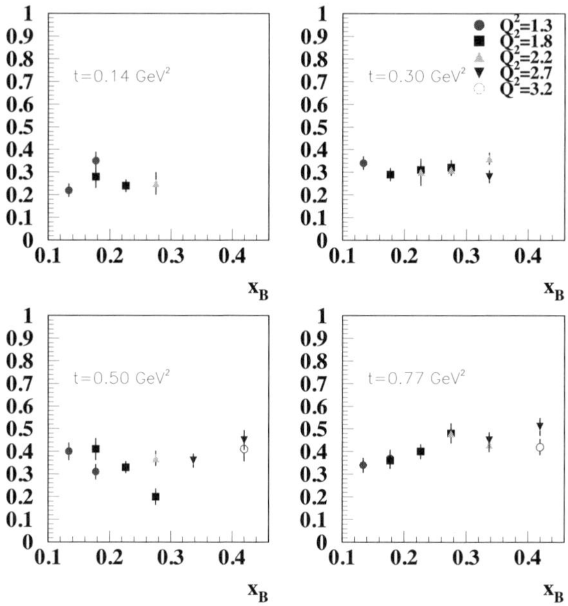

Fig. 5. η to π^0 cross sections ratio as a function of x_B for different values of Q^2 and t (preliminary).

Beam spin asymmetry.

The beam spin asymmetry (BSA) is defined by

$$A = \frac{\overrightarrow{\sigma} - \overleftarrow{\sigma}}{\overrightarrow{\sigma} + \overleftarrow{\sigma}} \sim \alpha sin\phi. \tag{2}$$

From Eq. (1) the beam spin asymmetry directly yields the $L - T$ interference structure function σ'_{LT}. Any observation of a non-zero BSA would be indicative of an L-T interference. If σ_L dominates, then σ_{LT}, σ_{TT}, and σ'_{LT} should be small. Sizable beam spin asymmetries for exclusive π^0 and η mesons electroproduction have been measured above the resonance region over a large number of kinematic bins for the first time.[9,10] An example of a t distribution of the BSA for π^0 and η production at several kinematic bins is shown in Fig. 6.

These large non-zero asymmetries imply that both transverse and longitudinal amplitudes participate in the process. The predictions of a GPD based model and Regge model[5] is shown in Fig. 6. It is not clear how well

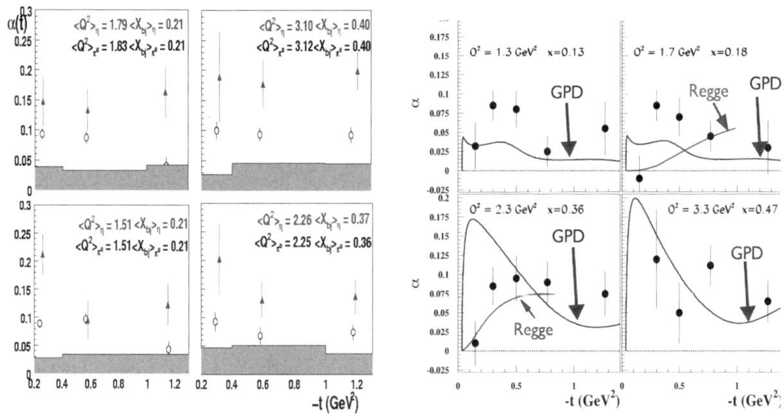

Fig. 6. Left: the t distribution of the beam spin asymmetry for π^0 (open circles) and η (solid triangles) at several values of Q^2 and x_B [9,10] (preliminary). The error bar is from statistics only, while the gray areas indicate the global systematic uncertainties for η. Right: the comparison of the π^0 BSA[9] with the GPD and Regge models[5] at several values of Q^2 and x_B .

these curves fit the data, but the GPD fit is better at $Q^2 = 2.3$ GeV^2 than at 1.7 GeV^2. Further calculations at higher Q^2 are anticipated. The results of a Regge model calculation at least qualitatively describe the experimental data .

Conclusions

Cross sections and asymmetries for the π^0 and η exclusive electroproduction in a wide kinematic range of Q^2, t and x_B have been measured and initial analyses already are showing remarkable results. These data will help us to better understand the transition from soft to hard mechanisms. Initial results show that both transverse and longitudinal amplitudes participate in the exclusive processes at currently accessible kinematics. We view the work presented here as leading into the program of the Jefferson Lab 12 GeV upgrade. The increased energy and luminosity will allow us to make the analysis presented here at much higher Q^2 and x_B as well as to perform Rosenbluth L/T separations.

We acknowledge the outstanding efforts of the staff of the Accelerator and Physics Divisions at Jefferson Lab that made this experiment possible. We also acknowledge useful discussions with H. Avakian, A. Afanasev, M. Burkardt, V. Burkert, R. De Masi, L. Elouadrhiri, M. Garcon,

F-X Girod, M. Guidal, G. Goldstein, K. Joo, J-M. Laget, S. Liuti, S. Niccolai, R. Niyazov, A. Radyushkin, S. Stepanyan, M. Strikman, I. Strakovski, M, Ungaro, C. Weiss, and B. Zhao. This work was supported by the U.S. Department of Energy and National Science Foundation. The Jefferson Science Associates (JSA) operates the Thomas Jefferson National Accelerator Facility for the United States Department of Energy under contract DE-AC05-06OR23177.

References

1. J. C. Collins, L. Frankfurt and M. Strikman, Phys. Rev. D**56**, 2982 (1997).
2. M. I. Eides, L. L. Frankfurt and M. I. Strikman, Phys. Rev. D**59**, 114025 (1999).
3. L. Mankiewicz, G. Piller and A. Radyushkin, Eur. Phys.J. C**10**:307-312 (1999) [hep-ph/9812467].
4. S.V. Goloskokov and P. Kroll, Eur. Phys. J. C**65**:137-151 (2010) [0906.0460].
5. G.R. Goldstein, and S. Liuti, arXiv:0908.2197 (2009);
   S. Ahmad, G.R. Goldstein, and S. Liuti, Phys. Rev. D**79**:054014 (2000).
6. A. Afanasev et al., Phys. Rev. D**66**:074004 (2002).
7. J-M. Laget, arXiv:1004.1949 (2010).
8. Mattias Burkardt, arXiv:0711.1881 (2007).
9. R. De Masi et al., Phys. Rev. C**77**:042201 (2008) [0711.4736].
10. K. Joo et al., AIP Conf.Proc.915:607-610 (2007), to be published.

PSEUDOSCALAR MESON ELECTROPRODUCTION AND TRANSVERSITY

GARY R. GOLDSTEIN*

Department of Physics and Astronomy, Tufts University,
Medford, MA 02155, USA
**E-mail: gary.goldstein@tufts.edu*
www.tufts.edu

SIMONETTA LIUTI

Department of Physics, University of Virginia,
Charlottesville, VA 22901, USA
E-mail: sl4y@virginia.edu

Exclusive meson leptoproduction from nucleons in the deeply virtual exchanged boson limit can be described by generalized parton distributions (GPDs). Including spin dependence in the description requires 8 independent quark-parton and gluon-parton functions. The chiral even subset of 4 quark-nucleon GPDs are related to nucleon form factors and to parton distribution functions. The chiral odd set of 4 quark-nucleon GPDs are related to transversity, the tensor charge, and other quantities related to transversity. Different meson or photon production processes access different combinations of GPDs. This is analyzed in terms of t-channel exchange quantum numbers, J^{PC} and it is shown that pseudoscalar production can isolate chiral odd GPDs. There is a sensitive dependence in various cross sections and asymmetries on the tensor charge of the nucleon and other transversity parameters. In a second section, analyticity and completeness are shown to limit the partonic interpret ation of the GPDs in the ERBL region.

Keywords: Exclusives; GPD; DVMP; π^0 production.

1. Introduction - Spin Dependent GPDs

We will report here on recently completed work [1,2] and work in progress. Deeply virtual exclusive leptoproduction of photons and mesons (DVCS and DVMP) can be described in terms of Generalized Parton Distributions (GPDs). These provide a holographic view into the nucleon structure. With measurements of polarization and angular asymmetries, a general parameterization requires 8 quark-nucleon spin-dependent GPDs and a corresponding number of gluon-nucleon GPDs (for a review, see Ref. [3]). The basic definition of the quark-nucleon GPDs is through off-forward matrix elements of quark field correlators,

$$\Phi_{ab} = \int \frac{dy^-}{2\pi} e^{iy^- X} \langle P'S' \mid \overline{\psi}_b(0)\psi_a(y^-) \mid PS\rangle \tag{1}$$

where we write the Dirac indices explicitly. Contracting with the Dirac matrices, γ^μ or $\gamma^\mu\gamma^5$ and integrating over the internal quark momenta gives rise to the Chiral even GPDs H, E or $\widetilde{H}, \widetilde{E}$, respectively. On the other hand, contracting with $\sigma^{\mu\nu}$ yields 4 chiral odd GPDs, which have been chosen in Ref. [3] to be $H_T, E_T, \widetilde{H}_T, \widetilde{E}_T$, through the spinor decomposition

$$\int dk^-\, d^2\mathbf{k}\ \mathrm{Tr}\left[\mathrm{i}\sigma^{+\mathrm{i}}\Phi\right]_{\mathrm{XP^+=k^+}}$$
$$= \frac{1}{2P^+}\overline{U}(P',S')\,[H_T^q\, i\sigma^{+i} + \widetilde{H}_T^q \frac{P^+\Delta^i - \Delta^+P^i}{M^2} + E_T^q \frac{\gamma^+\Delta^i - \Delta^+\gamma^i}{2M}$$
$$+ \widetilde{E}_T^q \frac{\gamma^+P^i - P^+\gamma^i}{M}]\, U(P,S) \tag{2}$$

The crucial connection of the 8 GPDs that enter the partonic description of electroproduction to spin dependent observables in DVCS and DVMP is through the helicity decomposition [3], where, for example, one of the chiral even helicity amplitudes is given by substituting explicit Dirac helicity state spinors for nucleons to yield

$$A_{++,++}(X,\xi,t) = \frac{\sqrt{1-\xi^2}}{2}(H^q + \tilde{H}^q - \frac{\xi^2}{1-\xi^2}(E^q + \tilde{E}^q)), \tag{3}$$

while one of the chiral odd amplitudes is obtained from Eq. 2,

$$A_{++,--}(X,\xi,t) = \sqrt{1-\xi^2}(H_T^q + \frac{t_0 - t}{4M^2}\tilde{H}_T^q - \frac{\xi}{1-\xi^2}(\xi E_T^q + \tilde{E}_T^q)). \tag{4}$$

We have constructed a robust model for the GPDs, extending previous work [5] that is based on the parameterization of diquark spectators and Regge behavior at small X. The GPD model parameters are constrained by their relations to PDFs, $H^q(X,0,0) = f_1^q(X)$, $\tilde{H}^q(X,0,0) = g_1^q(X)$, $H_T^q(X,0,0) = h_1^q(X)$ and to nucleon form factors $F_1(t)$, $F_2(t)$, $g_A(t)$, $g_P(t)$ through the first x moments of $H(X,\zeta,t), E(X,\zeta,t), \tilde{H}(X,\zeta,t), \tilde{E}(X,\zeta,t)$, respectively. These are all normalized to the corresponding charge, anomalous moment, axial charge and pseudoscalar "charge", respectively. For Chiral odd GPDS there are fewer constraints. $H_T(X,0,0) = h_1(X)$ can be fit to the loose constraints in Ref. [6] - the first moment of $H(X,\xi,t)$ is the "tensor form factor", called $g_T(t)$ by Hägler [8]. Further, it is conjectured that the first moment of $2\tilde{H}_T^q(X,0,0) + E_T^q(X,0,0)$ i s a "transverse anomalous moment", κ_T^q, with the latter defined by Burkardt [4].

With our *ansatz* many observables can be determined in parallel with corresponding Regge predictions. Since the initial work [1], we have undertaken a more extensive parameterization,and presented several new predictions. Here we show one example - the transversely polarized target asymmetry, in Fig. 1 that will be explained below.

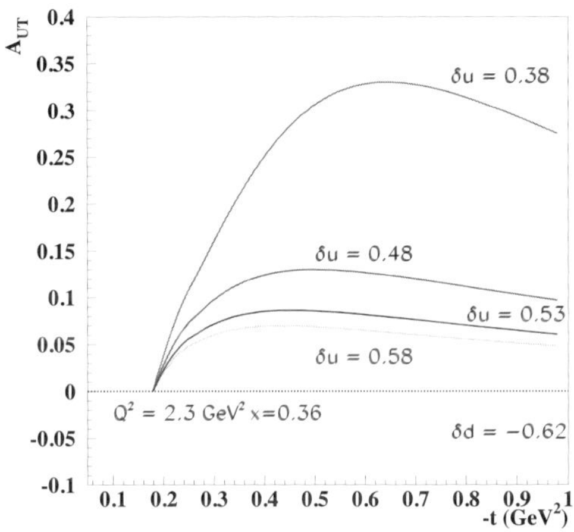

Fig. 1. Transverse spin asymmetry, A_{UT}, vs. $-t$, at $Q^2 = 2.3$ GeV^2, $x_{Bj} = 0.36$ for different values of the tensor charge, δu, with fixed $\delta d = -0.62$, *i.e.* equal to the central value extracted in a global fit [6].

In the case of π^0 production there are important constraints that restrict the GPDs that contribute. Consider the t-channel quantum numbers corresponding to combinations of GPDs. The x moments of the GPDs have expansions in terms of t-dependent form factors and polynomials in ξ. It has been shown by Lebed and Ji for pdfs [7] and Haegler for GPDs [8] (see also Ref. 9), that these moments have t-channel angular momentum decompositions, as appropriate for t-channel exchanges, as well as Regge poles. From the $t-$channel perspective $\gamma^* + \pi^0$, which has C-parity negative, goes into a $q\bar{q}$ pair (actually a non-local pair of field operators that have an operator product expansion and Mellin moments), which subsequently becomes an $N\bar{N}$ system. In that chain, each part has the same J^{PC}.

Consider first the chiral even GPDs. In the table we show the relevant crossing (anti-)symmetric combinations of the GPDs. The crossing odd $\widetilde{H}$ has contributions from $2^{--}, 4^{--}$ and higher. There are several different reasons that this GPD is not expected to contribute at leading order. There cannot be a 0^{--} coupling to $\gamma + \pi^0$ or $N\bar{N}$. The 2^{--} appears in the triplet spin with L=2. For simple resonance exchanges, there would be an angular momentum barrier compared to the $J = 1$ exchanges. In Regge language the trajectory with 2^{--} is non-leading even signature and the absence of 0^{--} would require a "nonsense" factor killing the pole, thereby suppressing the effect in the physical region. Furthermore, there is no well established 2^{--} isoscalar or isovector (ρ or ω) meson below mass of 2 GeV/c^2, although there is a nearly degenerate pair, $1^{--}\omega(1650)$ with a $3^{--}\omega(1670)$. Since these can be categorized as $S = 1 \,\&\, L = 2$, then the 2^{--} would lie in this region of masses. This puts the Regge trajectory going through $J = 2$ at $t \approx 2.8 GeV^2$, which will lie well below the ρ and lower than a_1 and b_1, minimizing its importance at small x as well.

The crossing odd $\widetilde{E}$ has contributions from $1^{+-}, 2^{--}, 3^{+-}$, etc., so it is the leading candidate for chiral even GPDs that contribute to π^0. Its first moment is the pseudoscalar form factor, for which the main contribution is

Table 1. Chiral even GPDs & J^{PC}

GPDs' t-channel $J^{P\,(C=-)}$ for π^0 production		
crossing symmetrized GPD	J^{PC}	Spin & (crossing)
$\widetilde{H}(x,\xi,t) - \widetilde{H}(-x,\xi,t)$	$2^{--}, 4^{--}...$	$S = 1$ (even)
$\widetilde{E}(x,\xi,t) - \widetilde{E}(-x,\xi,t)$	$1^{+-}, 2^{--}, 3^{+-}, ...$	$S = 0$ (odd) & 1 (even)
$H_T(x,\xi,t) + H_T(-x,\xi,t)$	$1^{+-}, 2^{--}, 3^{+-}, ...$	$S = 1$ (odd)

the π itself. However, for the neutral π this is not the case - there is no π pole because there is no $\gamma \to \pi^0 + \pi^0$ coupling. How does this effect the π^0 production? It should be said that the role of the charged pion pole in the GPD is not settled in any case, so its contribution to charged π production is not agreed upon. Our approach is that the pole would be included in the GPD, as in the review of Goeke, et al. [12], rather than providing a separate contribution to the amplitudes, as in Ref. [13]. That being said, in π^0 the non-pole contribution to the form factor is relevant. How do we estimate that?

2. Digression on Nucleon Form Factors

The electroweak form factors of the nucleon include $g_A(t)$, the axial vector form factor, and $g_P(t)$, the "induced pseudoscalar". Hence for the axial electroweak current,

$$\langle N(p') \mid J_A^\nu \mid N(p)\rangle = \bar{u}(p')[g_A(q^2)\gamma^\nu\gamma^5 + \frac{g_P(q^2)}{m_\mu}q^\nu\gamma^5]u(p), \quad (5)$$

where $q^2 = (p' - p)^2 = t$, and the choice of the muon mass to make the g_P dimensionless follows convention (the best experimental values of the induced pseudoscalar form factor come from μ capture). The divergence of the isovector part of the axial current is approximated in PCAC by the pion pole - the Goldberger-Treiman relation for $g_A(0)$ in terms of the π-nucleon coupling constant. For non-zero q^2 there is a relation between the two form factors,

$$g_P(q^2) = \frac{2m_\mu M}{m_\pi^2 - q^2} g_A(0). \quad (6)$$

Now $\widetilde{H}$ integrates to $g_A(t)$; $\widetilde{E}$ integrates to $h_A(t) = \frac{2M}{m_\mu} g_P(t)$, proportional to the above pseudoscalar form factor of the nucleon. Recent experimental determinations show that $g_A(0) = 1.267$ and $g_P(-0.88m_\mu^2) = 8.58$ [11]. For π^0 electroproduction on the nucleon, however, there is no π^0 exchange. The difference between g_P from Eq. 6 and experiment is a measure of the non-pole contribution, which is quite small [11]. Thence, the size of π^0 electroproduction cross sections would be expected to be considerably less than charged π's if $\widetilde{E}$ were the major contribution (see for example Ref. [12], where an estimate is based on a chiral-soliton model).

3. π^0 and pseudoscalar production

The measured cross section for π^0 is sizable and has large transverse γ^* contributions. This indicates that the main contributions should come from chiral odd GPDs, for which the t-channel decomposition is richer. In particular, because these GPDs arise from the Dirac matrices $\sigma^{\mu\nu}$, there are 2 series of J^{PC} values for each GPD [8] corresponding to space-space or time-space combinations - 1^{--} and 1^{+-} These series occur for 3 of the 4 chiral odd GPDs, the exception being $\widetilde{E}_T$. We are thus led to the conclusion that chiral odd GPDs will dominate the neutral pseudoscalar leptoproduction cross sections. This result has interesting consequences. For one thing, in a factorized handbag picture, these GPDs will couple to the hard part, the $\gamma^* +$ quark $\rightarrow \pi^0 +$ quark providing the π^0 couples through γ^5, which is naively twist 3, rather than the twist 2 $\gamma^+\gamma^5$. Nevertheless, the previous arguments support this choice. Secondly the vector 1^{--} and axial vector 1^{+-} in the t-channel, viewed as particles (ρ^0, ω and $\mathrm{b}_1^0, \mathrm{h}$), couple primarily to the transverse virtual photon. For Reggeons, the 1^{--} does not couple at all to the longitudinal photon, while the axial vector 1^{+-} does through helicity flip [10]. Guided by these observations [1], we assume the hard part depends on whether the exchange quantum numbers are in the vector or axial vector series, thereby introducing orbital angular momentum into the model. We use Q^2 dependent electromagnetic "transition" form factors for vector or axial vector quantum numbers going to a pion. We calculate these using PQCD for $q+\bar{q}+\gamma^*(Q^2) \to q+\bar{q}$ and a standard $z-$dependent pion wave function, convoluted in an impact parameter representation that allows orbital excitations to be easily implemented.

With our model for the chiral odd, spin-dependent GPDs and these transition form factors, we can obtain the full range of cross sections and asymmetries in kinematic regimes that coincide with ongoing JLab experiments. We are able to predict the important transverse photon contributions to the observables [1]. In figures 1 and 2 we show two examples of predictions that depend on the values of the tensor charges, thereby providing a means to narrow down those important quantities. Preliminary versions of this program have been presented and further details will soon appear. A similar emphasis on chiral odd contributions for π electroproduction has recently been proposed [13], although the details of that model are quite different from ours.

4. Dispersion Relations and Partonic Interpretation of GPDs

At the heart our understanding of the role of GPDs in exclusive leptoproduction reactions are the analyticity properties of the amplitudes. We have examined the applicability of Dispersion Relations (DRs) to the GPD formulation of DVCS. Unitarity and completeness are crucial ingredients in establishing analytic properties of the amplitudes. The amplitudes are analytic in energy variables, which allows the amplitudes ("Compton Form Factors" or CFF's) to satisfy DRs relating real and imaginary parts. The imaginary part of a CFF is given by the GPD evaluated at the kinematic point where the returning quark has only transverse momentum relative to the nucleon direction. Then the DR can determine the real part thereby. However, at non-zero momentum transfer the DRs require integration over

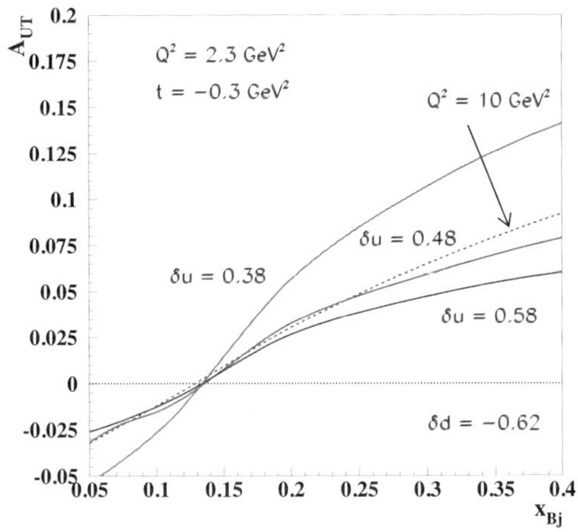

Fig. 2. Transverse spin asymmetry, A_{UT}, vs. x_{Bj}, at $Q^2 = 2.3$ GeV^2, $xt = -0.3GeV^2$ for different values of the tensor charge, δu, with fixed $\delta d = -0.62$, *i.e.* equal to the central value extracted in a global fit [6].

unphysical regions of the variables and that region is considerable - the real parts must still be measured by using interference with the Bethe-Heitler contribution [2].

We have also investigated the analyticity in the $X < \zeta$ region (the ERBL region), which conventionally is described as a quark-antiquark or meson distribution in the proton. We extended the derivation of the parton model from connected matrix elements for non-local quark and gluon field operators in inclusive hard processes [14] to the non-forward GPDs [15]. At leading twist, the kinematics require semi-disconnected amplitudes, *i.e.* vacuum fluctuations, that vitiate the partonic interpretation. In order to restore the sensible partonic picture it is necessary to include gluon exchange, appearing as an initial or final state interaction that "dresses" the struck or returning quark.

5. Conclusions

Spin dependent GPDs enable a QCD based look into the parton angular momentum distributions that constitute the nucleon structure. DVCS depends on chiral even GPDs that involve longitudinal asymmetries. On the other hand, π^0 electroproduction provides a window into transversity via Chiral odd GPDs, which are essential to understand the importance of transverse γ^*. The transverse photon cross sections and asymmetries are major components of electroproduction observables. A broad vista of spin phenomenology is opened up by our careful modeling of GPDs, as we have explained.

In a more general study of the analyticity of GPDs, we presented the problem with the partonic interpretation of the ERBL region of GPD kinematics. This work is ongoing, and suggests an important connection between partonic distributions and final state interactions.

Acknowledgments

The authors appreciate the work of the organizers of Exclusive Reactions 2010. This work is supported in part by U.S. Department of Energy Grant Nos. DE-FG02-92ER40702 (G. R. G.) and DE-FG02-01ER4120 (S. L).

References

1. S. Ahmad, G.R. Goldstein, S. Liuti, Phys. Rev. D **79**, 054014 (2009).
2. G.R. Goldstein and S. Liuti, Phys. Rev. D **80**, 071501 (2009).
3. M. Diehl, Phys. Rept. **388**, 41 (2003); *ibid* Eur. Phys. Jour. C **19**, 485 (2001).
4. M. Burkardt, Phys. Rev. D **72**, 094020 (2005); *ibid* Phys. Lett. **B** 639 (2006) 462.
5. S. Ahmad, *et al.* (AHLT), Phys. Rev. D **75**, 094003 (2007); *ibid* Eur. Jour. Phys. C **63**, 407 (2009).
6. M. Anselmino, *et al.*, Phys. Rev. D **75**, 054032 (2007).
7. X. Ji and R.F. Lebed, Phys. Rev. D **63**, 076005 (2001).
8. P. Hägler, Phys. Lett. B **594** 164 (2004).
9. M. Diehl and D.Yu. Ivanov, Eur. Phys. Jour. C **52**, 919 (2007).
10. G. R. Goldstein and J. F. Owens, Phys. Rev. **D7** 865 (1973).
11. T. Gorringe and H.W. Fearing, Rev. Mod. Phys. **76**, 31 (2004).
12. K. Goeke, M.V. Polyakov and M. Vanderhaeghen, Prog. Part. Nucl. Phys. **47** 401 (2001).
13. S. V. Goloskokov and P. Kroll, Eur. Phys. J. C **50**, 829 (2007); *ibid* arXiv:0906.0460.
14. R.L. Jaffe, Nucl. Phys. B **229**, 205 (1983).
15. G.R. Goldstein and S. Liuti, arXiv:1006.0213.

SINGLE TRANSVERSE-SPIN ASYMMETRY IN HARD EXCLUSIVE MESON ELECTROPRODUCTION IN THE BACKWARD REGION

J.P. LANSBERG[1]*, B. PIRE[1], L. SZYMANOWSKI[2]

[1] *Centre de Physique Théorique, École Polytechnique, CNRS, F-91128, Palaiseau, France*

[2] *Soltan Institute for Nuclear Studies, Warsaw, Poland*

We discuss the relevance of studying single transverse-spin asymmetry in hard-exclusive meson electroproduction in the backward region. Such an asymmetry could help us discriminate between contributions from a soft baryon exchange in the u-channel and a hard parton-induced scattering.

Keywords: Exclusive-Backward Reactions, Single Transverse-Spin Asymmetry

1. Introduction

In Refs. 1–3, we introduced the framework to study backward pion electroproduction, $\gamma^\star(q)N(p_1) \to N'(p_2)\pi(p_\pi)$, on a proton (or neutron) target, in the Bjorken regime (q^2 large and $q^2/(2p_1.q)$ fixed) in terms of a factorized amplitude (see Fig. 1) where a hard part is convoluted with Transition Distribution Amplitudes (TDAs), as well as the reaction, $N(p_1)\bar{N}(p_2) \to \gamma^\star(q)\pi(p_\pi)$, in the near forward region.[4–6] This extended the concept of Generalised Parton Distributions (GPDs). Such an extension of the GPD framework has already been advocated in the pioneering work of Ref. 7. The TDAs involved in the description of Deeply-Virtual Compton Scattering (DVCS) in the backward kinematics $\gamma^\star(q)N(p_1) \to N'(p_2)\gamma(p_\gamma)$ and the reaction $N(p_1)\bar{N}(p_2) \to \gamma^\star(q)\gamma(p_\gamma)$ in the near forward region were given in Ref. 8.

Recently, a study of the TDAs in the meson-cloud model[9] has become available.[10] Yet, more work is needed before being able to proceed to *quantitative* comparisons between different TDA models and between theory and experiments. For the time being, model independent analyses – looking for

*lansberg@cpht.polytechnique.fr

scaling or characteristic polarization effects – sound more expedient. In this context, we would like to argue here that the study of target transverse-spin asymmetry (which we will denote SSA) could be used as a test of the dominance of a hard parton-induced scattering in the backward region at large Q^2 rather than that of a soft baryon exchange in the u-channel. Such a reaction would only generate phases through final state interactions, expected to decrease for $W^2 \gg (M + m_\pi)^2$ and large Q^2. On the contrary, in reactions at the parton level, one expects an imaginary part to develop and to generate a SSA independently of whether W^2 and Q^2 are large or not. This will be explained later on.

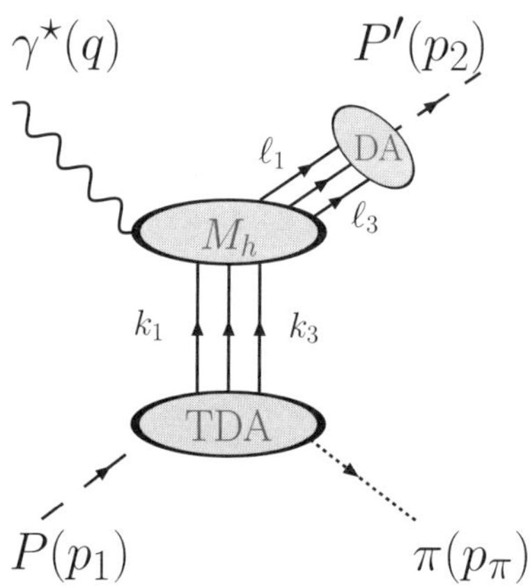

Fig. 1. Illustration of the factorisation for backward electroproduction of a pion.

2. Some definitions

The five-fold differential cross section for the process $eP \to e'P'\pi^0$ can be reduced to a two-fold one – expressible in the center-of-mass frame of the $P'\pi^0$ pair – multiplied by a flux factor Γ, $\frac{d^5\sigma}{dE_{e'}d^2\Omega_e d^2\Omega_\pi^*} = \Gamma \frac{d^2\sigma}{d^2\Omega_\pi^*}$, where Ω_e is the differential solid angle for the scattered electron in the lab frame, and Ω_π^* is the differential solid angle for the pion in the $P'\pi^0$ center-of-mass frame, such that $d\Omega_\pi^* = d\phi\, d\cos\theta_\pi^*$. θ_π^* is defined as the polar angle between the virtual photon and the pion in the latter system (see Fig. 2). ϕ is the azimuthal angle between the electron plane and the plane of the process $\gamma^\star P \to P'\pi^0$ (hadronic plane) ($\phi = 0$ when the pion is emitted in the half plane containing the outgoing electron).

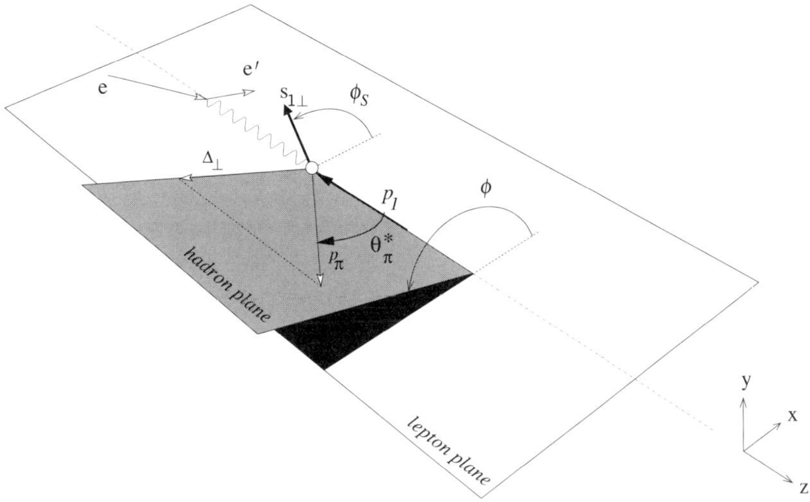

Fig. 2. Kinematics of electroproduction of a pion and definition of the angles ϕ and ϕ_S

In general, we have contributions from different polarisations of the photon. For that reason, we define four polarised cross sections, $d^2\sigma_{\rm T}$, $d^2\sigma_{\rm L}$, $d^2\sigma_{\rm TL}$ and $d^2\sigma_{\rm TT}$,[a] which do not depend on ϕ but only on W, Q^2 and θ^*_π. The ϕ dependence is written as[13]

$$\frac{d^2\sigma}{d\Omega^*_\pi} = \frac{d^2\sigma_{\rm T}}{d\Omega^*_\pi} + \epsilon\,\frac{d^2\sigma_{\rm L}}{d\Omega^*_\pi} + \sqrt{2\epsilon(1+\epsilon)}\,\frac{d^2\sigma_{\rm TL}}{d\Omega^*_\pi}\cos\phi + \epsilon\,\frac{d^2\sigma_{\rm TT}}{d\Omega^*_\pi}\cos 2\phi. \quad (1)$$

At the leading-twist accuracy, the QCD mechanism considered here contributes only to $\frac{d^2\sigma_{\rm T}}{d\Omega^*_\pi}$.

In the scaling regime, the amplitude for $\gamma^\star P(p_1) \to P'(p_2)\pi(p_\pi)$ in the backward kinematics – namely small $u = (p_\pi - p_1)^2 = \Delta^2$ or $\cos\theta^*_\pi$ close to -1 – involves the TDAs $T(x_i, \xi, \Delta^2)$, where x_i $(i = 1, 2, 3)$ denote the light-cone-momentum fractions carried by participant quarks and ξ is the skewness parameter such that $2\xi = x_1 + x_2 + x_3$.

The amplitude is then a convolution of the proton DAs, a perturbatively-calculable-hard-scattering amplitude and the TDAs, defined from the Fourier transform of a matrix element of a three-quark-light-cone operator between a proton and a meson state. We have shown that these TDAs

[a]For L, x, y, the linear polarisations of the virtual photon (for the definition of the x & y axis, see Fig. 1), one defines[12,13] $d^2\sigma_{\rm L} \propto \mathcal{M}^L(\mathcal{M}^L)^*$, $d^2\sigma_{\rm T} \propto 1/2[\mathcal{M}^x(\mathcal{M}^x)^* + \mathcal{M}^y(\mathcal{M}^y)^*]$, $d^2\sigma_{\rm TL} \propto \mathcal{M}^x(\mathcal{M}^L)^* + \mathcal{M}^L(\mathcal{M}^x)^*$ and $d^2\sigma_{\rm TT} \propto 1/2[\mathcal{M}^x(\mathcal{M}^x)^* - \mathcal{M}^y(\mathcal{M}^y)^*]$.

obey QCD evolution equations, which follow from the renormalisation-group equation of the three-quark operator. Their Q^2 dependence is thus completely under control.

The momenta of the process $\gamma^\star P \to P'\pi$ are defined as in Fig. 1 & 2. The z-axis is chosen along the initial nucleon and the virtual photon momenta and the $x-z$ plane is identified with the collision or hadronic plane (Fig. 2). Then, we define the light-cone vectors p and n ($p^2{=}n^2{=}0$) such that $2\,p.n = 1$, as well as $P = \frac{1}{2}(p_1 + p_\pi)$, $\Delta = p_\pi - p_1$ and its transverse component Δ_T ($\Delta_T.\Delta_T = \Delta_T^2 < 0$). From those, we define ξ in an usual way as

$$\xi = -\frac{\Delta.n}{2P.n}. \tag{2}$$

We can then express the momenta of the particles through their Sudakov decomposition and, keeping the first-order corrections in the masses and Δ_T^2, we have:

$$p_1 = (1+\xi)p + \frac{M^2}{1+\xi}n, q \simeq -2\xi\left(1 + \frac{(\Delta_T^2 - M^2)}{Q^2}\right)p + \frac{Q^2}{2\xi\left(1 + \frac{(\Delta_T^2 - M^2)}{Q^2}\right)}n,$$

$$p_\pi = (1-\xi)p + \frac{m_\pi^2 - \Delta_T^2}{1-\xi}n + \Delta_T, \Delta = -2\xi p + \left[\frac{m_\pi^2 - \Delta_T^2}{1-\xi} - \frac{M^2}{1+\xi}\right]n + \Delta_T$$

$$p_2 \simeq -2\xi\frac{(\Delta_T^2 - M^2)}{Q^2}p + \left[\frac{Q^2}{2\xi\left(1 + \frac{(\Delta_T^2 - M^2)}{Q^2}\right)} - \frac{m_\pi^2 - \Delta_T^2}{1-\xi} + \frac{M^2}{1+\xi}\right]n - \Delta_T, \tag{3}$$

For $\varepsilon_x = (0,1,0,0)$ and $\varepsilon_y = (0,0,1,0)$ with the axis definitions of Fig. 2, one may further specify that

$$\Delta_T = |\Delta_T|(\cos\phi\,\varepsilon_x + \sin\phi\,\varepsilon_y) \text{ and } s_{T,1} = s_1 = \cos\phi_S\,\varepsilon_x + \sin\phi_S\,\varepsilon_y, \tag{4}$$

for the transverse spin of the target ($s_1.p_1 = s_1.p = s_1.n = 0$).

In the following, we use the same definition of the $p \to \pi^0$ TDAs as given by Eq. (15) of Ref. [3].

3. Hard-amplitude calculation

At leading order in α_s, the amplitude $\mathcal{M}^{\lambda}_{s_1 s_2}$ for $\gamma^\star(q,\lambda)P(p_1,s_1) \to P'(p_2,s_2)\pi^0(p_\pi)$ reads

$$\mathcal{M}^{\lambda}_{s_1 s_2} = \underbrace{-i\frac{(4\pi\alpha_s)^2\sqrt{4\pi\alpha_{em}}f_N^2}{54 f_\pi}}_{\mathcal{C}}\frac{1}{Q^4}\Bigg[\underbrace{\bar{u}_2 \not\epsilon(\lambda)\gamma^5 u_1}_{\mathcal{S}^{\lambda}_{s_1 s_2}}\underbrace{\int\left(2\sum_{\alpha=1}^{7}T_\alpha+\sum_{\alpha=8}^{14}T_\alpha\right)}_{\mathcal{I}}$$
$$-\underbrace{\bar{u}_2\not\epsilon(\lambda)\frac{\not\Delta_T}{M}\gamma^5 u_1}_{\mathcal{S'}^{\lambda}_{s_1 s_2}}\underbrace{\int\left(2\sum_{\alpha=1}^{7}T'_\alpha+\sum_{\alpha=8}^{14}T'_\alpha\right)}_{\mathcal{I}'}\Bigg] , \qquad (5)$$

where

$$\int \equiv \int_{-1+\xi}^{1+\xi} dx_1 dx_2 dx_3 \delta(2\xi - x_1 - x_2 - x_3)\int_0^1 dy_1 dy_2 dy_3 \delta(1-y_1-y_2-y_3) ,$$

$u(p_1,s_1) \equiv u_1$, $\bar{u}(p_2,s_2) \equiv \bar{u}_2$ and the coefficients T_α and $T'_\alpha (\alpha = 1, ..., 14)$ are functions of x_i,y_j,ξ and Δ^2 and are given in Table 1 of Ref. [3].

The expression in Eq. (5) is to be compared with the leading-twist amplitude for the baryonic-form factor[14]

$$\mathcal{M}^\lambda \propto -i(\bar{u}_2\not\epsilon(\lambda)u_1)\frac{\alpha_s^2 f_N^2}{Q^4}\int\left(2\sum_{\alpha=1}^{7}T^p_\alpha(x_i,y_j,\xi,t)+\sum_{\alpha=8}^{14}T^p_\alpha(x_i,y_j,\xi,t)\right). \qquad (6)$$

The factors T^p_α are very similar to the T_α obtained here. However, the integration domain is different. In the form factor case

$$\int \text{stands for} \int_0^1 dx_1 dx_2 dx_3 \delta(1-\sum_i x_i)\int_0^1 dy_1 dy_2 dy_3 \delta(1-\sum_i y_i). \qquad (7)$$

Consequently, the integration of denominators in T^p_α such as $1/(x_i + \imath\varepsilon)$ do not generate any imaginary parts. On the contrary, the integrations of similar denominators in T_α and T'_α over the TDA integration domain will generate an imaginary part when passing from the ERBL region (all $x_i > 0$) to one of the DGLAP regions (one $x_i < 0$). This will be the source of the SSA as we will show later on.

4. Single Transverse Spin Asymmetry

In order to study a possible SSA we shall study the quantity $\sigma^{s_1} - \sigma^{-s_1}$ with the definition

$$\sigma^{s_1} = \sum_{\lambda} \sum_{s_2} (\mathcal{M}^{\lambda}_{s_1 s_2})(\mathcal{M}^{\lambda}_{s_1 s_2})^*. \tag{8}$$

Since we are interested in the leading twist contribution of this asymmetry, we can sum only over the transverse polarisation of the virtual photon using $\sum_{\lambda=x,y} \varepsilon(\lambda)^\mu (\varepsilon(\lambda)^\nu)^* = -g^{\mu\nu} + (p^\mu n^\nu + p^\nu n^\mu)/(p.n)$. The sum on the final-proton spin s_2 is done using $\sum_{s_2} u_\alpha(p_2, s_2) \bar{u}_\beta(p_2, s_2) = (\not{p}_2 + M)_{\alpha\beta}$. As regards the initial-proton spinor, one uses the following relation involving its transverse spin s_1, $u_\alpha(p_1, s_1)\bar{u}_\beta(p_1, s_1) = 1/2(1 + \gamma^5 \not{s}_1)(\not{p}_1 + M)_{\alpha\beta}$.

Hence, one has

$$\begin{aligned}
\sum_{\lambda=x,y} \sum_{s_2} (\mathcal{M}^{\lambda}_{s_1 s_2})(\mathcal{M}^{\lambda}_{s_1 s_2})^* &= \frac{|\mathcal{C}|^2}{Q^8}\Big(-g^{\mu\nu} + \frac{p^\mu n^\nu + p^\nu n^\mu}{p.n}\Big) \times \\
\Big[\bar{u}_2 \gamma^\mu \gamma^5 u_1 \mathcal{I} - \bar{u}_2 \gamma^\mu \frac{\not{\Delta}_T}{M} \gamma^5 u_1 \mathcal{I}'\Big] &\times \Big[-\bar{u}_1 \gamma^5 \gamma_\mu u_2 \mathcal{I}^* + \bar{u}_1 \gamma^5 \frac{\not{\Delta}_T}{M} \gamma_\mu u_1 \mathcal{I}'^* \Big] \\
&= \frac{|\mathcal{C}|^2}{Q^8}\Big(-g^{\mu\nu} + \frac{p^\mu n^\nu + p^\nu n^\mu}{p.n}\Big) \times \\
\Big[-\mathrm{Tr}\Big((\not{p}_2 + M)\gamma^\mu \gamma^5 (\not{p}_1 + M) &\frac{(1+\gamma^5 \not{s}_1)}{2} \gamma^5 \gamma_\mu\Big) \mathcal{I}\mathcal{I}^* \\
+\mathrm{Tr}\Big((\not{p}_2 + M)\gamma^\mu \frac{\not{\Delta}_T}{M} \gamma^5 (\not{p}_1 + M) &\frac{(1+\gamma^5 \not{s}_1)}{2} \gamma^5 \gamma_\mu\Big) \mathcal{I}'\mathcal{I}^* \\
+\mathrm{Tr}\Big((\not{p}_2 + M)\gamma^\mu \gamma^5 (\not{p}_1 + M) &\frac{(1+\gamma^5 \not{s}_1)}{2} \gamma^5 \frac{\not{\Delta}_T}{M} \gamma_\mu\Big) \mathcal{I}\mathcal{I}'^* \\
-\mathrm{Tr}\Big((\not{p}_2 + M)\gamma^\mu \frac{\not{\Delta}_T}{M} \gamma^5 (\not{p}_1 + M) &\frac{(1+\gamma^5 \not{s}_1)}{2} \gamma^5 \frac{\not{\Delta}_T}{M} \gamma_\mu\Big) \mathcal{I}'\mathcal{I}'^* \Big]
\end{aligned} \tag{9}$$

Dropping the contributions proportional to the proton mass, the spin asymmetry reads ($\epsilon_{0123} = -\epsilon^{0123} = +1$)

$$\begin{aligned}
\sigma^{s_1} - \sigma^{-s_1} =& 8 \frac{|\mathcal{C}|^2}{Q^6} \frac{1+\xi}{\xi} \frac{\epsilon^{n p s_1 \Delta_T}}{M} \Im m(\mathcal{I}'\mathcal{I}^*) \\
=& -4 \frac{|\mathcal{C}|^2}{Q^6} \frac{|\Delta_T|}{M} \frac{1+\xi}{\xi} \sin(\phi - \phi_S) \Im m(\mathcal{I}'\mathcal{I}^*).
\end{aligned} \tag{10}$$

Comparing with the expressions for the unpolarised cross section obtained in Ref. 3, one concludes that the asymmetry for the hard-parton induced contribution is leading-twist as soon as $\Delta_T \neq 0$ and $\mathcal{I}$ or $\mathcal{I}'$ are no longer pure real or pure imaginary numbers. This is precisely what one expects when DGLAP contributions are taken into account.[16]

5. Discussion and conclusion

Although the knowledge of *baryon to meson* TDAs has recently improved significantly thanks to a first study in the meson cloud model[10] and another one focused on their spectral representations,[15] model-independent observables aimed at studying the backward regime of meson electroproduction will still be the bread-and-butter of this field for the months to come.

In this context, we find it particularly relevant to emphasize that the study of the asymmetry of the target transverse spin would reveal unique information on the nature of the particles exchanged in the u channel, be it a "mere" baryon slightly off-shell, or three perturbative quarks. For non-vanishing transverse momenta (Δ_T), one expects in the latter case an asymmetry of the same order as the unpolarised cross section, while, in the former case, they would be most likely decreasing for increasing W^2 and Q^2.

Acknowledgments

We thank S.J. Brodsky, G. Huber, V. Kubarovsky, K.J. Park, B. Pasquini, K. Semenov, P. Stoler for useful and motivating discussions as well as the organizers of the workshop "Exclusive 2010" for their kind invitation to present our work. This work is partly supported by the ANR contract BLAN07-2-191986. L.Sz. acknowledges the support by the Polish Grant N202 249235.

References

1. B. Pire and L. Szymanowski, Phys. Lett. B **622** (2005) 83 [arXiv:hep-ph/0504255].
2. B. Pire and L. Szymanowski, PoS **HEP2005** (2006) 103 [arXiv:hep-ph/0509368].
3. J. P. Lansberg, B. Pire and L. Szymanowski, Phys. Rev. D **75** (2007) 074004 [Erratum-ibid. D **77** (2008) 019902] [arXiv:hep-ph/0701125]. Note that the differential cross-section plots are in nb/sr, not pb/sr as indicated.
4. B. Pire and L. Szymanowski, Phys. Rev. D **71** (2005) 111501 [arXiv:hep-ph/0411387].
5. J. P. Lansberg, B. Pire and L. Szymanowski, Phys. Rev. D **76** (2007) 111502(R) [arXiv:0710.1267 [hep-ph]].
6. M. F. Lutz, *et al.* [The PANDA Collaboration], arXiv:0903.3905 [hep-ex].

7. L. L. Frankfurt, P. V. Pobylitsa, M. V. Polyakov and M. Strikman, Phys. Rev. D **60** (1999) 014010 [arXiv:hep-ph/9901429]; L. Frankfurt, M. V. Polyakov, M. Strikman, D. Zhalov and M. Zhalov, in *Newport News 2002, Exclusive Processes at High Momentum Transfer* (edited by A. Radyushkin, P. Stoler; Singapore, World Scientific, 2002, pp 361-368), arXiv:hep-ph/0211263.
8. J. P. Lansberg, B. Pire and L. Szymanowski, Nucl. Phys. A **782** (2007) 16 [arXiv:hep-ph/0607130].
9. B. Pasquini and S. Boffi, Phys. Rev. D **73** (2006) 094001 [arXiv:hep-ph/0601177].
10. B. Pasquini, M. Pincetti and S. Boffi, Phys. Rev. D **80** (2009) 014017 [arXiv:0905.4018 [hep-ph]].
11. P. V. Pobylitsa, M. V. Polyakov and M. Strikman, Phys. Rev. Lett. **87** (2001) 022001 [arXiv:hep-ph/0101279].
12. P. J. Mulders, Phys. Rept. **185** (1990) 83.
13. K. Park *et al.* [CLAS Collaboration], Phys. Rev. C **77** (2008) 015208 [arXiv:0709.1946 [nucl-ex]].
14. V. L. Chernyak and A. R. Zhitnitsky, Phys. Rept. **112**, 173 (1984);
15. B. Pire, K. Semenov-Tian-Shansky and L. Szymanowski, arXiv:1008.0721 [hep-ph].
16. J.P. Lansberg, *et al.*, work in progress.

TRANSVERSE STRUCTURE OF HADRONS

M. BURKARDT

Department of Physics, New Mexico State University, Las Cruces, NM 88011
**E-mail: burkardt@nmsu.edu*

Parton distributions in impact parameter space, which are obtained by Fourier transforming GPDs, exhibit a significant deviation from axial symmetry when the target and/or quark are transversely polarized. Connections between this deformation and single-spin asymmetries as well as with quark-gluon correlations are discussed. The information content of the DVCS amplitude at fixed Q^2 can be condensed into GPDs along the 'diagonal' $x = \xi$ plus D-form factor.

1. Impact Parameter Dependent PDFs and SSAs

The Fourier transform of the GPD $H_q(x, 0, t)$ yields the distribution $q(x, \mathbf{b}_\perp)$ of unpolarized quarks and target, in impact parameter space [1]

$$q(x, \mathbf{b}_\perp) = \int \frac{d^2\mathbf{\Delta}_\perp}{(2\pi)^2} H_q(x, 0, -\mathbf{\Delta}_\perp^2)\, e^{-i\mathbf{b}_\perp \cdot \mathbf{\Delta}_\perp}, \tag{1}$$

with $\mathbf{\Delta}_\perp = \mathbf{p}'_\perp - \mathbf{p}_\perp$. For a transversely polarized target (e.g. polarized in the $+\hat{x}$-direction) the impact parameter dependent PDF $q_{+\hat{x}}(x, \mathbf{b}_\perp)$ is no longer axially symmetric and the transverse deformation is described by the gradient of the Fourier transform of the GPD $E_q(x, 0, t)$ [2]

$$q_{+\hat{x}}(x,\mathbf{b}_\perp) = q(x,\mathbf{b}_\perp) - \frac{1}{2M}\frac{\partial}{\partial b_y} \int \frac{d^2\mathbf{\Delta}_\perp}{(2\pi)^2} E_q(x, 0, -\mathbf{\Delta}_\perp^2)\, e^{-i\mathbf{b}_\perp \cdot \mathbf{\Delta}_\perp} \tag{2}$$

$E_q(x, 0, t)$ and hence the details of this deformation are not very well known, but its x-integral, the Pauli form factor F_2, is. This allows to relate the average transverse deformation resulting from Eq. (2) to the contribution from the corresponding quark flavor to the anomalous magnetic moment.

For example, u quarks in a proton contribute with a positive anomalous magnetic moment and d quarks (after factoring out the negative d quark charge) with a negative value. Eq. (2) thus implies that for a nucleon target

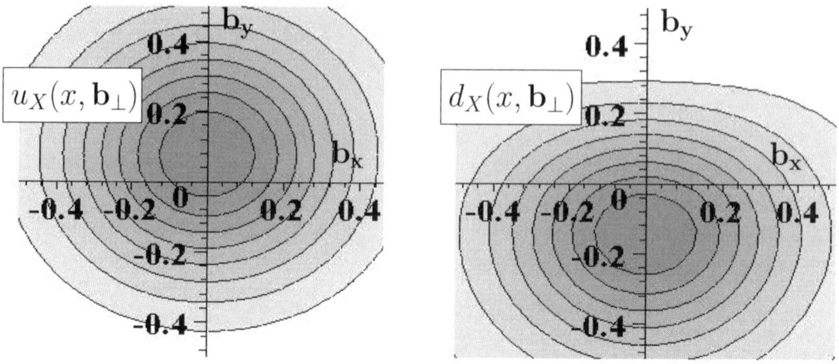

Fig. 1. Distribution of the j^+ density for u and d quarks in the $\perp$ plane ($x_{Bj} = 0.3$) for a proton polarized in the x direction in the model from Ref. [2]. For other values of x the distortion looks similar. The signs of the distortion are determined by the signs of the contribution from each quark flavor to the proton anomalous magnetic moment.

polarized in the $+\hat{x}$ direction, the leading twist distribution of u quarks is shifted in the $+\hat{y}$ direction while that of d quarks is shifted in the $-\hat{y}$ direction. This is important for understanding the sign of the Sivers function.

In a target that is polarized transversely (*e.g.* vertically), the quarks in the target nucleon can exhibit a (left/right) asymmetry of the distribution $f_{q/p^\uparrow}(x_{\rm B}, \boldsymbol{k}_T)$ in their transverse momentum $\boldsymbol{k}_T$ [3,4]

$$f_{q/p^\uparrow}(x_{\rm B}, \boldsymbol{k}_T) = f_1^q(x_{\rm B}, k_T^2) - f_{1T}^{\perp q}(x_{\rm B}, k_T^2)\frac{(\hat{\boldsymbol{P}} \times \boldsymbol{k}_T) \cdot \boldsymbol{S}}{M}, \tag{3}$$

where $\boldsymbol{S}$ is the spin of the target nucleon and $\hat{\boldsymbol{P}}$ is a unit vector opposite to the direction of the virtual photon momentum. The fact that such a term may be present in (3) is known as the Sivers effect and the function $f_{1T}^{\perp q}(x_{\rm B}, k_T^2)$ is known as the Sivers function. The latter vanishes in a naive parton picture since $(\hat{\boldsymbol{P}} \times \boldsymbol{k}_T) \cdot \boldsymbol{S}$ is odd under naive time reversal (a property known as naive-T-odd), where one merely reverses the direction of all momenta and spins without interchanging the initial and final states. The significant distortion of parton distributions in impact parameter space (Fig. 1) provides a natural mechanism for a Sivers effect. In semi-inclusive DIS, when the virtual photon strikes a u quark in a $\perp$ polarized proton, the u quark distribution is enhanced on the left side of the target (for a proton with spin pointing up when viewed from the virtual photon perspective). Although in general the final state interaction (FSI) is very complicated, we expect it to be on average attractive thus translating a position space distortion to the left into a momentum space asymmetry to the right and vice

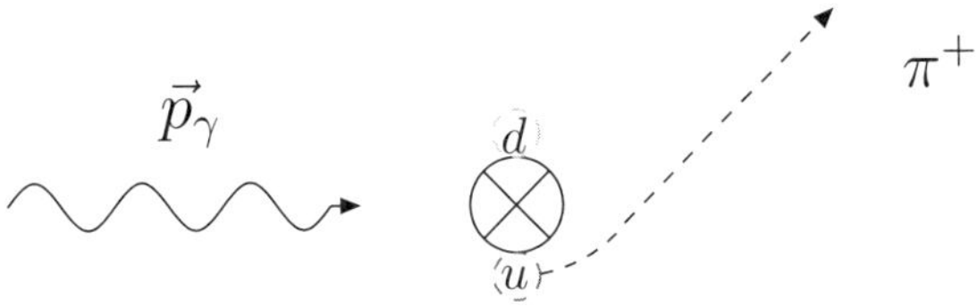

Fig. 2. The transverse distortion of the parton cloud for a proton that is polarized into the plane, in combination with attractive FSI, gives rise to a Sivers effect for u (d) quarks with a $\perp$ momentum that is on the average up (down).

versa (Fig. 2) [5]. Since this picture is very intuitive, a few words of caution are in order. First of all, such a reasoning is strictly valid only in mean field models for the FSI as well as in simple spectator models [6]. Furthermore, even in such mean field or spectator models there is in general no one-to-one correspondence between quark distributions in impact parameter space and unintegrated parton densities (e.g. Sivers function) (for a recent overview, see Ref. [7]). While both are connected by an overarching Wigner distribution [8], they are not Fourier transforms of each other. Nevertheless, since the primordial momentum distribution of the quarks (without FSI) must be symmetric, we find a qualitative connection between the primordial position space asymmetry and the momentum space asymmetry due to the FSI. Another issue concerns the x-dependence of the Sivers function. The x-dependence of the position space asymmetry is described by the GPD $E(x,0,-\Delta_\perp^2)$. Therefore, within the above mechanism, the x dependence of the Sivers function should be related to that of $E(x,0,-\Delta_\perp^2)$. However, the x dependence of E is not known yet and we only know the Pauli form factor $F_2 = \int \mathrm{d}xE$. Nevertheless, if one makes the additional assumption that E does not fluctuate as a function of x then the contribution from each quark flavor q to the anomalous magnetic moment κ determines the sign of $E^q(x,0,0)$ and hence of the Sivers function. With these assumptions, as well as the very plausible assumption that the FSI is on average attractive, one finds that $f_{1T}^{\perp u} < 0$, while $f_{1T}^{\perp d} > 0$. Both signs have been confirmed by a flavor analysis based on pions produced in a SIDIS experiment by the HERMES collaboration [9] and are consistent with a vanishing isoscalar Sivers function observed by COMPASS [10].

2. Transverse Force on Quarks in DIS

The chirally even spin-dependent twist-3 parton distribution $g_2(x) = g_T(x) - g_1(x)$ is defined as

$$\int \frac{d\lambda}{2\pi} e^{i\lambda x} \langle PS|\bar{\psi}(0)\gamma^\mu\gamma_5\psi(\lambda n)|_{Q^2}|PS\rangle$$
$$= 2\left[g_1(x,Q^2)p^\mu(S\cdot n) + g_T(x,Q^2)S_\perp^\mu + M^2 g_3(x,Q^2)n^\mu(S\cdot n)\right].$$

Neglecting m_q, one finds $g_2(x) = g_2^{WW}(x) + \bar{g}_2(x)$, with $g_2^{WW}(x) = -g_1(x) + \int_x^1 \frac{dy}{y} g_1(y)$ [11], where $\bar{g}_2(x)$ involves quark-gluon correlations, e.g. [12,13]

$$\int dx x^2 \bar{g}_2(x) = \frac{d_2}{3} \tag{4}$$

with

$$4MP^+P^+S^x d_2 = g\left\langle P,S\left|\bar{q}(0)G^{+y}(0)\gamma^+q(0)\right|P,S\right\rangle. \tag{5}$$

At low Q^2, g_2 has the physical interpretation of a spin polarizability, which is why the matrix elements (note that $\sqrt{2}G^{+y} = B^x - E^y$)

$$\chi_E 2M^2\vec{S} = \langle P,S|\, q^\dagger\vec{\alpha}\times g\vec{E}q\,|P,S\rangle \qquad \chi_B 2M^2\vec{S} = \langle P,S|\, q^\dagger g\vec{B}q\,|P,S\rangle \tag{6}$$

are sometimes called spin polarizabilities or color electric and magnetic polarizabilities [14]. In the following we will discuss that at high Q^2 a better interpretation for these matrix elements is that of a 'force' [15].

To see this we express the $\hat{y}$-component of the Lorentz force acting on a particle with charge g that is moving with (nearly) the speed of light $\vec{v} = (0,0,-1)$ along the $-\hat{z}$ direction in terms of light-cone variables, yielding

$$g\left[\vec{E} + \vec{v}\times\vec{B}\right]^y = g\left(E^y + B^x\right) = g\sqrt{2}G^{y+}, \tag{7}$$

which coincides with the component that appears in the twist-3 correlator above (5). Thus Eq. (5) represents the (twist 2) quark density correlated with the transverse color-Lorentz force that a quark would experience at that position if it moves with the velocity of light in the $-\hat{z}$ direction — which is exactly what the struck quark does after it has absobed the virtual photon in a DIS experiment in the Bjorken limit. Therefore the correct semi-classical interpretation of Eq. (5) is that of an average[a] transverse force

$$F^y(0) \equiv -\frac{\sqrt{2}}{2P^+}\langle P,S|\,\bar{q}(0)G^{+y}(0)\gamma^+q(0)\,|P,S\rangle \tag{8}$$
$$= -2\sqrt{2}MP^+S^x d_2 = -2M^2 d_2$$

[a]The average is meant as an ensemble average since the forward matrix element in plane wave states automatically provides an average over the nucleon volume.

acting on the active quark in the instant right after[b] it has been struck by the virtual photon.

Although the identification of $\langle p|\bar{q}\gamma^+ G^{+y} q|p\rangle$ as an average color Lorentz force due to the final state interactions (8) may be intuitively evident from the above discussion, it is also instructive to provide a more formal justification. For this purpose, we consider the time dependence of the transverse momentum of the 'good' component of the quark fields (the component relevant for DIS in the Bjorken limit) $q_+ \equiv \frac{1}{2}\gamma^-\gamma^+ q$

$$\begin{aligned} 2p^+ \frac{d}{dt}\langle p^y\rangle &\equiv \frac{d}{dt}\langle PS|\,\bar{q}\gamma^+\left(p^y - gA^y\right) q\,|PS\rangle \\ &= \frac{1}{\sqrt{2}}\frac{d}{dt}\langle PS|\,q_+^\dagger\left(p^y - gA^y\right) q_+\,|PS\rangle \\ &= 2p^+\langle PS|\left[\dot{\bar{q}}\gamma^+\left(p^y - gA^y\right) q + \bar{q}\gamma^+\left(p^y - gA^y\right)\dot{q}\right]|PS\rangle \\ &\quad - \langle PS|\,\bar{q}\gamma^+ g\dot{A}^y q\,|PS\rangle\,. \end{aligned} \tag{9}$$

Using the QCD equations of motion

$$\dot{q} = \left(igA^0 + \gamma^0\vec{\gamma}\cdot\vec{D}\right) q, \tag{10}$$

where $-iD^\mu = p^\mu - gA^\mu$, yields

$$\begin{aligned} 2p^+ \frac{d}{dt}\langle \mathbf{p}^y\rangle &= \langle PS|\,\bar{q}\gamma^+ g\left(G^{y0} + G^{yz}\right) q\,|PS\rangle + `\langle PS|\,\bar{q}\gamma^+\gamma^-\gamma^i D^i D^j q\,|PS\rangle' \\ &= \sqrt{2}\,\langle PS|\,\bar{q}\gamma^+ g G^{y+} q\,|PS\rangle + `\langle PS|\,\bar{q}\gamma^+\gamma^-\gamma^i D^i D^j q\,|PS\rangle'\,, \end{aligned} \tag{11}$$

where '$\langle PS|\,\bar{q}\gamma^+\gamma^-\gamma^i D^i D^j q\,|PS\rangle'$' stands symbolically for all terms that involve a product of $\gamma^+\gamma^-$ as well as a $\gamma^\perp$ and only $\perp$ derivatives D^i.

Now it is important to keep in mind that we are not interested in the average force on the 'original' quark fields (before the quark is struck by the virtual photon), but *after* absorbing the virtual photon and moving with (nearly) the speed of light in the $-\hat{z}$ direction. In this limit, the first term on the r.h.s. of (11) dominates, as it contains the largest number of '+' Lorentz indices. Dropping the other terms yields (8).

The identification of $2M^2 d_2$ with the average transverse force acting on the active quark in a SIDIS experiment is also consistent with the Qiu Sterman result [16] for the average transverse momentum of the ejected quark (also averaged over the momentum fraction x carried by the active quark)

$$\langle k_\perp^y\rangle = -\frac{1}{2P^+}\left\langle P,S\left|\bar{q}(0)\int_0^\infty dx^- G^{+y}(x^+ = 0, x^-)\gamma^+ q(0)\right|P,S\right\rangle \tag{12}$$

[b]'Right after', since the quark-gluon correlator in (8) is local!

The average transverse momentum is obtained by integrating the transverse component of the color Lorentz force along the trajectory of the active quark — which is an almost light-like trajectory along the $-\hat{z}$ direction, with $z = -t$. The local twist-3 matrix element describing the force at time=0 is the first integration point in the Qiu-Sterman integral (12).

Lattice calculations of the twist-3 matrix element yield [17]

$$d_2^{(u)} = 0.010 \pm 0.012 \qquad\qquad d_2^{(d)} = -0.0056 \pm 0.0050 \tag{13}$$

renormalized at a scale of $Q^2 = 5$ GeV^2 for the smallest lattice spacing in Ref. [17]. These numbers are also consistent with experimental studies [18]. Using (8) these (ancient) lattice results thus imply

$$F_{(u)} \approx -100\text{MeV/fm} \qquad\qquad F_{(d)} \approx 56\text{MeV/fm}. \tag{14}$$

In the chromodynamic lensing picture, one would have expected that $F_{(u)}$ and $F_{(d)}$ are of about the same magnitude and with opposite sign. The same holds in the large N_C limit. A vanishing Sivers effect for an isoscalar target would be more consistent with equal and opposite average forces. However, since the error bars for d_2 include only statistical errors, the lattice result may not be inconsistent with $d_2^{(d)} \sim -d_2^{(u)}$.

The average transverse momentum from the Sivers effect is obtained by integrating the transverse force to infinity (along a light-like trajectory) $\langle k^y \rangle = \int_0^\infty dt F^y(t)$. This motivates us to define an 'effective range'

$$R_{eff} \equiv \frac{\langle k^y \rangle}{F^y(0)}. \tag{15}$$

Note that R_{eff} depends on how rapidly the correlations fall off along a light-like direction and it may thus be larger than the (spacelike) radius of a hadron. Of cource, unless the functional form of the integrand is known, R_{eff} cannot really tell us about the range of the FSI, but if the integrand in (8) does not oscillate, (15) provides a reasonable estimate for the range over which the integrand in (8) is significantly nonzero.

Fits of the Sivers function to SIDIS data yield about $|\langle k^y \rangle| \sim 100$ MeV [19]. Together with the (average) value for $|d_2|$ from the lattice this translates into an effective range R_{eff} of about 1 fm. It would be interesting to compare R_{eff} for different quark flavors and as a function of Q^2, but this requires more precise values for d_2 as well as the Sivers function.

A relation similar to (8) can be derived for the x^2 moment of the twist-3 scalar PDF $e(x)$. For its interaction dependent twist-3 part $\bar{e}(x)$ one finds for an unpolarized target [20]

$$4MP^+P^+e_2 = g \left\langle p \right| \bar{q}\sigma^{+i}G^{+i}q \left| P \right\rangle, \tag{16}$$

where $e_2 \equiv \int_0^1 dx x^2 \bar{e}(x)$. The matrix element on the r.h.s. of Eq. (16) can be related to the average transverse force acting on a transversely polarized quark in an unpolarized target right after being struck by the virtual photon. Indeed, for the average transverse momentum in the $+\hat{y}$ direction, for a quark polarized in the $+\hat{x}$ direction, one finds

$$\langle k^y \rangle = \frac{1}{4P^+} \int_0^\infty dx^- g \langle p| \bar{q}(0)\sigma^{+y} G^{+y}(x^-) q(0) |p\rangle . \qquad (17)$$

A comparison with Eq. (16) shows that the average transverse force at $t = 0$ (right after being struck) on a quark polarized in the $+\hat{x}$ direction reads

$$F^y(0) = \frac{1}{2\sqrt{2}p^+} g \langle p| \bar{q}\sigma^{+y} G^{+y} q |p\rangle = \frac{1}{\sqrt{2}} M P^+ S^x e_2 = \frac{M^2}{2} e_2. \qquad (18)$$

The impact parameter distribution for quarks polarized in the $+\hat{x}$ direction [21] is shifted in the $+\hat{y}$ direction [22,23]. Applying the chromodynamic lensing model implies a force in the negative $\hat{y}$ direction for these quarks and one thus expects $e_2 < 0$ for both u and d quarks. Furthermore, since $\kappa_\perp > \kappa$, one would expect that in a SIDIS experiment the $\perp$ force on a $\perp$ polarized quark in an unpolarized target on average to be larger than that on unpolarized quarks in a $\perp$ polarized target, and thus $|e_2| > |d_2|$.

3. Information Content of the DVCS Amplitude

In the following we will focus on charge even GPDs $GPD^+(x, \xi, t) \equiv GPD(x, \xi, t) - GPD(-x, \xi, t)$. Furthermore, the Q^2 dependence will not be shown explicitly although all terms depend on Q^2.

Using dispersion relations one can show that [24]

$$\Re \mathcal{A}_{DVCS} \sim \int_{-1}^{1} dx \frac{GPD^+(x, \xi, t)}{x - \xi} = \int_{-1}^{1} dx \frac{GPD^+(x, x, t)}{x - \xi} + \Delta(t), \quad (19)$$

where $\Delta(t)$ is the D-form factor [25]. This remarkable relation also follows from polynomiality [26] and implies that the information content of the DVCS amplitude (at fixed Q^2) can be condensed to GPDs along the diagonal $x = \xi$ plus the D-form factor. It should be emphasized that $\Re \mathcal{A}_{DVCS}(\xi, t)$ still adds more information to $\Im \mathcal{A}_{DVCS}(\xi, t)$ than just $\Delta(t)$, since, for fixed t not the whole range $0 < \xi < 1$ is accessible (at very low ξ the Bjorken limit may not yet have been reached and high ξ is inaccessible since $t = -\frac{4\xi^2 M^2 + \Delta_\perp^2}{1-\xi^2}$), while the above integrals extend from $\xi = 0$ to $\xi = 1$. Nevertheless, (19) suggests to fit parameterizations for $GPD(\xi, \xi, t)$ and $D(t)$ to DVCS data rather than attempting to constrain parameters for

$GPD(x, \xi, t)$ over the whole $x - \xi$ range. $GPD(\xi, \xi, t)$ and $\Delta(t)$ could then be used as an interface between experimental data and models. In fact, due to (19), any model/parameterization of GPDs satisfying polynomiality that fits both $\Re\mathcal{A}_{DVCS}$ and $\Im\mathcal{A}_{DVCS}$, would also fit $GPD(\xi, \xi, t)$ and $\Delta(t)$ and vice versa. Moreover, fitting GPD-models to DVCS-data is not unique. For example, one can always fit DVCS data with the ansatz

$$GPD(x, \xi, t) = GPD_{DD}(x, \xi, t) + \Theta(|\xi| - |x|) D\left(\frac{x}{\xi}, t\right) \tag{20}$$

where for the 'double distribution' [27] one makes the specific ansatz $GPD_{DD}(x, \xi, t) = GPD(x, x, t)$, and $D(z, t)$ being an arbitrary function that satisfies $\int_{-1}^{1} \frac{dz}{z-1} D(z, t) = \Delta(t)$.

With the information from DVCS reduced to GPDs along the diagonal plus the D-form factor, the question arises what one can learn from this information. GPDs along the diagonal have been discussed e.g. in Ref. [28].

While GPDs for $x > \xi > 0$ are still diagonal in the absolute transverse positions of all partons, they appear off-diagonal when positions are measured relative to the $\perp$ center of momentum. However, as the momentum carried by the active quark changes between initial and final state, so does the location of the transverse center of momentum [2]. Therefore, even though the (absolute) $\perp$ positions of the active quark/spectators remain unchanged, their separation from the $\perp$ center of momentum changes since the latter does. For the physical interpretation of GPDs in the case of $\zeta \neq 0$, working with relative $\perp$ position coordinates (i.e. relative to each other) rather than impact parameter (measured relative to the $\perp$ center of momentum may thus be preferable. Indeed, the discussion above illustrates that, for nonzero ξ, the Fourier transform of GPDs w.r.t. the transverse momentum transfer $\mathbf{\Delta}_\perp$ yields information about the transition matrix element between the initial and final state, when the $\perp$ distance between the active quark and the center of momentum of the spectators is $\mathbf{r}_\perp$. More precisely, $\frac{1-\zeta}{1-X}\mathbf{r}_\perp$ [c] is Fourier conjugate to $\mathbf{\Delta}_\perp$, and for $x = \zeta$, the variable conjugate to the $\mathbf{\Delta}_\perp$ is just $\mathbf{r}_\perp$.

In the limit $x = \xi$ the Fourier transform of GPDs w.r.t. $\mathbf{\Delta}_\perp$ thus yields the dependence of the overlap matrix element on the separation $\mathbf{r}_\perp$ between the active quark and the center of momentum $\mathbf{R}_{\perp s}$ of the spectators. While for $\zeta = 0$ it is the separation from the center of momentum of the whole hadron that sets the scale, it is the separation from then center of momentum of the spectators that matters for $x = \zeta$. In order to

[c] Here X and ζ are the variables used in Ref. [27]. We note that $x = \xi$ when $X = \zeta$.

utilise the above observations in the interpretation of GPDs, we note that $-t = \frac{\zeta^2 M^2 + \mathbf{\Delta}_\perp^2}{1-\zeta}$. Therefore, if the t-dependence of GPDs is parameterized in the form $GPD \propto e^{Bt}$ one finds for the $\mathbf{\Delta}_\perp^2$-dependence $GPD \propto e^{-B_\perp \mathbf{\Delta}_\perp^2}$ with $B_\perp = \frac{1}{1-\zeta} B$. Thus, even if the $\mathbf{\Delta}_\perp^2$-slope (described by $B_\perp$) remains finite as $\zeta \to 1$, the t-slope (described by B) goes to zero.

4. Summary

The GPD $E^q(x, 0, -\mathbf{\Delta}_\perp^2)$, which appears in the '$x$-decomposition' of the contribution from quark flavor q to the Pauli form factor F_2^q describes the transverse deformation of the unpolarized quark distribution in impact parameter space. That deformation provides a very intuitive mechanism for transverse SSAs in SIDIS. As a result, the signs of SSAs can be related to the contribution from quark flavor q to the nucleon anomalous magnetic moment. Quark-gluon correlations appearing in the x^2-moment of the twist-3 part of the polarized parton distribution $g_2^q(x)$ have a semi-classical interpretation as the average (enemble average) transverse force acting on the struck quark in DIS from a transversely polarized target in the moment after it has absorbed the virtual photon. Since the direction of that force can be related to the transverse deformation of PDFs, one can thus also relate the sign of these quark-gluon correlations to the contribution from quark flavor q to the nucleon anomalous magnetic moment.

Such a correlation between observables that at first appear to have little in common also occurs in the chirally odd sector: the impact parameter space distribution of quarks with a given transversity in an unpolarized target can be related to the Boer-Mulders function discribing the left-right asymmetry of quarks with a given transversity in SIDIS from an unpolarized target. Furthermore, semi-classically, the quark-gluon correlations appearing in the x^2-moment of the twist-3 part of the scalar PDF $e(x)$ discribes the average transverse force acting on a quark with given transversity immediately after it has absorbed the virtual photon.

At leading twist, the information content of the DVCS amplitude at fixed Q^2 can be 'condensed' into GPDs along the diagonal $x = \xi$ plus the D-form factor. This does not render measurements of the real part of the DVCS amplitude redundant since only part of the diagonal $x = \xi$ is kinematically accessible plus there is the additional information contained in the D-form factor. Q^2 evolution provides further constraints for $x \neq \xi$.

For $x = \xi$, the variable Fourier conjugate to $\mathbf{\Delta}_\perp$ is $\mathbf{r}_\perp$ the separation between the active quark and the center of momentum of the spectators —

which does *not* have to go to zero for $x \to 1$ [28]. The t-slope must still go to zero for $x = \xi \to 1$ as it is smaller than the $\mathbf{\Delta}_\perp^2$-slope by a factor $1 - \zeta$.

Acknowledgements: I would like to thank A.Bacchetta, D. Boer, J.P. Chen, Y.Koike, and Z.-E. Mezziani for useful discussions. This work was supported by the DOE under grant number DE-FG03-95ER40965.

References

1. M.Burkardt, Phys. Rev. D **62**, 071503 (2000), Erratum-ibid. D **66**, 119903 (2002); M. Diehl, Eur. Phys. J. C **25**, 223 (2002); J.P. Ralston and B. Pire, Phys. Rev. D **66**, 111501 (2002).
2. M. Burkardt, Int. J. Mod. Phys. A **18**, 173 (2003).
3. D.W. Sivers, Phys. Rev. D **43**, 261 (1991).
4. A. Bacchetta et al., Phys. Rev. D **70**, 117504 (2004).
5. M. Burkardt, Phys. Rev. D **66**, 114005 (2002); Phys. Rev. D **69**, 057501 (2004).
6. S.J. Brodsky, D.S. Hwang, and I. Schmidt, Nucl. Phys. B **642**, 344 (2002); M. Burkardt and D.S. Hwang, Phys. Rev. **D69**, 074032 (2004); L.P. Gamberg et al., Phys. Rev. D **67**, 071504 (2003); D. Boer, S.J. Brodsky, and D.S. Hwang, Phys. Rev. D **67**, 054003 (2003); A. Bacchetta et al., Phys. Lett **B578**, 109 (2004); M. Radici et al., hep-ph/0708.0232; L.P. Gamberg et al. hep-ph/0708.0324; D. Boer et al., Phys. Rev. D **67**, 054003 (2003); L.P. Gamberg et al., Phys. Rev. D **67** (2003) 071504.
7. S. Meissner et al., Phys. Rev. D **76**, 034002 (2007)
8. A.V. Belitsky, X. Ji, and F. Yuan, Phys. Rev. D **69**, 074014 (2004).
9. A. Airapetian et al. (HERMES collab.), Phys. Rev. Lett. **94**, 012002 (2005).
10. A. Martin (COMPASS collab.), Czech. J. Phys. **56**, F33 (2006).
11. S. Wandzura and F. Wilczek, Phys. Lett. **72B**, 195 (1977).
12. E. Shuryak and A.I. Vainshtein, Nucl. Phys. B **201**, 141 (1982).
13. R.L. Jaffe, Comm. Nucl. Part. Phys. **19**, 239 (1990).
14. B.W. Filippone and X. Ji, Adv. Nucl. Phys. **26**, 1 (2001).
15. M. Burkardt, arXiv:0811.1206
16. J. Qiu and G. Sterman, Phys. Rev. Lett. **67**, 2264 (1991).
17. M. Göckeler et al., Phys. Rev. D **72**, 054507 (2005).
18. Z.-E. Mezziani et al., hep-ph/0404066.
19. M. Anselmino et al., Eur. Phys. J. A **39**, 89 (2009).
20. Y. Koike and K. Tanaka, Phys. Rev. D **51**, 6125 (1995).
21. M. Diehl and P.Hägler, Eur. Phys. J. **C44**, 87 (2005).
22. M. Göckeler et al. (QCDSF collaboration), Phys. Rev. Lett. **98**, 222001 (2007); Ph. Hägler et al. (LHPC collaboration), hep-lat/0705.4295.
23. M. Burkardt and B. Hannafious, Phys. Lett. B **658**, 130 (2008).
24. M. Diehl and D.Y. Ivanov, Eur. Phys. J. C **52**, 919 (2007).
25. M.V. Polyakov and C. Weiss, Phys. Rev. D **60**, 114017 (1999).
26. I.V. Anikin and O.V. Teryaev, Phys. Rev. D **76**, 056007 (2007).
27. A.V. Radyushkin, Phys. Rev. D **56**, 5524 (1997).
28. M. Burkardt, arXiv:0711.1881

EXCLUSIVE PION ELECTROPRODUCTION AND TRANSVERSITY

P. KROLL

Fachbereich Physik, Universität Wuppertal,
Wuppertal, D-42097, Germany
**E-mail: kroll@physik.uni-wuppertal.de*

In this talk it is reported on an analysis of hard exclusive π^+ electroproduction within the handbag approach. Particular emphasis is laid on single-spin asymmetries. It is argued that a recent HERMES measurement of asymmetries measured with a transversely polarized target clearly indicate the occurrence of strong contributions from transversely polarized photons. Within the handbag approach such $\gamma_T^* \to \pi$ transitions are described by the transversity GPDs accompanied by a twist-3 pion wave function. It is shown that this approach leads to results on cross sections and single-spin asymmetries in fair agreement with experiment.

Keywords: Handbag factorization, generalized parton distributions, transversity, electroproduction

1. Introduction

In this article it will be reported upon an analysis of hard exclusive electroproduction of positively charged pions[1] within the frame work of the so-called handbag approach which offers a partonic description of meson electroproduction provided the virtuality of the exchanged photon, Q^2, is sufficiently large. The theoretical basis of the handbag approach is the factorization of the process amplitudes in hard partonic subprocesses and soft hadronic matrix elements, parameterized as generalized parton distributions (GPDs), as well as wave functions for the produced mesons, see Fig. 1. In collinear approximation factorization has been shown[2,3] to hold rigorously for exclusive meson electroproduction in the limit $Q^2 \to \infty$. It has also been shown that the transitions from a longitudinally polarized photon to the pion, $\gamma_L^* \to \pi$, dominates at large Q^2. Transitions from transversely polarized photons to pions, $\gamma_T^* \to \pi$, are suppressed by inverse powers of the hard scale.

However, as has been argued in Ref. 1, $\gamma_T^* \to \pi$ transitions are quite large at experimentally accessible values of Q^2 which are typically of the order of a few GeV^2. This follows from data of asymmetries measured with a transversely polarized target[4] and is also seen in the transverse cross section measured by the $F_\pi - 2$ collaboration.[5] It is demonstrated in Ref. 1 that within the handbag approach, these $\gamma_T^* \to \pi$ transitions can be calculated as a twist-3 effect consisting of the leading-twist helicity-flip GPDs[6,7] combined with the twist-3 pion distribution amplitude.[8] In the following the main ideas of the approach advocated for in Ref. 1 will be briefly described and some of the results will be discussed and compared to experiment.

2. The handbag approach

Within the handbag approach the amplitudes for pion electroproduction through longitudinally polarized photons read

$$\mathcal{M}^{\pi^+}_{0+,0+} = \sqrt{1-\xi^2}\,\frac{e_0}{Q}\left[\langle\widetilde{H}^{(3)}\rangle - \frac{\xi^2}{1-\xi^2}\langle\widetilde{E}^{(3)}\rangle - \frac{2\xi m Q^2}{1-\xi^2}\frac{\rho_\pi}{t-m_\pi^2}\right],$$
$$\mathcal{M}^{\pi^+}_{0-,0+} = \frac{e_0}{Q}\,\frac{\sqrt{-t'}}{2m}\left[\xi\langle\widetilde{E}^{(3)}\rangle + 2mQ^2\frac{\rho_\pi}{t-m_\pi^2}\right]. \tag{1}$$

Here, the usual abbreviation $t' = t - t_0$ is employed where $t_0 = -4m^2\xi^2/(1-\xi^2)$ is the minimal value of t corresponding to forward scattering. The mass of the nucleon is denoted by m and the skewness parameter, ξ, is related to Bjorken-x by

$$\xi = \frac{x_{\rm Bj}}{2 - x_{Bj}}\,. \tag{2}$$

Helicity flips at the baryon vertex are taken into account since they are only suppressed by $\sqrt{-t'}/m$. In contrast to this, effects of order $\sqrt{-t'}/Q$ are neglected. The last term in each of the above amplitudes is the contribution from the pion pole (see Fig. 1). Its residue reads

$$\rho_\pi = \sqrt{2}g_{\pi NN}F_\pi(Q^2)F_{\pi NN}(t')\,, \tag{3}$$

where $g_{\pi NN}$ is the familiar pion-nucleon coupling constant. The structure of the pion and the nucleon is taken into account by form factors, the electromagnetic one for the pion, $F_\pi(Q^2)$, whereby the small virtuality of the exchanged pion is as usual ignored, and $F_{\pi NN}(t)$ for the π-nucleon vertex. The pion-pole term has been used in essentially this form in the measurement of the pion form factor.[5] In contrast to other work on hard exclusive pion electroproduction (an exception is Ref. 9) the full pion form factor is taken into account this way and not only its so-called perturbative contribution which only amounts to about a third of its experimental value. It is to be stressed that the pion pole also contributes to the amplitudes for transversely polarized photons. However, these contributions are very small.[1]

The convolutions $\langle F \rangle$ in (1) have been worked out in Ref. 1 with subprocess amplitudes calculated within the modified perturbative approach.[10] In this approach the quark transverse momenta are retained in the subprocess and Sudakov suppressions are taken into account. The partons are still emitted and re-absorbed by the nucleon collinearly, i.e. we still have collinear factorization in GPDs and hard subprocess amplitudes. It has been shown[11] that within this variant of the handbag approach the data on cross sections and spin density matrix elements for vector-meson production are well fitted for small values of skewness ($\xi \simeq x_{Bj}/2 \lesssim 0.1$).

3. $\gamma_T^* \to \pi$ transitions

The electroproduction cross sections measured with a transversely or longitudinally polarized target consist of many terms, each can be projected out by $\sin\varphi$ or $\cos\varphi$ moments where φ is a specific linear combination of ϕ, the azimuthal angle between the lepton and the hadron plane and ϕ_s, the orientation of the target spin vector. A number of these moments have been measured recently.[4,12] A particularly striking result is the $\sin\phi_S$ moment. The data on it, displayed in Fig. 2, exhibit a mild t-dependence and do not show any indication for a turnover towards zero for $t' \to 0$. This behavior of $A_{UT}^{\sin\phi_s}$ at small $-t'$ can only be produced by an interference term between the two helicity non-flip amplitudes $\mathcal{M}_{0+,0+}$ and $\mathcal{M}_{0-,++}$ which are not

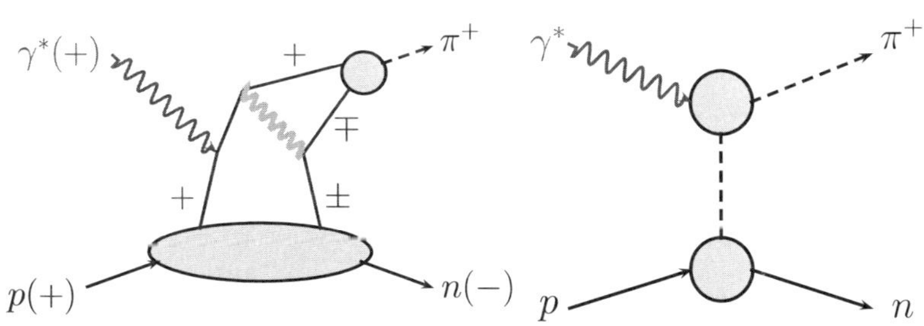

Fig. 1. The pion exchange graph and a typical lowest order Feynman graph for pion electroproduction. The signs indicate helicity labels for the contribution from transversity GPDs to the amplitude $\mathcal{M}_{0-,++}$, see text.

Table 1. Features of the asymmetries for a transversally and longitudinally polarized target. The photon polarization is denoted by L (longitudinal) and T (transversal). Asymmetries under control of TT interference terms are not shown in the table; they are very small. 1: There is a second contribution for which the helicities of the outgoing proton are interchanged.

observable	dominant interf. term	amplitudes	low t' behavior
$A_{UT}^{\sin(\phi-\phi_s)}$	LL	$\mathrm{Im}\big[\mathcal{M}^*_{0-,0+}\mathcal{M}_{0+,0+}\big]$	$\propto\sqrt{-t'}$
$A_{UT}^{\sin(\phi_s)}$	LT	$\mathrm{Im}\big[\mathcal{M}^*_{0-,++}\mathcal{M}_{0+,0+}\big]$	const.
$A_{UT}^{\sin(2\phi-\phi_s)}$	LT	$\mathrm{Im}\big[\mathcal{M}^*_{0-,-+}\mathcal{M}_{0+,0+}\big]$ 1)	$\propto t'$
$A_{UL}^{\sin(\phi)}$	LT	$\mathrm{Im}\big[\mathcal{M}^*_{0-,++}\mathcal{M}_{0-,0+}\big]$	$\propto\sqrt{-t'}$
$A_{LU}^{\sin(\phi)}$	LT	$\mathrm{Im}\big[\mathcal{M}^*_{0-,++}\mathcal{M}_{0-,0+}\big]$	$\propto\sqrt{-t'}$
$A_{LL}^{\cos(\phi)}$	LT	$\mathrm{Re}\big[\mathcal{M}^*_{0-,++}\mathcal{M}_{0-,0+}\big]$	$\propto\sqrt{-t'}$

forced to vanish in the forward direction by angular momentum conservation. The amplitude $\mathcal{M}_{0-,++}$ has to be sizeable because of the large size of the $\sin\phi_s$ moment. We therefore have to conclude that there are strong contributions from $\gamma_T^*\to\pi$ transitions at moderately large values of Q^2.

How can this amplitude be modeled in the frame work of the handbag approach? From Fig. 1 where the helicity configuration of the amplitude $\mathcal{M}_{0-,++}$ is shown, it is clear that contributions from the usual helicity non-flip GPDs, $\widetilde{H}$ and $\widetilde{E}$, to this amplitude do not have the properties required by the data on the $\sin\phi_s$ moment. For these GPDs the emitted and re-absorbed partons from the nucleon have the same helicity. Consequently, there are net helicity flips of one unit at both the parton-nucleon vertex and the subprocess. Angular momentum conservation therefore forces both parts to vanish as $\sqrt{-t'}$. Thus, a contribution from the ordinary GPDs to $\mathcal{M}_{0-,++}$ vanishes $\propto t'$. There is a second set of leading-twist GPDs, the helicity-flip or transversity ones $H_T, E_T,\ldots$ [6,7] for which the emitted and re-absorbed partons have opposite helicities. As an inspection of Fig. 1 reveals the parton-nucleon vertex as well as the subprocess amplitude $\mathcal{H}_{0-,++}$ are now of helicity non-flip nature and are therefore not forced to vanish in the forward direction. The prize to pay is that quark and anti-quark forming the pion have the same helicity. Therefore, the twist-3 pion wave function is needed instead of the familiar twist-2 one. The dynamical mechanism building up the amplitude $\mathcal{M}_{0-,++}$ is so of twist-3 accuracy. It has been first proposed in Ref. 13 for wide-angle photo- and electroproduction of mesons where $-t$ is considered to be the large scale.[14]

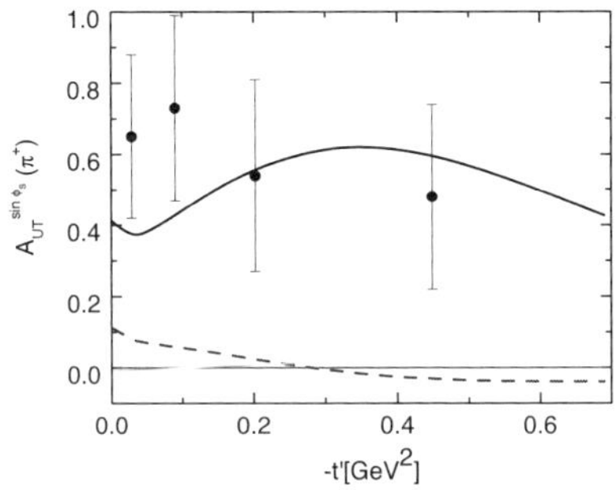

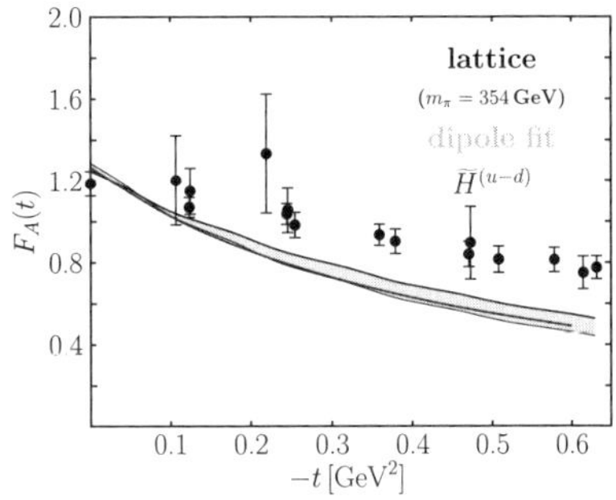

Fig. 2. (Left) The $\sin\phi_s$ moment for a transversely polarized target at $Q^2 \simeq 2.45\,\text{GeV}^2$ and $W = 3.99\,\text{GeV}$ for π^+ production. The predictions from the handbag approach[1] are shown as a solid line. The dashed line is obtained under neglect of the twist-3 contribution. Data are taken from Ref. 4.

Fig. 3. (Right) The axial-vector form factor. The band represents the dipole fit to the experimental data,[15] the solid circles the lattice results[16] for $m_\pi = 354\,\text{MeV}$. The thick curve is the form factor evaluated from $\widetilde{H}$ with the profile function of Ref. 20.

Allowing only for H_T as the only transversity GPD in an admittedly rough approximation the twist-3 mechanism only contributes to the amplitude

$$\mathcal{M}_{0-,++}^{\text{twist}-3} = e_0 \int_{-1}^{1} dx H_T^{(3)}(x,\xi,t)\mathcal{H}_{0-,++} \tag{4}$$

For the calculation of the subprocess amplitude $\mathcal{H}_{0-,++}$ the twist-3 pion wave function is taken from Ref. 8 with the three-particle Fock component neglected.[1] This wave function contains a pseudo-scalar and a tensor component. The latter one provides a contribution to $\mathcal{M}_{0-,++}$ which is proportional to t'/Q^2 and, hence, neglected. The contribution from the pseudo-scalar component to $\mathcal{M}_{0-,++}$ has the required properties. It is proportional to the parameter $\mu_\pi = m_\pi^2/(m_u + m_d)$ which appears as a consequence of the divergency of the axial-vector current. Since m_u and m_d are current quark masses μ_π is large, actually $\simeq 2\,\text{GeV}$ at the scale of $2\,\text{GeV}$. Thus, although parametrically suppressed by μ_π/Q as compared to the longitudinal amplitudes, the twist-3 effect is sizeable for Q of the order of a few GeV.

4. The GPDs at small skewness

For π^+ electroproduction the GPDs, namely $\widetilde{H}$, the non-pole part of $\widetilde{E}$ and the most important one of the transversity GPDs, H_T, contribute in the isovector combination

$$F_i^{(3)} = F_i^u - F_i^d \,. \tag{5}$$

The GPDs are constructed with the help of double distributions ansatz[17] consisting of the product of the zero-skewness GPDs and an appropriate weight function, actually parameterized as a power of the valence Fock state meson distribution amplitude. This weight function generates the skewness dependence of the GPD. It is important to note that other methods to generate the skewness dependence, namely the Shuvaev transform[18] or the dual parameterization[19] lead to very similar results for the GPDs at small skewness. The zero-skewness GPDs in the double distribution ansatz are assumed to be given by products of their respective forward limits and Regge-like t dependences, $\exp\left[f_i(x,t)t\right]$, with profile functions that read

$$f_i(x,t) = b_i - \alpha_i' \ln x \tag{6}$$

where α_i' is the slope of an appropriate Regge trajectory (pole or cut). These profile functions can be regarded as small-x approximations of more complicated versions used for the determination of the zero-skewness GPDs from the nucleon form factors[20]

$$f_i(x,t) = b_i(1-x)^3 - \alpha_i'(1-x)^3 \ln x + A_i x(1-x)^2 \tag{7}$$

which hold at all x. There is a strong correlation between t and x in this ansatz: the behavior of moments or convolutions of a GPD at small (large) $-t$ is determined by the small (large) x behavior of this GPD. It is to be stressed that the analysis performed in Ref. 20 as well as recent results from lattice QCD[16] clearly rule out a factorization of the zero-skewness GPDs in x and t.

The forward limit of $\widetilde{H}$ is given by the polarized parton distributions $\Delta q(x)$, that of H_T by the transversity distribution $\delta(x)$ for which the results of an analysis of the asymmetries in semi-inclusive electroproduction have been taken.[21] Finally, the forward limit of the non-pole part of $\widetilde{E}$ is parameterized as

$$\tilde{e}^{(3)}(x) = \widetilde{E}^{(3)}_{\text{n.p.}}(x, \xi = t = 0) = \widetilde{N}^{(3)}_{\tilde{e}} x^{-0.48}(1-x)^5 \,, \tag{8}$$

in analogy to the PDFs. The normalization $\widetilde{N}^{(3)}_{\tilde{e}}$ is fitted to experiment. The full set of parameters used in the analysis of π^+ electroproduction can be found in Ref. 1.

The full GPD $\widetilde{E}^{(3)}$ is the sum of the pole and non-pole contribution where the first one reads

$$\widetilde{E}^u_{\rm pole} = -\widetilde{E}^d_{\rm pole} = \Theta(|x| \le \xi)\frac{F_P^{\rm pole}(t)}{4\xi}\Phi_\pi((x+\xi)/(2\xi))\,. \tag{9}$$

Here, Φ_π is the pion's distribution amplitude and F_P is the pseudo-scalar form factor of the nucleon being related to $\widetilde{E}^{(3)}$ by the sum rule

$$\int_{-1}^{1} dx \widetilde{E}^{(3)}(x,\xi,t) = F_P(t)\,. \tag{10}$$

The evaluation of the pion electroproduction amplitude from the graph shown on the left hand side of Fig. 1 just using $\widetilde{E}_{\rm pole}$ leads to the pion-pole contribution as given in (1) but with only the perturbative contribution to the pion's electromagnetic form factor occurring in the residue (3). Other graphs have to be considered in addition for the pion pole, e.g. the Feyman mechanism. In order to avoid this complication the pion-pole contribution is simply worked out from the graph shown on the right hand side of Fig. 1.

In summary: the GPDs used in Ref. 1 are valid at small skewness ($\xi \lesssim 0.1$) and are probed by experiment for $x \lesssim 0.6$. Due to the double distribution ansatz they satisfy polynomiality and the reduction formulas. It has also been checked numerically that the lowest moments of the GPDs $\widetilde{H}$ and $\widetilde{E}$ are in agreement with the data on the axial-vector[15] and pseudo-scalar[22] form factors of the nucleon (see Fig. 3) and respect various positivity bounds.[23,24] Comparison with recent lattice QCD studies[16,25] reveals that there is good agreement with the relative strength of moments and their relative t dependences. At small t even the absolute values of the moments agree quite well but the t dependence of the moments obtained from lattice QCD are usually flatter than those from the GPDs and the form factor data. An exception is the lowest moment of H_T for u quarks for which we have a value that is about 25% smaller than the lattice result. Similar observation can be made for the GPDs H and E which have been constructed analogously and probed in vector meson electroproduction.[11] As an example the axial-vector form factor obtained from $\widetilde{H}$ as used in Ref. 1 is compared with experiment and with the lattice QCD results in Fig. 3.

5. Results

It is shown in Ref. 1 that with the described GPDs, the π^+ cross sections as measured by HERMES[26] are nicely fitted as well as the transverse target asymmetries.[4] This can be seen for instance from Fig. 1 where $A_{UT}^{\sin\phi_s}$

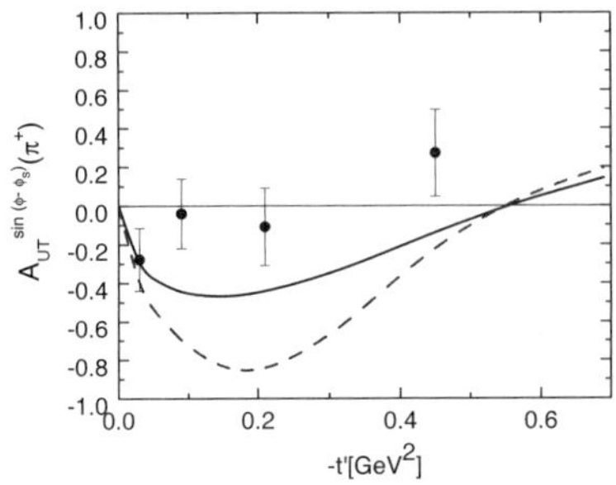

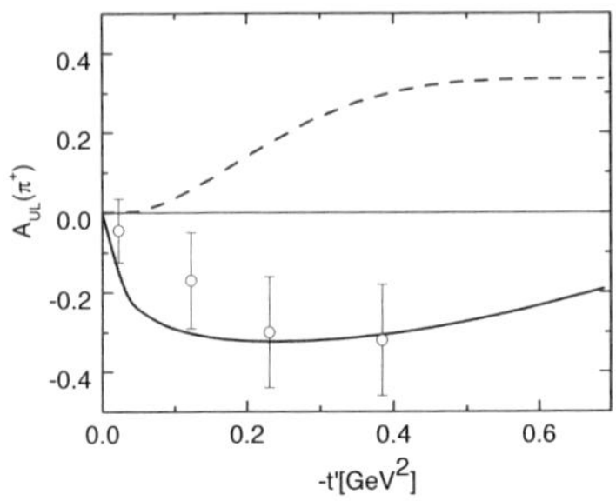

Fig. 4. (Left) Predictions for the $\sin(\phi - \phi_s)$ moment at $Q^2 = 2.45\,\mathrm{GeV}^2$ and $W = 3.99\,\mathrm{GeV}$ shown as solid lines.[1] The dashed line represents the longitudinal contribution to the $\sin(\phi - \phi_s)$ moment. Data are taken from Ref. 4.

Fig. 5. (Right) The asymmetry for a longitudinally polarized target at $Q^2 \simeq 2.4\,\mathrm{GeV}^2$ and $W \simeq 4.1\,\mathrm{GeV}$. The dashed line is obtained disregarding the twist-3 contribution. Data are taken from Ref. 12.

is displayed. Also the $\sin(\phi - \phi_s)$ moment which is dominantly fed by an interference term of the two amplitudes for longitudinally polarized photons, is fairly well described as is obvious from Fig. 4. Very interesting is also the asymmetry for a longitudinally polarized target. It is dominated by an interference term between $\mathcal{M}_{0-,++}$ which comprises the twist-3 effect, and the nucleon helicity-flip amplitude for $\gamma_L^* \to \pi$ transition, $\mathcal{M}_{0-,0+}$. Results for $A_{UL}^{\sin\phi}$ are displayed and compared to the data in Fig. 5. In both the cases, $A_{UT}^{\sin\phi_s}$ and $A_{UL}^{\sin\phi}$, the prominent role of the twist-3 mechanism is clearly visible. Switching it off one obtains the dashed lines which are significantly at variance with experiment. In this case the transverse amplitudes are only fed by the pion-pole contribution.

Although the main purpose of the work presented in Ref. 1 is focused on the analysis of the HERMES data one may also be interested in comparing this approach to the Jefferson Lab data on the cross sections.[5] With the GPDs $\widetilde{H}, \widetilde{E}$ and H_T in their present form the agreement with these data is poor. I remind the reader that the approach advocated for in Refs. 1 and 11 is optimized for small skewness. At larger values of it the parameterizations of the GPDs are perhaps to simple and may require improvements as for instance the replacement of the profile function (6) by (7). As mentioned above the GPDs are probed by the HERMES data only for x less than about 0.6. One may therefore change the GPDs at large x to some extent without changing much the results for cross sections and asymmetries in the kinematical region of small skewness. For Jefferson Lab kinematics, on the other hand, such changes of the GPDs may matter. Finally one should be aware that at larger values of skewness the other transversity GPDs may not be negligible. In a recent lattice study[27] the moments of the combination $2\widetilde{H}_T + E_T$ have been found to be rather large in comparison to those of H_T. Including this combination of GPDs into the analysis of pion electroproduction one would have

$$\mathcal{M}^{\text{twist}-3}_{0+,\mu+} = -e_0 \frac{\sqrt{-t'}}{4m} \int_{-1}^{1} dx \big[2\widetilde{H}_T^{(3)} + E_T^{(3)}\big] \mathcal{H}_{0-,++} \qquad (11)$$

in addition to (4). Here, μ (± 1) labels the photon helicity. The amplitude (11) holds up to corrections of order ξ.

6. Summary and outlook

In summary, there is strong evidence for transversity in hard exclusive electroproduction of pions. A most striking effect is seen in the target asymmetry $A_{UT}^{\sin\phi_s}$. The interpretation of this effect requires a large helicity non-flip amplitude $\mathcal{M}_{0-,++}$. Within the handbag approach this amplitude is generated by the helicity-flip or transversity GPDs in combination with a twist-3 pion wave function. This explanation establishes an interesting connection to transversity parton distributions measured in inclusive processes. Further studies of transversity in exclusive reactions are certainly demanded. Good data on π^0 electroproduction would also be welcome. They would not only allow for further tests of the twist-3 mechanism but also give the opportunity to verify the model GPDs $\widetilde{H}$ and $\widetilde{E}$ as used in Ref. 1. An intriguing issue is whether or not the handbag approach in its present form for pion electroproduction works for the kinematics presently accessible at Jlab. It is known that it cannot accommodate the CLAS data on ρ^0, ρ^+ and ω production.[28] Other applications and tests of the handbag approach including the twist-3 mechanism are pion electroproduction measured at the upgraded Jlab accelerator or by the COMPASS collaboration and the measurement of the time-like process $\pi^- p \to \mu^+\mu^- n$.[29] The extension of this approach to electroproduction of other pseudoscalar mesons, in particular the η and η', is also of interest. In principle this would give access to the GPDs for strange quarks. As has been shown in Ref. 30 there is no complication in the analysis of the electroproduction data due to the two-gluon Fock components of the η and η' since they are suppressed by t'/Q^2.

7. Acknowledgments

The author wishes to thank Anatoly Radyushkin and Paul Stoler for inviting him to present this talk at the workshop on Exclusive Reactions at High Momentum Transfer. This work is supported in part by the BMBF under contract 06RY258.

References

1. S.V. Goloskokov and P. Kroll, *Eur. Phys. J.* **C**65 (2010) 137.
2. A. V. Radyushkin, *Phys. Lett.* **B**385 (1996) 333.
3. J.C. Collins, L. Frankfurt and M. Strikman, *Phys. Rev.* **D**56 (1997) 2982.
4. A. Airapetian *et al.* [HERMES Collaboration], *Phys. Lett.* **B682**, 345 (2010).
5. H. P. Blok *et al.* [Jefferson Lab Collaboration], *Phys. Rev.* **C**78 (2008) 045202.
6. M. Diehl, *Eur. Phys. J.* **C**19 (2001) 485.
7. P. Hoodbhoy and X. Ji, *Phys. Rev.* **D**58 (1998) 054006.
8. V. M. Braun and I. E. Halperin, *Z. Phys.* **C**48 (1990) 239. [*Sov. J. Nucl. Phys.* **52** (1990 YAFIA,52,199-213.1990) 126].
9. C. Bechler and D. Mueller, arXiv:0906.2571.
10. J. Botts and G. Sterman, *Nucl. Phys.* **B**325 (1989) 62.
11. S. V. Goloskokov and P. Kroll, *Eur. Phys. J.* **C**42 (2005) 281; *ibid.* **C**53 (2008) 367.
12. A. Airapetian *et al.* [HERMES Collaboration], *Phys. Lett.* **B**535 (2002) 85.
13. H. W. Huang *et al.*, *Eur. Phys. J.* **C**33 (2004) 91.
14. H. W. Huang and P. Kroll, *Eur. Phys. J.* **C**17 (2000) 423.
15. T. Kitagaki *et al.*, *Phys. Rev.* **D**28 (1983) 436.
16. Ph. Hagler *et al.* [LHPC Collaborations], *Phys. Rev.* **D**77 (2008) 094502.
17. A. V. Radyushkin, *Phys. Lett.* **B**449 (1999) 81.
18. A. D. Martin, C. Nockles, M. G. Ryskin, A. G. Shuvaev and T. Teubner, *Eur. Phys. J.* **C**63 (2009) 57.
19. M. V. Polyakov and K. M. Semenov-Tian-Shansky, *Eur. Phys. J.* **A**40 (2009) 181.
20. M. Diehl, T. Feldmann, R. Jakob and P. Kroll, *Eur. Phys. J.* **C**39 (2005) 1.
21. M. Anselmino *et al.*, *Nucl. Phys. Proc. Suppl* **191** (2009) 98.
22. S. Choi *et al.*, *Phys. Rev. Lett.* **71** (1993) 3927.
23. P. V. Pobylitsa, *Phys. Rev.* **D**65 (2002) 114015.
24. M. Diehl and Ph. Hagler, *Eur. Phys. J.* **C**44 (2005) 87.
25. M. Gockeler *et al.* [QCDSF Collaboration and UKQCD Collaboration], *Phys. Lett.* **B**627 (2005) 113.
26. A. Airapetian *et al.* [HERMES Collaboration], *Phys. Lett.* **B**659 (2008) 486.
27. M. Gockeler *et al.* [QCDSF Collaboration and UKQCD Collaboration], *Phys. Rev. Lett.* **98** (2007) 222001.
28. S. A. Morrow *et al.* [CLAS Collaboration], *Eur. Phys. J.* **A**39 (2009) 5.
29. E. R. Berger, M. Diehl and B. Pire, *Phys. Lett.* **B**523 (2001) 265.
30. P. Kroll and K. Passek-Kumericki, *Phys. Rev.* **D**67 (2003) 054017.

EXCLUSIVE VECTOR MESON ELECTROPRODUCTION @ CLAS

A. FRADI
Institut de Physique Nucléaire d'Orsay
91405 Orsay, France
**E-mail: fradi@ipno.in2p3.fr*

We present the results of exclusive electroproduction of vector mesons on the proton at CLAS. We discuss the interpretation of these cross sections in terms of t-channel Reggeon exchanges and in terms of Generalized Parton Distributions (GPDs) formalism.

Keywords: Nucleon structure; Exclusive vector meson electroproduction; Generalized Parton Distribution

1. Introduction

The exclusive electroproduction of mesons on the nucleon is an important tool to better understand the nucleon structure and, more generally, the transition between the low energy hadronic and high energy partonic domains of the Quantum Chromodynamics (QCD) theory. The CEBAF ("Continous Electron Beam Accelerator Facility") of the Jefferson Laboratory (JLab) at Newport News (USA) with its high intensity electron beam and its large acceptance spectrometer CLAS[1] (CEBAF Large Acceptance Spectrometer) offers a great opportunity to study the exclusive electroproduction of mesons, in particular those that have multi-particle decays.
This proceeding presents the results for the exclusive electroproduction of the vector mesons ρ^0, ω and ϕ on the proton at CLAS and provides a first (preliminary) look at the ρ^+ channel. These results come mainly from two experiments: the e1-6 and the e1-dvcs experiments. For e1-6, the data were collected between October 2001 and January 2002. The beam energy was 5.754 GeV and the integrated luminosity was 28.5 fb^{-1}. The e1-dvcs data were collected between March and May 2005. The beam energy was 5.75 GeV and the integrated luminosity was $\sim$ 40 fb^{-1}. The cross sections for the ρ^0, ω and ϕ channels are published in Refs. [2–4]. For the ρ^+ channel,

it is the first ever measurement of its cross section. The presented results are still preliminary and more details can be found in Ref. [5].

2. Data analysis

2.1. *The reaction $ep \to ep\rho^0$*

The final state that is analyzed is $ep \to ep\pi^+\pi^-$. The final state particles e, p and π^+ were detected in CLAS and the exclusive process was identified by selecting a missing π^- with the missing mass technique. Fig. 1 shows the distributions of $M_X[epX]$ for some (Q^2, x_B) bins. One sees clearly the ρ^0 peak sitting on top of a background of a non-resonant two-pion continuum. One can also distinguish two bumps at masses around 980 MeV and 1270 MeV corresponding, respectively, to the scalar f_0 and tensor f_2 mesons. After weighting each event with the appropriate acceptance, we have used four contributions to fit these spectra: three skewed Breit-Wigner distributions to describe the ρ^0, f_0 and f_2 resonant structures and a parametrisation of the $M_{\pi^+\pi^-}$ non-resonant continuum determined from simulations (mostly $ep \to ep\pi^+\pi^-$ phase space). After the background subtraction procedure, total and differential cross sections of the reaction $ep \to ep\rho^0$ were extracted, by calculating for each (Q^2, x_B) bin the area of the ρ^0 Breit-Wigner. A separation of the longitudinal cross sections from the transverse

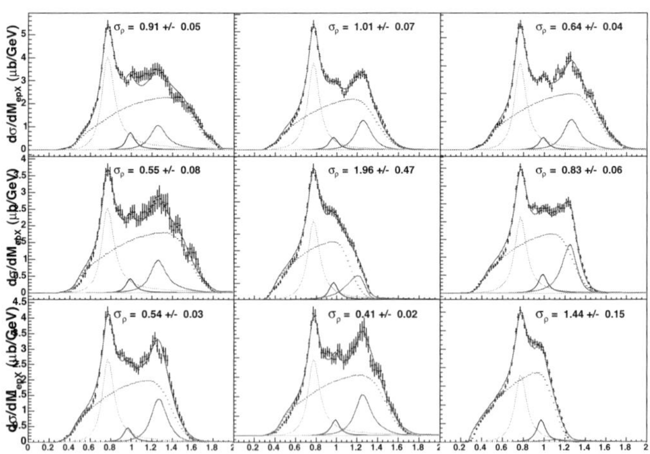

Fig. 1. Missing mass $M_X[epX]$ for some (Q^2, x_B) bins. See Ref. 2 for the kinematics of each bin.

ones was performed by analysing the pion decay angles of the ρ^0 and relying on the SCHC (s-channel helicity conservation) which was checked experimentally.

2.2. *The reaction $ep \to ep\omega$*

The final state that is analyzed is $ep \to e\pi^+\pi^-\pi^0$. The final state particles e, p and π^+ were detected in CLAS and the exclusive process was identified by the missing mass $M_X[epX]$. Fig. 2 (left) shows this acceptance-corrected missing mass distribution for a particular (Q^2, x_B) bin. The ω peak is fitted with a skewed Gaussian shape (taking into account the experimental resolution and radiative tail) and the background with a second-order polynomial.

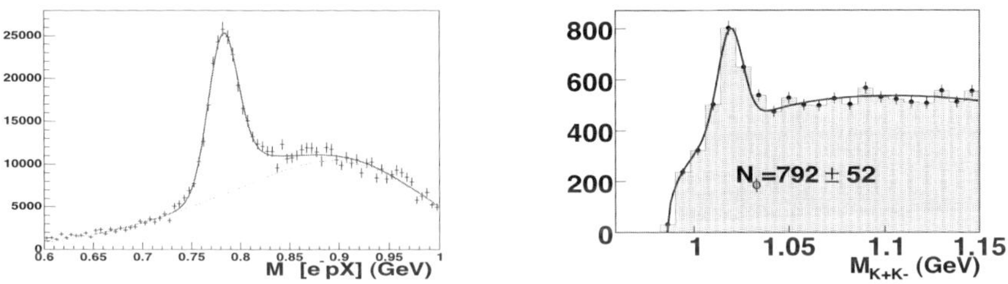

Fig. 2. Left: missing mass $M_X[epX]$ distribution for a particular (Q^2, x_B) bin. Right: invariant mass $M_{K^+K^-}$ for the whole data set.

2.3. *The reaction $ep \to ep\phi$*

The final state that is analyzed is $ep \to epK^+K^-$. The final state particles e, p and K^+ were detected in CLAS and the exclusive process was identified by selecting a missing K^- with the missing mass technique. Fig. 2 (right) shows the invariant mass $M_{K^+K^-}$ for the entire data set with a clear ϕ meson peak. This distribution is fitted with a Gaussian plus a polynomial function for the background. After weighting each event with the acceptance, the background was subtracted and the cross section of the exclusive electroproduction of ϕ was extracted.

2.4. *The reaction $ep \to en\rho^+$*

The final state that is analyzed is $ep \to en\pi^+\pi^0$. The final state particles e, π^+ and π^0 were detected in CLAS and the exclusive process was identified by selecting a missing neutron with the missing mass technique. Fig. 3 shows the acceptance-corrected invariant mass $M_{\pi^+\pi^0}$ distributions for each (Q^2, x_B) bin. One sees clearly the ρ^+ peaks around 775 MeV. They are however sitting on top of a non-negligeable non resonant two-pion background. We have used two contributions to fit these spectra: a skewed Breit-Wigner distribution to describe the resonant structure of the ρ^+ and a parametrization of the non-resonant two-pion continuum determined from simulations (mostly $ep \to en\pi^+\pi^0$ phase space). In order to extract the total cross section of the reaction $\gamma^* p \to n\rho^+$, we have calculated for each (Q^2, x_B) bin the area of the Breit-Wigner (peaked smooth curve in Fig. 3), the distributions being already normalized and corrected by the acceptance.

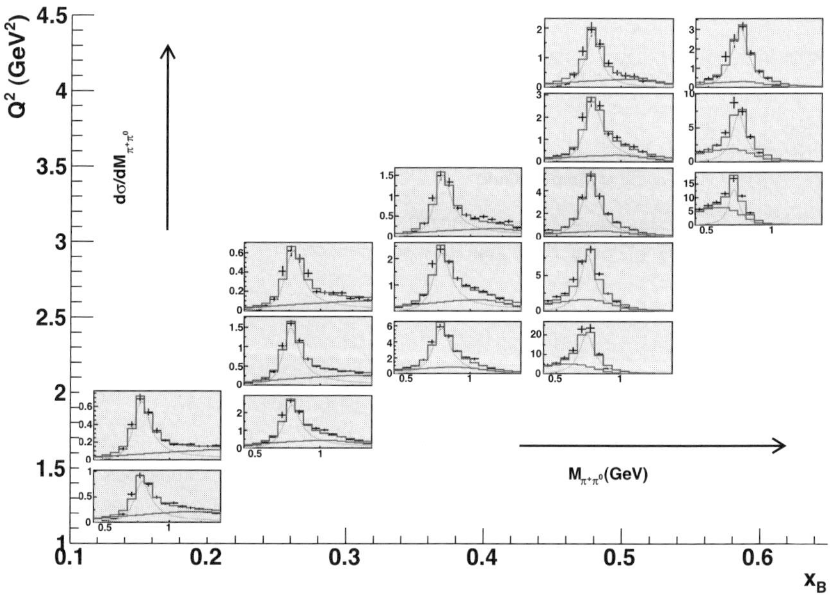

Fig. 3. $M_{\pi^+\pi^0}$ acceptance-corrected distributions, showing fits for the background subtraction. Crosses: experimental data; Smooth peaked curves: Breit-Wigner of ρ^+; Smooth low-lying curves: $M_{\pi^+\pi^0}$ projection of the the non-resonant continuum $\gamma^* p \to n\pi^+\pi^0$ reaction; Histogram: total fit result. In this proceedings contribution, units are arbitrary for the vertical scale.

3. Discussion

We now compare the extracted total and differential cross sections of exclusive electroproduction for these vector mesons with two theoretical approaches: on the one hand, the hadronic approach based on Regge theory and meson trajectory exchanges in the t-channel and on the other hand, the partonic approach based on the handbag diagram and GPDs. These two approaches are illustrated in Fig. 4. This comparison will be useful in order to better understand the domain of validity of these two approaches and constrain their inputs.

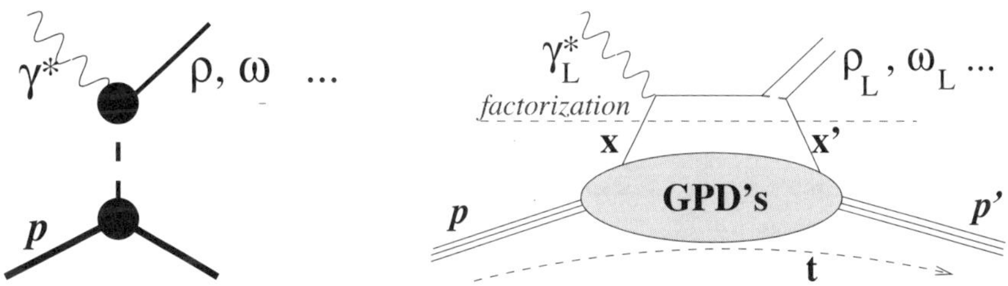

Fig. 4. Shematic representations of the Reggeon t-channel exchange (left) and of the handbag diagram (right) for exclusive vector meson electroproduction.

3.1. *The Regge "hadronic" approach*

The Regge approach is most appropriate above the resonance region and at forward angles where the cross section is the largest. It consists of exchanges of meson "trajectories" in the t-channel. In the following, we will use the "JML" acronym to refer to the specific model developed by J.-M. Laget and collaborators.[6]

For the ρ^0 channel, the model consists of t-channel exchanges of σ, f_2 and Pomeron trajectories. For the ω channel, the exchange particles are π^0 and Pomeron trajectories. For the ϕ channel, the only exchange particle is the Pomeron trajectory. For the ρ^+ channel, the exchange particles are the π^+ and the ρ^+ trajectories. The free parameters are the coupling constants at the hadronic vertices (most of them being well constrained) and the mass scales of the electromagnetic form factors.

Fig. 5 shows the total longitudinal cross section $\sigma_L(\gamma^* p \to p\rho^0)$ as a function of W for fixed Q^2. The results of the JML calculation is shown with the dot-dashed curve. The JML model is able to successfully reproduce the cross sections for almost all of our (Q^2, W) range. It can also reproduce

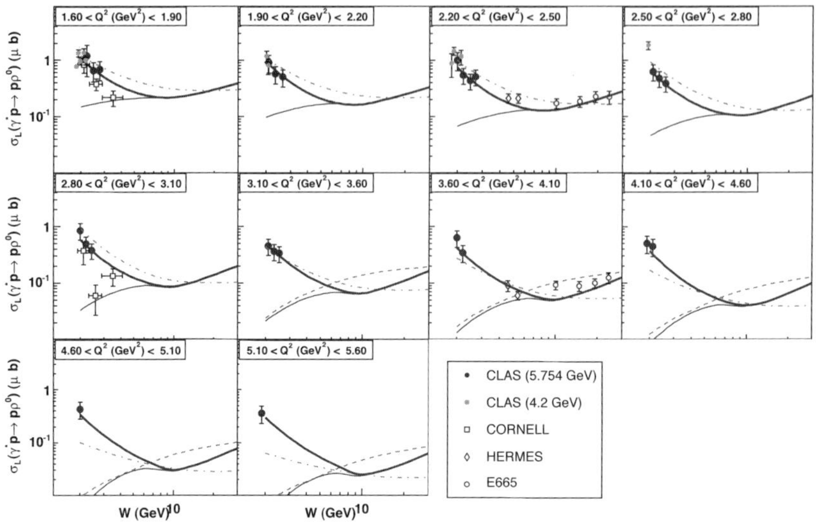

Fig. 5. Longitudinal cross section as a function of W at fixed Q^2, for the reaction $\gamma^* p \to p\rho^0$. The dashed curve shows the result of the GK calculation and the thin solid curve shows the result of the VGG calculation. The thick solid curve is the VGG calculation with the addition of the new t-channel meson exchange term. The dot-dashed curve shows the results of the Regge JML calculation. The references of the presented data can be found in Ref. [2].

successfully the main trends of the W, Q^2 and t dependences of the total and differential cross sections of the reactions $\gamma^* p \to p\omega$, $\gamma^* p \to p\phi$ and $\gamma^* p \to n\rho^+$ as presented in Refs. [3–5].

3.2. *The GPD "partonic" approach*

The formalism of GPDs is valid in the so-called Bjorken regime, i.e. $Q^2, \nu \to \infty$ with $x_B = \frac{Q^2}{2M\nu}$ finite. It was proven[7] that the dominant process for exclusive meson electroproduction, in the Bjorken limit, is given by the so-called handbag diagram represented in Fig. 4. The handbag diagram is based on the notion of factorization in leading-order/leading-twist pQCD between a hard scattering process, exactly calculable in pQCD, and a non-perturbative nucleon structure part that is parametrized by the GPDs. For mesons, the factorization of the handbag diagram is only valid for the longitudinal part of the cross section. For electroproduction of ρ^0 and ω we are sensitive to the sum of the quark handbag diagram and the gluon handbag diagram, while for the ρ^+ and the ϕ channels we are sensitive to, respectively, only the quark handbag diagram and the gluon handbag diagram.

In the following we discuss the two particular GK[8] and VGG[9] GPD-based calculations that provide quantitative results for the longitudinal part of the exclusive meson cross sections. In Fig. 5, the dashed line shows the results of the GK model, while the thin solid line shows the result of the VGG model. We see that they give a good description of the high and intermediate W region, down to $W \sim 5$ GeV. At high W the slow rise of the cross section is due to the gluon and sea contributions, while the valence quarks contribute only at small W. At lower W values, where the new CLAS data lie, both the GK and VGG models fail to reproduce the data. This discrepancy can reach an order of magnitude at the lowest W values. The trend of these particular GPD calculations is to decrease as W decreases, whereas the data increase. The same behavior was observed, in the low W region, for the exclusive electroproduction of the ρ^+ as one sees in Fig. 6. The results of the calculations of the GK and VGG models are shown, respectively, with the red and the blue curves.

An attempt to reconcile the GPD calculation with the low W ρ^0 cross sections is presented in Ref. [10]. Through a toy-model, t-channel meson exchanges are included in the GPDs and the result of this calculation (actually a fit) is illustrated by the thick blue curve in Fig. 5.

Fig. 7 shows the longitudinal cross section of ϕ electroproduction compared with the calculation of the GK model. One can see that the model reproduces well the data for all the presented W range: for low W data where the CLAS data lie and for higher W. As mentionned before, for the ϕ channel, we have only the contribution of the handbag diagram of gluons and this comparison shows that the GPD approach can work well for gluons.

From the results of the three mesons channels ρ^0, ρ^+ and ϕ, one can conclude that the GPD approach seems to work for gluons and sea quarks and largely fail for valence quarks. Many theorical efforts are now in progress to understand this transition between the small W region and the high W region.

For the ω channel, SCHC was found to be violated by a significant amount, thus preventing a longitudinal/transverse cross section separation. We therefore do not discuss here its comparison with GPD calculations.

3.3. *Comparison of the t slope for the ρ^0, ω, ϕ and ρ^+ channels*

Fig. 8 shows the slope of the differential cross section $d\sigma/dt$ for the ρ^0, ω, ϕ and ρ^+ channels as a function of W (on the top part) and as a function of Q^2 (in the bottom part). One can see the same trends of this slope b for

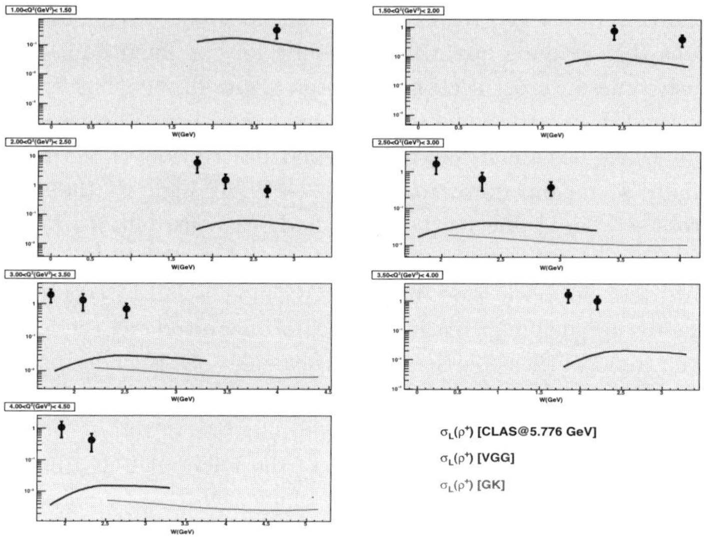

Fig. 6. PRELIMINARY longitudinal cross section as a function of W at fixed Q^2, for the reaction $\gamma^* p \to n\rho^+$. The curves are the results of the GK and VGG models. Units are arbitrary on the y axis.

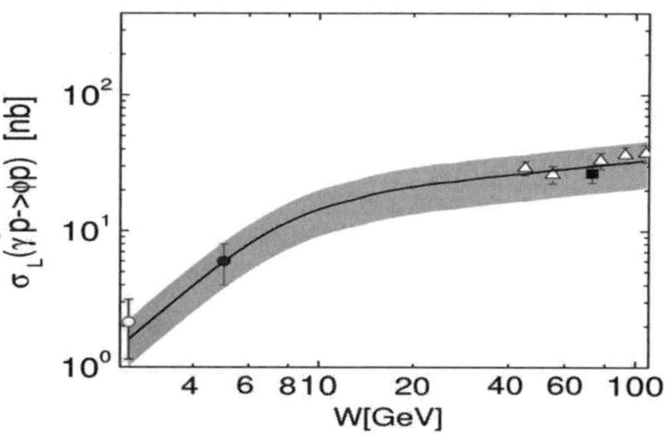

Fig. 7. Longitudinal cross section as a function of W, for the reaction $\gamma^* p \to p\phi$. The references of the presented data can be found in Ref. [4]. The black curve is the result of the GK model.

all mesons channels, which can be interpreted in simple and intuitive terms in the following way:

- b increases with W: the size of the nucleon increases as one probes the high W values (i.e. the sea quarks), which could mean that the sea quarks tend to extend to the periphery of the nucleon.

- b decreases with Q^2: as we go to large Q^2, the resolution of the probe increases and we tend to see smaller and smaller objects.

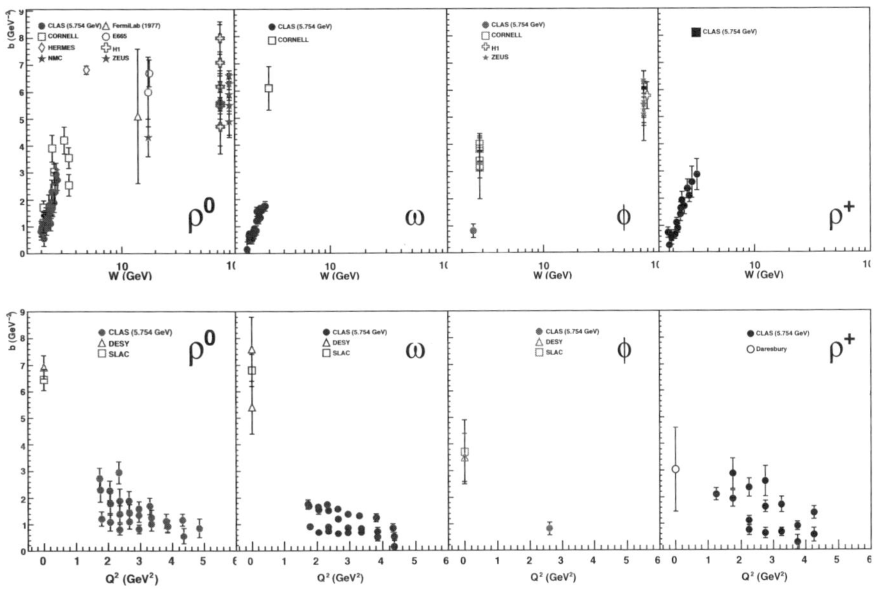

Fig. 8. The slope b as a function of W (on the top) and as a function of Q^2 (on the bottom) for the ρ^0, ω, ϕ and ρ^+ channels.

4. Summary

Using the CLAS detector at JLab, we have collected the largest ever set of data for the exclusive electroproduction of vector mesons on the proton in the valence region. We have presented the published results for ρ^0, ω and ϕ and given a first look at the ρ^+ (preliminary) cross sections. These data can be interpreted with two approaches:

- hadronic approach: the JML model describes well most of the features of the ρ^0, ω, ϕ and ρ^+ cross sections up to $Q^2 \sim 4.5$ GeV^2.
- partonic approach: GPD models describe well the data for $W > 5$ GeV but largely fail for $W < 5$ GeV.

A comparison between the features of the different mesons is in progress: we find for instance the same trends of the variation of the t slope as a function of W and as a function of Q^2 for all ρ^0, ω, ϕ and ρ^+ channels.

References

References

1. B.A. Mecking et al., Nucl. Instr. Meth. A **503**, (2003) 513.
2. S. Morrow et al., Eur. Phys. J. A **39**, (2009) 5.
3. L. Morand et al., Eur. Phys. J. A **24** (2005) 445.
4. J. P. Santoro et al., Phys. Rev. C **78** (2008) 025210.
5. A. Fradi, thesis Univ. Paris-Sud at Orsay, 2009.
6. J.-M. Laget, Phys. Lett. B **489** (2000) 313;
   F. Cano and J.-M. Laget, Phys. Rev. D **65** (2002) 074022;
   J. M. Laget, Phys. Rev. D **70** (2004) 054023;
   F. Cano and J. M. Laget, Phys. Lett. B **551** (2003) 317.
7. J.C. Collins, L. Frankfurt and M. Strikman, Phys.Rev.D **56** (1997) 2982.
8. S.V. Goloskokov and P. Kroll, Eur. Phys. J.C **42** (2005) 281; Eur.Phys. J.C **50** (2007) 829.
9. M. Vanderhaeghen, P.A.M. Guichon, M. Guidal, Phys. Rev. D **60**, 094017 (1999).
10. M. Guidal and S. Morrow, *Proceeding of the International Workshop "Exclusive reactions at high momentum transfer", Jefferson Laboratory, Newport-News, Virginia, USA, May 21-24 2007*, World Scientific, arXiv:0711.3743 (hep-ph).

EXCLUSIVE MESON PRODUCTION AT HIGH Q^2 AND FACTORIZATION

T. HORN*

Physics Department, The Catholic University of America,
Washington, DC 20064, USA

The p$(e, e'\pi^+)$n and the p$(e, e', K^+)\Lambda$ (Σ^0) reactions are important tools in the study of hadron structure. In particular, the flavor degree of freedom introduced with the addition of the strange quark provides important information on the reaction mechanism underlying strangeness production and the transition from hadronic to partonic degrees of freedom in exclusive processes.

Depending on the virtuality of the photon, Q^2, in the exclusive production process one can access different kinematic regimes. At low Q^2, where the process is described by hadronic fluctuations and their interaction with the target, the spatial structure of the meson is described by the meson form factor. A well studied example is the pion form factor.[1] At sufficiently large values of Q^2, the process becomes effectively pointlike and can be understood in terms of a virtual photon scattering from quark-gluon degrees of freedom. At fixed x_B and $-t$, *i.e.*, the longitudinal momentum fraction carried by the struck quark and the momentum transfer to the nucleon, a QCD factorization theorem allows one to calculate the electroproduction amplitude in terms of the Generalized Parton Distributions (GPDs), which have been shown to factorize from QCD perturbative (hard) processes for longitudinal photons.[2] The GPDs encode the long-distance (soft) physics. Detailed descriptions of the GPD formalism can be found, for instance, in Refs. 3–5. Though it is expected that the factorization theorem is valid for $Q^2 > 10$ GeV^2, to date it is unclear whether it may already be approximately valid at moderate values of Q^2 under certain conditions.[6]

Longitudinal/transverse (L/T) separated meson production cross sections are perhaps the best way to test the predictions of the factorization theorem for hard exclusive processes. To leading order, the meson electroproduction cross section for longitudinally polarized virtual photons scales like $\sigma_L \sim Q^{-6}$ at fixed $-t$ and x_B.[7] The contribution of transversely polarized photons is suppressed by at least an additional power of $1/Q$ in the amplitude.

While factorization is necessary for the interpretation of the GPDs, the approach to the factorization regime is interesting in its own right. Studies of the nonperturbative region and its transition to the hard regime hold important information for our understanding of the transverse spatial structure of hadrons. A comparison of systems with different quark flavors, for instance, strange and non-strange systems, also provides important information about the production mechanism and the transverse spatial distribution of strange and light quarks. Furthermore, the approach to the factorization regime for strange systems is thus far unexplored. The natural candidates for our study are the pion, which is the simplest two-quark system, and the kaon, which is the simplest system containing strangeness.

The goal of the present analysis is to investigate the meson production mechanism through the comparison of strange and non-strange systems, and in particular, determining which values of Q^2 are sufficient for factorization to apply in pion and kaon electroproduction. The Q^2 dependence of the π^+ and K^+ cross sections also allows for a quasi model independent comparison in the transition from soft to hard physics, which is more robust than calculations of cross sections based on QCD factorization and present GPD models.

Before testing the Q^2-dependence of the cross section it is beneficial to first look into the agreement with the well known expectation for the $-t$ dependence. A feature of π^+ production, and to some extent K^+ production, is the existence of a "pole term", which gives the cross section a characteristic exponential rise as t decreases. The applicability of the t-pole dominance over the full dynamic range can be tested through the $-t$ dependence of the cross sections for different values of x_B. Furthermore, by comparing the t-dependences of π^+ and K^+ production, one can also get an indication of the transition to the region where the pole term is believed not to be dominant.

In our analysis, we use the previously published unseparated and L/T separated $^1\text{H}(e,e'\pi^+)n$ and $^1\text{H}(e,e'K^+)\Lambda$ cross sections,[10–14] as well as newly available pion production results from Hall C[15,16] and kaon production results from Hall A [17] at JLab. The recent data were acquired for values of Q^2 up to about 4 GeV^2, for values of $-t$ ranging between 0.1 and 0.5 GeV^2 for pions and 0.7 GeV^2 for kaons, at a center of mass energy of up to W=2.2 GeV. A more detailed description of the experiments can be found in Ref. [18].

To combine and compare the $-t$ dependence of the results from recent JLab measurements with those of earlier experiments we scaled the experimental cross sections according to the empirical form $(W^2 - M^2)^{-2}$, where M_p is the proton mass.[11] This W dependence provides a good parameterization of the photo- and electroproduction data.[11,19]

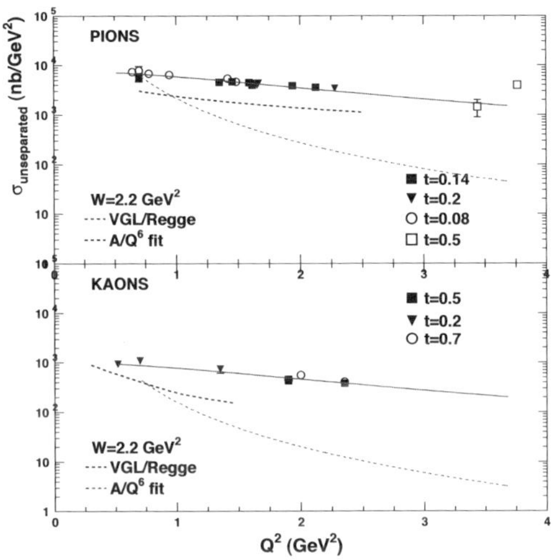

Fig. 1. The Q^2 dependence of the unseparated pion and kaon production cross section at fixed W=2.2 GeV. The error bars include the statistical and systematic uncertainty combined in quadrature.

The smooth Q^2 dependence of the data allows to also scale to a common value of Q^2. The cross sections lie on a universal curve when scaled with $Q^2\ F_\pi^2(Q^2)$ as shown in figure 1. Here, the monopole form was used for the form factor. The form $(1 + Q^2)^{-1}$ gives a reasonably good fit.

Figure 2 shows the $-t$ dependence of the pion and kaon cross sections scaled to W=2.2 GeV using the functional form $(W^2 - M^2)^{-2}$, and to Q^2=1.6 GeV^2 according to the form $\frac{1}{1+Q^2}$. The data are binned in x_B with bin centers at x_B=0.15, 0.25, and 0.4. Although overall the cross section data follow an almost exponential t dependence, upon close inspection one can see that the data deviate from that curve, *i.e.*, the Q^2, W, and t-dependences do not factorize completely. The t dependence shows the

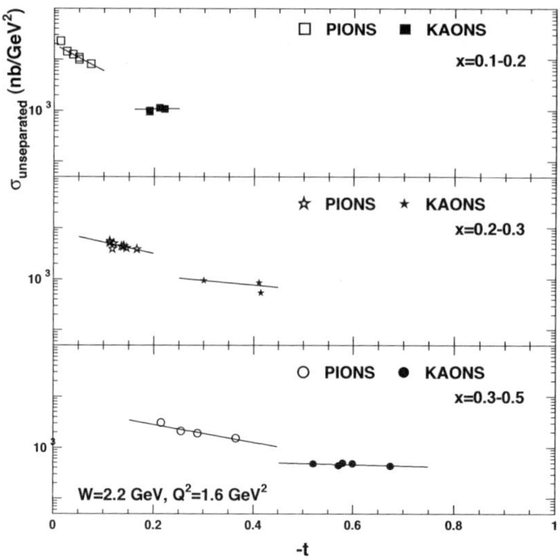

Fig. 2. The $-t$ dependence of the unseparated pion and kaon production cross sections for three bins in x_B at fixed Q^2 and W. The error bars include the statistical and systematic uncertainty combined in quadrature.

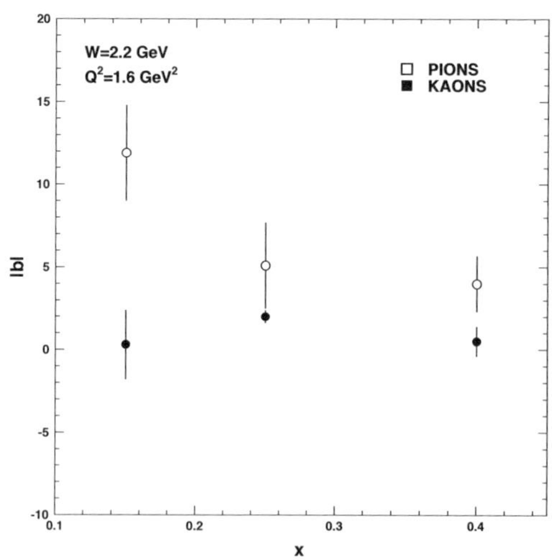

Fig. 3. The $-t$ slope for kaons and pions as a function of x_B. The slopes are plotted at the center of each x_B bin. The error bars denote the statistical uncertainty.

characteristic fall-off associated with t pole dominance. Furthermore, the t slope of the pion cross section is steeper at low x_B than the one of the kaons. This seems to suggest that the kaon pole factor $(m^2 - t)^{-2}$ is less prominent than theoretically predicted in reference 4. The $-t$ dependence decreases at larger values of x_B. This suggests that the t pole process in general becomes less dominant as x_B increases.

Looking at the t dependence over the full kinematic range in x_B allows for probing the hadron structure beyond the t pole. A priori it is not clear if this mechanism will be dominant everywhere and if trends in non-strange and strange systems should be similar. For instance, u-quark scattering deep inside the nucleon would predict no difference in the t-dependence for pions and kaons. Fitting the unseparated cross section data with an exponential $Be^{-b|t|}$ results in the slope parameters illustrated in Figure 3. At values of x_B ~0.1, the t-slopes seem to be different by a factor of about ten.

As mentioned above, the pion t slope agrees with a dominant t pole in that region. The kaon pole is less prominent. At values of x_B >0.2 the pion and kaon t slopes seem to become similar. This would suggest that the dominant production mechanism is no longer the t pole process. However, recall that the cross sections shown are unseparated and so may make it more difficult to identify the pole term, which can be best isolated through the separated longitudinal cross section.

The t-dependence can also be interpreted as a measure of the overall transverse size of the interaction region. The pole term in the GPD in which the QCD operator measuring the quark density is connected to the nucleon by the t-channel meson exchange. A priori it is not clear if strange quarks will show similar trends as those that are non-strange. However, recall that to access the physics contained in the GPDs one needs to show that hard-soft factorization applies.

A stringent and model independent test of QCD factorization is the leading order Q^2 power law scaling of the separated cross sections. The Q^2-dependence of the cross section ratio σ_L/σ_T, where we have only used L/T separated data, is shown in Figure 4. The data were scaled to constant values of W=2.2 GeV using the functional form $(W^2 - M^2)^{-2}$, and to Q^2=1.6 GeV^2 according to the form $\frac{1}{1+Q^2}$. The "hard scattering" predictions for σ_L/σ_T ($\sim Q^2$) are indicated by the red lines in Figure 4.

Our data are also compared to a Regge based calculation by Vanderhaeghen, Guidal and Laget,[20] in which the t dependence is determined by the exchange of Regge trajectories for π- and ρ-like particles. Such calculations have been well established over a wide kinematic range in $-t$ and W. Most parameters are fixed by pion photoproduction data with the pion and pion transition form factors as the only free parameters. Both form factors are parameterized by a monopole form, $[1+Q^2/\Lambda_i^2]^{-1}$, where $\Lambda_{\pi,\rho}^2$ is the π (ρ) mass scale. The VGL Regge model underestimates the magnitude of the Q^2 dependence of the data. There are suggestions that this is due to the limited predictive power of this model for σ_T.[21] For pion production, these recent models suggest that σ_T could be described as the limit of semi-inclusive production via the fragmentation mechanism. Unfortunately, separated kaon cross section data above the resonance region are extremely rare. The difficulty in describing the exisiting pion data and the lack of suitable kaon data emphasizes the need for new L/T separation experiments in these kinematics.

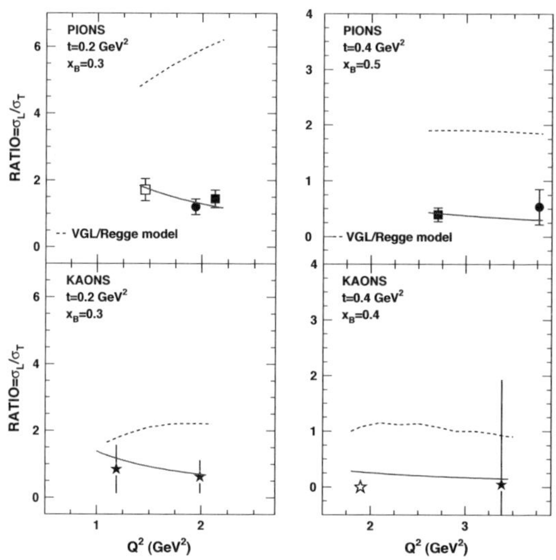

Fig. 4. The Q^2 dependence of the separated cross section ratio at fixed values of $-t$ and x_B. The error bars denote the statistical and systematic uncertainty combined in quadrature. The solid curve shows a fit of the form Q^2. The dashed line is a VGL/Regge calculation using Λ_π^2 from a global fit to F_π.

An interesting observation is that, at first glance, σ_L/σ_T appears to scale in a manner consistent with the Q^2 hard scattering prediction. However, an observation of the "correct" power law behavior of the cross section is not a proof that QCD factorization is already applicable, and it is necessary to examine several characteristics of the reaction before drawing conclusions about the reaction mechanism. Recent π^+ data indicate a $1/Q^6$-scaling of σ_L that is consistent with the hard scattering mechanism aready at values of $Q^2 > 1$ GeV^2, but σ_T does not show a corresponding $1/Q^8$ behavior. The Q^2 dependence of the pion form factor is also consistent with the $1/Q^2$ scaling expectation for $Q^2 > 1$ GeV^2, even though the observed magnitude is much larger than the hard QCD prediction.[15] The latter puzzle may, for instance, be due to QCD factorization not being applicable in this regime, or due to insufficient knowledge about additional soft contributions from the wave function in meson production. Recent results from the ρ^0 and ω channels[22] seem to support the former, while large angle Compton scattering data[23] suggest that higher order corrections are dominant at currently available energies. It would thus be of great interest to determine whether the scaling observed for σ_L and the pion form factor holds over an extended kinematic range and also manifests itself in other similar systems. The kaon, where one of the light quarks is replaced by a heavier strange one, makes a natural comparison. The JLab 12 GeV upgrade will provide the experimental facilities needed to study these puzzles of the meson reaction mechanism. The pion factorization experiment (E12-07-105)[24] will search for the onset of factorization with an extended kinematic reach, and also compare π^+ and π^- production to check possibilities to extract GPDs without performing explicit L/T separations. The kaon factorization experiment (E12-09-011)[25] will provide the first L/T separated kaon data above the resonance region. It will not only provide information about the onset of factorization in strange systems, but also add a new flavor for understanding meson production. A new experiment on π° production will shed light on pole and non-pole contributions.

In summary, the p$(e, e'\pi^+)$n and the p$(e, e', K^+)\Lambda$ (Σ^0) reactions play an important role in our understanding of hadronic structure. The t dependence of the cross section provides important information about the dominance of the t channel exchange, which is needed for the extraction of the meson form factors. Current unseparated data suggest a dominance of the pion exchange at low t, decreasing as t increases. The relative role of σ_L and σ_T is, however, not well understood due to the lack of sufficiently precise data. This needs to be addressed, as the Q^2 dependence of the separated pion electroproduction cross sections plays a large role in investigations of the transition from hadronic to partonic degrees of freedom in exclusive processes. The Q^2-dependence of separated pion and kaon longitudinal to transverse cross section ratios is in good agreement with the Q^2-scaling prediction within the experimental uncertainty. This would suggest that GPDs can be accessed at relatively low values of Q^2. However, higher order contributions may still be significant in this kinematic regime and mimic the expected Q^2-scaling behavior. JLab 12 GeV will allow rigorous tests of factorization in meson production, extending the kinematic reach and make studies of additional systems possible.

References

1. T. Horn et. al., Phys. Rev. Lett **97** 192001 (2006).
2. J.C. Collins, L. Frankfurt and M. Strikman, Phys. Rev. **D56**, 2982 (1997).
3. K. Goeke, V. Polyakov and M. Vanderhaeghen, Prog. Part. Nucl. Phys. **47**, 401 (2001).
4. M. Diehl, Phys. Rep. **388**, 41 (2003).
5. A. Belitsky and A. V. Radyushkin, Phys. Rep. **418**, 1 (2005).
6. L. Frankfurt, P.V. Pobylitsa, V. Polyakov and M. Strikman, Phys. Rev. **D60**, 014010 (1999).
7. S.R. Brodsky and G.R. Farrar, Phys. Rev. Lett. **31**, 1153 (1973);
8. M. Penttinen, M.V. Polyakov, K. Goeke, Phys. Rev. D**62**, 014024 (2000).
9. M. Diehl, W. Kugler, A. Schaefer, C. Weiss, Phys. Rev. D**72**, 034034 (2005).
10. A. Airapetian et al., Phys. Lett.**B659**, 486 (2008).
11. P. Brauel et al., Phys. Lett. **B69**, 253 (1977); P. Brauel et al., Z. Phys. **C3**, 101 (1979).
12. C.J. Bebek et al., Phys. Rev. D**17**, 1693 (1978).
13. C.J. Bebek et al., Phys. Rev. D**15**, 594 (1977).
14. R. Mohring et al., Phys. Rev. C**67** 055205 (2003).
15. T. Horn et al., Phys. Rev. C **78**, 058201 (2008);
16. X. Qian et al., Phys. Rev. C **81**, 055209 (2010);
17. M. Coman et al., Phys. Rev. C **81**, 055201 (2010);
18. B. Clasie, Ph.D. thesis, Massachussets Institute of Technology (2006); M. Coman, Ph.D. thesis, Florida International University (2006).
19. H.P. Blok et al., Phys. Rev. C **78**, 045202 (2008);
20. M. Vanderhaeghen, M. Guidal and J.-M. Laget, Phys. Rev. **C57**, 1454 (1998); Nucl. Phys. **A627** 645 (1997).
21. M.M. Kaskulov, K. Gallmeister, U. Mosel, Phys. Rev. D **78**, 114022 (2008);
22. C. Hadjidakis et al., Phys. Lett. B**605**, 256-264 (2005); M. Guidal et al., arXiv:0711.3743.v1 (2008); L. Morand et al., Eur. Phys. J**A24**, 445-458 (2005).
23. M.A. Shupe et al., Phys. Rev. D**19**, 1921 (1979); A. Danagoulian et al., Phys. Rev. Lett.**98**, 152001 (2007).
24. T. Horn et al., E12-07-105, *Scaling Study of the L-T Separated Pion Electroproduction Cross Section at 11 GeV*, approved at JLab PAC32 (2007).
25. T. Horn et al., E12-09-011, *Studies of the L-T Separated Kaon Electroproduction Cross Section from 5-11 GeV*, approved at JLab PAC34 (2009).

HARD EXCLUSIVE NEUTRAL PION PRODUCTION IN THE DEEPLY VIRTUAL REGIME

E. FUCHEY*

Clermont Université, Université Blaise Pascal, CNRS/IN2P3, Laboratoire de Physique Corpusculaire, 63000 Clermont-Ferrand, France
**E-mail: efuchey@jlab.org*

We present measurements of the $ep \to ep\pi^0$ cross section extracted at two values of four-momentum transfer $Q^2 = 1.9$ GeV^2 and $Q^2 = 2.3$ GeV^2 at Jefferson Lab Hall A. The kinematic range allows to study the evolution of the extracted hadronic tensor as a function of Q^2 and W. Results will be confronted with Regge inspired calculations and GPD predictions. An intepretation of our data within the framework of semi-inclusive deep inelastic scattering has also been attempted.

Keywords: Deeply virtual pion production; Regge; Generalized parton distributions; Semi-inclusive DIS.

1. Introduction

The past decade has shown a strong evolution of the study of hadron structure through exclusive processes (such as Deeply Virtual Compton Scattering (DVCS) and Deeply Virtual Meson Production (DVMP)), allowing access to the three-dimensional structure of hadrons. This document focuses on deeply virtual neutral pion production.

We present measurements of the differential cross section for the forward exclusive electroproduction reaction $ep \to ep\pi^0$. Results will be presented for four kinematics, two of them defined by the same value of $x_{Bj} = 0.36$ and two values of Q^2 (1.9 and 2.3 GeV^2), and the two other defined by the same value of $Q^2 = 2.1$ GeV^2 and two values of x_{Bj} (0.33 and 0.4). The cross sections will be compared to different models that are available to describe π^0 electroproduction, including the Regge Model and the Generalized Parton Distribution (GPD) framework.

Forward photo-production at asymptotically high energies can be described by Regge theory, which exploits the analytic properties of the scattering amplitude in the limit $t/s \to 0$ [1]. Recent computations with Regge-inspired models exist for our kinematics. Among these, there is the t-channel meson exchange (TME) model by J.M. Laget *et al.* [2,3]. This model is able to reproduce π^0 photoproduction data, low Q^2 π^0 electroproduction data and π^0 beam spin asymmetries for $Q^2 > 1$ GeV^2 [4], as well as the longitudinal charged pion electroproduction cross section for $Q^2 > 1$ GeV^2 [5].

In the Bjorken limit $Q^2 \to \infty$, and $t/Q^2 \ll 1$ at fixed x_{Bj}, the scattering amplitude is dominated by the leading twist amplitude of GPDs and the pion Distribution Amplitude (DA) [6–8]. The GPDs are light-cone matrix elements of non-local bilinear quark and gluon operators [9–11], unifying the elastic electro-weak form factors with the forward parton distributions of deep inelastic lepton scattering. With the definitions of [6–8], the cross sections are predicted to scale as $\sigma_L \sim Q^{-6}$ and $\sigma_T \sim Q^{-8}$ in the GPD framework.

An interpretation of exclusive data with semi-inclusive mechanisms also exists to explain transverse cross sections of hard exclusive charged pion electroproduction [12].

2. Experiment

The present data were acquired as part of Jefferson Lab Hall A experiment E00-110 [13]. Additional details about the experimental configuration, calibrations, and analysis can be found in 14. This document reports on the analysis of the triple coincidence $\mathrm{H}(e, e'\gamma\gamma)X$ events. A 5.75 GeV electron beam was incident on a liquid hydrogen target Electrons were detected in a High Resolution Spectrometer (HRS), and the photons in an 132 element PbF_2 calorimeter. The validation threshold for the data acquisition trigger was set to about 1 GeV for each photon cluster. Thanks to the resolution of the spectrometer and the calorimeter, one can use the missing-mass squared to ensure exclusivity. The exclusive sample is selected by applying a cut on the missing-mass squared at $(M_p + m_\pi)^2$. The pion selection relies on the invariant mass of the two photons. We select all events which invariant mass is within $\pm$ 30 MeV of the pion mass.

3. Cross-section analysis

In order to extract the differential cross section, it is advantageous to incorporate all model-independent kinematic dependences of the differential cross section into the experimental simulation. To this end, we express the differential cross section in terms of structure functions as described in the paper of Drechsel and Tiator [15]. We define the differential phase-space element $d^5\Phi = dQ^2 dx_{Bj} d\phi_e d[t_{min} - t] d\phi_\pi$ and the equivalent real photon energy in the CM frame $k_\gamma^{CM} = (W^2 - M_p^2)/2W$.

To lowest order in the fine structure constant α, the differential cross section for an electron of helicity h is:

$$\frac{d^5\sigma(h)}{d^5\Phi} = \frac{\alpha}{2\pi^2}\frac{k'}{k}\frac{k_\gamma}{Q^2}\frac{1}{1-\epsilon}\frac{d^2\sigma_v(h)}{dt d\phi_\pi}, \tag{1}$$

with $k_\gamma = (W^2 - M_p^2)/2M_p$, and $\epsilon = 1/[1 + 2(q^2/Q^2)\tan^2\theta_e/2]$ the degree of linear polarization of the virtual photons. The virtual photo-absorption cross section is expanded, with $\epsilon_L/\epsilon = 4M_p^2 x_{Bj}^2/Q^2$, as

$$\begin{aligned}\frac{d^2\sigma_v(h)}{dt d\phi_\pi} &= \frac{1}{2\pi}\left\{\frac{d\sigma_T}{dt} + \epsilon_L\frac{d\sigma_L}{dt} + \sqrt{2\epsilon_L(1+\epsilon)}\frac{d\sigma_{TL}}{dt}\cos\phi_\pi\right.\\ &\left. + \epsilon\frac{d\sigma_{TT}}{dt}\cos 2\phi_\pi + h\sqrt{2\epsilon_L(1-\epsilon)}\frac{d\sigma_{TL'}}{dt}\sin\phi_\pi\right\};\\ &= \frac{1}{2q^{CM}k_\gamma^{CM}}\left\{\frac{W_{xx}+W_{yy}}{2} + \epsilon_L W_{zz} - \sqrt{2\epsilon_L(1+\epsilon)}\Re\mathrm{e}W_{xz}\right.\\ &\left. + \epsilon\frac{W_{xx}-W_{yy}}{2} - h\sqrt{2\epsilon_L(1-\epsilon)}\Im\mathrm{m}W_{yz}\right\}\end{aligned} \tag{2}$$

in terms of the usual hadronic tensor $W^{\mu\nu}$. The interference terms σ_{TL} and $\sigma_{TL'}$ have a leading $\sin\theta_\pi^{CM}$ dependence, and the linear polarization interference term σ_{TT} has a leading $\sin^2\theta_\pi^{CM}$ dependence. For this reason, we define reduced structure functions r_Λ which are equal to the hadronic tensors divided by their leading order angular dependence.

We divide the acceptance into 24 equal bins in $\phi_\pi \in [0, 2\pi]$ and 8 bins in $t_{min} - t \in [0, 0.3]$ GeV^2 for both the helicity dependent and independent parts of the cross section. We compare our experimental yields N_d in these bins, noted j_d, with yields obtained from a full simulation of the detectors acceptance and response, N_s which depend on the r_Λ values averaged over the bin. The values of the reduced structure functions r_Λ are those which minimize (σ_d the statistical error on the bin j_d)

$$\chi^2 = \sum_{j_d}\frac{(N_d - N_s(r_\Lambda))^2}{\sigma_d^2}. \tag{3}$$

The results will be presented as the four separated cross sections following the usual decomposition available in Eq. (2).

4. Radiative Corrections and Systematic errors

The external radiative effects on the incident electron, and internal real radiative effects at the vertex are treated in the equivalent radiator approximation [16,17]. Pre-radiation is modeled by generating an event-by-event energy loss ΔE_{in} of the incident electron (E_0) following a distribution ($b \simeq 4/3$):

$$I_{in}(E_0, \Delta E_{in}, t_{in}) = \frac{bt_{in} + \delta_S/2}{\Delta E_{in}} \left[\frac{\Delta E_{in}}{E_0}\right]^{bt_{in}+\delta_S/2} \tag{4}$$

with

$$\delta_S = \frac{2\alpha}{\pi}\left[\ln\frac{Q^2}{m_e} - 1\right] \tag{5}$$

where t_{in} is the event-by-event target thickness (in radiation lengths) traversed by the electron before the scattering vertex. The Schwinger term δ_S models the internal pre-radiation. The scattered energy at the vertex is $E'_v = E_0 - \Delta E_{in} - Q^2/(2M_p x_{Bj})$. A similar distribution of the post radiated energy ΔE_{out} is used to model internal post-radiation.

External post-radiation by the scattering electron is modeled with the GEANT3 simulation. Internal virtual radiative correction factors, including vertex renormalization, vacuum polarization and the cut-off independent real radiation effects are calculated according to [18].

Two sources of systematic errors lie on the missing mass squared cut and in calorimeter threshold. The influence of the missing mass squared cut comes from the competition of the inclusive channels with the exclusive $\mathrm{H}(e, e'\pi^0)p$ reaction which cannot be reproduced in the simulation. The cross section stability relative to this cut indicates that these inclusive channels are not contributing. The systematic error from the missing mass cut is 1 % for ($x_{Bj} = 0.36$, $Q^2 = 2.3$ GeV^2) and ($x_{Bj} = 0.33$, $Q^2 = 2.1$ GeV^2), and 3 % for ($x_{Bj} = 0.36$, $Q^2 = 1.9$ GeV^2) and ($x_{Bj} = 0.4$, $Q^2 = 2.1$ GeV^2). The influence of the calorimeter threshold comes from the competition of the hardware threshold, which determines the event trigger, with the software analysis threshold. The cross section stability relative to the threshold indicates that the software threshold is set safely above the hardware threshold. The systematic error from the calorimeter threshold is 1 % for all kinematics.

The other sources of kinematic errors have experimental or instrumental origins. Among these sources of systematic errors, the most significative contributions come from the HRS acceptence for 2.2 %, and the radiative corrections for 1.5 %. For the helicity dependent cross section, the contribution form the beam polarization (2.0 %) has also to be mentionned. All sources of systematic errors are independent and are added quadratically. The total quadratic systematic error is given for each kinematics in Tab. 1.

Table 1. Systematic errors for both helicity dependent and independent cross sections for all kinematics.

$x_{Bj} = 0.36$ $Q^2 = 2.1$ GeV^2	$Q^2 = 2.3$ GeV^2 $x_{Bj} = 0.33$	$Q^2 = 1.9$ GeV^2 $x_{Bj} = 0.4$
helicity-independent	3.0 %	4.2 %
helicity-dependent	3.7 %	4.6 %

5. Results and discussions

The exclusive π^0 electroproduction cross section were extracted for $Q^2 =$ 1.9 and 2.3 GeV^2 at $x_{Bj} = 0.36$, and $x_{Bj} = 0.33$ and 0.4 at $Q^2 = 2.1$ GeV^2. Our statistics allowed us to achieve, for the ϕ_π-independent cross section, a statistical precision of 2.5% for $x_{Bj} = 0.36$, and of 5% for $Q^2 = 2.1$ GeV^2.

Fig. 1 (respectively 2) shows $\sigma_T + \epsilon_L \sigma_L$ plotted as a function of $t_{min} - t$, for $x_{Bj} = 0.36$ (respectively $Q^2 = 2.1$ GeV^2). It also shows the Q^2-dependence (respectively W-dependence) for the total cross section ($\sigma_T + \epsilon_L \sigma_L$, which behaves as $1/Q^{4.5}$ (respectively $1/W^{3.5}$). The ϕ_π-independent results are rather flat in $t_{min} - t$. The t-channel meson exchange model by J.M. Laget [3] gives a good agreement with $\sigma_T + \epsilon_L \sigma_L$ and $\sigma_{TL'}$ for both Q^2 values at x_{Bj} fixed. However, it shows a disagreement in order of magnitude for σ_{TT} and disagreement in sign for σ_{TL} for both of those kinematics.

The Q^2-dependence of the hadronic tensor implies $\sigma_T + \epsilon_L \sigma_L$ behaves roughly as $1/Q^{4.5}$. Assuming $\sigma_L \gg \sigma_T$, σ_L would behave roughly as $1/Q^3$, which is far from the QCD leading twist predicted by GPD formalism.

The data we extracted yield two conclusions with regard to the available models.

- The model by J.M. Laget is able to describe $\sigma_T + \epsilon_L \sigma_L$ and $\sigma_{TL'}$, but neither σ_{TL} nor σ_{TT}.
- The Q^2-dependence of the cross sections (Fig. 1) demonstrates that we are far from the QCD leading twist prediction of $d\sigma_L/dt$, which behaves as $1/Q^6$. On the other hand, it is similar to the Q^2-

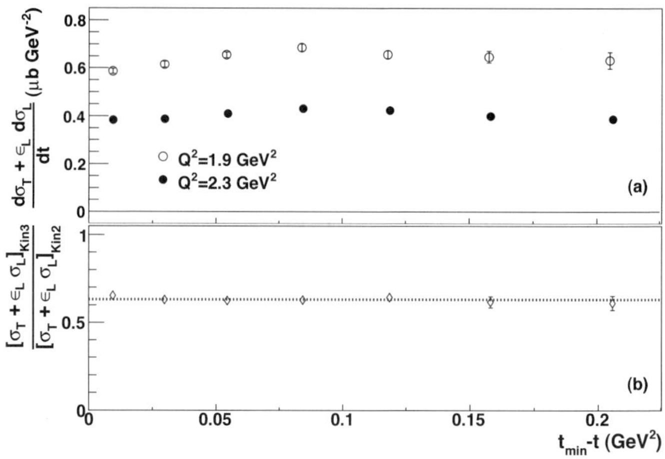

Fig. 1. (a) separated $H(e, e'\pi^0)p$ cross section $\sigma_T + \epsilon_L\sigma_L$ as a function of $t_{min} - t$ for $x_{Bj} = 0.36$. Error bars represent statistical errors only. (b): ratio of $\sigma_T + \epsilon_L\sigma_L$ for the two kinematics as a function of $t_{min} - t$. The fit of this ratio (dashed line) indicates the Q^2-dependence of the cross section.

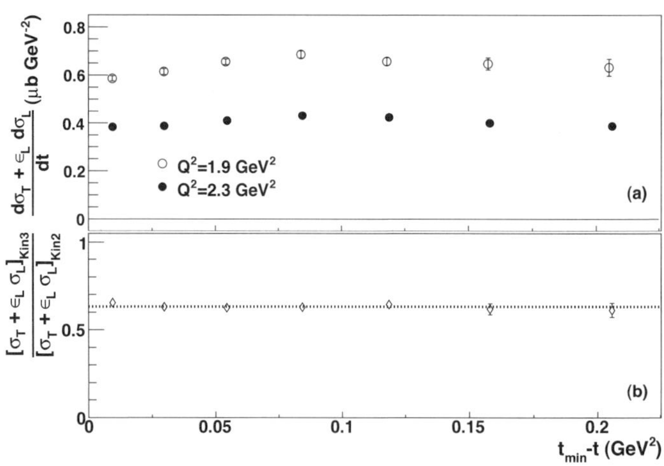

Fig. 2. (a): separated $H(e, e'\pi^0)p$ cross section $\sigma_T + \epsilon_L\sigma_L$ as a function of $t_{min} - t$ for $Q^2 = 2.1$ GeV^2. Error bars represent statistical errors only. (b) ratio of $\sigma_T + \epsilon_L\sigma_L$ for the two kinematics as a function of $t_{min} - t$. The fit of this ratio (dashed line) indicates the W-dependence of the cross section.

dependence of the transverse cross section for charged pion electroproduction published by Hall C.

Moreover, the π^0 has no charge, no spin, so a direct coupling with a virtual photon is suppressed, which removes the pion-pole contribution to the longitudinal cross section. This suggests that the transverse $ep \to ep\pi^0$ cross section is likely to dominate, and transverse $ep \to en\pi^+$ cross sections have already been described by quark fragmentation mechanisms usually used to describe semi inclusive processes.

T. Horn *et al.* measured the exclusive π^+ electroproduction cross section at $Q^2 = 1.60$ and 2.45 GeV^2, with σ_T and σ_L separation [5]. The t-channel meson exchange model by J.M. Laget reproduces the σ_L component. However, the σ_T component does not follow the TME model prediction. M. Kaskulov *et al.* performed PYTHIA-JETSET calculations using the Lund model applied to π^+ transverse cross sections at Hall C kinematics [12]. These calculations applied to Hall C π^+ transverse cross sections are in excellent agreement with the data. This gives evidence that the π^+ transverse cross sections at $Q^2 > 1$ GeV^2 above the resonance region is described by a partonic process. This suggests that the present π^0 data could similarly be described by incoherent scattering on the partonic structure of the nucleon target.

For these reasons, we consider our data within the context of Semi-Inclusive Deep Inelastic Scattering (SIDIS). We attempted to fit our data with a SIDIS formalism written by Anselmino *et al.* [19]. The equation (38) of [19] gives the cross section at $\mathcal{O}(k_\perp/\sqrt{Q^2})$ for semi-inclusive production for any hadron. It also provides the order of magnitude for the $\cos\phi_\pi$ contribution of a SIDIS process (Cahn effect). The idea is to adjust the ratio of the $\cos\phi_\pi$ contribution (identified to σ_{TL}) over the constant term (identified to $\sigma_T + \epsilon_L\sigma_L$) by adjusting only the parameter $\langle p_\perp^2\rangle/\langle k_\perp^2\rangle$:

$$\frac{\sqrt{2\epsilon_L(1+\epsilon)}\sigma_{TL}}{\sigma_T + \epsilon_L\sigma_L} = \frac{4(2-y)\sqrt{1-y}z_\pi p_{\pi\perp}}{\left(\frac{\langle p_\perp^2\rangle}{\langle k_\perp^2\rangle} + z_\pi^2\right)\sqrt{Q^2}(1+(1-y)^2)} \tag{6}$$

This adjustment is shown in Figure 3. Our exclusive data (where particle multiplicity is unity) is reproduced by this model if $\langle p_\perp^2\rangle/\langle k_\perp^2\rangle \sim 4.0$, whereas at high particle multiplicity (inclusive data), $\langle p_\perp^2\rangle/\langle k_\perp^2\rangle \sim 0.8$.

6. Conclusions

We extracted the separated differential π^0 cross section at Jefferson Lab, Hall A, at four kinematic settings, with a 2.5 % statistical precision at x_{Bj} fixed, and with a 5 % statistical precision at Q^2 fixed. We studied the Q^2- and W-dependence of the hadronic tensor.

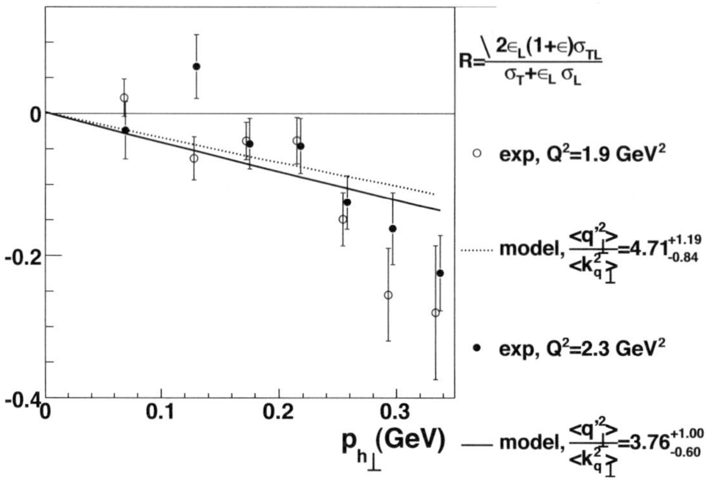

Fig. 3. Ratio $\frac{\sqrt{2\epsilon_L(1+\epsilon)}\sigma_{TL}}{\sigma_T+\epsilon_L\sigma_L}$ for $Q^2 = 1.9$ GeV^2 (open circles) and $Q^2 = 2.3$ GeV^2 (solid circles) plotted as a function of $p_{\pi\perp}$. Error bars represent statistical errors only. The fits to these quantities are represented by a dotted curve (solid curve) for $Q^2 = 1.9$ GeV^2 ($Q^2 = 2.3$ GeV^2).

The model of J.M. Laget is able to describe the total cross section, but still needs improvement. There is a contradiction between our data and the leading twist QCD prediction for high Q^2 ($\sigma_T + \epsilon_L\sigma_L \propto 1/Q^{4.5}$ instead of $1/Q^8$). The $\cos\phi_\pi$ cross section contribution is compatible in sign and shape with the Cahn effect of SIDIS.

To improve the understanding of our data, we plan to run another π^0 experiment in Fall 2010, at two beam energies, allowing us to disentangle σ_L from σ_T.

References

1. G. R. Goldstein and J. F. Owens, *Phys. Rev.* **D7**, 865 (1973).
2. J. M. Laget, *Phys. Rev.* **C76**, p. 052201 (2007).
3. J. M. Laget (2010).
4. R. De Masi *et al.*, *Phys. Rev.* **C77**, p. 042201 (2008).
5. T. Horn *et al.*, *Phys. Rev.* **C78**, p. 058201 (2008).
6. J. C. Collins, L. Frankfurt and M. Strikman, *Phys. Rev.* **D56**, 2982 (1997).
7. K. Goeke, M. V. Polyakov and M. Vanderhaeghen, *Prog. Part. Nucl. Phys.* **47**, 401 (2001).
8. M. Vanderhaeghen, P. A. M. Guichon and M. Guidal, *Phys. Rev.* **D60**, p. 094017 (1999).

9. D. Mueller, D. Robaschik, B. Geyer, F. M. Dittes and J. Horejsi, *Fortschr. Phys.* **42**, p. 101 (1994).
10. X.-D. Ji, *Phys. Rev. Lett.* **78**, 610 (1997).
11. A. V. Radyushkin, *Phys. Rev.* **D56**, 5524 (1997).
12. M. M. Kaskulov, K. Gallmeister and U. Mosel, *Phys. Rev.* **D78**, p. 114022 (2008).
13. P. Y. Bertin, C. E. Hyde, R. Ransome and F. Sabatié, Jefferson Lab Proposal, E00-110, (2000).
14. C. Munoz Camacho *et al.*, *Phys. Rev. Lett.* **97**, p. 262002 (2006).
15. D. Drechsel and L. Tiator, *J. Phys.* **G18**, 449 (1992).
16. L. W. Mo and Y.-S. Tsai, *Rev. Mod. Phys.* **41**, 205 (1969).
17. Y.-S. Tsai, *Rev. Mod. Phys.* **46**, p. 815 (1974).
18. M. Vanderhaeghen *et al.*, *Phys. Rev.* **C62**, p. 025501 (2000).
19. M. Anselmino *et al.*, *Phys. Rev.* **D71**, p. 074006 (2005).

HARD PHOTODISINTEGRATION OF ^{3}He

CARLOS GRANADOS*

Florida International University
Miami, Florida 33199, U.S.
**E-mail: cgran005@fiu.edu*
www.fiu.edu

Large angle photodisintegration of two nucleons from the ^{3}He nucleus is studied within the framework of the hard rescattering model (HRM). In the HRM the incoming photon is absorbed by one nucleon's valence quark that then undergoes a hard rescattering reaction with a valence quark from the second nucleon producing two nucleons emerging at large transverse momentum . Parameter free cross sections for pp and pn break up channels are calculated through the input of experimental cross sections on *pp* and *pn* elastic scattering. The calculated cross section for *pp* breakup and its predicted energy dependency are in good agreement with recent experimental data. Predictions on spectator momentum distributions and helicity transfer are also presented.

Keywords: Hard photodisintegration.

1. Introduction

Breakup of nucleon-nucleon (NN) systems at large energy and momentum transfer is an important tool in probing QCD descriptions of nuclear processes at this kinematic regime. In NN breakup, large NN center of mass energies where QCD degrees of freedom become evident, are reached with moderate beam energies in comparison with nucleon nucleon collisions. The simplest of these processes, the photodisintegration of the deuteron has been intensively studied through experiments at kinematics where the hadronic description is no longer applicable (see, e.g., Refs. [1, 2]). It has been observed that at 90° center of mass angle and beam energies of few GeV, the measured differential cross section scales in accordance with the constituent counting rule for exclusive processes [3], signaling the onset of quark degrees of freedom in the reaction.

Several QCD models of NN breakup have been developed in order to described features observed in the experimental data. Among these, models such as the reduced nuclear amplitude (RNA) approach [4], the quark gluon string (QGS) model [5], and the hard rescattering model (HRM) [6] rely on different approaches in dealing with the nonperturbative factors that enter the calculations. For the RNA approach, these factors are accounted for by introducing parameterizations of nucleon and nuclear form factors. The QGS model introduces a reggeization of the scattering amplitude, while through the HRM, breakup cross sections are calculated through the input of experimental data of cross sections of the corresponding nucleon nucleon elastic scattering process.

Calculations of cross sections of deuteron photodisintegration in the RNA and the QGS approaches require a normalization to the data. In contrast, in the HRM these cross sections are obtained without the need of a normalization parameter. In order to further test the validity of these competing approaches, it was proposed to extend photodisintegration studies to the breakup of proton proton (pp) and proton neutron (pn) systems in the ^{3}He nucleus [7]. For these reactions, numerical predictions for the cross section differ considerably from one approach to another. Recent experimental data [8] shows a better agreement with HRM calculations [9].

Additional constrains on the analysis of NN breakup are set in ^{3}He photodisintegration by studying observables such as the spectator momentum distributions and polarization transfer asymmetries. Features on these observables predicted here within the HRM are expected to be put to the test in future experimental analysis.

2. HRM of ^{3}He Photodisintegration

Fig.1 illustrates the HRM picture of the breakup of a NN pair in ^{3}He photodisintegration. It differs graphically from that of the deuteron breakup by the presence of a spectator nucleon. In Fig.1, by absorbing the incoming photon, a valence quark is knocked out from one of the nucleons in the initial NN system. This quark then re scatters off a valence quark from the second nucleon. Both quarks then recombine with the spectators systems to form two nucleons with large transverse momentum in the final state.

Within the HRM, The scattering amplitude is obtained using Feynman rules on the diagram in Fig.1. The only calculable terms are the QED vertex corresponding to the photon absorption by the valence quark and the component of the nuclear wave function represented by the ^{3}He vertex. The scattering amplitude is then reduced to the form in Eq.(1) through two main assumptions (see Ref. [9] for details). First, by working in a $q^+ = 0$

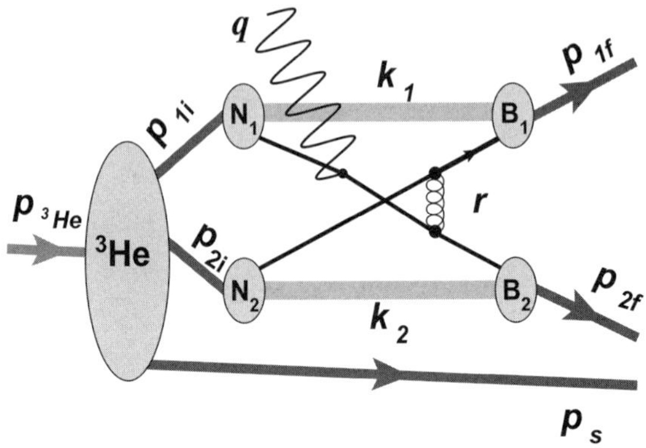

Fig. 1. ^{3}He photodisintegration according to the HRM

reference frame, we approximate the struck nucleon's propagator to its on shell component. Second, we assume that the nucleons in the NN system have an equal share of the NN light cone momentum, which maximizes the nuclear wave function. The result as shown in Eq.(1) is a convolution of a nuclear wave function $\Psi_{^3\mathrm{He}}$ and the scattering amplitude T of a kinematically corresponding NN elastic scattering reaction.

$$
\begin{aligned}
&\langle \lambda_{1f}, \lambda_{2f}, \lambda_s \mid \mathcal{M} \mid \lambda_\gamma, \lambda_A \rangle \\
&= ie[\lambda_\gamma] \Bigg\{ \sum_{i \in N_1} \sum_{\lambda_{2i}} \int \frac{Q_i^{N_1}}{\sqrt{2s'}} \langle \lambda_{2f}; \lambda_{1f} \mid T^{QI}_{NN,i}(s, t_N) \mid \lambda_\gamma; \lambda_{2i} \rangle \\
&\qquad\qquad \times \Psi^{\lambda_A}_{^3\mathrm{He}}(p_1, \lambda_\gamma; p_2, \lambda_{2i}; p_s, \lambda_s) \frac{d^2 p_\perp}{(2\pi)^2} \\
&+ \sum_{i \in N_2} \sum_{\lambda_{1i}} \int \frac{Q_i^{N_2}}{\sqrt{2s'}} \langle \lambda_{2f}; \lambda_{1f} \mid T^{QI}_{NN,i}(s, t_N) \mid \lambda_{1i}; \lambda_\gamma \rangle \\
&\qquad\qquad \times \Psi^{\lambda_A}_{^3\mathrm{He}}(p_1, \lambda_{1i}; p_2, \lambda_\gamma; p_s, \lambda_s) \frac{d^2 p_\perp}{(2\pi)^2} \Bigg\}, \qquad (1)
\end{aligned}
$$

where $\lambda_1, \lambda_2, \lambda_s$ are the helicities of the nucleons from ^{3}He. $\lambda_A, \lambda_\gamma$ are the helicities of the nucleus and of the incoming photon respectively, and Q_i^N is the charge in e units of the quark struck by the photon. A further reduction is achieved if $-t_N >> m_N^2$. Under this condition, the matrix elements of T can be factored out of the $p_\perp$ integral.

Using the following notation for the independent helicity amplitudes in

NN elastic scattering,

$$\begin{aligned}
\langle +,+|T|+,+\rangle &= \phi_1 \\
\langle +,-|T|+,-\rangle &= \phi_3 \\
\langle -,+|T|+,-\rangle &= \phi_4 \\
\langle -,-|T|+,+\rangle &= \phi_2 \\
\langle -,+|T|+,+\rangle &= \phi_5,
\end{aligned} \tag{2}$$

and from explicitly working the matrix elements in Eq.(1), we can write the averaged of the squared of the scattering amplitudes $\mathcal{M}$ as follows,

$$\begin{aligned}
|\bar{\mathcal{M}}|^2 = \frac{1}{2}\frac{1}{3}\frac{e^2}{2s'}\Big[& S_{12}\left\{|(\hat{Q}^{N_1}+\hat{Q}^{N_2})\phi_1|^2 + |(\hat{Q}^{N_1}+\hat{Q}^{N_2})\phi_2|^2\right\} \\
& + S_{34}\left\{|\hat{Q}^{N_1}\phi_3 + \hat{Q}^{N_2}\phi_4|^2 + |\hat{Q}^{N_1}\phi_4 + \hat{Q}^{N_2}\phi_3|^2\right\} \\
& + 2S_0|(\hat{Q}^{N_1}+\hat{Q}^{N_2})\phi_5|^2\Big],
\end{aligned} \tag{3}$$

where the light-cone spectral functions of the ^{3}He nucleus are defined as follows:

$$\begin{aligned}
S_{12} &= N_{NN} \sum_{(\lambda_1=\lambda_2=-\frac{1}{2})}^{\frac{1}{2}} \sum_{(\lambda_s=-\frac{1}{2})}^{\frac{1}{2}} \left| \int \Psi^{\frac{1}{2}}_{^3\mathrm{He}}(p_1,\lambda_1;p_2,\lambda_2;p_s,\lambda_s)\frac{d^2p_\perp}{(2\pi)^2}\right|^2, \\
S_{34} &= N_{NN} \sum_{(\lambda_1=-\lambda_2=-\frac{1}{2})}^{\frac{1}{2}} \sum_{(\lambda_s=-\frac{1}{2})}^{\frac{1}{2}} \left| \int \Psi^{\frac{1}{2}}_{^3\mathrm{He}}(p_1,\lambda_1;p_2,\lambda_2;p_s,\lambda_s)\frac{d^2p_\perp}{(2\pi)^2}\right|^2, \\
S_0 &= S_{12} + S_{34},
\end{aligned} \tag{4}$$

and the charge factors $\hat{Q}^N$ account for all possible quark photon interactions contributing to the reaction's amplitude. For two nucleon breakup in ^{3}He, the unpolarized cross section is given by

$$\frac{d\sigma}{dtd^3p_s/(2E_s(2\pi)^3)} = \frac{|\bar{\mathcal{M}}|^2}{16\pi(s-M_A^2)(s_{NN}-M_{NN}^2)} \tag{5}$$

where $s = (q+p_A)^2$, and $s_{NN} = (q+p_{NN})^2$. Then according to Eq.(3), the cross section in Eq.(5) can be computed from the knowledge of the ϕ scattering amplitudes, and of the charge factors $\hat{Q}$. Because there is not enough experimental data in polarized NN scattering to extract the helicity amplitudes ϕ, Eq.(5) can instead be computed through general assumptions regarding these amplitudes . This is further discussed in the coming sections.

Alternatively, the ϕ amplitudes can be modeled according to a NN scattering mechanism. Experiments have demonstrated the dominance of the quark interchange mechanism (QIM) in hard exclusive reactions between baryons of common quark flavors such as pp and pn elastic scattering [10]. Within a (QIM) approach, the ϕ amplitudes are obtained by assuming a symmetry structure of the valence quark wave functions of the interacting baryons (see e.g., Refs.[11]). A starting point is using SU(6) symmetric wave functions [11], but it has been suggested that a diquark model better describes general features in pn elastic scattering such as its strength relative to pp scattering, and the asymmetry observed in its angular distributions (see, e.g., Ref.[12]).

Finally, an important feature of the HRM is that at large values of s and $-t$ it predicts an energy distribution of Eq.(5) in accordance with the constituent counting rule [3]. That is $\frac{d\sigma}{dt} \sim s^{-11}$ [6,7,9], which comes from the s^{-4} dependence of the ϕ amplitudes entering Eq.(3). This behavior has already been observed in deuteron photodisintegration [1,2], and in recent experiments on pp breakup in ^{3}He [8].

3. *pn* and *pp* breakup

As mentioned in the previous section $\frac{d\sigma}{dt}$ in Eq.(5) can be computed through general assumptions on pn and pp elastic scattering.

3.1. $\gamma\ {}^3He \rightarrow pn + p$

In pn breakup, pn scattering amplitudes ϕ enter Eq.(3). Using that for hard pn elastic scattering $\phi_3 \sim \phi_4$ (see, e.g., Ref. [13]), and that at low transverse momentum $p_\perp << m_N$, $\Psi_{^3\text{He}} \approx \sqrt{2}(2\pi)^3 m_N \Psi_{^3\text{He,NR}}$ [14], the cross section of pn breakup can be approximated to a simple product of a spectral function and a cross section of a kinematically corresponding pn elastic reaction,

$$\frac{d\sigma^{\gamma^3He\rightarrow(pn)p}}{dt\frac{d^3p_s}{E_s}} = \alpha Q^2_{F,pn} 16\pi^4 \frac{S^{pn}_{0,NR}(\alpha_c = \frac{1}{2}, \vec{p}_s)}{2} \frac{s_{NN}(s_{NN} - 4m_N^2)}{(s')(s - M^2_{^3\text{He}})} \times \frac{d\sigma^{pn\rightarrow pn}(s_{NN}, t_N)}{dt_N}, \tag{6}$$

where $Q_F = 1/3$, $\alpha = 1/137$ and $s' = s_{NN} - M^2_{NN}$.

3.1.1. $\gamma^3 He \to pp + n$

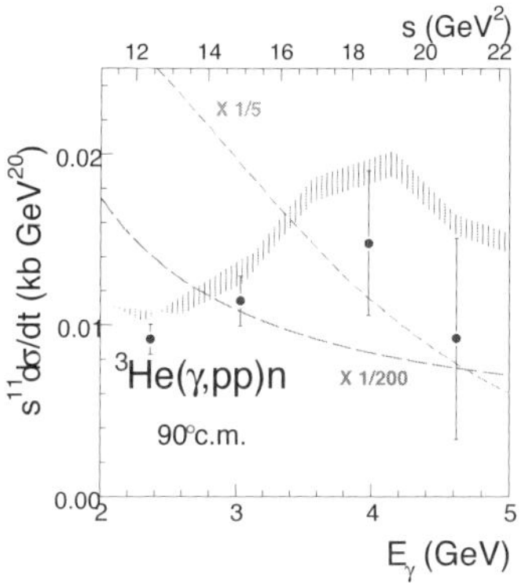

Fig. 2. Cross section data for pp breakup in ^{3}He photodisintegration at 90° c.m. angle of the $\gamma + pp$ system [8]. (Dashed) QGS prediction times 1/5. (Long dashed) RNA prediction times 1/200. (Shaded) HRM prediction.

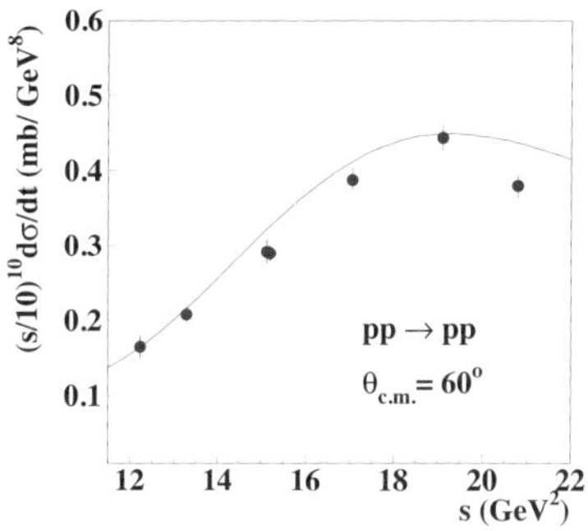

Fig. 3. pp elastic scattering data at 60° of center of mass angle of scattering. These data is used in Eq.(7) to calculate the HRM cross section of pp breakup shown in Fig.2.

At large center of mass angle of scattering in pp elastic scattering, it follows from Pauli's principle that $\phi_4 \sim -\phi_3$ (see, e.g., Ref.[15]). Also, assuming that the S states dominates the ground state nuclear wave function, from the exclusion principle, the probability of two protons having the same helicity is negligible. Therefore, $S_{12} << S_{34}$. These assumptions allow to write the cross section for pp breakup in ^{3}He photodisintegration as follows:

$$\frac{d\sigma^{\gamma^3 He \to (pp)n}}{dt \frac{d^3 p_s}{E_s}} = \alpha Q^2_{F,pp} 16\pi^4 S^{pp}_{34}(\alpha = \frac{1}{2}, \vec{p}_s) \frac{s_{NN}(s_{NN} - 4m^2_N)}{s'(s - M^2_{^3\text{He}})} \times \frac{2\beta^2}{1 + 2C^2} \frac{d\sigma^{pp\to pp}(s_{NN}, t_N)}{dt}, \tag{7}$$

where, $C^2 = (\phi_3^2/\phi_1^2) \approx (\phi_4^2/\phi_1^2)$, $\beta = (|\phi_3| - |\phi_4|)/|\phi_1|$ and Q_F=5/3. Because the pp breakup cross section is proportional to β and having that $|\phi_4| \sim |\phi_3|$, we predict a large suppression of the pp breakup relative to the pn breakup case.

Eq.(7) shows that just as in the pn breakup case, the pp breakup cross section is proportional to the cross section of the corresponding pp elastic scattering. This is illustrated in Figs. 2 and 3 that show that the shape of the scaled energy distribution for pp breakup mimics that of the scaled energy distribution for the corresponding pp elastic scattering. Fig.2 also shows that the HRM calculation is in very good agreement with the experimental data.

3.2. *Spectator momentum distributions*

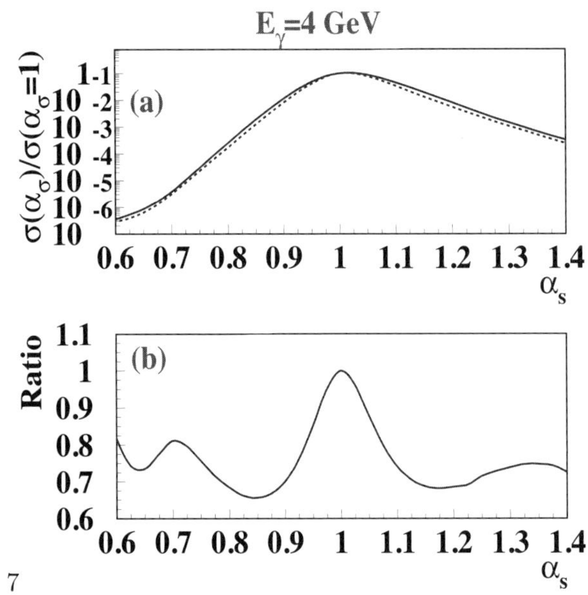

Fig. 4. (a) pp breakup (solid line) and pn breakup (dashed line) s^{11}-weighted HRM cross sections as functions of the spectator nucleon's light cone momentum fraction (α_s) and normalized to their values at $\alpha_s = 1$. (b) Ratio (R) of the pn to pp breakup HRM cross sections normalized to its value at $\alpha_s = 1$.

Additional observables can be extracted from ^3He photodisintegration in order to test or constrain the validity of the HRM. We consider for instance the spectator momentum distributions plotted in Fig.4. In Fig.4a we noted that the asymmetry in both distributions around $\alpha_s = 1$ is due to the s^{-11} dependence of both cross sections. As less energy is taken by the spectator nucleon, more energy is taken by the NN system, hence the sharper drop towards $\alpha_s = 0$. The broader pp breakup distribution in Fig.4a, as well as the sharp drop of R around $\alpha_s = 1$ are due to the suppression of the S_{12}^{pp} spectral function in the low NN momentum component of the nuclear wave function.

3.3. *Polarization transfer observables*

Another observable of interest is related to the effectiveness with which the incoming photon transfers its helicity to an outgoing proton. This is measured through the asymmetry $C_{z'}$ (see, e.g., [9]) that in the HRM takes the form

$$C_{z'} = \frac{(|\phi_1|^2 - |\phi_2|^2)S^{12} + (|\phi_3|^2 - |\phi_4|^2)S_{34}}{2|\phi_5|^2 S_0 + (|\phi_1|^2 + |\phi_2|^2)S_{12} + (|\phi_3|^2 + |\phi_4|^2)S_{34}}, \tag{8}$$

Noting that for pp breakup $S_{12} \approx 0$ and $\phi_4 \approx -\phi_3$, we have that $C_{z'}^{pp} \approx 0$, while for pn breakup $C_{z'}^{pn} \approx 2/3$.

4. Summary

We studied the hard breakup of a nucleon pair in hard a ^{3}He nucleus by introducing quark degrees of freedom through the hard rescattering model (HRM). Cross sections for proton neutron pn and proton proton breakup are calculated in the (HRM) without the need for introducing a parameter to normalize to data. The obtained cross section for pp breakup shows good agreement with recent experimental data. We also predict distinctive spectator momentum distributions for both pp and pn breakup, and a cancellation of the $C_{z'}$ asymmetry in the pp breakup channel. Mounting evidence in favor of the rescattering picture of breakup processes motivates an extension of these study to different baryonic channels such as $\Delta\Delta$-isobar in deuteron breakup (see, e.g., Ref.[16]) for which experiments are already been proposed.

References

1. C. Bochna *et al.*, *Phys. Rev. Lett.* **81**, 4576 (1998).
2. E. C. Schulte *et al.*, *Phys. Rev. Lett.* **87**, p. 102302 (2001).
3. S. J. Brodsky and G. R. Farrar, *Phys. Rev. Lett.* **31**, 1153 (1973).
4. S. J. Brodsky and J. R. Hiller, *Phys. Rev.* **C28**, p. 475 (1983).
5. V. Y. Grishina *et al.*, *Eur. Phys. J.* **A10**, 355 (2001).
6. L. L. Frankfurt, G. A. Miller, M. M. Sargsian and M. I. Strikman, *Phys. Rev. Lett.* **84**, 3045 (2000).
7. S. J. Brodsky *et al.*, *Phys. Lett.* **B578**, 69 (2004).
8. I. Pomerantz *et al.*, *Phys. Lett.* **B684**, 106 (2010).
9. M. M. Sargsian and C. Granados, *Phys. Rev.* **C80**, p. 014612 (2009).
10. C. G. White *et al.*, *Phys. Rev.* **D49**, 58 (1994).
11. G. R. Farrar, S. A. Gottlieb, D. W. Sivers and G. H. Thomas, *Phys. Rev.* **D20**, p. 202 (1979).
12. C. G. Granados and M. M. Sargsian, *Phys. Rev. Lett.* **103**, p. 212001 (2009).

13. G. P. Ramsey and D. W. Sivers, *Phys. Rev.* **D45**, 79 (1992).
14. L. L. Frankfurt and M. I. Strikman, *Phys. Rept.* **160**, 235 (1988).
15. M. Jacob and G. C. Wick, *Ann. Phys.* **7**, 404 (1959).
16. C. G. Granados and M. M. Sargsian, *[arXiv: hep-ph 1007.4705]* (2010).

MEASUREMENT OF THE GENERALIZED FORM FACTORS NEAR PION THRESHOLD VIA $\gamma^* p \to n\pi^+$ IN CLAS

KIJUN PARK

Thomas Jefferson National Accelerator Facility,
Newport News, Virginia 23606, USA
**E-mail: parkkj@jlab.org*
http://www.jlab.org/~parkkj

We extracted the multipoles near pion threshold for the first time at high momentum transfers (Q^2) in the $n\pi^+$ final state channel. The dominance of the S-wave transverse multipole (E_{0+}) is expected in this region, which allows us to access the new generalized form factors G_1, G_2 within the Light-Cone-Sum-Rule (LCSR) framework. In this analysis, we used one of the recent CLAS experimental data sets which utilized the high-energy polarized electron beam on unpolarized proton target using the 4π detector coverage. We measured the differential cross sections and extracted the multipoles using two methods, LCSR and multipole fit. Both methods show consistent results, which are almost Q^2 independent.

Keywords: CLAS; Threshold Pion; S-wave; Generalized Form Factors; Differential Cross Section.

1. Introduction

We report the extraction of the multipole E_{0+} near pion threshold in the charged single pion electroproduction channel ($\vec{e}p \to e'n\pi^+$) with a high-energetic polarized electron beam on the unpolarized proton target. The recent CLAS experimental data set (*e1-6*) allows us to study near threshold pion production at high energy and in particular at photon virtualities Q^2 up to ~ 6 GeV^2. The recent and future CLAS polarization measurements are especially helpful in this respect as well. We believe that such studies are feasible with CLAS data and the task of this analysis is to provide the necessary theoretical guidance and experimental extraction techniques.

The LCSR has been addressed that a realistic QCD-motivated model for the Q^2 dependence of both transverse E_{0+} and longitudinal L_{0+}, $S-$wave multipoles at threshold in the region $Q^2 \sim 1-10$ GeV^2 can be accessed by present experimental data [1]. This approach is attractive because in LCSR soft contributions to the form factors are calculated in terms of the same nucleon Distribution Amplitudes (DAs) that enter the pQCD calculation and there is no double counting. The same technique can be applied to pion electroproduction. The relevant generalized form factors were estimated in the LCSR approach [2] for the range of momentum transfers $Q^2 \sim 5-10$ GeV^2. For this work, the sum rules have been re-derived in [2] taking into account the semi-disconnected pion-nucleon contributions in the intermediate state.

Two generalized form factors that contribute to pion electroproduction at the kinematic threshold, were introduced and explained the relation to $S-$wave multipoles and suggested a model for their Q^2 dependence based on LCSR. As a simple approach in LCSR for the electroproduction close to threshold, the correlation function shows the complementing the $S-$wave form factor-like contributions by $P-$wave terms corresponding to pion emission in the final state that can be expressed in terms of the nucleon electromagnetic form factors. Under the one-photon-exchange approximation, the single pion electroproduction cross section factorizes:

$$
\begin{aligned}
\frac{k_\gamma^*}{p_\pi^*}\frac{d^2\sigma_u}{d\Omega_\pi^*} =& \sigma_T + \epsilon_L\sigma_L + \epsilon\sigma_{TT}\sin^2\theta_\pi^*\cos 2\phi_\pi^* \\
&+ \frac{1}{\sqrt{2\epsilon_L(1+\epsilon)}}\sigma_{LT}\sin\theta_\pi^*\cos\phi_\pi^* + \frac{h}{\sqrt{2\epsilon_L(1-\epsilon)}}\sigma_{LT'}\sin\theta_\pi^*\sin\phi_\pi^* \ ,
\end{aligned}
$$

where p_π^*, θ_π^* are the π^+ momentum and polar angle in the center-of-mass frame, ϕ_π^* is the azimuthal rotation of the $\pi^+ n$ plane with respect to the electron scattering plane (e, e'), h is the electron beam helicity, and k_γ^* is the equivalent photon energy.

Once we got the four structure functions from the fit of ϕ-dependent differential cross section, the partial wave decomposition by legendre polynomials of the structure functions in the limit of angular momentum $l_\pi = 1$ is given by

$$
\begin{aligned}
\sigma_T + \epsilon_L\sigma_L &= D_0^{T+L} + D_1^{T+L}P_1(\cos\,\theta_\pi^*) + D_2^{T+L}P_2(\cos\theta_\pi^*) \\
\sigma_{TT} &= D_0^{TT} \quad ; \quad \sigma_{LT} = D_0^{LT} + D_1^{LT}P_1(\cos\,\theta_\pi^*) \ ,
\end{aligned}
$$

where the coefficients (D_0^{T+L}, D_1^{T+L}, D_2^{T+L}, D_0^{TT}, D_0^{LT} and D_1^{LT}) are composite of 7 complex multipoles. If we take LCSR calculation in pion threshold region [1] into account, the Legendre coefficients consist of the generalized form factors as shown,

$$D_0^{T+L} = \frac{1}{f_\pi^2}\left[\frac{4\vec{k_i}^2 Q^2}{m_N^2}|G_1^{\pi N}|^2 + \frac{c^2{}_\pi g_A{}^2 \vec{k_f}^2}{W^2 - m_N^2} Q^2 m_N^2 G_M^2\right.$$
$$\left. + \epsilon_L \left(\vec{k_i}^2 |G_2^{\pi N}|^2 + \frac{4c^2{}_\pi g_A{}^2 \vec{k_f}^2}{W^2 - m_N^2} m_N^4 G_E^2\right)\right],$$
$$D_1^{T+L} = \frac{1}{f_\pi^2}\frac{4c_\pi g_A |k_i||k_f|}{W^2 - m_N^2}\left(Q^2 G_M Re(G_1^{\pi N}) - \epsilon_L m_N^2 G_E Re(G_2^{\pi N})\right),$$
$$D_0^{LT} = -\frac{1}{f_\pi^2}\frac{c_\pi g_A |k_i||k_f|}{W^2 - m_N^2} Q m_N \left(G_M Re(G_2^{\pi N}) + 4 G_E Re(G_1^{\pi N})\right),$$

where G_M, G_E are magnetic and electric Sachs form factors due to the pion emission off the initial proton state and for charged pions there is an additional contribution corresponding to the chiral rotation of the electromagnetic current. A consequence of the absence of $D-$waves is that $D_0^{TT} = 0$. The isospin factor $c_\pi = \sqrt{2}$, $f_\pi = 93$ MeV, and $g_A = 1.267$. We use the following parametrization for electric/magnetic form factor of proton/neutron; G_M^p, G_E^p [3][4], G_M^n and G_E^n [5].

2. Data analysis

We used the *e1-6* experimental data set for this analysis. Data were accumulated with at the beam charge of 21 mC using a liquid hydrogen target. The incident beam had an energy of 5.754 GeV with an averaged 70% beam polarization. The kinematic range bins were set to satisfy the required statistics in covered kinematic range. Table 1 shows the covered kinematics and bin size.

Table 1. Covered kinematic range.

Variable	Num. Bin	Range	Bin Size
W	3	1.11 GeV $\sim$ 1.15 GeV	20 MeV
Q^2	5	2.05 GeV^2 $\sim$ 4.16 GeV^2	variable
$cos\theta^*{}_\pi$	10	$-1.0 \sim 1.0$	0.2
$\phi^*{}_\pi$	12	$0.0^\circ \sim 360^\circ$	30°

where W is invariant energy, $\theta_\pi{}^*$, $\phi_\pi{}^*$ are the polar and azimuthal angles of the detected positive pion in the center-of-mass frame respectively, Q^2 is the four momentum transfer squared ($q = p_e - p_{e'}, Q^2 = -q^2$. The acceptance and radiative corrections were calculated with the exactly same size and range of kinematic bins.

The particle identification (PID) and kinematic corrections are applied for near threshold physics regime. The CLAS electron identification at the trigger level is done by requiring a minimum amount of energy in the electromagnetic calorimeter (EC) in coincidence with a signal in the Čerenkov counter (CC). In the off-line analysis, some additional requirements were used to select events with a good electron, which have been already described in [6]. To improve the electron selection we applied the cuts in terms of energy released in the electro-magnetic calorimeter, z-vertex, fiducial volume, charge, and the number of CC photo-electrons [7]. Pions (π^+) are identified by a coincidence of DC and SC. Pions are basically identified by a 2σ cut in the β_h distribution. In addition to the particle identification done by SEB a few more conditions were imposed on the π^+ candidates. We made proper geometrical fiducial cuts, which were applied to both simulation and experimental data, where they chose the areas of uniform detector response that can be reproduced by GSIM [7]. The background contribution under the neutron missing mass is very small on the pion threshold region, hence no background subtraction was performed.

We calculate the detector acceptance from simulation to extract cross sections from experimental yields. We used the two different event generators for systematic error estimation, which are AAO-RAD (physics model; Unitary isobar(MAID03) based event generator), GENEV-EG (phase space event generator). We generate exclusive events including radiative effects for our kinematics. The acceptance is defined by including the efficiency of the detector and radiative effects:

$$Acceptance_i = \frac{\#ofREC_i}{\#ofGEN_i},$$

where GEN is the GENerated event and REC is the REConstructed event after *user_ana* and all cuts. Figure 1 shows some examples of ϕ^*_π dependent acceptances for $W = 1.13\text{GeV}$, $Q^2 = 2.05\text{GeV}^2$.

The radiative corrections were calculated by ExcluRad [8]. We studied the model dependence of ExcluRad using Sato-Lee2004 (dynamic model) and MAID2003 (unitary isobar model) to verify whether ExcluRad gives us consistent correction factors independent of the physics models. Figure 2 shows examples of the radiative-correction comparison (left-axis label) and cross section (right-axis label) between two models as a function of cos θ^*_π at fixed $W = 1.15$ GeV and $Q^2 = 2.91$ GeV^2 for different $\phi^*_\pi = 105^o - 345^o$.

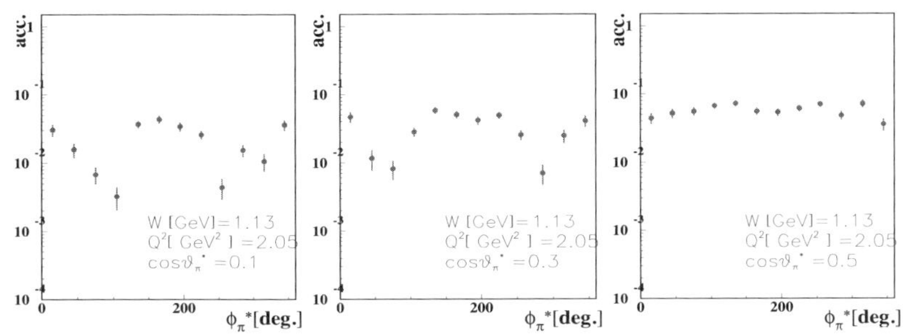

Fig. 1. Some examples of the ϕ_π^*-dependent acceptances for the single-pion electroproduction ($\gamma^* p \rightarrow \pi^+ n$) at given $W = 1.13$ GeV, $Q^2 = 2.05$ GeV^2 and cos $\theta_\pi^* = 0.1,\ 0.3$ and 0.5. based on AAO_RAD.

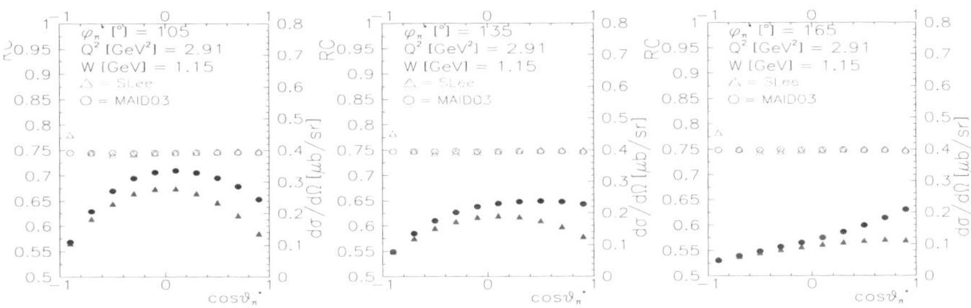

Fig. 2. Some examples of the radiative correction factor from ExcluRad as function of *cos* θ_π^* at fixed $W = 1.15$ GeV and $Q^2 = 2.91$ GeV^2 for different $\phi_\pi^* = 105^o - 345^o$. Each plot shows the RC factor on the left y-axis and differential cross section on the right y-axis. The circles represent RC and differential cross sections from MAID2003 and triangles from Sato-Lee2004 respectively.

3. Results

In order to extract the differential cross sections, all previously described corrections and cuts have to be applied to the data. Figures 3 shows examples of the ϕ_π^*-dependent differential cross sections in $W = 1.11$ GeV bin near pion threshold. We compared the cross section with various physics models such as Dubna-Mainz-Taipei(DMT) [9], Sato-Lee04 [10], and MAID03 [11]. Each data point in the plots contains only statistical error, and the shaded bars below show the systematic uncertainties. The systematic uncertainties studies were done for various sources. The averaged total systematic error is around 7%. The major source of systematic error come from particle ID for electrons ($\sim$ 4%), pions ($\sim$ 3%) and accpetnace with different physics event generators ($\sim$ 4%). The radiaitve correction shows small systematic uncertainty for two different models (dynamic model and unitary model), which is about $\sim 0.5\%$.

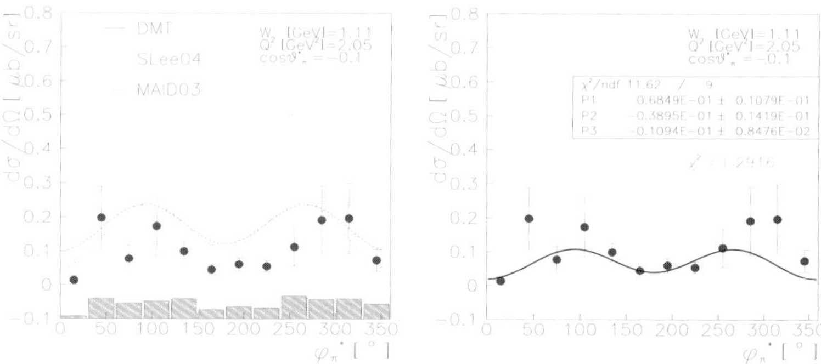

Fig. 3. An example of differential cross sections as function of ϕ_π^* at $W = 1.11$ GeV, $Q^2 = 2.05$ GeV^2, and cos $\theta_\pi^* = -0.1$. The error bars of the data include only statistical errors and the shaded bars show the systematic uncertainties. Solid, dashed and dash-dotted lines are the DMT, MAID2003 and Sato-Lee2004 model predictions respectively.

The fit of ϕ_π^*-dependent cross sections allows us to obtain the structure functions. The fit function for $\frac{d\sigma}{d\Omega_\pi}$ has three fit parameters, $\frac{d\sigma}{d\Omega_\pi} = A + C\cos 2\phi_\pi^* + D\cos\phi_\pi^*$ which corresponds to structure functions $A = \sigma_{T+L}$, $D = \sigma_{LT}$ and $C = \sigma_{TT}$ respectively. Once we obtain the Legendre moments from fitting the structure functions, the observables have only W and Q^2 degrees of freedom. These observables can be directly used for extracting the generalize form factors depending on parametrization of the electric form factor of neutron (G_E^n) and other kinematic constants. Figure 4 shows the preliminary results of the E_{0+} multipole divided by the dipole form factor (G_D) as function of Q^2. Three plots in Figure 4 show the comparison of E_{0+} multipoles extracted by using different electric form factor parametrization for three W bins. The calculations from MAID2007 [12] and LCSR are also presented. The experimental results for E_{0+}/G_D are about $0.2 - 0.3$ and almost flat as function of Q^2. The amplitude is bigger than predicted by MAID and similar (or a bit smaller) than LCSR calculation, although LCSR has a steeper Q^2 dependence, which may be caused by present LCSR calculations for the form factors in the chiral limit $m_\pi \to 0$. The plots show also the comparison of multipole extraction with two different G_E^n form factor parametrizations [13] (open circles) and [14] (full circles) under the assumption of vanishing pion mass.

We also extracted multipole using the fitting method. In general, multipoles are complex numbers and we have seven multipoles (E_{0+}, E_{1+}, M_{1+},

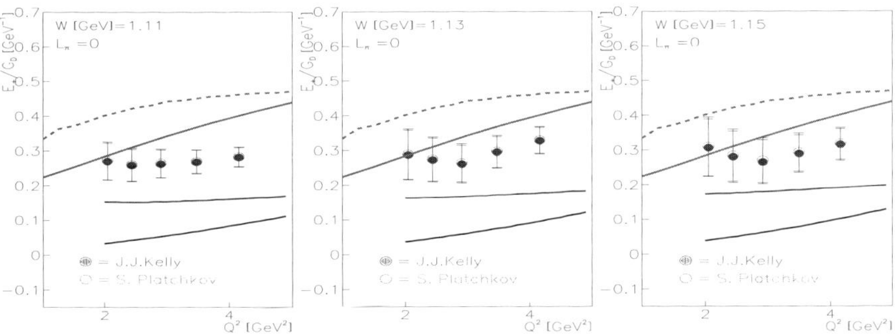

Fig. 4. Preliminary $n\pi^+$ results of the E_{0+} multipole divided by dipole form factor (G_D) as function of Q^2. Three plots show E_{0+}/G_D multipole by ignoring the pion mass with different neutron electric form factor parametrization, open circles are from [13] and full circles are from [14]; middle solid line: MAID2007 for E_{0+}/G_D, lower solid line: MAID2007 for L_{0+}/G_D. Upper solid line: LCSR calculation using experimental electromagnetic form factors as input, dashed line is pure LCSR.

M_{1-}, S_{0+}, S_{1+}, and S_{1-}) near threshold with $l \leq 1$ and hence 14 quantities should be determined [15]. Since we are interested in threshold behavior of multipoles E_{0+} and S_{0+}. However, because the measurements can not be done precisely at the threshold, the multipoles M_{1-},S_{1-},M_{1+},E_{1+}, and S_{1+} also contribute. The multipoles $M_{1+}(3/2)$, $E_{1+}(3/2)$, $S_{1+}(3/2)$ have large resonance contributions, which are determined very well from the data at the resonance region. They should be kept at the values found at the resonance positions. These are values in the convenient analytical form, i.e. in the form of parametrization over Q^2. W-dependence of the whole $M_{1+}(3/2)$ should be taken the same as W-dependence of the SAID $M_{1+}(3/2)$ multipole. W-dependence of $E_{1+}(3/2)$ and $S_{1+}(3/2)$ is more complicated. However, these are small multipoles. Other multipoles M_{1-}, S_{1-} and $M_{1+}(1/2)$, $E_{1+}(1/2)$, $S_{1+}(1/2)$ should be fitted along with E_{0+},S_{0+}. In addition, we use the some assumptions [15]. Using multipole relations and amplitudes, we are able to fit the measure cross section to extract the E_{0+} multipole. Figure 5 shows the preliminary results of E_{0+} multipole extraction in terms of Q^2 between 2.0 GeV^2 and 4.5 GeV^2 using the multipole fit method for near the pion threshold three W bins. The averaged χ^2 of fit is 1.03 over all Q^2.

4. Summary

We extracted the E_{0+} multipole near pion threshold ($W = 1.11 - 1.15$ GeV) at high $Q^2 = 2.05 - 4.16$ GeV^2 with different G_E^n form factor parametriza-

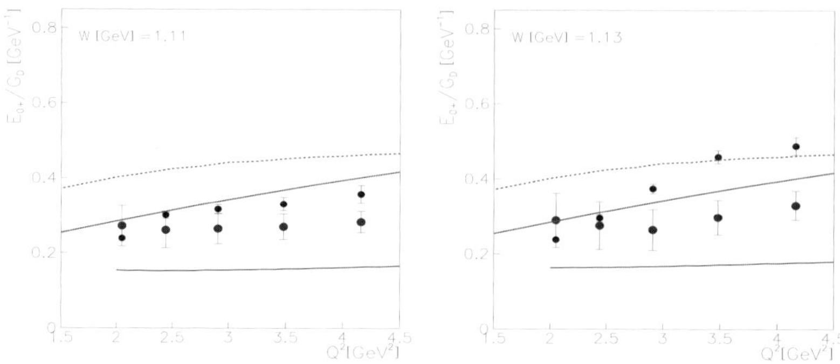

Fig. 5. Q^2 dependence of the E_{0+} multipole normalized by the dipole form factor from the multipole fit. The lower circles show the E_{0+}/G_D extraction based on LCSR without taking the pion mass into account. The upper circles are the results of the multipole method. Curves as in Figure 4.

tions. The preliminary results for vanishing pion mass show $E_{0+}/G_D \sim 0.3$ and almost flat as function of Q^2. This amplitude is bigger than the MAID prediction and similar with LCSR predictions. From this analysis, the $n\pi^+$ channel shows dominance of transverse S-wave multipole E_{0+} along with independence of pion mass considering and G_E^n form factor parametrization. The LCSR calculations predict ϕ^*-independent differential cross sections caused by a complete cancellation of the contribution to σ_{LT} and σ_{TT} from G_1 and G_2 at particular Q^2. This analysis shows the θ^*_π dependence for the longitudinal-transverse, σ_{LT} interference term, σ_{LT} is especially strong for large angle and relatively small Q^2. However, this is still quite consistent result with LCSR prediction if we take the highest Q^2 bin into account because the ϕ-independence is only expected for large Q^2 in LCSR [1]. The experimental data (σ_{LT}) shows small deviation from zero over all cos θ^*_π for all three W bins at $Q^2 = 4.16$ GeV^2, which is possibly caused by the mentioned cancellation of the G_1 and G_2 contributions.

References

1. V.M. Braun *et al.*, *Phys. Rev.* *D*77, 034016(2008).
2. V.M. Braun *et al.*, *Phys. Rev.* *D*75, 014021(2007).
3. E.J. Brash *et al.*, *Phys. Rev.* *C* **65**, 051001 (2002).
4. O. Gayou *et al.*, *Phys. Rev. Lett.* **88**, 092301 (2002).
5. P. E. Bosted, *Phys. Rev.* *C* **51**, 409 (1995).
6. K. Park and R. W. Gothe, CLAS-Analysis Note 2010-101 (2010).
7. K. Park *et al.*, *Phys. Rev.* *C* **77**, 015208 (2008).
8. A. Afanasev *et al.*, *Phys. Rev.* *D* **66**, 074004, (2002).
9. S. S. Kamalov and S. N. Yang, *Phys. Rev. Lett.* **83**, 4494 (1999).

10. T.-S. Lee, Private communication.
11. D. Drechsel *et al.*, *Nucl. Phys.* **A645**,145-174,(1999)
12. D. Drechsel, S.S. Kamalov, L. Tiator, *Eur. Phys. J.***A34**, 69,(2007)
13. S. Platchkov *et al.*, *Nucl. Phys. A***510**, 740 (1990).
14. J. J. Kelly *et al.*, *Phys. Rev. C* **70**, 068202 (2004).
15. I. G. Aznauryan, *Phys. Rev. D* **57**, 2727 (1998).

NEAR THRESHOLD NEUTRAL PION ELECTROPRODUCTION AT HIGH MOMENTUM TRANSFERS

P. KHETARPAL

Physics Department, Rensselaer Polytechnic Institute,
Troy, NY 12180, USA
**E-mail: khetap@rpi.edu*
www.rpi.edu

Near threshold pion electroproduction as a function of Q^2 can be written in terms of new form factors. These new generalized form factors $G_1^{\pi N}$ and $G_2^{\pi N}$ have been predicted using light cone sum rules in the chiral limit for the reactions $ep \to eN\pi$ in the Q^2 1-10 GeV^2 range. An experiment at Jefferson Lab has been conducted using the CLAS spectrometer to measure near threshold neutral pion electroproduction as a function of Q^2. Preliminary cross sections, S-wave multipoles and generalized form factors are presented.

Keywords: Threshold; Neutral Pion; Generalized Form Factors

1. Introduction

Pion photo- and electroproduction $\gamma N \to \pi N$, $\gamma^* N \to \pi N$ close to threshold has been studied extensively since the 1950s both experimentally and theoretically. Exact predictions for the threshold cross sections and the axial form factor were pioneered by Kroll and Ruderman in 1954 for photoproduction[1] known as the low energy theorem (LET). The predictions of this LET were restricted only to charged pions and the π^0 contribution was shown to vanish in the 'soft pion' limit ($m_\pi \sim p_\pi$). In later years, using vanishing pion mass chiral symmetry ($m_\pi \to 0$) these predictions were extended to pion electroproduction for both charged and neutral pions.[2,3]

These soft-pion theorems or LETs[1–5] are not applicable for $Q^2 \gg \Lambda^3_{QCD}/m_\pi$, where $\Lambda_{QCD} \sim 200 - 300$ MeV is a typical hadronic scale. In the case of asymptotically large momentum transfers ($Q^2 \to \infty$) perturbative QCD (pQCD) factorization techniques[6–8] have been used to obtain predictions for cross section amplitudes and axial form factors near threshold.

Recently, Braun *et al*[9,10] suggest a method to extract new baryon form factors, $G_1^{\pi N}(Q^2)$ and $G_2^{\pi N}(Q^2)$, for $Q^2 \sim 1 - 10\,\mathrm{GeV}^2$ using light cone sum rules (LCSR). The transition matrix elements of the electromagnetic interaction, J_μ, can be written in terms of these form factors at threshold:

$$\langle N(P')\pi(k)|J_\mu|p(P)\rangle = \\ -\frac{i}{f_\pi}\bar{N}\gamma_5\left[(\gamma_\mu q^2 - q_\mu \not{q})\frac{G_1^{\pi N}(Q^2)}{m_N^2} - \frac{i\sigma_{\mu\nu}q^\nu}{2m_N}G_2^{\pi N}(Q^2)\right]p. \quad (1)$$

Here, $N(P')$ and $p(P)$ are spinors for the final and initial nucleons with momenta P' and P, respectively, m_N is the mass of the nucleon, f_π is the pion decay constant and $q \equiv P - P'$. Since the pion is a negative parity object and the electromagnetic current is parity conserving, the γ_5 object is present to conserve the overall parity of the reaction. These form factors are directly related to the S-wave multipoles E_{0+} and L_{0+}. As one moves away from threshold, the P-wave starts to contribute to the transition amplitude in addition to the S-wave multipoles, which can be given in terms of the elastic electromagnetic form factors, $F_1(Q^2)$ and $F_2(Q^2)$.

Additionally, the sum rules can be extended to the $Q^2 < 1\,\mathrm{GeV}^2$ regime and the LET are recovered to $O(m_\pi)$ accuracy. This approach provides a bridge between the low and high Q^2 regimes. Predictions for the axial form factor and the generalized form factors are also obtained in this approach.

An experiment at Jefferson Lab has been conducted to measure the cross section amplitudes for the neutral pion electroproduction process, $ep \rightarrow ep\pi^0$ using Hall-B's CLAS[a] detector. From these cross sections, the generalized form factors $G_1^{\pi^0 p}(Q^2)$ and $G_2^{\pi^0 p}(Q^2)$ are obtained.

2. Kinematic Definitions and Notations

The neutral pion reaction

$$e(l) + p(P) \rightarrow e(l') + p(P') + \pi^0(k') \quad (2)$$

is shown schematically in the virtual photon-proton center of mass frame in Fig. 1. Here, $l = (E_e, E_e\hat{\mathbf{z}})$, $l' = (E'_e, \mathbf{p'_e})$, $P = (M_p, \mathbf{0})$ and $P' = (E'_p, \mathbf{p'_p})$ are the initial and final electron and proton 4-momenta of the reaction in the lab frame. Also, $k = (E_\pi, \mathbf{p}_\pi)$ is the 4 momentum of the emitted pion. In this experiment, it is assumed that the incident electron interacts with the target proton via exchange of a single virtual photon with 4-momentum $q_\mu = l_\mu - l'_\mu = (\omega, \mathbf{q})$. In this approximation, we also assume that the

[a]Continuous electron beam accelerator facility's Large Acceptance Calorimeter

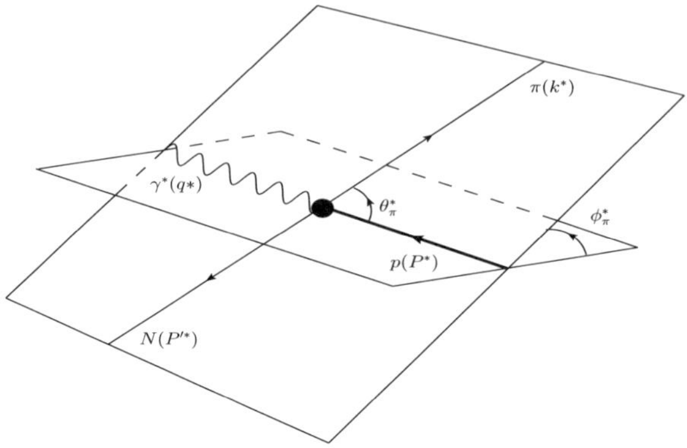

Fig. 1. Neutral pion electroproduction in the center of mass frame.

electron mass is negligible ($m_e \approx 0$) compared to the overall kinematics of the reaction. The two important kinematic invariants of interest are

$$Q^2 \equiv -q^2 = -\omega^2 + |\mathbf{q}|^2 = 4E_e E'_e \sin^2(\theta'_e/2)$$
$$s = W^2 = (q+p)^2 = M_p^2 + 2\omega M_p - Q^2. \tag{3}$$

Here, M_p is the mass of the proton and θ'_e is the polar angle of the scattered electron in the lab frame.

The transition amplitude for the reaction can be separated in terms of the leptonic $L_{\mu\nu}$ and hadronic $M_{\mu\nu}$ currents in the one photon exchange approximation:

$$\mathcal{M} = eL_{\mu\nu}M^{\mu\nu}. \tag{4}$$

The five-fold differential cross section for the reaction can be written in terms of the cross section for the subprocess $\gamma^* p \to p\pi^0$,[11] which depends only on the matrix elements of the hadronic interaction:

$$\frac{d^5\sigma}{dE'_e d\Omega'_e d\Omega^*_\pi} = \Gamma \frac{d^2\sigma_{\gamma^* p \to p\pi^0}}{d\Omega^*_\pi}. \tag{5}$$

Here, $d\Omega'_e = d\cos\theta'_e d\phi'_e$ is the differential solid angle for the scattered electron in the lab frame; $d\Omega^*_\pi = d\cos\theta^*_\pi d\phi^*_\pi$ is the differential solid angle for the pion in the virtual photon-proton ($\gamma^* p$) center of mass frame. On the other side

$$\Gamma = \frac{\alpha}{2\pi^2}\frac{E'_e}{E_e}\frac{W^2 - M_p^2}{2M_p Q^2}\frac{1}{1-\varepsilon}, \tag{6}$$

which is the *virtual photon flux* and depends entirely on the matrix elements of the leptonic interaction and contains the *transverse polarization* of the virtual photon

$$\varepsilon = \left(1 + 2\frac{|\mathbf{q}|^2}{Q^2}\tan^2\frac{\theta_e}{2}\right)^{-1}. \tag{7}$$

Similarly, the *longitudinal polarization* of the virtual photon can be defined as $\varepsilon_L = \frac{Q^2}{\omega^2}\varepsilon$.

For unpolarized beam and target the reduced cross section from Eq. 5 can be expanded in terms of the hadronic structure functions:

$$\frac{d\sigma_{\gamma^* p\to p\pi^0}}{d\Omega_\pi^*} = \frac{2p_\pi^* W}{W^2 - M_p^2}\left(\frac{d\sigma_T}{d\Omega_\pi^*} + \varepsilon_L\frac{d\sigma_L}{d\Omega_\pi^*} + \varepsilon\frac{d\sigma_{TT}}{d\Omega_\pi^*}\cos 2\phi_\pi^* + \sqrt{2\varepsilon_L(\varepsilon+1)}\frac{d\sigma_{LT}}{d\Omega_\pi^*}\cos\phi_\pi^*\right). \tag{8}$$

Here, p_π^* is the pion momentum in the CM frame and $\sigma_T + \varepsilon_L\sigma_L$, σ_{LT} and σ_{TT} are the structure functions that describe the transverse, longitudinal, longitudinal-transverse interference, and transverse-transverse interference components of the virtual photon-proton interaction differential cross section.

Each of these structure functions contain the $\cos\theta_\pi^*$ dependence and can be parameterized in terms of the CGLN[12] multipole amplitudes that are a set of complex coefficients defined by derivatives of Legendre polynomials. These amplitudes $E_{l\pm}$, $M_{l\pm}$ and $S_{l\pm}$ describe the electric, magnetic and scalar multipoles respectively, where l describes the total orbital angular momentum of the pion relative to the nucleon and $\pm$ describe the spin $\pm\frac{1}{2}$ of the nucleon in the final state.

3. Experimental Overview

The near threshold reaction was studied using the CLAS detector that is housed in Jefferson Lab's Hall-B. The description of the CLAS detector system is outside the scope of this article and is discussed elsewhere.[13]

There are several difficulties in the analysis of this experiment in the near threshold region. In this region, the pion electroproduction cross section goes to zero; so, the statistics are very low. Also, a major source of contamination to the neutral pion signal near threshold are the elastic Bethe-Heitler events. This contamination is suppressed by subtracting these events from the pion signal. Additionally, since the π^0 comes out approximately in the same direction as the γ^* in the lab frame, the resolution of the ϕ_π^*

is poor. This resolution worsens as one approaches threshold. The analysis for the experiment is done carefully to maximize the yield, to minimize the elastic radiative contamination, and to minimize resolution effects of the kinematic variables by using reasonable binning.

The scattered electrons in the final state of the reaction are detected by requiring a geometrical coincidence between the Cerenkov Counters and the Electromagnetic Calorimeter of the CLAS detector. The recoiled protons are detected by looking at the timing information obtained from the Scintillator Counters. The neutral pion in the final state is selected using conservation of 4-momentum and missing mass squared, $M_X^2 = (q + P - P')^2$. The pions are selected by requiring the M_X^2 to be around the expected PDG[14] value of the mass of the neutral pion, *i.e.*, 0.018 GeV^2.

4. Differential Cross Sections

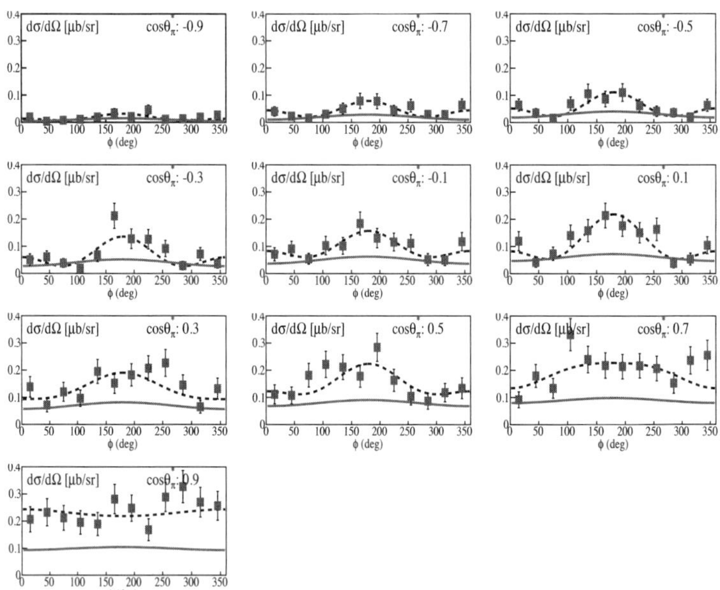

Fig. 2. The preliminary differential cross sections in μb/sr for $W = 1.11$ GeV and $Q^2 = 2.75\,\text{GeV}^2$. Each subplot shows cross section for a $\cos\theta_\pi^*$ bin as a function of ϕ_π^*. Experimental points (red) are shown with statistical errors only. The black dashed curve is a preliminary fit to the cross section as in Eq. 9. Predictions from Braun *et al* are shown in magenta.

The kinematic coverage of the experiment spans over W of 1.08 to 1.16 GeV and Q^2 of ~ 2 to $5\,\mathrm{GeV}^2$. The cross sections are reported over the entire solid angle of the neutral pion in the CM frame.

The reduced differential cross section for the reaction is computed for each kinematic bin. Fig. 2 shows the differential cross section for the kinematic bin $W = 1.11$ GeV and $Q^2 = 2.75\,\mathrm{GeV}^2$. The results from experiment are shown as red points with statistical errors. One can observe that the cross section is more-or-less symmetric in ϕ_π^* but is reduced for backward angles in θ_π^*. Some bins show large errors because of small statistical yield from experiment.

The predictions from Braun *et al*[10] are shown as the magenta curves in the figure. One can observe a large discrepancy with the predictions of Braun near $\phi_\pi^* \sim 180°$. This is because the predictions do not include enough P-wave contribution to the cross section. Using Eq. 8, the differential cross section for each kinematic bin can be fitted with the following parameterization:

$$\frac{d\sigma}{d\Omega_\pi^*} = \frac{p_\pi^*}{k_\gamma^*}\left(A + \varepsilon B \cos 2\phi_\pi^* + \sqrt{2\varepsilon_L(\varepsilon+1)}C\cos\phi_\pi^*\right). \tag{9}$$

Here, A, B and C are the parameters of the fit and are equivalent to $\sigma_T + \varepsilon_L\sigma_L$, σ_{TT} and σ_{LT}, respectively. The result of the fit is shown as the black dashed curves in Fig. 2.

5. S-Wave Multipoles and Generalized Form Factors

The parameters of the fit for the differential cross sections (Eq. 9) are functions of $\cos\theta_\pi^*$ and can be expressed in terms of Legendre polynomials:

$$\begin{aligned}
A(\cos\theta_\pi^*) &= A_0 + A_1 P_1(\cos\theta_\pi^*) + A_2 P_2(\cos\theta_\pi^*)\\
B(\cos\theta_\pi^*) &= B_0 \sin^2\theta_\pi^*\\
C(\cos\theta_\pi^*) &= (C_0 + C_1\cos\theta_\pi^*)\sin\theta_\pi^*
\end{aligned} \tag{10}$$

The polynomials are expanded up to the second order to include terms only from $l = 0$ and $l = 1$ multipoles since these are the only multipoles expected to contribute in the near threshold region. The coefficients of the polynomials from Eq. 10 can be associated with these multipoles.

We are interested in extracting the E_{0+} and S_{0+} ($l = 0$) multipoles. The S-wave multipoles were extracted by I. Aznauryan.[15] Fig. 3(a) and Fig. 3(b) show the extracted S-wave multipoles E_{0+} and S_{0+} as a function of Q^2 using three different methods. These multipoles have been normalized to the dipole formula $G_D(Q^2) = \left(1 + \frac{Q^2}{.71}\right)^{-2}$.

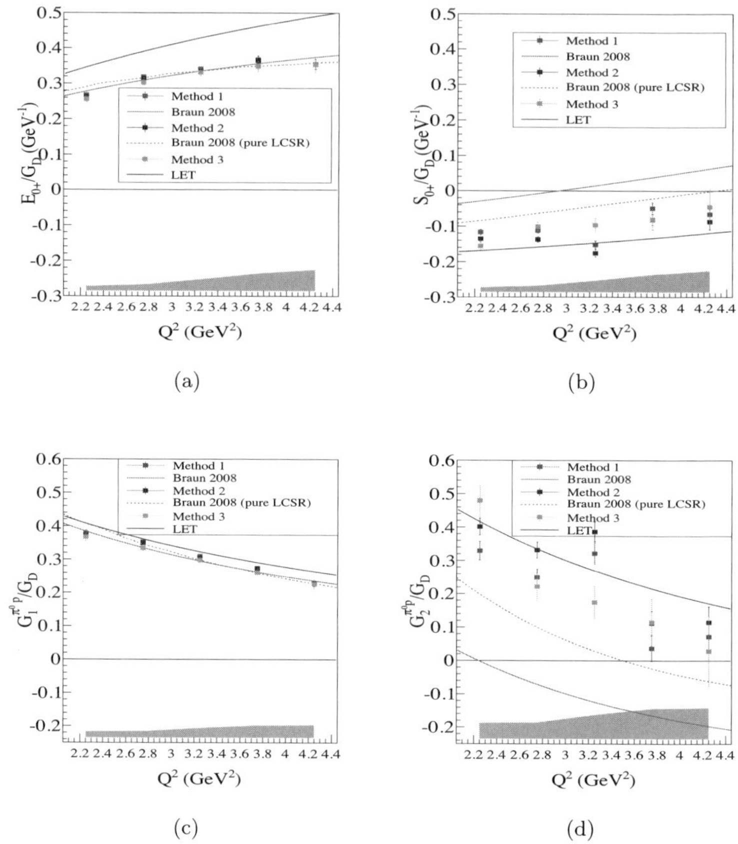

Fig. 3. The preliminary S-wave multipoles (a) E_{0+} and (b) S_{0+} and generalized form factors (c) $G_1^{\pi^0 p}$ and (d) $G_2^{\pi^0 p}$ normalized to the dipole formula G_D are plotted as a function of Q^2. The multipoles are extracted using three methods (red, blue and green points) as indicated in the text. The form factors are calculated from these multipoles. The error bars include systematic and statistical uncertainties added in quadrature. The systematic uncertainties are shown separately as a gray band at the bottom. The LCSR based model predictions from Braun *et al* are shown (magenta-solid) along with estimated uncertainties (magenta-dashed) in calculations. The LET predictions are also shown (blue).

The three different fit methods include the following prescriptions. Only real parts of E_{0+}, S_{0+}, M_{1-}, S_{1-}, E_{1+} and S_{1+} are involved in the fitting. The first method (red points) includes a fit at all W ranges ($W = 1.08-1.16$ GeV of the current experiment) but M_{1-}, S_{1-}, M_{1+}, E_{1+}, and S_{1+} were fitted to world data at higher W. The second method (blue points) includes a fit over all the W range as in the first method and M_{1-}, M_{1+} and S_{1+} were similarly obtained by fit to higher W data. However, the ranges of the amplitudes S_{1-} and E_{1+} are increased by a factor of two to see their effect on the current results. The third method (green points) is the same as the second one except that the fit is made only for $W = 1.09$ and 1.11 GeV, *i.e.*, very close to threshold. All three fit methods show consistent results.

Fig. 3(c) and (d) show the obtained form factors, G_1 and G_2, as a function of Q^2. The red, blue and green data points correspond to the three methods of fitting the multipoles from which these form factors are extracted.

The error bars on the points include systematic and statistical uncertainties added in quadrature. The systematic uncertainties are shown separately at the bottom of the plots as a gray band, which assume all systematic errors for all data points are entirely correlated. So, this gives an upper limit on the overall systematic uncertainty. The plots also show predictions from Braun *et al* as magenta curves. Braun *et al* have tried to minimize the uncertainties in their LCSR based model calculations by including electromagnetic form factor values known from experiment for the P-wave. The predictions from these calculations are shown as solid curves in the figure. The "pure" LCSR based models where the calculation uncertainties have not been minimized and the values for all form factors are obtained entirely from theoretical calculations are shown as dashed curves in the figure. One can observe that the experimental results for E_{0+} are consistent with the predictions of Braun *et al* but there is discrepancy with the predictions for S_{0+}.

Additionally, the LET predictions in the chiral limit are also shown for comparison in blue where existing parameterizations for the elastic form factors[16,17] have been used.

One can observe that the experimental results for G_1 are consistent with the predictions from Braun *et al*, but the results for G_2 show some discrepancy. In particular, one can observe a faster fall off than the dipole form for both G_1 and G_2. However, the zero crossing of G_2 is not observed as predicted. This faster than dipole form falling of the generalized form factors suggests a broadening of the spatial distribution of the correlated pion-nucleon system. It suggests that the correlated pion-nucleon system is broader than the nucleon itself because the nucleon follows the behavior of the dipole form factor.

6. Acknowledgements

The author acknowledges the CLAS collaboration, the Jefferson Lab technical and accelerator staff for making the experiment possible. This work is supported in part by the U.S. National Science Foundation and the U.S. Department of Energy.

References

1. N. M. Kroll and M. A. Ruderman, *Phys. Rev.* **93**, p. 233 (1954).
2. Y. Nambu and D. Lurié, *Phys. Rev.* **125**, p. 1429 (1962).
3. Y. Nambu and E. Shrauner, *Phys. Rev.* **128**, p. 862 (1962).
4. A. I. Vainshtein and V. I. Zakharov, *Nucl. Phys. B* **36**, p. 589 (1972).
5. S. Scherer and J. H. Koch, *Nucl. Phys. A* **534**, p. 461 (1991).
6. P. V. Pobylitsa, M. V. Polyakov and M. Strikman, *Phys. Rev. Lett.* **87**, p. 022001(Jun 2001).
7. A. V. Efremov and A. V. Radyushkin, *Phys. Lett. B* **94**, 245(Jul 1980).
8. G. P. Lepage and S. J. Brodsky, *Phys. Rev. D* **22**, 2157(Nov 1980).
9. V. M. Braun, D. Y. Ivanov, A. Lenz and A. Peters, *Phys. Rev. D* **75**, p. 014021 (2007).
10. V. M. Braun, D. Y. Ivanov and A. Peters, *Phys. Rev. D* **77**, p. 034016 (2008).
11. E. Amaldi, S. Fubini and G. Furlan, *Pion-Electroproduction* (Springer-Verlag, New York, 1979).
12. G. F. Chew, M. L. Goldberger, F. E. Low and Y. Nambu, *Phys. Rev.* **106**, 1345(Jun 1957).
13. B. A. Mecking *et al.*, *Nucl. Instrum. Meth.* **A503**, 513 (2003).
14. C. Amsler *et al.*, *Phys. Lett.* **B667**, p. 1 ((2008) and 2009 partial update for the 2010 edition).
15. I. G. Aznauryan *et al.*, *Phys. Rev. C* **80**, p. 055203(Nov 2009).
16. E. J. Brash, A. Kozlov, S. Li and G. M. Huber, *Phys. Rev. C* **65**, p. 051001(Apr 2002).
17. O. Gayou *et al.*, *Phys. Rev. Lett.* **88**, p. 092301(Feb 2002).

THE ADS/QCD CORRESPONDENCE AND EXCLUSIVE PROCESSES

STANLEY J. BRODSKY*

Stanford Linear Accelerator Center
Stanford University, Stanford, California 94309
and CP^3-Origins, Southern Denmark University
Odense, Denmark
**E-mail: sjbth@slac.stanford.edu*

GUY F. DE TÉRAMOND*

Universidad de Costa Rica
San José, Costa Rica
**E-mail: gdt@asterix.crnet.cr*

ALEXANDRE DEUR*

Thomas Jefferson National Accelerator Facility
Newport News, VA 23606
**E-mail:deurpam@jlab.org*

The AdS/CFT correspondence between theories in AdS space and conformal field theories in physical space-time provides an analytic, semi-classical, color-confining model for strongly-coupled QCD. The soft-wall AdS/QCD model, modified by a positive-sign dilaton metric, leads to a remarkable one-parameter description of nonperturbative hadron dynamics at zero quark mass, including a zero-mass pion and meson and baryon Regge spectra of linear trajectories with the same slope in orbital angular momentum L and radial quantum number n. One also predicts the form of the non-perturbative effective coupling $\alpha_s^{AdS}(Q)$ and its β-function which agrees with the effective coupling α_{g_1} extracted from the Bjorken sum rule. Light-front holography, which connects the fifth-dimensional coordinate of AdS space z to an invariant impact separation variable ζ, allows one to compute the analytic form of the frame-independent light-front wavefunctions, the fundamental entities which encode hadron properties as well as decay constants, form factors, deeply-virtual Compton scattering, exclusive heavy hadron decays, and other exclusive scattering amplitudes. One thus obtains a relativistic description of hadrons in QCD at the amplitude level with dimensional counting for exclusive reactions at high momentum transfer. As specific examples, we discuss the behavior of the pion and nucleon form factors in the space-like and time-like regions. We also review the phenomenology of exclusive processes including some anomalous empirical results.

1. Introduction

Exclusive processes play a key role in quantum chromodynamics, testing the primary quark and gluon interactions of QCD and the structure of hadrons at the amplitude level. Two basic pictures have emerged based on perturbative QCD (pQCD) and nonperturbative AdS/QCD. In pQCD hard scattering amplitudes at a high scale $Q^2 >> \Lambda^2_{\rm QCD}$ factorize as a convolution of gauge-invariant hadron distribution amplitudes $\phi_H(x_i, Q)$ with the underlying hard scattering quark-gluon subprocess amplitude T_H. The leading power fall-off of the hard scattering amplitude follows from the conformal scaling of the underlying hard-scattering amplitude: $T_H \sim 1/Q^{n-4}$, where n is the total number of fields (quarks, leptons, or gauge fields) participating in the hard scattering.[1,2] Thus the reaction is dominated by subprocesses and Fock states involving the minimum number of interacting fields. In the case of $2 \to 2$ scattering processes, this implies the dimensional counting rules $d\sigma/dt(AB \to CD) = F_{AB\to CD}(t/s)/s^{n-2}$, where $n = N_A + N_B + N_C + N_D$ and N_H is the minimum number of constituents of H. The result is modified by the ERBL evolution[3,4] of the distribution amplitudes and the running of the QCD coupling.

It is striking that the dimensional counting rules are also a key feature of nonperturbative AdS/QCD models.[5] Although the mechanisms are different, both the pQCD and AdS/QCD approaches depend on the leading twist interpolating operators of the hadron and their structure at short distances. In both theories, hadronic form factors at high Q^2 are dominated by the wavefunctions at small impact separation. This in turn leads to the color transparency phenomena.[6,7] For example, measurements of pion photoproduction are consistent with dimensional counting $s^7 d\sigma/dt(\gamma p \to \pi^+ n) \sim$ constant at fixed CM angle for $s > 7$ GeV. The angular distributions seen in hard large CM angle scattering reactions are consistent with quark interchange,[8] a result also predicted by the hard wall AdS/QCD model. Reviews are given in Refs. [9] and [10]. One sees the onset of perturbative QCD scaling behavior even for exclusive nuclear amplitudes such as deuteron photodisintegration (Here $n = 1 + 6 + 3 + 3 = 13$) and $s^{11} d\sigma/dt(\gamma d \to pn) \sim$ constant at fixed CM angle.[11–13] The measured deuteron form factor[14] also appears to follow the leading-twist QCD predictions[15] at large momentum transfers in the few GeV region. The six color-triplet quarks of the valence Fock state of the deuteron can be arranged as a sum of five different color-singlet states, only one of which can be identified with the neutron-proton state and can account for the large magnitude of the deuteron form factor at high scales. A measurement of $d\sigma/dt(\gamma d \to \Delta^{++}\Delta)$ in the scaling region can establish the role of "hidden-color" degrees of freedom[16] of the nuclear wavefunction in hard deuteron reactions.

In the case of pQCD, the near-constancy of the effective QCD coupling at small scales helps explain the general empirical success of the dimensional counting rules for the near-conformal power law fall-off of form factors and fixed-angle scaling.[17]

Color transparency[6,7] is a key property of color gauge theory, and it thus stands at the foundations of QCD. Color transparency has been confirmed in diffractive dijet production,[18] pion photoproduction[19] and vector meson electroproduction,[20] but it is very important to also systematically validate it in large-angle hadron scattering processes. Color transparency and higher-twist subprocesses [21–25] where the trigger hadron is produced directly, such as $uu \to p\bar{d}$, can account for the anomalous growth of the baryon-to-meson ratio with increasing centrality observed in heavy ion collisions at RHIC.[26]

2. Anomalies in Exclusive Processes

Some exceptions to the general success of dimensional counting are known:

The transition form factor $F(Q^2)_{\gamma\to\pi^0}$ between a real photon and a pion has been recently measured at BaBar to high $Q^2 \simeq 10$ GeV^2, falling at high photon virtuality roughly as $1/Q$ rather than the predicted $1/Q^2$ fall-off. In contrast, preliminary measurements from BaBar[27] indicate that the transition form factors $F(Q^2)_{\gamma\to\eta}$ and $F(Q^2)_{\gamma\to\eta'}$ are consistent with the pQCD expectations. The photon to meson transition form factor is the simplest QCD hadronic exclusive amplitude, and thus it is critical to understand this discrepancy. As we shall discuss below, AdS/QCD predicts a broad distribution amplitude $\phi_\pi(x,Q)$ in the nonperturbative domain, but since ERBL evolution leads to a narrower distribution in the high Q domain, it cannot explain the BaBar anomaly. It is hard to imagine that the pion distribution amplitude is very flat,[28–30] since this corresponds to a pointlike non-composite hadron. It is crucial to measure $\frac{d\sigma}{dt}(\gamma\gamma \to \pi^0\pi^0)$ since the CM angular distribution is very sensitive to the shape of $\phi_\pi(x,Q)$.[31]

The Hall A collaboration[32] at JLab has reported another significant exception to the general empirical success of dimensional counting in fixed-CM-angle Compton scattering $d\sigma/dt(\gamma p \to \gamma p) \sim F(\theta_{CM})/s^8$ instead of the predicted $1/s^6$ scaling. The deviations from fixed-angle conformal scaling may be due to corrections from resonance contributions in the JLab energy range. It is interesting that the hadron form factor $R_V(t)$,[33] which multiplies the $\gamma q \to \gamma q$ amplitude is found by Hall A to scale as $1/t^2$, in agreement with the pQCD and AdS/QCD prediction. In addition, the Belle measurement[34] of the timelike two-photon cross section $d\sigma/dt(\gamma\gamma \to p\bar{p})$ is consistent with $1/s^6$ scaling.

Although large-angle proton-proton elastic scattering is well described by dimensional scaling $s^{10} d\sigma/dt(pp \to pp) \sim$ constant at fixed CM angle, extraordinarily large spin-spin correlations are observed.[35] The ratio of scattering cross sections for spin-parallel and normal to the scattering plane versus spin-antiparallel reaches $R_{NN} \simeq 4$ in large angle $pp \to pp$ at $\sqrt{s} \simeq 5$ GeV; this is a remarkable example of "exclusive transversity". Color transparency is observed at lower energies but it fails[36] at the same energy where R_{NN} becomes large. In fact, these anomalies have a natural explanation[37] as a resonance effect related to the charm threshold in pp scattering. Alternative explanations of the large spin correlation are discussed and reviewed in Ref. [38]. Resonance formation is a natural phenomenon when all constituents are relatively at rest. For example, a resonance effect can occur due to the intermediate state $uuduudc\bar{c}$ at the charm threshold $\sqrt{s} = 5$ GeV in pp collisions. Since the c and $\bar{c}$ have opposite intrinsic parity, the resonance appears in the $L = J = S = 1$ partial wave for $pp \to pp$ which is only allowed for spin-parallel and normal to the scattering plane $A_{NN} = 1$.[37] Resonance formation at the charm threshold also explains the dramatic quenching of color transparency seen in quasielastic pn scattering by the EVA BNL experiment[36] in the same kinematic region. The reason why these effects are apparent in $pp \to pp$ scattering is that the amplitude for the formation of an $uuduudc\bar{c}$ s-channel resonance in the intermediate state is of the same magnitude as the fast-falling background $pp \to pp$ pQCD amplitude from quark interchange at large CM angles: $M(pp \to pp) \sim 1/u^2t^2$. The open charm cross section in pp scattering is predicted by unitarity to be of order of 1 μb at threshold.[37] One also expects similar novel QCD phenomena in large-angle photoproduction $\gamma p \to \pi N$ near the charm threshold, including the breakdown of color transparency and strong spin-spin correlations. These effects can be tested by measurements at the new JLab 12 GeV facility, which would confirm resonance formation in a low partial wave in $\gamma p \to \pi N$ at $\sqrt{s} \simeq 4$ GeV due to attractive forces in the $uud\bar{c}c$ channel.

Another difficulty for the application of pQCD to exclusive processes is the famous $J/\psi \to \rho\pi$ puzzle; the observed unusually large branching ratio for $J/\psi \to \rho\pi$. In contrast, the branching ratio for $\Psi' \to \rho\pi$ is very small. Such decays into pseudoscalar plus vector mesons require light-quark helicity suppression or internal orbital angular momentum and thus should be suppressed by hadron helicity conservation in pQCD. However, the $J/\psi \to \rho\pi$ puzzle can be explained by the presence of intrinsic charm Fock states in the outgoing mesons.[39]

3. Light-Front Quantization and Exclusive Processes

Light-front (LF) quantization is the ideal framework for describing the structure of hadrons in terms of their quark and gluon degrees of freedom. The light-front wavefunctions (LFWFs) of bound states in QCD are relativistic generalizations of the Schrödinger wavefunctions, but they are determined at fixed light-front time $\tau = x^+ = x^0 + x^3$, the time marked by the front of a light wave,[40] rather than at fixed ordinary time t. They play the same role in hadron physics that Schrödinger wavefunctions play in atomic physics. In addition, the simple structure of the LF vacuum provides an unambiguous definition of the partonic content of a hadron in QCD. In the light-front formalism, one sets boundary conditions at fixed τ and then evolves the system using the LF Hamiltonian $P^- = P^0 - P^3 = id/d\tau$. The invariant Hamiltonian $H_{LF} = P_\mu P^\mu = P^+P^- - P_\perp^2$ has eigenvalues $\mathcal{M}^2$ where $\mathcal{M}$ is the physical hadron mass. The Heisenberg equation for QCD on the light-front thus takes the form $H_{LF}|\Psi_H\rangle = \mathcal{M}_H^2|\Psi_H\rangle$, where H_{LF} is determined canonically from the QCD Lagrangian. Its eigenfunctions are the light-front eigenstates which define the frame-independent light-front wavefunctions, and its eigenvalues yield the hadronic spectrum, the bound states as well as the continuum. The projection of the eigensolutions on the free Fock basis gives the n-parton LF wavefunctions $\psi_{n/H} = \langle n|\Psi_H\rangle$ needed for phenomenology. Heisenberg's problem on the light-front can be solved numerically using discretized light-front quantization (DLCQ)[41] by applying anti-periodic boundary conditions in $\sigma = x^0 - x^3$. This method has been used successfully to solve many lower dimension quantum field theories.[42]

The light-front Fock state wavefunctions $\psi_{n/H}(x_i, \vec{k}_\perp, \lambda_i)$ are functions of LF momentum fractions $x_i = \frac{k_i^+}{P^+} = \frac{k_i^0+k_i^3}{P^0+P^3}$ with $\sum_{i=1}^n x_i = 1$, relative transverse momenta satisfying $\sum_{i=1}^n \vec{k}_{\perp i} = 0$, and spin projections λ_i. Remarkably, the LFWFs are independent of the hadron's total momentum $P^+ = P^0 + P^3$, so that once they are known in one frame, they are known in all frames; Wigner transformations and Melosh rotations are not required. The light-front formalism for gauge theories in light-cone gauge is particularly useful in that there are no ghosts and one has a direct physical interpretation of orbital angular momentum. They also allow one to formulate hadronization in inclusive and exclusive reactions at the amplitude level.

A key example of the utility of the light-front is the Drell-Yan West formula[43,44] for the spacelike form factors of electromagnetic currents given as overlaps of initial and final LFWFs. At high momentum where one can iterate the hard scattering kernel, this yields the dimensional counting rules, factorization theorems, and ERBL evolution of the distribution amplitudes. The gauge-invariant distribution amplitudes $\phi_H(x_i, Q)$ defined from the integral over the transverse momenta $\mathbf{k}^2_{\perp i} \leq Q^2$ of the valence (smallest n) Fock state provide a fundamental measure of the hadron at the amplitude level;[3,4] they are the nonperturbative inputs to the factorized form of hard exclusive amplitudes and exclusive heavy hadron decays in pQCD.

Given the light-front wavefunctions $\psi_{n/H}$ one can compute a large range of other hadron observables. For example, the valence and sea quark and gluon distributions which are measured in deep inelastic lepton scattering are defined from the squares of the LFWFs summed over all Fock states n. Exclusive weak transition amplitudes[45] such as $B \to \ell\nu\pi$, and the generalized parton distributions[46] measured in deeply virtual Compton scattering $\gamma^* p \to \gamma p$ are (assuming the "handbag" approximation) overlaps of the initial and final LFWFs with $n = n'$ and $n = n' + 2$. The resulting distributions obey the DGLAP and ERBL evolution equations as a function of the maximal invariant mass, thus providing a physical factorization scheme.[47] In each case, the derived quantities satisfy the appropriate operator product expansions, sum rules, and evolution equations. At large x where the struck quark is far-off shell, DGLAP evolution is quenched,[48] so that the fall-off of the DIS cross sections in Q^2 satisfies Bloom-Gilman inclusive-exclusive duality at fixed W^2.

The simple features of the light-front contrast with the conventional instant form where one quantizes at $t = 0$. For example, calculating a hadronic form factor requires boosting the hadron's wavefunction from the initial to final state, a dynamical problem as difficult as solving QCD itself. Moreover current matrix elements require computing the interaction of the probe with all of connected currents fluctuating in the QCD vacuum. Each contributing diagram is frame-dependent.

A fundamental theorem for gravity can be derived from the equivalence principle: the anomalous gravitomagnetic moment defined from the spin-flip matrix element of the energy-momentum tensor is identically zero $B(0) = 0$.[49] This theorem can be proven in the light-front formalism Fock state by Fock state.[50] The LF vacuum is trivial up to zero modes in the front form, thus eliminating contributions to the cosmological constant from

QED or QCD.[51]

4. AdS/QCD

One of the most significant theoretical advances in recent years has been the application of the AdS/CFT correspondence[52] between string theories defined in 5-dimensional Anti–de Sitter (AdS) space-time and conformal field theories in physical space-time, to study the dynamics of strongly coupled quantum field theories. The essential principle underlying the AdS/CFT approach to conformal gauge theories is the isomorphism of the group of Poincaré and conformal transformations $SO(4,2)$ to the group of isometries of Anti-de Sitter space. The AdS metric is

$$ds^2 = \frac{R^2}{z^2}(\eta_{\mu\nu}dx^\mu dx^\nu - dz^2), \tag{1}$$

which is invariant under scale changes of the coordinate in the fifth dimension $z \to \lambda z$ and $x_\mu \to \lambda x_\mu$. Thus one can match scale transformations of the theory in $3+1$ physical space-time to scale transformations in the fifth dimension z. In the AdS/CFT duality, the amplitude $\Phi(z)$ represents the extension of the hadron into the additional fifth dimension. The behavior of $\Phi(z) \to z^\tau$ at $z \to 0$ matches the twist-dimension τ of the hadron at short distances.

QCD is not conformal but there is in fact much empirical evidence from lattice gauge theory,[53] Dyson Schwinger theory,[54] and empirical effective charges,[55] that the QCD β-function vanishes in the infrared.[56] The QCD infrared fixed point arises since the propagators of the confined quarks and gluons in the loop integrals contributing to the β-function have a maximal wavelength.[57] The decoupling of quantum loops in the infrared is analogous to QED where vacuum polarization corrections to the photon propagator decouple at $Q^2 \to 0$. Since there is a window where the QCD coupling is large and approximately constant, QCD resembles a conformal theory for massless quarks. Thus, even though QCD is not conformally invariant, one can use the mathematical representation of the conformal group in five-dimensional Anti-de Sitter space to construct an analytic first approximation to the theory.

The AdS/QCD correspondence is now providing important insight into the properties of QCD needed to compute exclusive reactions. In particular, the soft-wall AdS/QCD model modified by a positive sign dilaton metric, which represents color confinement, leads to a remarkable one-parameter description of nonperturbative hadron dynamics, including successful predictions for the meson and baryon spectrum for zero quark mass, including a zero-mass pion, a Regge spectrum of linear trajectories with the same slope in orbital angular L and the principal quantum number n, as well as dynamical form factors. The theory predicts dimensional counting for form factors and other fixed CM angle exclusive reactions. Moreover, as we shall review, light-front holography allows one to map the hadronic amplitudes $\phi_H(z)$ determined in the AdS fifth dimension z to the valence LFWFs of each hadron as a function of a covariant impact variable ζ. Moreover, the same techniques provide a prediction for the QCD coupling $\alpha_s(Q^2)$ and its β-function which reflects the dynamics of confinement.

5. AdS/QCD Models

We thus begin with a conformal approximation to QCD to model an effective dual gravity description in AdS space. The five-dimensional AdS_5 geometrical representation of the conformal group represents scale transformations within the conformal window. Confinement can be introduced with a sharp cut-off in the infrared region of AdS space, as in the "hard-wall" model,[5] or, more successfully, using a dilaton background in the fifth dimension to produce a smooth cutoff at large distances as in the "soft-wall" model.[58] We assume a dilaton profile $\exp(+\kappa^2 z^2)$,[59–62] with opposite sign to that of Ref. [58]. The soft-wall AdS/QCD model with a positive-sign dilaton-modified AdS metric,[60] $ds^2 = (R^2/z^2)e^{+\kappa^2 z^2}(\eta_{\mu\nu}dx^\mu dx^\nu - dz^2)$, leads to the harmonic potential[59] $U(z) = \kappa^4 z^2 + 2\kappa^2(L+S-1)$ in the fifth dimension coordinate z. the resulting spectrum reproduces linear Regge trajectories. $\mathcal{M}^2(S,L,n) = 4\kappa^2(n+L+S/2)$, where $\mathcal{M}^2(S,L,n)$ is proportional to the internal spin, orbital angular momentum L and the principal radial quantum number n.

The modified metric induced by the dilaton can be interpreted in AdS space as a gravitational potential for an object of mass m in the fifth dimension: $V(z) = mc^2\sqrt{g_{00}} = mc^2 R\, e^{\pm\kappa^2 z^2/2}/z$. In the case of the negative solution, the potential decreases monotonically, and thus an object in AdS will fall to infinitely large values of z. For the positive solution, the potential is non-monotonic and has an absolute minimum at $z_0 = 1/\kappa$. Furthermore, for large values of z the gravitational potential increases exponentially, confining any object to distances $\langle z \rangle \sim 1/\kappa$.[59] We thus will choose the confining positive sign dilaton solution[59,60] with opposite sign to that of Ref. [58]. This additional warp factor leads to a well-defined scale-dependent effective coupling.

Glazek and Schaden[63] have shown that a harmonic oscillator confining potential naturally arises as an effective potential between heavy quark states when one stochastically eliminates higher gluonic Fock states. Also, Hoyer[64] has argued that the Coulomb and linear potentials are uniquely allowed in the Dirac equation at the classical level. The linear potential becomes a harmonic oscillator potential in the corresponding Klein-Gordon equation.

6. Light-Front Holography

Light-front holography[65–69] connects the equations of motion in AdS space and the Hamiltonian formulation of QCD in physical space-time quantized on the light front at fixed LF time. This correspondence provides a direct connection between the hadronic amplitudes $\Phi(z)$ in AdS space with LF wavefunctions $\phi(\zeta)$ describing the quark and gluon constituent structure of hadrons in physical space-time. In the case of a meson, $\zeta = \sqrt{x(1-x)\mathbf{b}_\perp^2}$ is a Lorentz invariant coordinate which measures the distance between the quark and antiquark; it is analogous to the radial coordinate r in the Schrödinger equation. Here $\vec{b}_\perp$ is the Fourier conjugate of the transverse momentum $\vec{k}_\perp$. In effect ζ represents the off-light-front energy shell and invariant mass dependence of the bound state; it allows the separation of the dynamics of quark and gluon binding from the kinematics of constituent spin and internal orbital angular momentum.[65] Light-front holography thus provides a connection between the description of hadronic modes in AdS space and the Hamiltonian formulation of QCD in physical space-time quantized on the light-front at fixed LF time τ. The resulting equation for the mesonic $q\bar{q}$ bound states at fixed light-front time has the form of a single-variable relativistic Lorentz invariant Schrödinger equation[65]

$$\left(-\frac{d^2}{d\zeta^2} - \frac{1-4L^2}{4\zeta^2} + U(\zeta)\right)\phi(\zeta) = \mathcal{M}^2\phi(\zeta), \tag{2}$$

where the confining potential is $U(\zeta) = \kappa^4\zeta^2 + 2\kappa^2(L+S-1)$ in the soft-wall model with a positive-sign dilaton-modified AdS metric.[59] Its eigenvalues determine the hadronic spectra and its eigenfunctions are related to the light-front wavefunctions of hadrons for general spin and orbital angular momentum. This LF wave equation serves as a semiclassical first approximation to QCD, and it is equivalent to the equations of motion which describe the propagation of spin-J modes in AdS space. The resulting light-front wavefunctions provide a fundamental description of the structure and internal dynamics of hadronic states in terms of their constituent quark and gluons. There is only one parameter, the mass scale $\kappa \sim 1/2$ GeV, which enters the confinement potential. In the case of mesons $S = 0, 1$ is the combined spin of the q and $\overline{q}$ state, L is their relative orbital angular momentum as determined by the hadronic light-front wavefunctions.

The mapping between the LF invariant variable ζ and the fifth-dimension AdS coordinate z was originally obtained by matching the expression for electromagnetic current matrix elements in AdS space with the corresponding expression for the current matrix element, using LF theory in physical space time.[66] It has also been shown that one obtains the identical holographic mapping using the matrix elements of the energy-momentum tensor,[68,70] thus verifying the consistency of the holographic mapping from AdS to physical observables defined on the light front.

7. The Hadron Spectrum and Form Factors in Light-Front AdS/QCD

The meson spectrum predicted by Eq. 2 has a string-theory Regge form $\mathcal{M}^2 = 4\kappa^2(n + L + S/2)$; *i.e.*, the square of the eigenmasses are linear in both the orbital angular momentum L and n, where n counts the number of nodes of the wavefunction in the radial variable ζ. The spectrum also depends on the internal spin S. This is illustrated for the pseudoscalar and vector meson spectra in Fig. 1, where the data are from Ref. [71]. The pion $(S = 0, n = 0, L = 0)$ is massless for zero quark mass, consistent with the chiral invariance of massless quarks in QCD. Thus one can compute the hadron spectrum by simply adding $4\kappa^2$ for a unit change in the radial quantum number, $4\kappa^2$ for a change in one unit in the orbital quantum number L and $2\kappa^2$ for a change of one unit of spin S. Remarkably, the same rule holds for three-quark baryons as we shall show below.

The eigensolutions of Eq. 2 provide the light-front wavefunctions of the valence Fock state of the hadrons $\psi(x, \vec{b}_\perp)$ as illustrated for the pion in Fig. 2 for the soft-wall (a) and hard-wall (b) models. The resulting distribution amplitude has a broad form $\phi_\pi(x) \sim \sqrt{x(1-x)}$ which is compatible with moments determined from lattice gauge theory. One can then immediately compute observables such as hadronic form factors (overlaps of LFWFs), structure functions (squares of LFWFs), as well as the generalized parton distributions and distribution amplitudes which underly hard exclusive reactions. For example, hadronic form factors can be predicted from the overlap of LFWFs in the Drell-Yan West formula. The prediction for the space-like pion form factor is shown in Fig. 2 (c). The pion form factor and the vector meson poles residing in the dressed current in the soft wall model require choosing a value of κ smaller by a factor of $1/\sqrt{2}$ than the canonical value of κ which determines the mass scale of the hadronic spectra. This

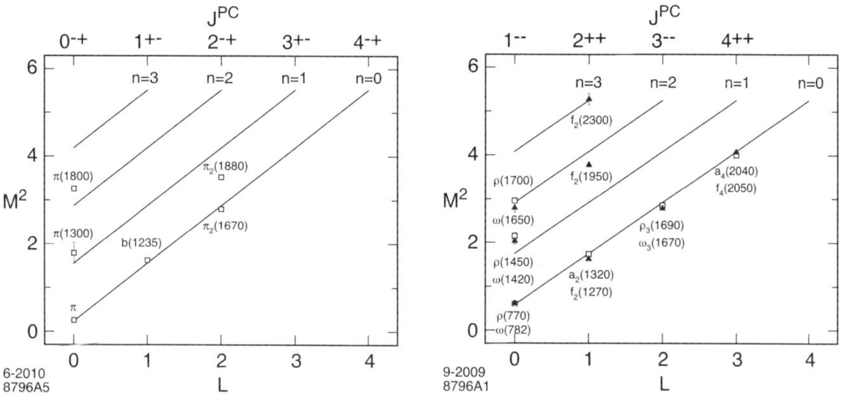

Fig. 1. Parent and daughter Regge trajectories for (a) the π-meson family with $\kappa = 0.6$ GeV; and (b) the $I=1$ ρ-meson and $I=0$ ω-meson families with $\kappa = 0.54$ GeV.

shift is apparently due to the fact that the transverse current in $e^+e^- \to q\bar{q}$ creates a quark pair with $L^z = \pm 1$ instead of the $L^z = 0$ $q\bar{q}$ composition of the vector mesons in the spectrum.

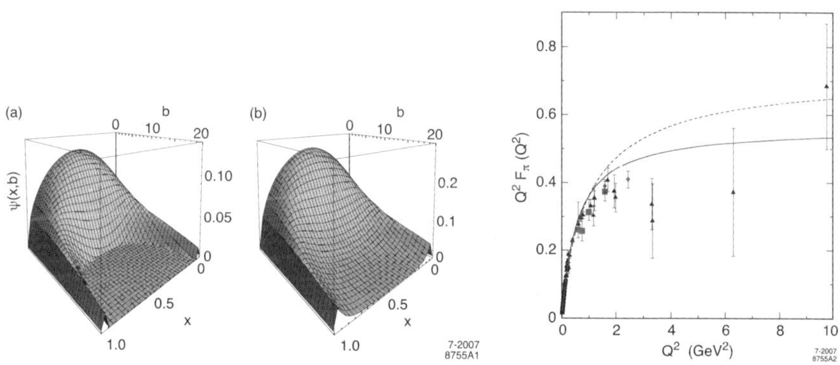

Fig. 2. Pion LF wavefunction $\psi_\pi(x, \vec{b}_\perp)$ for the AdS/QCD (a) hard-wall ($\Lambda_{QCD} = 0.32$ GeV) and (b) soft-wall ($\kappa = 0.375$ GeV) models. (c) Space-like scaling behavior for $Q^2 F_\pi(Q^2)$. The continuous line is the prediction of the soft-wall model for $\kappa = 0.375$ GeV. The dashed line is the prediction of the hard-wall model for $\Lambda_{QCD} = 0.22$ GeV. The triangles are the data compilation of Baldini *et al.*,[72] the boxes are JLAB 1 data[73] and the diamonds are JLAB 2 data.[74]

Individual hadrons in AdS/QCD are identified by matching the power behavior of the hadronic amplitude at the AdS boundary at small z to the twist τ of its interpolating operator at short distances as required by the AdS/CFT dictionary. The twist also equals the dimension of fields appearing in chiral super-multiplets;[75] thus the twist of a hadron equals the number of constituents plus the relative orbital angular momentum. One then can apply light-front holography to relate the amplitude eigensolutions in the fifth-dimension coordinate z to the LF wavefunctions in the physical space-time variable ζ.

Equation (2) was derived by taking the LF bound-state Hamiltonian equation of motion as the starting point.[65] The term L^2/ζ^2 in the LF equation of motion (2) is derived from the reduction of the LF kinetic energy when one transforms to two-dimensional cylindrical coordinates (ζ, φ), in analogy to the $\ell(\ell+1)/r^2$ Casimir term in Schrödinger theory. One thus establishes the interpretation of L in the AdS equations of motion. The interaction terms build confinement corresponding to the dilaton modification of AdS space.[65] The duality between these two methods provides a direct connection between the description of hadronic modes in AdS space and the Hamiltonian formulation of QCD in physical space-time quantized on the light-front at fixed LF time τ.

The identification of orbital angular momentum of the constituents is a key element in the description of the internal structure of hadrons using holographic principles. In our approach quark and gluon degrees of freedom are explicitly introduced in the gauge/gravity correspondence,[76] in contrast with the usual AdS/QCD framework[77,78] where axial and vector currents become the primary entities as in effective chiral theory. Unlike the top-down string theory approach, one is not limited to hadrons of maximum spin $J \leq 2$, and one can study baryons with finite color $N_C = 3$. Higher spin modes follow from shifting dimensions in the AdS wave equations. In the soft-wall model the usual Regge behavior is found $\mathcal{M}^2 \sim n + L$, predicting the same multiplicity of states for mesons and baryons as observed experimentally.[79] It is possible to extend the model to hadrons with heavy quark constituents by introducing nonzero quark masses and short-range Coulomb corrections. For other recent calculations of the hadronic spectrum based on AdS/QCD, see Refs. [80–94]. Other recent computations of the pion form factor are given in Refs. [95–97].

For baryons, the light-front wave equation is a linear equation determined by the LF transformation properties of spin 1/2 states. A linear confining potential $U(\zeta) \sim \kappa^2 \zeta$ in the LF Dirac equation leads to linear Regge trajectories.[98] For fermionic modes the light-front matrix Hamiltonian eigenvalue equation $D_{LF}|\psi\rangle = \mathcal{M}|\psi\rangle$, $H_{LF} = D_{LF}^2$, in a 2×2 spinor component representation is equivalent to the system of coupled linear equations

$$-\frac{d}{d\zeta}\psi_- - \frac{\nu+\frac{1}{2}}{\zeta}\psi_- - \kappa^2\zeta\psi_- = \mathcal{M}\psi_+,$$
$$\frac{d}{d\zeta}\psi_+ - \frac{\nu+\frac{1}{2}}{\zeta}\psi_+ - \kappa^2\zeta\psi_+ = \mathcal{M}\psi_-, \tag{3}$$

with eigenfunctions

$$\psi_+(\zeta) \sim z^{\frac{1}{2}+\nu} e^{-\kappa^2\zeta^2/2} L_n^{\nu}(\kappa^2\zeta^2),$$
$$\psi_-(\zeta) \sim z^{\frac{3}{2}+\nu} e^{-\kappa^2\zeta^2/2} L_n^{\nu+1}(\kappa^2\zeta^2), \tag{4}$$

and eigenvalues $\mathcal{M}^2 = 4\kappa^2(n+\nu+1)$.

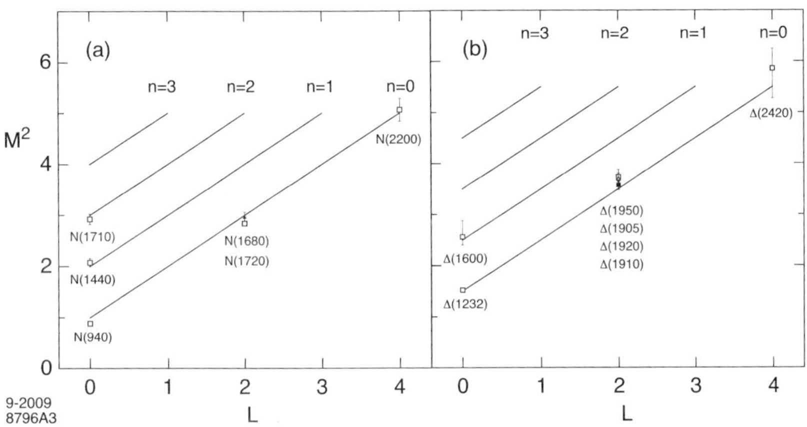

Fig. 3. **56**-plet Regge trajectories for the N and Δ baryon families for $\kappa = 0.5$ GeV.

The baryon interpolating operator

$$\mathcal{O}_{3+L} = \psi D_{\{\ell_1} \dots D_{\ell_q} \psi D_{\ell_{q+1}} \dots D_{\ell_m\}} \psi$$

with $L = \sum_{i=1}^{m} \ell_i$, is a twist-3, dimension-$9/2 + L$ operator with scaling behavior given by its twist-dimension $3 + L$. We thus require $\nu = L + 1$ to match the short-distance scaling behavior. Higher spin modes are obtained by shifting dimensions for the fields. Thus, as in the meson sector, the increase in the mass squared for higher baryonic state is $\Delta n = 4\kappa^2$, $\Delta L = 4\kappa^2$ and $\Delta S = 2\kappa^2$, relative to the lowest ground state, the proton. Since our starting point to find the bound state equation of motion for baryons is the light-front, we fix the overall energy scale identical for mesons and baryons by imposing chiral symmetry to the pion[69] in the LF Hamiltonian equations. By contrast, if we start with a five-dimensional action for a scalar field in presence of a positive sign dilaton, the pion is automatically massless.

The predictions for the **56**-plet of light baryons under the $SU(6)$ flavor group are shown in Fig. 3. As for the predictions for mesons in Fig. 1, only confirmed PDG[71] states are shown. The Roper state $N(1440)$ and the $N(1710)$ are well accounted for in this model as the first and second radial states. Likewise the $\Delta(1660)$ corresponds to the first radial state of the Δ family. The model is successful in explaining the important parity degeneracy observed in the light baryon spectrum, such as the $L = 2$, $N(1680) - N(1720)$ degenerate pair and the $L = 2$, $\Delta(1905), \Delta(1910), \Delta(1920), \Delta(1950)$ states which are degenerate within error bars. Parity degeneracy of baryons is also a property of the hard wall model, but radial states are not well described in this model.[99]

As an example of the scaling behavior of a twist $\tau = 3$ hadron, we compute the spin non-flip nucleon form factor in the soft wall model.[98] The proton and neutron Dirac form factors are given by

$$F_1^p(Q^2) = \int d\zeta \, J(Q,\zeta) \, |\psi_+(\zeta)|^2, \tag{5}$$

$$F_1^n(Q^2) = -\frac{1}{3}\int d\zeta \, J(Q,\zeta) \left[|\psi_+(\zeta)|^2 - |\psi_-(\zeta)|^2\right], \tag{6}$$

where $F_1^p(0) = 1$, $F_1^n(0) = 0$. The non-normalizable mode $J(Q,z)$ is the solution of the AdS wave equation for the external electromagnetic current in presence of a dilaton background $\exp(\pm\kappa^2 z^2)$.[67,100] Plus and minus components of the twist 3 nucleon LFWF are

$$\psi_+(\zeta) = \sqrt{2}\kappa^2 \, \zeta^{3/2} e^{-\kappa^2\zeta^2/2}, \quad \Psi_-(\zeta) = \kappa^3 \, \zeta^{5/2} e^{-\kappa^2\zeta^2/2}. \tag{7}$$

The results for $Q^4 F_1^p(Q^2)$ and $Q^4 F_1^n(Q^2)$ and are shown in Fig. 4

8. Nonperturbative Running Coupling from Light-Front Holography

The concept of a running coupling $\alpha_s(Q^2)$ in QCD is usually restricted to the perturbative domain. However, as in QED, it is useful to define the coupling as an analytic function valid over the full space-like and time-like domains. The study of the non-Abelian QCD coupling at small momentum transfer is a complex problem because of gluonic self-coupling and color confinement.

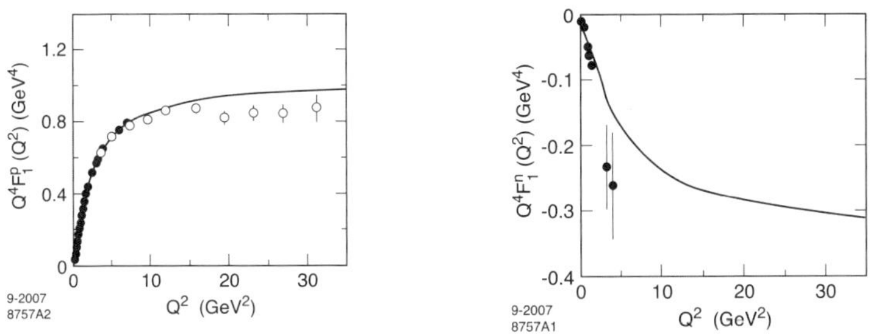

Fig. 4. Predictions for $Q^4 F_1^p(Q^2)$ and $Q^4 F_1^n(Q^2)$ in the soft wall model for $\kappa = 0.424$ GeV.[101]

The definition of the running coupling in perturbative quantum field theory is scheme-dependent. As discussed by Grunberg,[102] an effective coupling or charge can be defined directly from physical observables. Effective charges defined from different observables can be related to each other in the leading-twist domain using commensurate scale relations (CSR).[103] The potential between infinitely heavy quarks can be defined analytically in momentum transfer space as the product of the running coupling times the Born gluon propagator: $V(q) = -4\pi C_F \alpha_V(q)/q^2$. This effective charge defines a renormalization scheme – the α_V scheme of Appelquist, Dine, and Muzinich.[104] In fact, the holographic coupling $\alpha_s^{AdS}(Q^2)$ can be considered to be the nonperturbative extension of the α_V effective charge defined in Ref. [104]. We can also make use of the g_1 scheme, where the strong coupling $\alpha_{g_1}(Q^2)$ is determined from the Bjorken sum rule.[105] The coupling $\alpha_{g_1}(Q^2)$ has the advantage that it is the best-measured effective charge, and it can be used to extrapolate the definition of the effective coupling to large distances.[106] Since α_{g_1} has been measured at intermediate energies, it is particularly useful for studying the transition from small to large distances.

We have recently shown[107] how the LF holographic mapping of effective classical gravity in AdS space, modified by a positive-sign dilaton background, can be used to identify an analytically simple color-confining nonperturbative effective coupling $\alpha_s^{AdS}(Q^2)$ as a function of the space-like momentum transfer $Q^2 = -q^2$. This coupling incorporates confinement and agrees well with effective charge observables and lattice simulations. It also exhibits an infrared fixed point at small Q^2 and asymptotic freedom at large Q^2. However, the fall-off of $\alpha_s^{AdS}(Q^2)$ at large Q^2 is exponential: $\alpha_s^{AdS}(Q^2) \sim e^{-Q^2/\kappa^2}$, rather than the pQCD logarithmic fall-off. It agrees with hadron physics data extracted phenomenologically from different observables, as well as with the predictions of models with built-in confinement and lattice simulations. We also show that a phenomenological extended coupling can be defined which implements the pQCD behavior. The β-function derived from light-front holography becomes significantly negative in the non-perturbative regime $Q^2 \sim \kappa^2$, where it reaches a minimum, signaling the transition region from the infrared (IR) conformal region, characterized by hadronic degrees of freedom, to a pQCD conformal ultraviolet (UV) regime where the relevant degrees of freedom are the quark and gluon constituents. The β-function vanishes at large Q^2 consistent with asymptotic freedom, and it vanishes at small Q^2 consistent with an infrared fixed point.[57,108]

Let us consider a five-dimensional gauge field F propagating in AdS$_5$ space in presence of a dilaton background $\varphi(z)$ which introduces the energy scale κ in the five-dimensional action. At quadratic order in the field strength the action is

$$S = -\frac{1}{4}\int d^5x \sqrt{g}\, e^{\varphi(z)} \frac{1}{g_5^2}\, F^2, \tag{8}$$

where the metric determinant of AdS$_5$ is $\sqrt{g} = (R/z)^5$, $\varphi = \kappa^2 z^2$ and the square of the coupling g_5 has dimensions of length. We can identify the prefactor

$$g_5^{-2}(z) = e^{\varphi(z)} g_5^{-2}, \tag{9}$$

in the AdS action (8) as the effective coupling of the theory at the length scale z. The coupling $g_5(z)$ then incorporates the non-conformal dynamics of confinement. The five-dimensional coupling $g_5(z)$ is mapped, modulo a constant, into the Yang-Mills (YM) coupling g_{YM} of the confining theory in physical space-time using light-front holography. One identifies z with the invariant impact separation variable ζ which appears in the LF Hamiltonian: $g_5(z) \to g_{YM}(\zeta)$. Thus

$$\alpha_s^{AdS}(\zeta) = g_{YM}^2(\zeta)/4\pi \propto e^{-\kappa^2\zeta^2}. \tag{10}$$

In contrast with the 3-dimensional radial coordinates of the non-relativistic Schrö-dinger theory, the natural light-front variables are the two-dimensional cylindrical coordinates (ζ, ϕ) and the light-cone fraction x. The physical coupling measured at the scale Q is the two-dimensional Fourier transform of the LF transverse coupling $\alpha_s^{AdS}(\zeta)$ (10). Integration over the azimuthal angle ϕ gives the Bessel transform

$$\alpha_s^{AdS}(Q^2) \sim \int_0^\infty \zeta d\zeta \, J_0(\zeta Q) \, \alpha_s^{AdS}(\zeta), \tag{11}$$

in the $q^+ = 0$ light-front frame where $Q^2 = -q^2 = -\mathbf{q}_\perp^2 > 0$ is the square of the space-like four-momentum transferred to the hadronic bound state. Using this ansatz we then have from Eq. (11)

$$\alpha_s^{AdS}(Q^2) = \alpha_s^{AdS}(0) \, e^{-Q^2/4\kappa^2}. \tag{12}$$

In contrast, the negative dilaton solution $\varphi = -\kappa^2 z^2$ leads to an integral which diverges at large ζ. We identify $\alpha_s^{AdS}(Q^2)$ with the physical QCD running coupling in its nonperturbative domain.

The flow equation (9) from the scale dependent measure for the gauge fields can be understood as a consequence of field-strength renormalization. In physical QCD we can rescale the non-Abelian gluon field $A^\mu \to \lambda A^\mu$ and field strength $G^{\mu\nu} \to \lambda G^{\mu\nu}$ in the QCD Lagrangian density $\mathcal{L}_{\rm QCD}$ by a compensating rescaling of the coupling strength $g \to \lambda^{-1} g$. The renormalization of the coupling $g_{phys} = Z_3^{1/2} g_0$, where g_0 is the bare coupling in the Lagrangian in the UV-regulated theory, is thus equivalent to the renormalization of the vector potential and field strength: $A^\mu_{ren} = Z_3^{-1/2} A^\mu_0$, $G^{\mu\nu}_{ren} = Z_3^{-1/2} G^{\mu\nu}_0$ with a rescaled Lagrangian density $\mathcal{L}^{ren}_{\rm QCD} = Z_3^{-1} \mathcal{L}^0_{\rm QCD} = (g_{phys}/g_0)^{-2} \mathcal{L}_0$. In lattice gauge theory, the lattice spacing a serves as the UV regulator, and the renormalized QCD coupling is determined from the normalization of the gluon field strength as it appears in the gluon propagator. The inverse of the lattice size L sets the mass scale of the resulting running coupling. As is the case in lattice gauge theory, color confinement in AdS/QCD reflects nonperturbative dynamics at large distances. The QCD couplings defined from lattice gauge theory and the soft wall holographic model are thus similar in concept, and both schemes are expected to have similar properties in the nonperturbative domain, up to a rescaling of their respective momentum scales.

8.1. *Comparison of the Holographic Coupling with Other Effective Charges*

The effective coupling $\alpha^{AdS}(Q^2)$ (solid line) is compared in Fig. 5 with experimental and lattice data. For this comparison to be meaningful, we have to impose the same normalization on the AdS coupling as the g_1 coupling. This defines α_s^{AdS} normalized to the g_1 scheme: $\alpha_{g_1}^{AdS}\left(Q^2 = 0\right) = \pi$. Details on the comparison with other effective charges are given in Ref. [55].

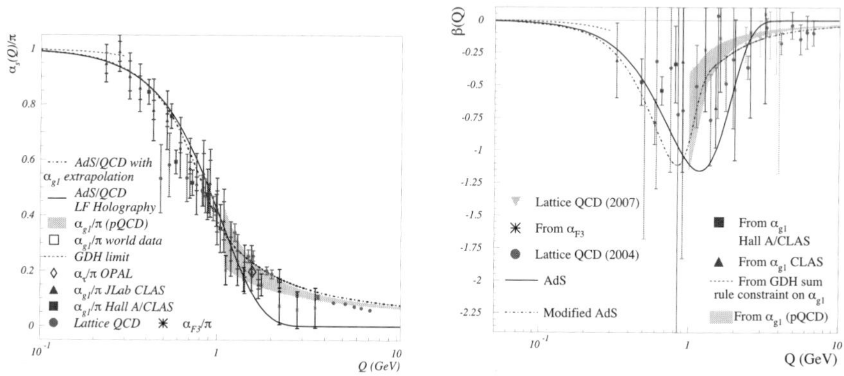

Fig. 5. (a) Effective coupling from LF holography for $\kappa = 0.54$ GeV compared with effective QCD couplings extracted from different observables and lattice results. (b) Prediction for the β-function compared to lattice simulations, JLab and CCFR results for the Bjorken sum rule effective charge.

The couplings in Fig. 5(a) agree well in the strong coupling regime up to $Q \sim 1$ GeV. The value $\kappa = 0.54$ GeV is determined from the vector meson Regge trajectory.[59] The lattice results shown in Fig. 5 from Ref. [53] have been scaled to match the perturbative UV domain. The effective charge α_{g_1} has been determined in Ref. [55] from several experiments. Fig. 5 also displays other couplings from different observables as well as α_{g_1} which is computed from the Bjorken sum rule[105] over a large range of momentum transfer (cyan band). At $Q^2 = 0$ one has the constraint on the slope of α_{g_1} from the Gerasimov-Drell-Hearn (GDH) sum rule[109] which is also shown in the figure. The results show no sign of a phase transition, cusp, or other non-analytical behavior, a fact which allows us to extend the functional dependence of the coupling to large distances. As discussed below, the smooth behavior of the AdS strong coupling also allows us to extrapolate its form to the perturbative domain.[107]

The hadronic model obtained from the dilaton-modified AdS space provides a semi-classical first approximation to QCD. Color confinement is introduced by the harmonic oscillator potential, but effects from gluon creation and absorption are not included in this effective theory. The nonperturbative confining effects vanish exponentially at large momentum transfer (Eq. (12)), and thus the logarithmic fall-off from pQCD quantum loops will dominate in this regime. The running coupling α_s^{AdS} given by Eq. (12) is obtained from a color-confining potential. Since the strong coupling is an analytical function of the momentum transfer at all scales, we can extend the range of applicability of α_s^{AdS} by matching to a perturbative coupling at the transition scale $Q \sim 1$ GeV, where pQCD contributions become important, as described in Ref. [107]. The smoothly extrapolated result (dot-dashed line) for α_s is also shown on Fig. 5. In order to have a fully analytical model, we write

$$\alpha^{AdS}_{Modified,g_1}(Q^2) = \alpha^{AdS}_{g_1}(Q^2) g_+(Q^2) + \alpha^{fit}_{g_1}(Q^2) g_-(Q^2), \tag{13}$$

where $g_\pm(Q^2) = 1/(1 + e^{\pm(Q^2 - Q_0^2)/\tau^2})$ are smeared step functions which match the two regimes. The parameter τ represents the width of the transition region. Here $\alpha^{AdS}_{g_1}$ is given by Eq. (12) with the normalization $\alpha^{AdS}_{g_1}(0) = \pi$ – the plain black line in Fig. 5 - and $\alpha^{fit}_{g_1}$ in Eq. (13) is the analytical fit to the measured coupling α_{g_1}.[55] The couplings are chosen to have the same normalization at $Q^2 = 0$. The smoothly extrapolated result (dot-dashed line) for α_s is also shown on Fig. 5. We use the parameters $Q_0^2 = 0.8$ GeV^2 and $\tau^2 = 0.3$ GeV^2.

8.2. *The β-Function from AdS/QCD*

The β-function for the nonperturbative effective coupling obtained from the LF holographic mapping in a positive dilaton modified AdS background is

$$\beta^{AdS}(Q^2) = \frac{d}{d \log Q^2} \alpha^{AdS}(Q^2) = -\frac{\pi Q^2}{4\kappa^2} e^{-Q^2/(4\kappa^2)}. \tag{14}$$

The solid line in Fig. 5(b) corresponds to the light-front holographic result Eq. (14). Near $Q_0 \simeq 2\kappa \simeq 1$ GeV, we can interpret the results as a transition from the nonperturbative IR domain to the quark and gluon degrees of freedom in the perturbative UV regime. The transition momentum scale Q_0 is compatible with the momentum transfer for the onset of scaling behavior in exclusive reactions where quark counting rules are observed.[1] For example, in deuteron photo-disintegration the onset of scaling corresponds to momentum transfer of 1.0 GeV to the nucleon involved.[110] Dimensional counting is built into the AdS/QCD soft and hard wall models since the AdS amplitudes $\Phi(z)$ are governed by their twist scaling behavior z^τ at short distances, $z \to 0$.[5]

Also shown on Fig. 5 (b) are the β-functions obtained from phenomenology and lattice calculations. For clarity, we present only the LF holographic predictions, the lattice results from Ref. [53] and the experimental data supplemented by the relevant sum rules. The dot-dashed curve corresponds to the extrapolated approximation obtained by matching to AdS results to the perturbative coupling[107] given by Eq. (13). The β-function extracted from LF holography, as well as the forms obtained from the works of Cornwall,,[108] Bloch, Fisher *et al.*[111] Burkert and Ioffe[112] and Furui and Nakajima[53] are seen to have a similar shape and magnitude.

Judging from these results, we infer that the actual β-function of QCD will extrapolate between the non-perturbative results for $Q < 1$ GeV and the pQCD results for $Q > 1$ GeV. We also observe that the general conditions

$$\beta(Q \to 0) = \beta(Q \to \infty) = 0, \tag{15}$$

$$\beta(Q) < 0, \text{ for } Q > 0, \tag{16}$$

$$\tfrac{d\beta}{dQ}\Big|_{Q=Q_0} = 0, \tag{17}$$

$$\tfrac{d\beta}{dQ} < 0, \text{ for } Q < Q_0, \ \tfrac{d\beta}{dQ} > 0, \text{ for } Q > Q_0. \tag{18}$$

are satisfied by our model β-function obtained from LF holography.

Eq. (15) expresses the fact that QCD approaches a conformal theory in both the far ultraviolet and deep infrared regions. In the semiclassical approximation to QCD without particle creation or absorption, the β-function is zero and the approximate theory is scale invariant in the limit of massless quarks.[113] When quantum corrections are included, the conformal behavior is preserved at very large Q because of asymptotic freedom and near $Q \to 0$ because the theory develops a fixed point. An infrared fixed point is in fact a natural consequence of color confinement:[108] since the propagators of the colored fields have a maximum wavelength, all loop integrals in the computation of the gluon self-energy decouple at $Q^2 \to 0$.[57] Condition (16) for Q^2 large, expresses the basic anti-screening behavior of QCD where the strong coupling vanishes. The β-function in QCD is essentially negative, thus the coupling increases monotonically from the UV to the IR where it reaches its maximum value: it has a finite value for a theory with a mass gap. Equation (17) defines the transition region at Q_0 where the β-function has a minimum. Since there is only one hadronic-partonic transition, the minimum is an absolute minimum; thus the additional conditions expressed in Eq (18) follow immediately from Eqs. (15-17). The conditions given by Eqs. (15-18) describe the essential behavior of the full β-function for an effective QCD coupling whose scheme/definition is similar to that of the V-scheme.

9. Conclusions

The combination of Anti-de Sitter space methods with light-front holography provides an accurate first approximation for the spectra and wavefunctions of meson and baryon light-quark bound states. One also obtains an elegant connection between a semiclassical first approximation to QCD, quantized on the light-front, with hadronic modes propagating on a fixed AdS background. The resulting bound-state Hamiltonian equation of motion in QCD leads to relativistic light-front wave equations in the invariant impact variable ζ which measures the separation of the quark and gluonic constituents within the hadron at equal light-front time. This corresponds to the effective single-variable relativistic Schrödinger-like equation in the AdS fifth dimension coordinate z, Eq. (2). The eigenvalues give the hadronic spectrum, and the eigenmodes represent the probability distributions of the hadronic constituents at a given scale. In particular, the light-front holographic mapping of effective classical gravity in AdS space, modified by a positive-sign dilaton background, provides a very good description of the spectrum and form factors of light mesons and baryons. We have also shown that the light-front holographic mapping of effective classical gravity in AdS space, modified by the positive-sign dilaton background predicts the form of a non-perturbative effective coupling $\alpha_s^{AdS}(Q)$ and its β-function. The AdS/QCD running coupling is in very good agreement with the effective coupling α_{g_1} extracted from the Bjorken sum rule. The holographic β-function displays a transition from nonperturbative to perturbative regimes at a momentum scale $Q \sim 1$ GeV. Our analysis indicates that light-front holography captures the characteristics of the full β-function of QCD and the essential dynamics of confinement, thus giving further support to the application of the gauge/gravity duality to the confining dynamics of strongly coupled QCD.

There are many phenomenological applications where detailed knowledge of the QCD coupling and the renormalized gluon propagator at relatively soft momentum transfer are essential. This includes exclusive and semi-exclusive processes as well as the rescattering interactions which create the leading-twist Sivers single-spin correlations in semi-inclusive deep inelastic scattering,[114,115] the Boer-Mulders functions which lead to anomalous $\cos 2\phi$ contributions to the lepton pair angular distribution in the unpolarized Drell-Yan reaction,[116] and the Sommerfeld-Sakharov-Schwinger correction to heavy quark production at threshold.[117] The confining AdS/QCD coupling from light-front holography thus can provide a quantitative understanding of this factorization-breaking physics.[118]

Acknowledgments

We thank Volker Burkert, John Cornwall, Sadataka Furui, Philipp Hägler, Wolfgang Korsch, G. Peter Lepage, Takemichi Okui, Joannis Papavassiliou, Anatoly Radyushkin, Craig Roberts, Robert Shrock, and Peter Tandy for helpful comments. This research was supported by the Department of Energy contracts DE–AC02–76SF00515 and DE-AC05-84ER40150. Invited talk, presented by SJB at the Workshop on Exclusive Reactions at High Momentum Transfer (IV), May 18-21, 2010, Thomas Jefferson National Accelerator Facility, Newport News, VA. SLAC-PUB-14208, CP3-Origins-2010-3, and Jlab-PHY-10-1135.

References

1. S. J. Brodsky and G. R. Farrar, Phys. Rev. Lett. **31**, 1153 (1973); Phys. Rev. D **11**, 1309 (1975).
2. V. A. Matveev, R. M. Muradian and A. N. Tavkhelidze, Lett. Nuovo Cim. **7**, 719 (1973).
3. G. P. Lepage and S. J. Brodsky, Phys. Lett. B **87**, 359 (1979).
4. A. V. Efremov and A. V. Radyushkin, Phys. Lett. B **94**, 245 (1980).
5. J. Polchinski and M. J. Strassler, Phys. Rev. Lett. **88**, 031601 (2002) [arXiv:hep-th/0109174].
6. S. J. Brodsky and A. H. Mueller, Phys. Lett. B **206**, 685 (1988).
7. G. Bertsch, S. J. Brodsky, A. S. Goldhaber and J. F. Gunion, Phys. Rev. Lett. **47**, 297 (1981).
8. J. F. Gunion, S. J. Brodsky, R. Blankenbecler, Phys. Rev. **D8**, 287 (1973).
9. D. W. Sivers, S. J. Brodsky and R. Blankenbecler, Phys. Rept. **23**, 1 (1976).
10. C. G. White *et al.*, Phys. Rev. D **49**, 58 (1994).
11. R. J. Holt, Phys. Rev. C **41**, 2400 (1990).
12. C. Bochna *et al.* [E89-012 Collaboration], Phys. Rev. Lett. **81**, 4576 (1998) [arXiv:nucl-ex/9808001].
13. P. Rossi *et al.* [CLAS Collaboration], Phys. Rev. Lett. **94**, 012301 (2005) [arXiv:hep-ph/0405207].
14. S. Rock *et al.*, Phys. Rev. D **46**, 24 (1992).
15. S. J. Brodsky and B. T. Chertok, Phys. Rev. D **14**, 3003 (1976).
16. S. J. Brodsky, C. R. Ji and G. P. Lepage, Phys. Rev. Lett. **51**, 83 (1983).
17. S. J. Brodsky and G. P. Lepage, Adv. Ser. Direct. High Energy Phys. **5**, 93 (1989).
18. E. M. Aitala *et al.* [E791 Collaboration], Phys. Rev. Lett. **86**, 4773 (2001) [arXiv:hep-ex/0010044].
19. B. Clasie *et al.*, arXiv:0707.1481 [nucl-ex].
20. A. Airapetian *et al.* [HERMES Collaboration], Phys. Rev. Lett. **90**, 052501 (2003) [arXiv:hep-ex/0209072].
21. E. L. Berger, S. J. Brodsky, Phys. Rev. Lett. **42**, 940-944 (1979).
22. E. L. Berger, T. Gottschalk, D. W. Sivers, Phys. Rev. **D23**, 99 (1981).

23. E. L. Berger, S. J. Brodsky, Phys. Rev. **D24**, 2428 (1981).
24. F. Arleo, S. J. Brodsky, D. S. Hwang *et al.*, [arXiv:1006.4045 [hep-ph]].
25. F. Arleo, S. J. Brodsky, D. S. Hwang *et al.*, [arXiv:0911.4604 [hep-ph]].
26. S. J. Brodsky, A. Sickles, Phys. Lett. **B668**, 111-115 (2008) [arXiv:0804.4608 [hep-ph]].
27. V. P. Druzhinin, PoS **EPS-HEP2009**, 051 (2009) [arXiv:0909.3148 [hep-ex]] and preliminary results presented at IHCEP2010.
28. A. V. Radyushkin, Phys. Rev. **D80**, 094009 (2009) [arXiv:0906.0323 [hep-ph]].
29. M. V. Polyakov, JETP Lett. **90**, 228-231 (2009) [arXiv:0906.0538 [hep-ph]].
30. S. V. Mikhailov, N. G. Stefanis, Mod. Phys. Lett. **A24**, 2858-2867 (2009) [arXiv:0910.3498 [hep-ph]].
31. S. J. Brodsky, G. P. Lepage, Phys. Rev. **D24**, 1808 (1981).
32. A. Danagoulian *et al.* [Hall A Collaboration], Phys. Rev. Lett. **98**, 152001 (2007) [arXiv:nucl-ex/0701068].
33. M. Diehl, T. Feldmann, R. Jakob and P. Kroll, Eur. Phys. J. C **8**, 409 (1999) [arXiv:hep-ph/9811253].
34. A. Chen, *International Conference on the Structure and Interactions of the Photon and 14th International Workshop on Photon-Photon Collisions (Photon 2001), Ascona, Switzerland, 2-7 Sep 2001.*
35. G. R. Court *et al.*, Phys. Rev. Lett. **57**, 507 (1986).
36. I. Mardor *et al.*, Phys. Rev. Lett. **81**, 5085 (1998).
37. S. J. Brodsky and G. F. de Teramond, Phys. Rev. Lett. **60**, 1924 (1988).
38. D. Dutta and H. Gao, Phys. Rev. C **71**, 032201 (2005) [arXiv:hep-ph/0411267].
39. S. J. Brodsky and M. Karliner, Phys. Rev. Lett. **78**, 4682 (1997) [arXiv:hep-ph/9704379].
40. P. A. M. Dirac, Rev. Mod. Phys. **21**, 392 (1949).
41. H. C. Pauli and S. J. Brodsky, Phys. Rev. D **32**, 2001 (1985).
42. S. J. Brodsky, H. C. Pauli and S. S. Pinsky, Phys. Rept. **301**, 299 (1998) [arXiv:hep-ph/9705477].
43. S. D. Drell and T. M. Yan, Phys. Rev. Lett. **24**, 181 (1970).
44. G. B. West, Phys. Rev. Lett. **24**, 1206 (1970).
45. S. J. Brodsky and D. S. Hwang, Nucl. Phys. B **543**, 239 (1999) [arXiv:hep-ph/9806358].
46. S. J. Brodsky, M. Diehl and D. S. Hwang, Nucl. Phys. B **596**, 99 (2001) [arXiv:hep-ph/0009254].
47. G. P. Lepage and S. J. Brodsky, Phys. Rev. D **22**, 2157 (1980).
48. S. J. Brodsky and G. P. Lepage, SLAC-PUB-2294; Workshop on Current Topics in High Energy Physics, Cal Tech., Pasadena, Calif., Feb 13-17, 1979.
49. O. V. Teryaev, arXiv:hep-ph/9904376.
50. S. J. Brodsky, D. S. Hwang, B. Q. Ma and I. Schmidt, Nucl. Phys. B **593**, 311 (2001) [arXiv:hep-th/0003082].
51. S. J. Brodsky and R. Shrock, arXiv:0905.1151 [hep-th]. S. J. Brodsky, C. D. Roberts, R. Shrock, and P. Tandy [arXiv:1005.4610 [nucl-th]].
52. J. M. Maldacena, Adv. Theor. Math. Phys. **2**, 231 (1998) [Int. J. Theor.

Phys. **38**, 1113 (1999)] [arXiv:hep-th/9711200].
53. S. Furui and H. Nakajima, Phys. Rev. D **76**, 054509 (2007) [arXiv:hep-lat/0612009].
54. L. von Smekal, R. Alkofer and A. Hauck, Phys. Rev. Lett. **79**, 3591 (1997) [arXiv:hep-ph/9705242].
55. A. Deur, V. Burkert, J. P. Chen and W. Korsch, Phys. Lett. B **650**, 244 (2007) [arXiv:hep-ph/0509113].
56. A. Deur, V. Burkert, J. P. Chen and W. Korsch, Phys. Lett. B **665**, 349 (2008) [arXiv:0803.4119 [hep-ph]].
57. S. J. Brodsky and R. Shrock, Phys. Lett. B **666**, 95 (2008) [arXiv:0806.1535 [hep-th]].
58. A. Karch, E. Katz, D. T. Son and M. A. Stephanov, Phys. Rev. D **74**, 015005 (2006) [arXiv:hep-ph/0602229].
59. G. F. de Teramond and S. J. Brodsky, Nucl. Phys. Proc. Suppl. **199**, 89 (2010) [arXiv:0909.3900 [hep-ph]].
60. O. Andreev and V. I. Zakharov, Phys. Rev. D **74**, 025023 (2006) [arXiv:hep-ph/0604204].
61. F. Zuo, arXiv:0909.4240 [hep-ph].
62. S. S. Afonin, arXiv:1001.3105 [hep-ph].
63. S. D. Glazek and M. Schaden, Phys. Lett. B **198**, 42 (1987).
64. P. Hoyer, arXiv:0909.3045 [hep-ph].
65. G. F. de Teramond and S. J. Brodsky, Phys. Rev. Lett. **102**, 081601 (2009) [arXiv:0809.4899 [hep-ph]].
66. S. J. Brodsky and G. F. de Teramond, Phys. Rev. Lett. **96**, 201601 (2006) [arXiv:hep-ph/0602252].
67. S. J. Brodsky and G. F. de Teramond, Phys. Rev. D **77**, 056007 (2008) [arXiv:0707.3859 [hep-ph]].
68. S. J. Brodsky and G. F. de Teramond, Phys. Rev. D **78**, 025032 (2008) [arXiv:0804.0452 [hep-ph]].
69. G. F. de Teramond and S. J. Brodsky, arXiv:1001.5193 [hep-ph].
70. Z. Abidin and C. E. Carlson, Phys. Rev. D **77**, 095007 (2008) [arXiv:0801.3839 [hep-ph]].
71. C. Amsler *et al.* (Particle Data Group), Phys. Lett. B **667**, 1, (2008).
72. R. Baldini, S. Dubnicka, P. Gauzzi, S. Pacetti, E. Pasqualucci and Y. Srivastava, Eur. Phys. J. C **11**, 709 (1999).
73. V. Tadevosyan *et al.* [Jefferson Lab F(pi) Collaboration], Phys. Rev. C **75**, 055205 (2007) [arXiv:nucl-ex/0607007].
74. T. Horn *et al.* [Fpi2 Collaboration], Phys. Rev. Lett. **97**, 192001 (2006) [arXiv:nucl-ex/0607005].
75. N. J. Craig and D. Green, JHEP **0909**, 113 (2009) [arXiv:0905.4088 [hep-ph]].
76. S. J. Brodsky and G. F. de Teramond, Phys. Lett. B **582**, 211 (2004) [arXiv:hep-th/0310227].
77. J. Erlich, E. Katz, D. T. Son and M. A. Stephanov, Phys. Rev. Lett. **95**, 261602 (2005) [arXiv:hep-ph/0501128].
78. L. Da Rold and A. Pomarol, Nucl. Phys. B **721**, 79 (2005) [arXiv:hep-ph/0501218]; JHEP **0601**, 157 (2006) [arXiv:hep-ph/0510268].

79. E. Klempt and A. Zaitsev, Phys. Rept. **454**, 1 (2007) [arXiv:0708.4016 [hep-ph]].
80. H. Boschi-Filho, N. R. F. Braga and H. L. Carrion, Phys. Rev. D **73**, 047901 (2006) [arXiv:hep-th/0507063].
81. N. Evans and A. Tedder, Phys. Lett. B **642**, 546 (2006) [arXiv:hep-ph/0609112].
82. D. K. Hong, T. Inami and H. U. Yee, Phys. Lett. B **646**, 165 (2007) [arXiv:hep-ph/0609270].
83. P. Colangelo, F. De Fazio, F. Jugeau and S. Nicotri, Phys. Lett. B **652**, 73 (2007) [arXiv:hep-ph/0703316].
84. H. Forkel, Phys. Rev. D **78**, 025001 (2008) [arXiv:0711.1179 [hep-ph]].
85. A. Vega and I. Schmidt, Phys. Rev. D **78**, 017703 (2008) [arXiv:0806.2267 [hep-ph]].
86. K. Nawa, H. Suganuma and T. Kojo, Mod. Phys. Lett. A **23**, 2364 (2008) [arXiv:0806.3040 [hep-th].
87. W. de Paula, T. Frederico, H. Forkel and M. Beyer, Phys. Rev. D **79**, 075019 (2009) [arXiv:0806.3830 [hep-ph]].
88. P. Colangelo, F. De Fazio, F. Giannuzzi, F. Jugeau and S. Nicotri, Phys. Rev. D **78**, 055009 (2008) [arXiv:0807.1054 [hep-ph]].
89. H. Forkel and E. Klempt, Phys. Lett. B **679**, 77 (2009) [arXiv:0810.2959 [hep-ph]].
90. H. C. Ahn, D. K. Hong, C. Park and S. Siwach, Phys. Rev. D **80**, 054001 (2009) [arXiv:0904.3731 [hep-ph]].
91. Y. Q. Sui, Y. L. Wu, Z. F. Xie and Y. B. Yang, Phys. Rev. D **81**, 014024 (2010) [arXiv:0909.3887 [hep-ph]].
92. J. I. Kapusta and T. Springer, Phys. Rev. D **81**, 086009 (2010) [arXiv:1001.4799 [hep-ph]];
93. P. Zhang, Phys. Rev. D **81** (2010) 114029 [arXiv:1002.4352 [hep-ph]]; JHEP **1005** (2010) 039 [arXiv:1003.0558 [hep-ph]]; arXiv:1007.2163 [hep-ph].
94. See also: S. J. Wang, J. Tao, X. B. Guo and L. Li, arXiv:1007.2462 [hep-th].
95. H. J. Kwee and R. F. Lebed, JHEP **0801**, 027 (2008) [arXiv:0708.4054 [hep-ph]].
96. H. R. Grigoryan and A. V. Radyushkin, Phys. Rev. D **76**, 115007 (2007) [arXiv:0709.0500 [hep-ph]].
97. C. A. B. Bayona, H. Boschi-Filho, M. Ihl and M. A. C. Torres, arXiv:1006.2363 [hep-th].
98. S. J. Brodsky and G. F. de Teramond, arXiv:0802.0514 [hep-ph].
99. G. F. de Teramond and S. J. Brodsky, Phys. Rev. Lett. **94**, 201601 (2005) [arXiv:hep-th/0501022].
100. H. R. Grigoryan and A. V. Radyushkin, Phys. Rev. D **76**, 095007 (2007) [arXiv:0706.1543 [hep-ph]].
101. The data compilation is from M. Diehl, Nucl. Phys. Proc. Suppl. **161**, 49 (2006) [arXiv:hep-ph/0510221].
102. G. Grunberg, Phys. Lett. B **95**, 70 (1980); Phys. Rev. D **29**, 2315 (1984); Phys. Rev. D **40**, 680 (1989).
103. S. J. Brodsky and H. J. Lu, Phys. Rev. D **51**, 3652 (1995) [arXiv:hep-

ph/9405218]; S. J. Brodsky, G. T. Gabadadze, A. L. Kataev and H. J. Lu, Phys. Lett. B **372**, 133 (1996) [arXiv:hep-ph/9512367].
104. T. Appelquist, M. Dine and I. J. Muzinich, Phys. Lett. B **69**, 231 (1977).
105. J. D. Bjorken, Phys. Rev. **148**, 1467 (1966).
106. A. Deur, arXiv:0907.3385 [nucl-ex].
107. S. J. Brodsky, G. F. de Teramond, A. Deur, Phys. Rev. **D81**, 096010 (2010) [arXiv:1002.3948 [hep-ph]].
108. J. M. Cornwall, Phys. Rev. D **26**, 1453 (1982).
109. S. D. Drell and A. C. Hearn, Phys. Rev. Lett. **16**, 908 (1966); S. B. Gerasimov, Sov. J. Nucl. Phys. **2** (1966) 430 [Yad. Fiz. **2** (1966) 598].
110. H. Gao and L. Zhu, AIP Conf. Proc. **747**, 179 (2005) [arXiv:nucl-ex/0411014].
111. J. C. R. Bloch, Phys. Rev. D **66**, 034032 (2002) [arXiv:hep-ph/0202073]; P. Maris and P. C. Tandy, Phys. Rev. C **60**, 055214 (1999) [arXiv:nucl-th/9905056]; C. S. Fischer and R. Alkofer, Phys. Lett. B **536**, 177 (2002) [arXiv:hep-ph/0202202]; C. S. Fischer, R. Alkofer and H. Reinhardt, Phys. Rev. D **65**, 094008 (2002) [arXiv:hep-ph/0202195]; R. Alkofer, C. S. Fischer and L. von Smekal, Acta Phys. Slov. **52**, 191 (2002) [arXiv:hep-ph/0205125]; M. S. Bhagwat, M. A. Pichowsky, C. D. Roberts and P. C. Tandy, Phys. Rev. C **68**, 015203 (2003) [arXiv:nucl-th/0304003].
112. V. D. Burkert and B. L. Ioffe, Phys. Lett. B **296**, 223 (1992); J. Exp. Theor. Phys. **78**, 619 (1994) [Zh. Eksp. Teor. Fiz. **105**, 1153 (1994)].
113. G. Parisi, Phys. Lett. B **39**, 643 (1972).
114. S. J. Brodsky, D. S. Hwang and I. Schmidt, Phys. Lett. B **530**, 99 (2002) [arXiv:hep-ph/0201296].
115. J. C. Collins, Phys. Lett. B **536**, 43 (2002) [arXiv:hep-ph/0204004].
116. D. Boer, S. J. Brodsky and D. S. Hwang, Phys. Rev. D **67**, 054003 (2003) [arXiv:hep-ph/0211110].
117. S. J. Brodsky, A. H. Hoang, J. H. Kuhn and T. Teubner, Phys. Lett. B **359**, 355 (1995) [arXiv:hep-ph/9508274].
118. J. Collins and J. W. Qiu, Phys. Rev. D **75**, 114014 (2007) [arXiv:0705.2141 [hep-ph]].

EXPOSING THE DRESSED QUARK'S MASS

H. L. L. ROBERTS,[1,2] L. CHANG,[3] I. C. CLOËT,[4] and C. D. ROBERTS[1,2,5]

[1] *Physics Division, Argonne National Laboratory*
Argonne, Illinois 60439, USA
[2] *Institut für Kernphysik, Forschungszentrum Jülich*
D-52425 Jülich, Germany
[3] *Institute of Applied Physics and Computational Mathematics*
Beijing 100094, China
[4] *Department of Physics, University of Washington*
Seattle WA 98195, USA
[5] *Department of Physics, Peking University*
Beijing 100871, China

This snapshot of recent progress in hadron physics made in connection with QCD's Dyson-Schwinger equations includes: a perspective on confinement and dynamical chiral symmetry breaking (DCSB); a précis on the physics of in-hadron condensates; results on the hadron spectrum, including dressed-quark-core masses for the nucleon and Δ, their first radial excitations, and the parity-partners of these states; an illustration of the impact of DCSB on the electromagnetic pion form factor, thereby exemplifying how data can be used to chart the momentum-dependence of the dressed-quark mass function; and a prediction that $F_1^{p,d}/F_1^{p,u}$ passes through zero at $Q^2 \approx 5m_N^2$ owing to the presence of nonpointlike scalar and axial-vector diquark correlations in the nucleon.

Keywords: Confinement, dynamical chiral symmetry breaking, Dyson-Schwinger equations, light-front methods; hadron form factors; hadron spectrum

1. Confinement, DCSB and in-hadron condensates

Quantum chromodynamics is a theory whose elementary excitations are not those degrees-of-freedom readily accessible via experiment; i.e., whose elementary excitations are confined. Moreover, less than 2% of a nucleon's mass can be attributed to the so-called current-quark masses that appear in QCD's Lagrangian, with the remainder being generated through dynamical chiral symmetry breaking (DCSB).[1] Neither confinement nor DCSB is apparent in QCD's Lagrangian and yet they play the dominant role in determining the observable characteristics of real-world QCD. The physics

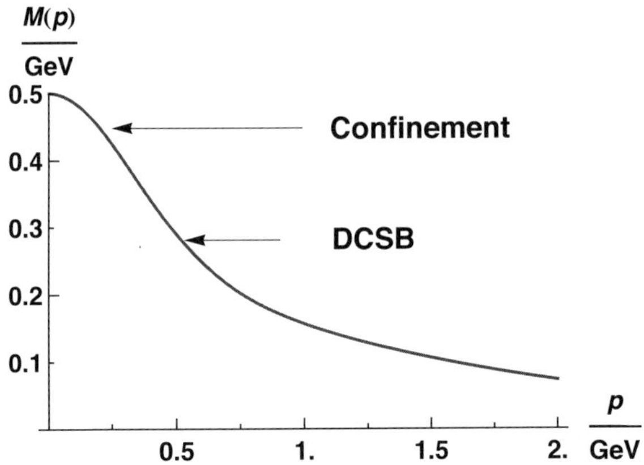

Fig. 1. Dressed-quark mass function used, e.g., in an extensive study of nucleon electromagnetic form factors.[2] Dynamical chiral symmetry breaking is evident in the enhancement around $p = 0.5\,\text{GeV}$; and confinement is signalled by the inflexion point at around $p = 0.25\,\text{GeV}$.

of hadrons is ruled by such *emergent phenomena*, which can only be elucidated via nonperturbative methods in quantum field theory. This is both the greatest novelty and the greatest challenge within the Standard Model.

Both confinement and DCSB can be seen in the dressed-quark mass function, illustrated by Fig. 1. The behaviour of $M(p)$ can be calculated in perturbation theory for $p \gtrsim 1.5\,\text{GeV}$. However, the evolution of $M(p)$ is essentially nonperturbative for infrared momenta. The existence of DCSB is signalled by the rapid increase in magnitude as p decreases from $1 \to 0.5\,\text{GeV}$. This longstanding prediction of Dyson-Schwinger equation (DSE) studies in QCD, see e.g. Refs. [3,4], is confirmed by numerical simulations of lattice-QCD.[5] The reality of DCSB means that the Higgs mechanism is largely irrelevant to the bulk of normal matter in the universe.

In analysing large momentum transfer processes it is common to employ a light-front formulation of QCD. This has merit but there appears to be a serious drawback; viz., DCSB has not yet been realised on the light-front. However, progress has recently been made toward resolving this conundrum,[12] by arguing for a shift in paradigm so that, in particular, DCSB is understood as being expressed in properties of hadrons rather than of the vacuum. It is contended therein that: the measurable impact of all condensates is entirely expressed in the properties of QCD's asymptotically realisable states; there is no evidence that the so-called vacuum condensates are anything more than mass-dimensioned parameters in one or another

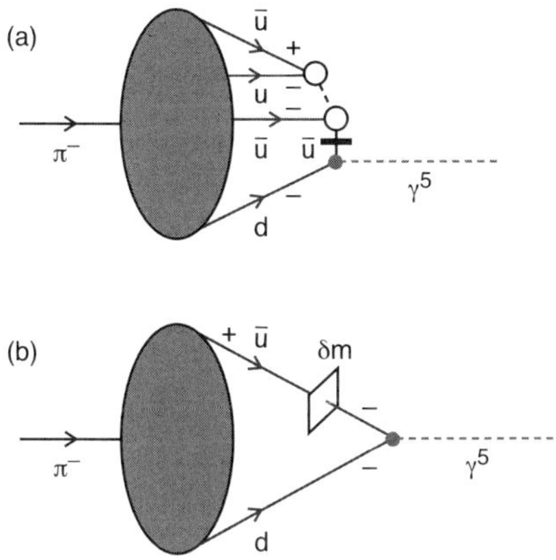

Fig. 2. Light-front contributions to $\rho_\pi = -\langle 0|\bar{q}\gamma_5 q|\pi\rangle$. *Upper panel* – A non-valence piece of the meson's light-front wavefunction, whose contribution to ρ_π is mediated by the light-front instantaneous quark propagator (vertical crossed-line). The "±" denote parton helicity. *Lower panel* – There are infinitely many such diagrams, which can introduce chiral symmetry breaking in the light-front wavefunction in the absence of a current-quark mass. (The case of f_π is analogous.)

theoretical truncation scheme; and condensates do not describe measurable spacetime-independent configurations of QCD's elementary degrees-of-freedom in a hadron-less ground state. This position assumes confinement, from which follows quark-hadron duality and hence that all observable consequences of QCD can, in principle, be computed using a hadronic basis.

ModelMp From this perspective a possible solution to the problem of DCSB in the light-front formulation is illustrated in Fig. 2. The light-front-instantaneous quark propagator can mediate a contribution from higher Fock state components to the matrix elements that define the pion's pseudoscalar (ρ_π) and pseudovector (f_π) decay constants. Such diagrams connect dynamically-generated chiral-symmetry breaking components of the meson's light-front wavefunction to these matrix elements. There are infinitely many contributions of this type and they do not depend sensitively on the current-quark mass in the neighborhood of the chiral limit. This leads to a conjecture that DCSB in the light-front formulation is expressed via in-hadron condensates and is connected with sea-quarks derived from higher Fock states.

Confinement is signalled in Fig. 1 by the inflexion point possessed by $M(p)$ at $p \approx 0.25\,\text{GeV}$, a relationship which is explained, e.g., in Refs. [6–8]. Confinement and DCSB are intimately connected in QCD, and it is natural to ask whether the connection is accidental or causal. There are models with DCSB but not confinement, however, a model with confinement but lacking DCSB has not yet been identified (see, e.g., Secs. 2.1 and 2.2 of Ref. [9]). This leads to a conjecture that DCSB is a necessary consequence of confinement. It is notable that there are numerous models and theories which exhibit both confinement and DCSB, and possess an external control parameter such that deconfinement and chiral symmetry restoration occur simultaneously at some critical value of this parameter; e.g., quantum electrodynamics in three dimensions with N_f electrons.[10,11]

These observations highlight that confinement can be related to the analytic properties of QCD's Schwinger functions,[6–8] which are often called Euclidean-space Green functions. From this standpoint the question of light-quark confinement can be translated into the challenge of charting the infrared behavior of QCD's *universal* β-function. This is a well-posed problem whose solution is an elemental goal of modern hadron physics and which can be addressed in any framework enabling the nonperturbative evaluation of renormalisation constants.

2. DCSB and the hadron spectrum

Through the gap and Bethe-Salpeter equations (BSEs) the pointwise behaviour of the β-function determines the pattern of chiral symmetry breaking; e.g., the behaviour in Fig. 1. Moreover, the fact that these and other DSEs connect the β-function to experimental observables entails, e.g., that comparison between computations and observations of the hadron mass spectrum, and hadron elastic and transition form factors, can be used to chart the β-function's long-range behaviour.[13] In order to realise this goal, a nonperturbative symmetry-preserving DSE truncation is necessary. Steady quantitative progress can be made with a scheme that is systematically improvable.[14,15] In fact, the mere existence of such a scheme has enabled the proof of exact nonperturbative results in QCD.[16–18]

On the other hand, significant qualitative advances in understanding the essence of QCD could be made with symmetry-preserving kernel *Ansätze* that express important additional nonperturbative effects, which are impossible to capture in any finite sum of contributions. Such an approach is now available.[19] It begins with a novel form for the axial-vector BSE, which is valid when the quark-gluon vertex is fully dressed. Therefrom, a Ward-Takahashi identity for the Bethe-Salpeter kernel is derived and solved for a class of dressed-quark-gluon-vertex models. The solution yields a symmetry-preserving closed system of gap and vertex equations. As the analysis can readily be extended to the vector equation, a comparison is possible between the responses of parity-partners in the meson spectrum to nonperturbatively dressing the quark-gluon vertex. The results indicate that: spin-orbit splitting in the ground-state meson spectrum is dramatically enhanced by DCSB;[19] DCSB generates a large anomalous chromomag-

Table 1. Meson and diquark masses (in GeV) computed using a contact-interaction DSE kernel,[20] which produces a momentum-independent dressed-quark mass $M = 0.41\,\mathrm{GeV}$ from a current-quark mass of $m = 8\,\mathrm{MeV}$. "RL" denotes rainbow-ladder truncation.[15] Row-2 is obtained by augmenting the RL kernel with spin-orbit repulsion, as described in the text. For reference, experimental masses are:[21] $m_\pi = 0.140$, $m_\sigma = 0.4 - 1.2$, $m_\rho = 0.775$, $m_{a_1} = 1.243$. NB. We implement isospin symmetry so, e.g., $m_\omega = m_\rho$, $m_{f_1} = m_{a_1}$, etc.

	m_π	m_σ	m_ρ	m_{a_1}	$m^{qq}_{0^+}$	$m^{qq}_{0^-}$	$m^{qq}_{1^+}$	$m^{qq}_{1^-}$
RL	0.141	0.825	0.798	1.073	0.723	0.975	1.006	1.173
RL $+ g_{\rm SO}$	0.141	1.079	0.798	1.243	0.723	1.189	1.006	1.323

netic moment term as an essential part of the dressed-quark-gluon vertex;[8] and, importantly, owing to the symmetry-preserving nature of the truncation procedure, the $M(p)$-driven vertex corrections alter the Bethe-Salpeter kernel in such a way as to leave ground-state pseudoscalar and vector meson masses almost unchanged.[8]

Motivated by this ongoing work, we employed the DSE model formulated in Ref. [20] to compute a spectrum of ground-state $u\,\&\,d$-quark mesons, with the results presented in Table 1. The first row lists results obtained in rainbow-ladder truncation. Of course, given the symmetry preserving nature of the truncation, in the chiral-limit $m_\pi = 0$ and $m_\sigma = 2M$ (which follows from Eq. (14) in Ref. [20]), where $M = 0.40\,\mathrm{GeV}$ is chiral-limit value of the model's momentum-independent dressed-quark mass. The second row lists values obtained with spin-orbit repulsion added to the scalar and axial-vector channels, the strength of which is described by a single parameter, $g_{\rm SO} = 0.32$, fixed so as to yield $m_{a_1} = 1.23\,\mathrm{GeV}$. (We emphasise that introducing $g_{\rm SO} \neq 1$ is an expedient that mimics those effects of a *momentum-dependent* mass function exposed and quantified in Refs. [8,19].) It will be noted that m_σ increases to a value which matches an estimate for the $\bar{q}q$-component of this meson obtained using unitarised chiral perturbation theory.[22]

In quantum field theory an analogue for baryons of the BSE for mesons is a Poincaré covariant Faddeev equation that sums all possible exchanges and interactions that can take place between three dressed-quarks. A tractable Faddeev equation[23] is founded on the observation that an interaction which describes colour-singlet mesons also generates quark-quark (diquark) correlations in the colour-$\bar{3}$ (antitriplet) channel.[24] In QCD these correlations are *nonpointlike*:[2] $r_{qq} \approx 0.8\,\mathrm{fm}$.

Within the model of Ref. [20] we have formulated the Faddeev equation for $u\,\&\,d$-quark octet and decuplet baryons. The kernel involves the

Table 2. *Row-1*: Dressed-quark-core masses for nucleon and Δ, their first radial excitations (denoted by $*$), and the parity-partners of these states, computed with $g_N = 1.33$, $g_\Delta = 1.58$. *Row-2*: Bare-masses inferred from a coupled-channels analysis at the Excited Baryon Analysis Center (EBAC).[27] EBAC's method does not provide a bare nucleon mass and "..." indicates states not found in their analysis.

	m_N	m_{N^*}	$m_{N\frac{1}{2}^-}$	$m_{N^*\frac{1}{2}^-}$	m_Δ	m_{Δ^*}	$m_{\Delta\frac{3}{2}^-}$	$m_{\Delta^*\frac{3}{2}^-}$
This work	1.05	1.73	1.89	2.10	1.33	1.85	2.01	2.18
EBAC		1.76	1.80	...	1.39	...	1.98	...

propagation of a dressed-quark, which originates at the breakup of one diquark and ends at the formation of another (see Fig. 1 of Ref. [2]). The equations are particularly simple if one uses an additional truncation: $S_{\rm exchange} \to (g_B/M)$, based on the so-called *static approximation*,[25] but here with a parameter g_B introduced to restore attraction that is lost through this expedient. Now the Faddeev equation for the ground-state Δ reduces to the following eigenvalue problem for m_Δ:

$$2\pi^2 = \frac{g_B}{M}\frac{1}{m_{1^+}^2}\int_0^1 d\alpha\, 4\int_q^\Lambda \Gamma_{1^+}^2 \frac{(m_{1^+}^2 + (1-\alpha)^2 m_\Delta^2)(\alpha\, m_\Delta + M)}{(q^2 + (1-\alpha)M^2 + \alpha m_{1^+}^2 - \alpha(1-\alpha)m_\Delta^2)^2}, \tag{1}$$

where Γ_{1^+} is the canonically normalised Bethe-Salpeter amplitude for the 1^+-diquark and $\int_q^\Lambda := \int d^4q/(4\pi^2)$, regularised as explained in Ref. [20]. The Δ is particularly simple because it only involves an axial-vector diquark. It could conceivably also contain a vector diquark component. However, that has conflicting parity and hence can only appear in combination with material orbital angular momentum between the quark and vector diquark, which is impossible in our model and strongly suppressed in Nature. The nucleon equation is more complicated than Eq. (1) because one must combine scalar and axial-vector diquark components but it remains quite simple: in general, a 5×5-matrix eigenvalue problem for m_N.

In Table 2 we report preliminary results for the dressed-quark-core masses of the ground-state nucleon and Δ, their first radial excitations, and the parity-partners of these states. This is a first step toward the theory goals described in Ref. [13]. The result $m_\Delta \approx 3M$, which is a good approximation over a wide range of current-quark masses and interaction strengths, emphasises the simplicity of the Δ resonance. Naturally, in quantum field theory as in quantum mechanics, the bound-state amplitude for a first radial excitation must possess a single zero.[16] We implement that feature via the method employed in Ref. [26]; i.e., for instance, by introducing a factor $(1 - d_{\rm r} q^2)$ on the right-hand-side in Eq. (1) with $d_{\rm r} = 1/[2\Lambda_{\rm ir}^2]$

where $1/\Lambda_{\rm ir} = 0.83\,{\rm fm}$ is the model's confinement length-scale.[20] We view the agreement between our dressed-quark-core masses and EBAC's bare masses as significant, since no attempt was made to ensure this. Our results hint that $N(2090)\,S_{11}$ is the first radial excitation of $N(1535)\,S_{11}$.

3. Elastic form factors

Once one has bound-state masses and Bethe-Salpeter or Faddeev amplitudes, the computation of hadron elastic and transition form factors can proceed. The DSE calculations of $F_\pi^{\rm em}(Q^2)$ in Refs. [28,29] are an archetype (Q^2 is the squared-momentum-transfer). The most systematic of these[29] predicted the measured form factor.[30] An elucidation of the sensitivity of $F_\pi^{\rm em}(Q^2)$ to the pointwise behaviour of the interaction between quarks is the main theme of Ref. [20]; and in Fig. 3 we compare the form factor computed using a contact-interaction with the QCD-based DSE prediction[29] and contemporary experimental data.[30–32]

Both the QCD-based result and that obtained from the momentum-in-

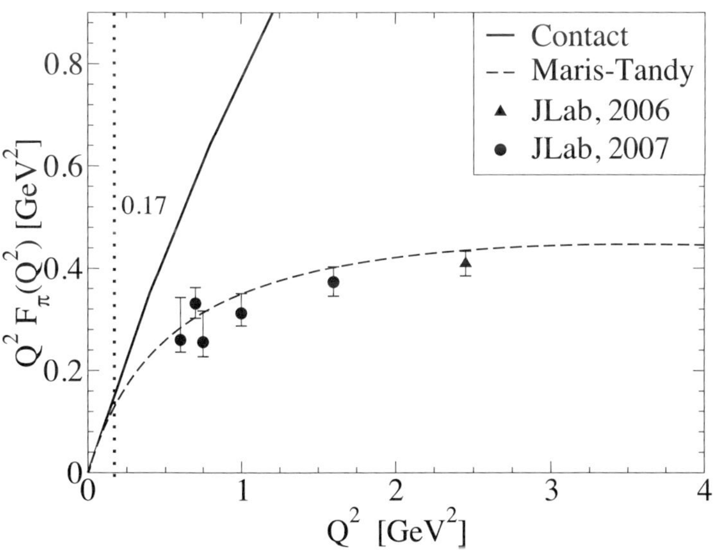

Fig. 3. *Solid curve*: Result obtained for $Q^2F_\pi(Q^2)$ using a contact interaction.[20] *Dashed curve*: DSE prediction,[29] which employed a momentum-dependent renormalisation-group-improved gluon exchange interaction. For $Q^2 > 0.17\,{\rm GeV}^2 \approx M^2$, marked by the vertical *dotted line*, the contact interaction result for $F_\pi(Q^2)$ differs from that in Ref. [29] by more than 20%. The data are from Refs. [31,32].

dependent interaction yield the same values for the pion's static properties. However, for $Q^2 > 0$ the form factor computed using $\sim 1/k^2$ vector boson exchange is immediately distinguishable empirically from that produced by a momentum-independent interaction. Indeed, the figure shows that for $F_\pi^{\rm em}$, existing experiments can already distinguish between different possibilities for the quark-quark interaction.

The analysis of Ref. [20] demonstrates that when a momentum-independent vector-exchange interaction is regularised in a symmetry-preserving manner, the results are directly comparable with experiment, computations based on well-defined and systematically-improvable truncations of QCD's DSEs,[29] and perturbative QCD. However, a contact interaction, whilst capable of describing pion static properties well, generates a form factor whose evolution with Q^2 deviates markedly from experiment for $Q^2 > 0.17\,{\rm GeV}^2 \approx M^2$ and produces asymptotic power-law behaviour: $Q^2 F_\pi(Q^2) \approx$ constant, in serious conflict with perturbative-QCD.[33–35]

It is noteworthy that the contact interaction produces a momentum-independent dressed-quark mass function, in contrast to QCD-based DSE studies[3,4,7] and lattice-QCD.[5] This is the origin of the marked discrepancy between the form factor it produces and extant experiment. Hence Ref. [20] highlights that form factor observables, measured at an upgraded Jefferson laboratory, e.g., are capable of mapping the running of the dressed-quark mass function. Efforts are underway to establish the signals of the running mass in baryon elastic and transition form factors. In this connection one statement can readily be made. Faddeev and Bethe-Salpeter amplitudes are peaked at zero relative momentum. Hence, as demonstrated explicitly in Ref. [36], the domain of greatest support in the impulse approximation calculation of elastic form factors is that with each quark absorbing: mesons, $Q/2$; and baryons, $Q/3$. The dressed-quark mass, $M(p^2)$, is perturbative for $p^2 > 2\,{\rm GeV}^2$. It can be argued from these observations that pQCD behaviour will not be observed in meson form factors unless $Q^2 > 8\,{\rm GeV}^2$ and not in baryon form factors unless $Q^2 > 18\,{\rm GeV}^2$.

Numerous nucleon properties have been computed via a more sophisticated version of the Faddeev equation than that used above to produce Table 2.[1,2,37–39] Herein we draw only one example from a comprehensive analysis of nucleon electromagnetic form factors[2] that incorporates the momentum-dependent mass function illustrated in Fig. 1. In Fig. 4 we depict a ratio of flavour-separated contributions to the proton's Dirac form factor. (NB. When isospin is a good symmetry, $F_1^{n,u}/F_1^{n,d} = F_1^{p,d}/F_1^{p,u}$.) The predicted Q^2-dependence owes to the presence of axial-vector diquark correla-

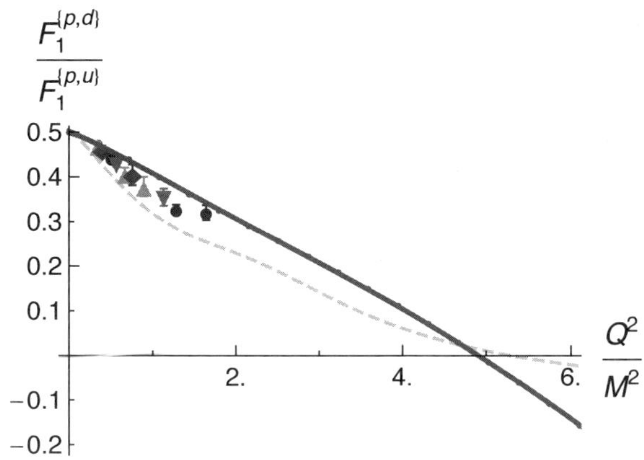

Fig. 4. Computed ratio of flavour-separated contributions to proton's Dirac form factor (M =nucleon mass) *solid curve.* Data, reconstructed from selected neutron electric form factor data using parametrisations of Ref. [40] when necessary:[41] Ref. [42], *up-triangles*; Ref. [43], *diamonds*; Ref. [44], *circles*; and Ref.[45], *down-triangles. Dashed-curve* – This ratio computed from a recent parametrisation of nucleon form factor data.[46]

tions in the nucleon. It has been found[2] that the proton's singly-represented d-quark is more likely to be struck in association with an axial-vector diquark correlation than with a scalar, and form factor contributions involving an axial-vector diquark are soft. On the other hand, the doubly-represented u-quark is predominantly linked with harder scalar-diquark contributions. This produces a d-quark Dirac form factor which is softer than that of u-quark and a ratio $F_1^{p,d}/F_1^{p,u}$ that passes through zero. The location of the zero depends on the relative probability of finding 1^+ and 0^+ diquarks in the proton. The same physics explains the $x = 1$ value of the $d_v(x)/u_v(x)$ ratio of valence-quark distribution functions in the proton.[47]

4. Epilogue

Herein we have exemplified the dramatic impact that DCSB has upon observables. The behaviour of the dressed-quark mass function heralds DCSB; and the momentum dependence manifest in Fig. 1 is an essentially quantum field theoretical effect. Exposing and elucidating its consequences therefore requires a nonperturbative and symmetry-preserving approach, where the latter means preserving Poincaré covariance, chiral and electromagnetic current-conservation, etc. The Dyson-Schwinger equations (DSEs) provide such a framework. Experimental and theoretical studies are underway that will identify observable signals of $M(p^2)$ and thereby explain the most important mass-generating mechanism for visible matter in the Standard Model.

This is an exciting time in hadron physics. Through the DSEs, we are positioned to unify phenomena as apparently diverse as the: hadron spectrum; hadron elastic and transition form factors, from small- to large-Q^2; and parton distribution functions.[47] The key is an understanding of both the fundamental origin of nuclear mass and the far-reaching consequences of the mechanism responsible; namely, DCSB. These things might lead us to an explanation of confinement, the phenomenon that makes nonperturbative QCD the most interesting piece of the Standard Model.

5. Acknowledgments

. We acknowledge valuable discussions with C. Hanhart, T.-S. H. Lee, V. Mokeev, S. Riordan, S. M. Schmidt, P. C. Tandy and B. Wojtsekhowski. This work was supported by: Forschungszentrum Jülich GmbH (HLLR, CDR); the National Natural Science Foundation of China, contract no. 10705002 (LC); the U. S. Department of Energy, Office of Nuclear Physics, contract nos. DE-AC02-06CH11357 (HLLR, CDR) and DE-FG03-97ER4014 (ICC); and the Department of Energy's Science Undergraduate Laboratory Internship programme (HLLR).

References

1. V. V. Flambaum *et al.*, *Few Body Syst.* **38**, 31 (2006).
2. I. C. Cloët *et al.*, Roberts, *Few Body Syst.* **46**, 1 (2009).
3. M. S. Bhagwat, M. A. Pichowsky, C. D. Roberts and P. C. Tandy, *Phys. Rev.* **C68**, 015203 (2003).
4. M. S. Bhagwat and P. C. Tandy, *AIP Conf. Proc.* **842**, 225 (2006).
5. P. O. Bowman *et al.*, *Phys. Rev.* **D71**, 054507 (2005).
6. G. Krein, C. D. Roberts and A. G. Williams, *Int. J. Mod. Phys.* **A7**, 5607 (1992).
7. C. D. Roberts, *Prog. Part. Nucl. Phys.* **61**, 50 (2008).
8. L. Chang and C. D. Roberts, "Hadron Physics: The Essence of Matter," arXiv:1003.5006 [nucl-th].
9. C. D. Roberts, M. S. Bhagwat, A. Höll and S. V. Wright, *Eur. Phys. J. ST* **140**, 53 (2007).
10. A. Bashir, A. Raya, I. C. Cloët and C. D. Roberts, *Phys. Rev.* **C78**, 055201 (2008).
11. A. Bashir, A. Raya, S. Sanchez-Madrigal and C. D. Roberts, *Few Body Syst.* **46**, 229 (2009).
12. S. J. Brodsky, C. D. Roberts, R. Shrock and P. C. Tandy, "Essence of the vacuum quark condensate," arXiv:1005.4610 [nucl-th].

13. I. Aznauryan *et al.*, "Theory Support for the Excited Baryon Program at the JLab 12 GeV Upgrade," arXiv:0907.1901 [nucl-th].
14. H. J. Munczek, *Phys. Rev.* **D52**, 4736 (1995).
15. A. Bender, C. D. Roberts and L. Von Smekal, *Phys. Lett.* **B380**, 7 (1996).
16. A. Höll, A. Krassnigg, C. D. Roberts, *Phys. Rev.* **C70**, 042203(R) (2004).
17. A. Höll *et al.*, *Phys. Rev.* **C71**, 065204 (2005).
18. M. S. Bhagwat *et al.*, *Phys. Rev.* C **76**, 045203 (2007).
19. L. Chang and C. D. Roberts, *Phys. Rev. Lett.* **103**, p. 081601 (2009).
20. L. X. Gutiérrez-Guerrero, A. Bashir, I. C. Cloët and C. D. Roberts, *Phys. Rev.* **C81**, 065202 (2010).
21. C. Amsler *et al.* [Particle Data Group], *Phys. Lett.* B **667**, 1 (2008).
22. J. R. Peláez and G. Ríos, *Phys. Rev. Lett.* **97**, 242002 (2006)
23. R. T. Cahill, C. D. Roberts and J. Praschifka, *Austral. J. Phys.* **42**, 129 (1989).
24. R. T. Cahill, C. D. Roberts and J. Praschifka, *Phys. Rev.* **D36**, 2804 (1987).
25. A. Buck, R. Alkofer and H. Reinhardt, *Phys. Lett.* **B286**, 29 (1992).
26. M. K. Volkov and V. L. Yudichev, *Phys. Part. Nucl.* **31**, 282 (2000) [*Fiz. Elem. Chast. Atom. Yadra* **31**, 576 (2000)].
27. N. Suzuki *et al.*, *Phys. Rev. Lett.* **104**, 042302 (2010).
28. P. Maris and C. D. Roberts, *Phys. Rev.* **C58**, 3659 (1998).
29. P. Maris and P. C. Tandy, *Phys. Rev.* **C62**, 055204 (2000).
30. J. Volmer *et al.*, *Phys. Rev. Lett.* **86**, 1713 (2001).
31. T. Horn *et al.*, *Phys. Rev. Lett.* **97**, 192001 (2006).
32. V. Tadevosyan *et al.*, *Phys. Rev.* **C75**, 055205 (2007).
33. G. R. Farrar and D. R. Jackson, *Phys. Rev. Lett.* **43**, 246 (1979).
34. A. V. Efremov and A. V. Radyushkin, *Phys. Lett.* **B94**, 245 (1980).
35. G. P. Lepage and S. J. Brodsky, *Phys. Rev.* **D22**, 2157 (1980).
36. C. D. Roberts, *Nucl. Phys.* A **605**, 475 (1996).
37. I. C. Cloët et al., *Few Body Syst.* **42**, 91 (2008).
38. I. C. Cloët and C. D. Roberts, *PoS* **LC2008**, 047 (2008).
39. L. Chang *et al.*, *Chin. Phys.* **C33**, 1189 (2009).
40. J. J. Kelly, *Phys. Rev.* **C70** (2004) 068202.
41. S. Riordan, private communication.
42. D. I. Glazier *et al.*, *Eur. Phys. J.* **A24**, 101 (2005).
43. J. Bermuth *et al.*, *Phys. Lett.* **B564**, 199 (2003).
44. B. Plaster *et al.*, *Phys. Rev.* **C73**, 025205 (2006).
45. G. Warren *et al.*, *Phys. Rev. Lett.* **92**, 042301 (2004).
46. R. Bradford, A. Bodek, H. S. Budd and J. Arrington, *Nucl. Phys. Proc. Suppl.* **159**, 127 (2006).
47. R. J. Holt and C. D. Roberts, "Distribution Functions of the Nucleon and Pion in the Valence Region," arXiv:1002.4666 [nucl-th].

ELECTROPRODUCTION OF RESONANCES AT LARGE MOMENTUM TRANSFERS

V. M. BRAUN

Institut für Theoretische Physik, Universität Regensburg, 93040 Regensburg, Germany

E-mail: vladimir.braun@physik.uni-regensburg.de

Measurements of transition form factors for the electro-excitation of nucleon resonances in the $Q^2 = 5 - 14$ GeV^2 region can provide one with the information on quark wave functions at small transverse separations. In particular a comparison of form factors for the states of opposite parity can give insight in the mechanisms of chiral symmetry breaking. I discuss perspectives of the theoretical description of such reactions using a combination of lattice calculations and light-cone sum rules.

Keywords: electroproduction; nucleon resonances; lattice QCD

1. Introduction

Electroproduction of nucleon resonances has long been recognized as an important tool in the exploration of the nucleon structure at different scales. There is a growing consensus that perturbative QCD (pQCD) factorization based on hard gluon exchange is not reached at present energies; however, the emergence of quarks and gluons as the adequate degrees of freedom is expected to happen earlier, at $Q^2 \sim$ a few GeV^2. Measurements of the form factors in this transition region are planned at Jefferson Lab[1,2] using the CLAS12 detector. In this talk such reactions are addressed from the theory perspective. I try to formulate the physics goals and explain an approach to the electroproduction of resonances that combines lattice calculations of wave functions at small transverse separations with dispersion relations and quark-hadron duality, known as light-cone sum rules (LCSR).

Quantum chromodynamics (QCD) predicts[3–6] that at large momentum transfer the transition form factors become increasingly dominated by the contribution of the valence Fock state with small transverse separation between the partons. In reality the dominance of valence configurations is likely to be achieved but "shrinkage" to small transverse separations does not seem to occur and also power counting rules based on helicity conservation do not work at moderate Q^2 that we are able to study experimentally.

To be more specific, the nucleon valence state contains contributions with different orbital angular momentum $\ell_z = 0, \pm 1$. The leading contribution to form factors at large Q^2 comes from the $\ell_z = 0$ state:[6]

$$|P\uparrow\rangle^{\ell_z=0} = \int \frac{[dx][d^2\vec{k}]}{6\sqrt{x_1x_2x_3}}\,\psi^{\ell_z=0}(x_i,\vec{k}_i)\times$$
$$\times\Big\{\big|u^\uparrow(x_1,\vec{k}_1)u^\downarrow(x_2,\vec{k}_2)d^\uparrow(x_3,\vec{k}_3)\big\rangle - \big|u^\uparrow(x_1,\vec{k}_1)d^\downarrow(x_2,\vec{k}_2)u^\uparrow(x_3,\vec{k}_3)\big\rangle\Big\}$$

where $\psi^{\ell_z=0}(x_i,\vec{k}_i)$ is the three-quark light-cone wave function that depends on quark momentum fractions x_i and transverse momenta $\vec{k}_i$[a].

The simplification that occurs at asymptotically large Q^2 is that the $\vec{k}$–dependence of wave functions becomes irrelevant and all necessary (non-perturbative) information is contained in the integral over transverse momenta

$$\Phi_3(x_i;\mu) = \int^{\mu} [d^2\vec{k}]\;\psi^{\ell_z=0}(x_i,\vec{k}_i)$$

where the cutoff $\mu \sim Q$ has to be imposed to make the integral converge.

The function $\Phi_3(x_i;\mu)$ is called the leading-twist distribution amplitude (DA). It can be studied using the operator product expansion[7]

$$\begin{aligned}\Phi_3(x_i;\mu) = 120 f_N(\mu)x_1x_2x_3\Big\{&1 + c_{10}(\mu)(x_1 - 2x_2 + x_3) + c_{11}(\mu)(x_1 - x_3)\\ &+ c_{20}(\mu)\left[1 + 7(x_2 - 2x_1x_3 - 2x_2^2)\right] + c_{21}(\mu)\,(1 - 4x_2)\,(x_1 - x_3)\\ &+ c_{22}(\mu)\left[3 - 9x_2 + 8x_2^2 - 12x_1x_3\right] + \ldots\Big\}\end{aligned} \tag{1}$$

where $f_N(\mu)$ (wave function at the origin) and $c_{ik}(\mu)$ (shape parameters) are scale-dependent coefficients which can be defined as matrix elements of (multiplicatively renormalizable) local operators. They can be calculated using QCD sum rules[8] or lattice QCD.[9] The DA $\Phi_3(x_i;\mu)$ is thus a much simpler object compared to the full light-cone wave function $\psi^{\ell_z=0}(x_i,\vec{k}_i)$; unfortunately this reduction does not seem to work in the $Q^2 = 5 - 15$ GeV^2 range.

Another problem is that the standard pQCD factorization approach does not take into account contributions of states with non-vanishing orbital

[a]Here and below we do not show contributions which vanish in the limit of small transverse separations, cf. Ref. [10].

angular momentum. For example[10]

$$|P\uparrow\rangle^{\ell_z=1} = \int \frac{[dx][d^2\vec{k}]}{6\sqrt{x_1x_2x_3}} \left[k_1^+\psi_1^{\ell_z=1}(x_i,\vec{k}_i) + k_2^+\psi_2^{\ell_z=1}(x_i,\vec{k}_i)\right] \times$$
$$\times\left\{\left|u^\uparrow(x_1,\vec{k}_1)u^\downarrow(x_2,\vec{k}_2)d^\downarrow(x_3,\vec{k}_3)\right\rangle - \left|d^\uparrow(x_1,\vec{k}_1)u^\downarrow(x_2,\vec{k}_2)u^\downarrow(x_3,\vec{k}_3)\right\rangle\right\}$$

where $k^\pm = k_x \pm ik_y$. The new light-cone wave functions $\psi_1^{\ell_z=1}(x_i,\vec{k}_i)$ and $\psi_2^{\ell_z=1}(x_i,\vec{k}_i)$ are reduced to next-to-leading twist-4 nucleon DAs[12,13]

$$\Phi_4(x_i;\mu) = \int^\mu \frac{[d^2\vec{k}]}{m_N x_3}\, k_3^-\left[k_1^+\psi_1^{L=1} + k_2^+\psi_2^{L=1}\right](x_i,\vec{k}_i)$$
$$\Psi_4(x_i;\mu) = \int^\mu \frac{[d^2\vec{k}]}{m_N x_2}\, k_2^-\left[k_1^+\psi_1^{L=1} + k_2^+\psi_2^{L=1}\right](x_i,\vec{k}_i)$$

and, again, can be studied using OPE[7,12]

$$\Phi_4(x_i;\mu) = 12\lambda_1(\mu)x_1x_2 + 12f_N(\mu)x_1x_2\left[1+\frac{2}{3}(1-5x_3)\right]+\dots$$
$$\Psi_4(x_i;\mu) = 12\lambda_1(\mu)x_1x_3 + 12f_N(\mu)x_1x_3\left[1+\frac{2}{3}(1-5x_2)\right]+\dots \qquad (2)$$

Note that to this accuracy twist-4 DAs include one new parameter only, λ_1. A similar expansion and the reduction to DAs can be worked out for nucleon resonances. Electroproduction of nucleon resonances at high Q^2 gives access to three-quark wave functions, more precisely to the overlap integrals between the wave functions of the nucleon and the resonance. These overlap integrals are in general too complicated to be calculated in QCD directly and our strategy will be to reduce these integrals to convolutions of distribution amplitudes which can be constrained using lattice QCD.

We believe that one important physics goal for such studies will be to compare DAs (alias light-cone wave functions at small transverse separations) of baryon states with opposite parity, e.g. $J^P = 1/2^+$ and $J^P = 1/2^-$. It is well known that such "parity-doublets" are non-degenerate in QCD because of spontaneous breaking of chiral symmetry. It is not known, however, whether this difference mainly affects the "pion cloud" or it is present at short distances already and affects momentum fraction distributions of valence quarks. Moreover, there are indications[14] that chiral symmetry is effectively restored in the spectra of higher-mass resonances and it would be extremely interesting to compare the corresponding wave functions.

The general strategy of combining the constraints on DAs from a lattice calculation with LCSRs to calculate the form factors is suggested in Ref. [15] where, as the first demonstration, we considered the electroproduction of

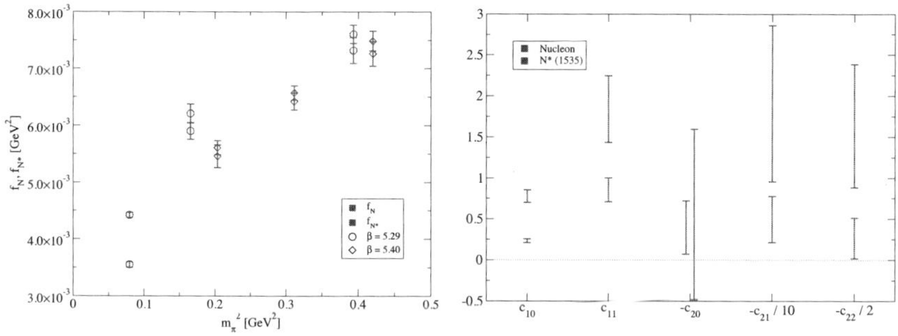

Fig. 1. Wave functions at the origin (left panel) and shape parameters of the nucleon and $N^*(1535)$ distribution amplitudes (right panel). Only statistical errors are shown.

$N^*(1535)$, the parity partner of the nucleon. In what follows I describe an ongoing work in this direction by the Regensburg theory group and QCDSF collaboration.

2. Nucleon and $N^*(1535)$ Distribution Amplitudes from Lattice QCD

Baryon states of different parity can be identified in a lattice calculation as those propagating forward or backward in (imaginary) time,[16,17] so in fact the results for $N^*(1535)$ reported in Ref. [15] are essentially a byproduct of our calculation of the nucleon DAs.[9] These results were obtained using QCDSF/DIK gauge configurations with two flavors of clover fermions for two different β values and several quark masses on $24^3 \times 48$ lattices. The calculation was done using nonperturbatively renormalized three-quark operators[18] with up to two derivatives imposing a RI'-MOM–like renormalization condition and converting the results to the $\overline{MS}$ scheme. In this way the mixing with "total derivatives" is automatically taken into account.

This work will be continued, using larger lattices and smaller pion masses in order to minimize effects of the chiral extrapolation. In Fig. 1 I present preliminary results[19] of the new calculation using ca. 600 $N_f = 2$ gauge configurations on a $32^3 \times 64$ lattice with $\beta = 5.29$ (a = 0.0753 fm) and pion mass $m_\pi \simeq 282$ MeV ($m_\pi L = 3.44$)). The left panel shows the comparison of nucleon and $N^*(1535)$ wave functions at the origin, f_N and f_{N^*}, as a function of m_π^2. The left-most points at $m_\pi^2 \simeq 0.08$ GeV^2 are new. They are much closer to the physical point and show, for the first time, that for small pion masses f_{N^*} becomes smaller than f_N. This phenomenon still has to be understood.

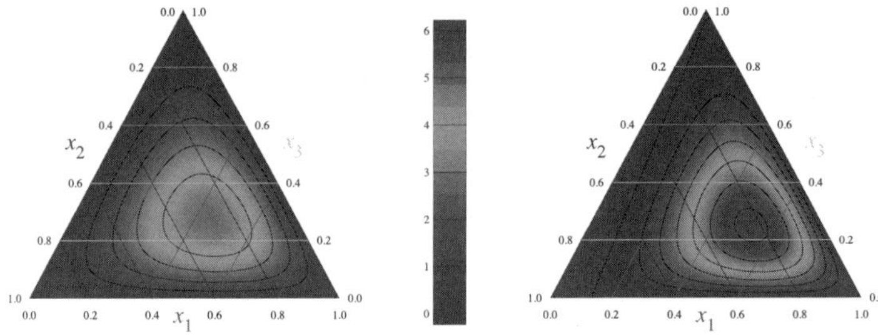

Fig. 2. Barycentric plots for the nucleon (left) and $N^*(1535)$ (right) distribution amplitudes.

The results for shape parameters of the nucleon distribution amplitudes (1) are shown on the right panel. One sees that coefficients of first-order polynomials c_{10} and c_{11} can be quantified, whereas contributions of second order polynomials are less constrained. The main reason for this are the $\mathcal{O}(a)$ discretization errors in the chain rule for derivatives $D(A \cdot B) = (DA) \cdot B + A \cdot (DB) + \mathcal{O}(a)$ which spoil energy conservation: after the (nonperturbative) renormalization we obtain for the sum of the quark momentum fractions $x_1 + x_2 + x_3 \simeq 0.94$ instead of unity. This is one of the issues that have to be addressed in future studies.

The main result so far is that the accuracy of modern lattice calculations is sufficient to detect differences in quark momentum fraction distributions in the nucleon and its parity partner state. Our calculations support a qualitative picture suggested by QCD sum rules[8] that the valence quark with the spin parallel to that of the nucleon carries most of its momentum, and for $N^*(1535)$ this effect appears to be even stronger. As an illustration we show in Fig. 2 the DAs of the nucleon and $N^*(1535)$ in barycentric coordinates.[b] Note that the $N^*(1535)$ DA appears to be more narrow and shifted towards the lower-right corner.

All these results are preliminary and the study will be continued using larger lattices and smaller pion masses. The comparison of the nucleon and $N^*(1535)$ will remain our primary goal for some time, but the calculations will also be extended to the full $J^P = 1/2^+$ and $J^P = 1/2^-$ baryon octets and to the decuplets. The main problem that has to be addressed in fu-

[b]In difference to the similar plot in Ref. [15] we only include the terms in c_{10} and c_{11} and discard contributions of second order polynomials which have large errors.

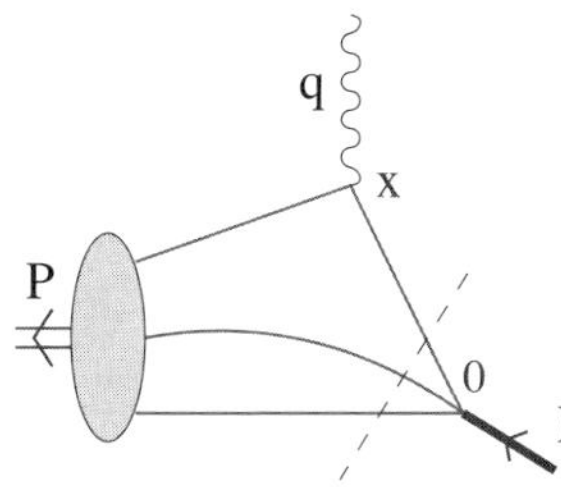

Fig. 3. Schematic structure of the light-cone sum rule for baryon form factors.

ture is the identification of resonance contributions for small quark (pion) masses such that strong decays, e.g. $N^*(1535) \to N\pi, N\eta$, are allowed. The separation of $N^*(1535)$ and $N^*(1650)$ may prove to be difficult. In particular the large decay width of $N^*(1535)$ in the $N\eta$ channel has to be understood and it can only be addressed in lattice calculations with three flavors of dynamic fermions. Discretization errors $\mathcal{O}(a)$ in the lattice definition of the relevant operators become a serious issue for second moments and have to be reduced in order that the results for c_{2k} coefficients become fully quantitative.

3. Transition form factors from Light-Cone Sum Rules

The matrix element of the electromagnetic current $j_\nu^{\rm em}$ between spin-1/2 states of opposite parity can be parametrized in terms of two independent form factors, which can be chosen as

$$\begin{aligned} \langle N^*(P')|j_\nu^{\rm em}|N(P)\rangle &= \bar{u}_{N^*}(P')\gamma_5\Gamma_\nu u_N(P)\,, \\ \Gamma_\nu &= \frac{G_1(q^2)}{m_N^2}(\not{q}q_\nu - q^2\gamma_\nu) - i\frac{G_2(q^2)}{m_N}\sigma_{\nu\rho}q^\rho\,, \end{aligned} \tag{3}$$

where $q = P' - P$ is the momentum transfer. The LCSRs are derived from the correlation function

$$\int dx\, e^{-iqx}\langle N^*(P)|T\{\eta(0)j_\mu^{\rm em}(x)\}|0\rangle\,,$$

where η is a suitable operator with nucleon quantum numbers, e.g. the Ioffe current.[20] Making use of the duality of QCD quark-gluon and hadronic degrees of freedom through dispersion relations one can write a representation for the form factors appearing in (3) in terms of the DAs of N^*. Schematically, the sum rules take the form

$$G_{1,2}(Q^2) = \sum_k \int [dx]\, C_{1,2;k}(x_i, Q^2, s_0, M^2, \mu, \alpha_s(\mu))\, \Phi_k(x_i, \mu)$$

where the sum goes over contributions of nucleon DAs of increasing twist and C_k are the coefficient functions that can be calculated in QCD perturbation theory and modified using dispersion relations to include two nonperturbative parameters: s_0, the interval of duality, and M^2, the Borel parameter which specifies the distance (in imaginary time) on which matching between hadronic and quark representations for the correlation function is being done. The dependence on M^2 is unphysical in the same sense as the factorization scale μ dependence in truncated perturbative expansions, but it is usually weak. The pQCD limit[3–6] corresponds to the leading part of the contribution of leading-twist DAs at large momentum transfer, and the main difference is that higher-twist contributions are *not* suppressed by powers of $\Lambda^2_{\rm QCD}/Q^2$, but rather of $\Lambda^2_{\rm QCD}/s_0$ with $s_0 \sim (1.5\ {\rm GeV})^2$. The reason for this is that LCSRs take into account "soft" contributions to the form factors coming from large transverse separations. The attractive feature of this approach is that such terms are calculated in terms of the DAs, thus avoiding the need to know (model) the full nonperturbative $k_\perp$ dependence of wave functions.

In leading order, the sum rules for $Q^2G_1(Q^2)/(m_N m_{N^*})$ and $-2G_2(Q^2)$ have the same functional form as the similar sum rules[21,22] for the Dirac and Pauli electromagnetic form factors of the proton, with the replacement $m_N \to m_{N^*}$ in the light-cone expansion part, and different DAs.

The experimental results for the electroproduction of $N^*(1535)$ are usually presented for helicity amplitudes $A_{1/2}(Q^2)$ and $S_{1/2}(Q^2)$ which can be expressed in terms of the form factors:[23]

$$\begin{aligned} A_{1/2} &= e\,B\left[Q^2G_1(Q^2) + m_N(m_{N^*} - m_N)G_2(Q^2)\right], \\ S_{1/2} &= \frac{e}{\sqrt{2}}B\,C\left[(m_N - m_{N^*})G_1(Q^2) + m_N G_2(Q^2)\right]. \end{aligned}$$

Here e is the elementary charge and B, C are kinematic factors defined as

$$B = \sqrt{\frac{Q^2 + (m_{N*} + m_N)^2}{2m_N^5(m_{N^*}^2 - m_N^2)}}, \qquad C = \sqrt{1 + \frac{(Q^2 - m_{N^*}^2 + m_N^2)^2}{4Q^2 m_{N^*}^2}}.$$

The results of the LCSR calculation of the form factors (normalized to the dipole) and helicity amplitudes using lattice-constrained N^* DAs from Ref. [15] is presented in Fig. 4. The shaded areas show the estimated uncertainties. The Q^2-dependence of the form factors $Q^2G_1(Q^2)$ and $G_2(Q^2)$ is predicted to be similar to the Dirac and Pauli nucleon electromagnetic form factors, respectively. Only the normalization is different: The observed negative amplitude $S_{1/2}$ implies a small value of G_2.

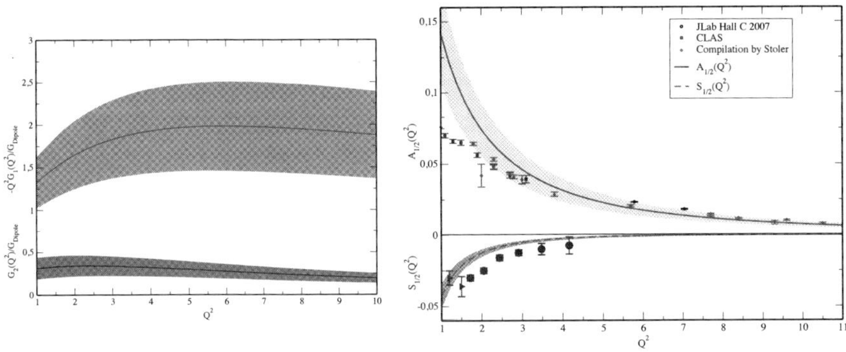

Fig. 4. Form factors (left panel) and helicity amplitudes (right panel) for the electroexcitation of the $N^*(1535)$ resonance calculated using the DAs in Ref. [15]. The experimental data are from Refs. [24–27].

These results have to be viewed as exploratory and further work is needed to make them quantitative. First of all, the LCSRs for baryon form factors have to be extended to the next-to-leading (NLO) accuracy, i.e. including the $\mathcal{O}(\alpha_s)$ corrections. This calculation is rather cumbersome and not straightforward because of contributions of evanescent operators. As the first step in this direction, the NLO corrections to contributions of leading-twist DAs in the LCSRs for the electromagnetic nucleon form factors F_1 and F_2 have been calculated in Ref. [28], see Fig. 5. One sees that such corrections are significant, especially in the G_E/G_M ratio. The extension of this calculation to contributions of sub-leading twist-4 DAs and the detailed study of baryon mass corrections to the sum rules is planned for the coming years.

Second, one has to remember that the LCSRs suffer from irreducible uncertainties due to the duality assumption for contributions of higher resonances and the continuum which for meson form factors is estimated to be of order 10%. (see e.g. Ref. [29]). The experience with applying this method to baryons is much less than for mesons, so that this "systematic error" cannot be estimated reliably. It is therefore imperative to apply the same technique to a maximally broad class of reactions. At present applications include nucleon electromagnetic and axial form factors,[21,22,30] $N\gamma\Delta$ transitions,[33] pseudoscalar- and vector-meson couplings to octet and decuplet baryons,[34] weak decays of the type $\Lambda_b \to p\ell\nu_\ell$[31] and $\Lambda_b \to \Lambda\ell^+\ell^-$,[32] and pion electroproduction at threshold.[35,36] I expect that this list will continue to grow, giving us confidence in the whole program.

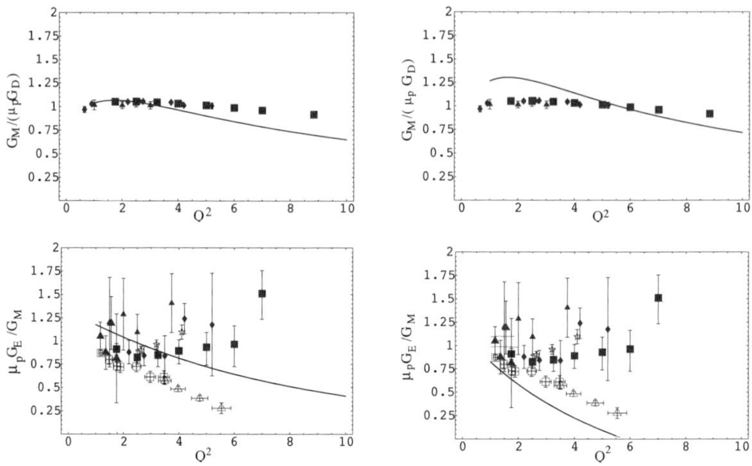

Fig. 5. LCSR results for the electromagnetic proton form factors for a realistic model of nucleon distribution amplitudes. Left panel: Leading order (LO); right panel: next-to-leading order (NLO) for twist-three contributions. Figure adapted from Ref. [28].

Acknowledgments

The author thanks M. Göckeler, R. Horsley, T. Kaltenbrunner, A. Lenz, Y. Nakamura, D. Pleiter, P. E. L. Rakow, J. Rohrwild, A. Schäfer, R. Schiel, G. Schierholz, H. Stüben, N. Warkentin, J. M. Zanotti for the collaboration on this project. Special thanks are due to Rainer Schiel for the possibility to use the results in Figs. 1,2 prior their presentation at LATTICE 2010.[19] This work is supported by the DFG through SFB/TR55 (Hadron Physics from Lattice QCD) and grants 9209070, 9209475.

References

1. V. D. Burkert, *Electromagnetic Transition Form Factors of Nucleon Resonances,* AIP Conf. Proc. **1056**, 348 (2008) [arXiv:0808.2326 [nucl-ex]].
2. I. Aznauryan *et al.*, *Theory Support for the Excited Baryon Program at the Jlab 12 GeV Upgrade,* arXiv:0907.1901 [nucl-th].
3. V. L. Chernyak and A. R. Zhitnitsky, JETP Lett. **25**, 510 (1977); V. L. Chernyak, A. R. Zhitnitsky and V. G. Serbo, JETP Lett. **26**, 594 (1977).
4. A. V. Efremov and A. V. Radyushkin, Phys. Lett. B **94**, 245 (1980); Theor. Math. Phys. **42**, 97 (1980).
5. G. P. Lepage and S. J. Brodsky, Phys. Rev. Lett. **43**, 545 (1979) [Erratum-ibid. **43**, 1625 (1979)].

6. G. P. Lepage and S. J. Brodsky, Phys. Rev. D **22**, 2157 (1980).
7. V. M. Braun, A. N. Manashov and J. Rohrwild, Nucl. Phys. B **807**, 89 (2009).
8. V. L. Chernyak and A. R. Zhitnitsky, Phys. Rept. **112**, 173 (1984).
9. V. M. Braun *et al.* [QCDSF Collaboration], Phys. Rev. D **79**, 034504 (2009).
10. X. d. Ji, J. P. Ma and F. Yuan, Nucl. Phys. B **652**,383 (2003);
11. X. d. Ji, J. P. Ma and F. Yuan, Eur. Phys. J. C **33**,75 (2004).
12. V. Braun, R. J. Fries, N. Mahnke and E. Stein, Nucl. Phys. B **589**, 381 (2000) [Erratum-ibid. B **607**, 433 (2001)].
13. A. V. Belitsky, X. d. Ji and F. Yuan, Phys. Rev. Lett. **91**, 092003 (2003).
14. L. Y. Glozman, Phys. Rept. **444**, 1 (2007).
15. V. M. Braun *et al.*, Phys. Rev. Lett. **103**, 072001 (2009).
16. F. X. Lee and D. B. Leinweber, Nucl. Phys. Proc. Suppl. **73**, 258 (1999).
17. S. Sasaki, T. Blum and S. Ohta, Phys. Rev. D **65**, 074503 (2002).
18. M. Göckeler *et al.* [QCDSF Collaboration and UKQCD Collaboration], Nucl. Phys. B **812**, 205 (2009).
19. R. Schiel *et al.* [QCDSF Collaboration], *presented at the XXVIII International Symposium on Lattice Field Theory, Sardinia, June 14-19, 2010.*
20. B. L. Ioffe, Nucl. Phys. B **188**, 317 (1981) [Erratum-ibid. B **191**, 591 (1981)].
21. V. M. Braun, A. Lenz, N. Mahnke and E. Stein, Phys. Rev. D **65**, 074011 (2002).
22. V. M. Braun, A. Lenz and M. Wittmann, Phys. Rev. D **73**, 094019 (2006).
23. I. G. Aznauryan, V. D. Burkert and T. S. Lee, arXiv:0810.0997 [nucl-th].
24. I. G. Aznauryan *et al.* [CLAS Collaboration], Phys. Rev. C **80**, 055203 (2009).
25. M. M. Dalton *et al.*, arXiv:0804.3509 [hep-ex].
26. H. Denizli *et al.* [CLAS Collaboration], Phys. Rev. C **76**, 015204 (2007).
27. P. Stoler, Phys. Rept. **226**, 103 (1993);
F. W. Brasse *et al.*, Nucl. Phys. B **110**, 413 (1976).
28. K. Passek-Kumericki and G. Peters, Phys. Rev. D **78**, 033009 (2008).
29. V. M. Braun, A. Khodjamirian and M. Maul, Phys. Rev. D **61**, 073004 (2000).
30. Z. G. Wang, S. L. Wan and W. M. Yang, Phys. Rev. D **73**, 094011 (2006); Eur. Phys. J. C **47**, 375 (2006).
31. M. q. Huang and D. W. Wang, Phys. Rev. D **69**, 094003 (2004).
32. T. M. Aliev, K. Azizi and M. Savci, Phys. Rev. D **81**, 056006 (2010).
33. V. M. Braun, A. Lenz, G. Peters and A.V. Radyushkin, Phys. Rev. D **73**, 034020 (2006).
34. T. M. Aliev, K. Azizi, A. Ozpineci and M. Savci, Phys. Rev. D **80**, 096003 (2009); T. M. Aliev, A. Ozpineci, M. Savci and V. S. Zamiralov, Phys. Rev. D **81**, 056004 (2010).
35. V. M. Braun, D. Y. Ivanov, A. Lenz and A. Peters, Phys. Rev. D **75**, 014021 (2007).
36. V. M. Braun, D. Y. Ivanov and A. Peters, Phys. Rev. D **77**, 034016 (2008).

FINAL RESULTS OF THE GEp-III EXPERIMENT AND THE STATUS OF THE PROTON FORM FACTORS

A. J. R. PUCKETT*

Los Alamos National Laboratory,
Los Alamos, NM 87544, U. S. A.
**E-mail: puckett@jlab.org*
www.lanl.gov

The GEp-III Collaboration

Thomas Jefferson National Accelerator Facility,
Newport News, VA 23606

The recently published final results of experiment E04-108 in Jefferson Lab Hall C extend the recoil polarization measurements of the ratio of the proton electric and magnetic form factors to $Q^2 = 8.5$ GeV^2, an increase in Q^2 coverage of more than 50%. A global fit of G_E^p and G_M^p to selected data for electron-proton elastic scattering cross sections and polarization observables is presented, illustrating the statistical impact of the new results.

Keywords: Proton Form Factors, Recoil Polarization Method

The elastic electromagnetic form factors of the nucleon are among the most fundamental quantities describing its structure. In the one-photon-exchange or Born approximation, the helicity-conserving Dirac form factor $F_1(Q^2)$ and the helicity-flip Pauli form factor $F_2(Q^2)$ fully characterize the amplitude for elastic electron-nucleon scattering at a spacelike momentum transfer $Q^2 = -q^2 > 0$. These form factors are accessible experimentally through both cross section and double-polarization measurements, and are often expressed in terms of the equivalent linear combinations $G_E = F_1 - \tau F_2$ (electric) and $G_M = F_1 + F_2$ (magnetic), which are more readily extracted from experimental observables. The Born differential cross section for elastic ep

scattering is given by[1]

$$\frac{d\sigma}{d\Omega_e} = \frac{\alpha^2}{Q^2}\left(\frac{E'_e}{E_e}\right)^2 \cot^2\frac{\theta_e}{2}\,\frac{G_E^2 + \frac{\tau}{\epsilon}G_M^2}{1+\tau}$$
$$= \frac{\sigma_{Mott}}{\epsilon(1+\tau)}\left[\epsilon G_E^2 + \tau G_M^2\right] \quad (1)$$

where α is the fine-structure constant of electromagnetism, E_e and E'_e are the lab incident and scattered electron energies, θ_e is the lab electron scattering angle, $\tau = Q^2/4M^2$, M is the nucleon mass, and $\epsilon = \left[1 + 2(1+\tau)\tan^2\frac{\theta_e}{2}\right]^{-1}$ corresponds to the longitudinal polarization of the virtual photon in the single-photon-exchange picture.

The polarization of protons scattered elastically by longitudinally polarized electrons has nonvanishing longitudinal and transverse components relative to the momentum transfer in the scattering plane given by[2,3]

$$I_0 P_l = \sqrt{\tau(1+\tau)}\tan^2\frac{\theta_e}{2}\,\frac{E_e + E'_e}{M} G_M^2, \quad (2)$$

$$I_0 P_t = -2\sqrt{\tau(1+\tau)}\tan\frac{\theta_e}{2} G_E G_M, \quad (3)$$

where $I_0 = G_E^2 + \frac{\tau}{\epsilon}G_M^2$. The component P_n normal to the scattering plane is exactly zero in the Born approximation. The ratio P_t/P_l is directly proportional to G_E/G_M:

$$R = \mu_p \frac{G_E^p}{G_M^p} = -\mu_p \frac{P_t}{P_l}\,\frac{E_e + E'_e}{2M}\tan\frac{\theta_e}{2} \quad (4)$$

The ratio (4) provides enhanced sensitivity to G_E at large Q^2, where G_M dominates the cross section (see equation (1)). Precise recoil polarization experiments[4–6] first established that R decreases rapidly with $Q^2 \geq 1$ GeV^2, in strong disagreement with cross section measurements. The extraction of a small G_E^2 term compared to a dominant $\frac{\tau}{\epsilon}G_M^2$ term from L/T separation measurements becomes highly sensitive to incompletely understood higher-order QED effects at large Q^2, including two-photon exchange (TPEX).[7] The enhanced sensitivity of the recoil polarization method to G_E and the typically smaller relative importance of radiative corrections and TPEX to the ratio (4) have led to a general consensus that the polarization data most reliably determine G_E^p at large Q^2.

Extending the accurate measurements of the proton and neutron electric and magnetic form factors to the highest accessible Q^2 provides crucial experimental input to understanding the transition between the nonperturbative and perturbative regimes of QCD. In addition to constraining

QCD-inspired phenomenological models, precise form factor data provide important model-independent information about the structure of the nucleon. The recently published results of experiment E04-108[8] in Jefferson Lab's Hall C extended the recoil polarization data for R to $Q^2 = 8.5$ GeV^2. In this paper we illustrate the statistical impact of the new data in a global fit to elastic electron-proton scattering data, including both cross section and polarization measurements, using the Kelly parametrization[9] of G_E^p and G_M^p.

The Kelly parametrization of the nucleon form factors[9] is given by the ratio of a polynomial of order n to a polynomial of order $n+2$ in $\tau = Q^2/4M^2$:

$$G(Q^2) = \frac{\sum_{k=0}^{n} a_k \tau^k}{1 + \sum_{k=1}^{n+2} b_k \tau^k} \tag{5}$$

This parametrization satisfies $G(Q^2) \propto Q^{-4}$ as $Q^2 \to \infty$ and gives the appropriate static ($Q^2 \to 0$) limit if one fixes $a_0 = 1$ for G_E^p and G_M^p/μ_p. This parametrization is chosen to have the large-Q^2 behavior expected from dimensional scaling laws in perturbative QCD.[10,11] The original 2004 analysis by Kelly using this parametrization found that a four-parameter fit with $n = 1$ could describe the data for G_E^p and $G_M^{p,n}$ with a reasonable χ^2. For simplicity's sake, the results presented here use the same choice.

The published elastic ep scattering data chosen for the fit include 421 differential cross section measurements spanning 0.005 GeV$^2 \leq Q^2 \leq$ 31.2 GeV^2 and 53 polarization measurements spanning 0.162 GeV$^2 \leq Q^2 \leq$ 8.49 GeV^2. The differential cross section data were taken from Refs. [12–23], while the polarization data were taken from Refs. [5,6,8,24–30]. While by no means comprehensive, the chosen data are sufficiently representative of the precision and Q^2 coverage of all elastic ep scattering data. All data included in this analysis were obtained from hydrogen targets.

The best fit parameters for G_E^p and G_M^p were obtained in a simultaneous global analysis of cross section and polarization data by minimizing the χ^2 function defined as

$$\chi^2 = \sum_{i=1}^{N_\sigma} \left[\frac{\sigma_{R,data}^{(i)} - \left(\epsilon G_E^2 + \tau G_M^2\right)^{(i)}}{\Delta\sigma_{R,data}^{(i)}} \right]^2 + \sum_{i=1}^{N_{pol.}} \left[\frac{R_{p,data}^{(i)} - R_p^{(i)}}{\Delta R_{p,data}^{(i)}} \right]^2, \tag{6}$$

where $R_p = \mu_p G_E^p / G_M^p$, G_E^p and G_M^p/μ_p are given by the formula (5) with $n = 1$ and $a_0 = 1$, $a_1^{E,M}$ and $b_{1,2,3}^{E,M}$ are eight adjustable parameters to be determined, $\sigma_{R,data} = \epsilon(1+\tau)\sigma_{data}/\sigma_{Mott}$ (see equation (1)), and $\Delta\sigma_{R,data}$ and $\Delta R_{p,data}$ are the uncertainties in the experimental data. For the cross section data, statistical uncertainties and overall normalization uncertainties, when quoted with the published results, were included in $\Delta\sigma_{R,data}$. For the polarization data, only statistical uncertainties were included.

The inconsistency between the Rosenbluth and polarization data for $Q^2 \geq 1$ GeV^2 was handled using an iterative procedure. Three iterations of the fit were performed. For all cross section data with $Q^2 \geq 1$ GeV^2, the value of G_E^p in equation (6) was replaced by a fixed value of G_E^p calculated from the results of the previous fit, or, in the case of the first iteration, using the results of Kelly's 2004 analysis.[9] This replacement has the effect that for $Q^2 \geq 1$ GeV^2, G_E^p is entirely determined by polarization data, while G_M^p is determined by cross section data, with G_E^p fixed by the polarization data for R_p. For $Q^2 < 1$ GeV^2, cross section and polarization data are treated on an equal footing for both G_E^p and G_M^p. The numerical minimization was carried out using the TMinuit class within the ROOT libraries.[31] Typically, no significant improvement of the fit was found after two iterations of the starting function for G_E^p.

This fit procedure was performed for two data sets identical in every respect except that the new JLab Hall C data for R_p were excluded from the first data set and included in the second set. Table 1 shows the results

Table 1. Fit results for G_E^p and G_M^p/μ_p, with ("Old") and without ("New") the data of Ref. [8]. $G\infty/G_D = 16M_p^4/\Lambda^2\,(a_1/b_3)$ is the asymptotic value of G_E^p/G_D and $G_M^p/(\mu_p G_D)$. The "standard" dipole form factor is defined as $G_D = \left(1 + Q^2/\Lambda^2\right)^{-2}$, with $\Lambda^2 = 0.71$ GeV^2.

	Old G_E^p	Old G_M^p	New G_E^p	New G_M^p
a_1	-.390 ± .081	.075 ± .023	-.299 ± .072	.081 ± .023
b_1	11.01 ± .10	11.14 ± .08	11.11 ± .09	11.15 ± .08
b_2	13.57 ± .84	18.44 ± .16	14.11 ± .83	18.45 ± .16
b_3	11.4 ± 4.5	5.12 ± .65	15.7 ± 4.2	5.31 ± .66
G_∞/G_D	-.84 ± .49	.360 ± .064	-.47 ± .23	.376 ± .061
$Q^2(G_E^p = 0)$	9.0		11.8	

of the fit with and without the new Jefferson Lab results. The parameter errors quoted are the standard errors calculated from the diagonal elements of the covariance matrix. Significant correlations among the parameters for each form factor were found. These correlations imply that the uncertainties in the asymptotic form factor values and the uncertainties in G_E^p and G_M^p at any given Q^2 cannot be calculated directly from the parameter errors, but instead require the full covariance matrix. On the other hand, the correlation coefficients between the parameters describing G_E^p and those describing G_M^p were generally small, at the level of a few percent. Table 2 shows the asymptotic values of $Q^6F_2^p$ and $Q^4F_1^p$ for the fits with and without the new high Q^2 data.

Table 2. Asymptotic values of Q^4F_1 and Q^6F_2, with and without data of Ref. [8].

Old $Q^6F_{2,\infty}^p$	New $Q^6F_{2,\infty}^p$	Old $Q^4F_{1,\infty}^p$	New $Q^4F_{1,\infty}^p$
$3.28 \pm .93$	$2.69 \pm .51$	$.507 \pm .090$	$.529 \pm .086$

Table 3 illustrates the quality of the fit in terms of χ^2. Although the overall χ^2 of the fit $\chi^2/n.d.f. = 828.4/466 = 1.78$ might be regarded as relatively poor, the extent to which the reduced χ^2 exceeds one per datum reflects the inconsistency between the Rosenbluth and polarization methods, as shown in table 3. The polarization data and the cross section data for $Q^2 \leq 1$ GeV^2 are consistent with the fit results for G_E^p and G_M^p. Relative to the number of data points, the cross section data for $1 \leq Q^2 \leq 10$ GeV^2 contribute disproportionately to χ^2, since the ϵ-dependence of the measured cross sections disagrees with the slope predicted by the polarization data.

Table 3. χ^2 of the final global fit with new data included.

Observable	N_{data}	χ^2
σ_R, $Q^2 \leq 1$ GeV^2	269	293.8
σ_R, $1 < Q^2 \leq 10$ GeV^2	142	483.3
σ_R, $Q^2 > 10$ GeV^2	10	3.96
$R_p^{pol.}$, all Q^2	53	47.4
All data	474	828.4

Figures 1 and 2 show the fit results together with the experimental data. In figure 1(a), polarization data for R_p were converted to G_E^p data using the fit result for G_M^p and appropriate error propagation. The best fit curves are shown with standard 1σ uncertainty bands calculated from the full covariance matrix. As shown in figure 1(a), the new G_E^p data[8] favor a slowing rate of decrease of G_E^p with Q^2 compared to previous data, and shrink the uncertainty in G_E^p for $Q^2 \geq 5$ GeV^2 by roughly a factor of two. The improved constraint on F_2^p in the high-Q^2 region shown in figure 1(b) is similarly dramatic.

Figure 2 shows the results for $G_M^p/\mu_p G_D$ and $Q^4F_1^p$. The interesting feature of the G_M^p results is that the fit is systematically higher than the

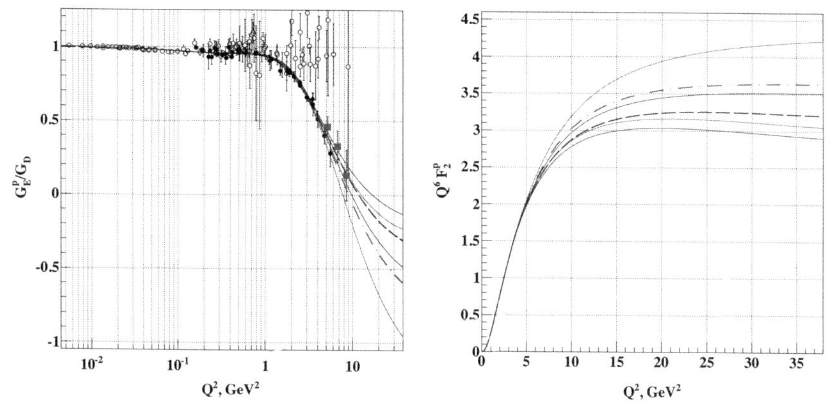

(a) G_E^p/G_D

(b) $Q^6 F_2^p$

Fig. 1. Fit results for G_E^p/G_D(a), and $Q^6 F_2^p$(b), before (dot-dashed) and after (dashed) including the results of Ref. 8, with 1σ uncertainty bands. Published values of G_E^p extracted from cross section (open circles) and polarization (filled circles) measurements. Results of Ref. [8] are shown by squares.

published data in the region of the discrepancy. Since the cross section measurements either extracted or assumed $R_p \approx 1$, the contribution of G_E^p to σ_R was overestimated, lowering the extracted values of G_M^p. This result for G_M^p is quantitatively and qualitatively similar to the result of a recent global analysis including TPEX corrections to the cross section data.[32] In contrast to this analysis, which neglects TPEX completely, the authors of Ref. [32] used the Kelly parametrization with $n = 3$ and achieved a substantially better χ^2, since the applied TPEX corrections largely reconciled the discrepancy. Although the new R_p data barely affect G_M^p and its uncertainty, the effect on F_1^p is more significant, an entirely expected result given the dramatic improvement in F_2^p and the definition $G_M = F_1 + F_2$. For yet larger Q^2, the fit result for G_M^p again agrees with the published data, reflecting the dominance of σ_R by G_M^p, shown in figure 3.

In conclusion, new high Q^2 measurements of $R = \mu_p G_E^p/G_M^p$ in Jefferson Lab's Hall C have significantly extended the range of Q^2 for which G_E^p is accurately determined. The impact of these new data on the world database of proton form factors was studied in the context of an empirical parametrization with physically reasonable assumptions for low and high Q^2 asymptotic behavior; i.e., $G_E^p(0) = 1$ and $G_M^p(0) = \mu_p$, and $F_{1(2)}^p \propto Q^{-4(6)}$ as $Q^2 \to \infty$. In this context, the new data for R_p reduce the uncertainty in

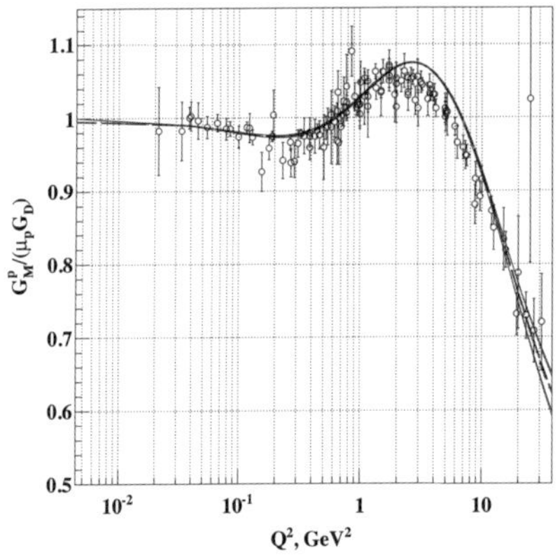

(a) $G^p_M/(\mu_p G_D)$

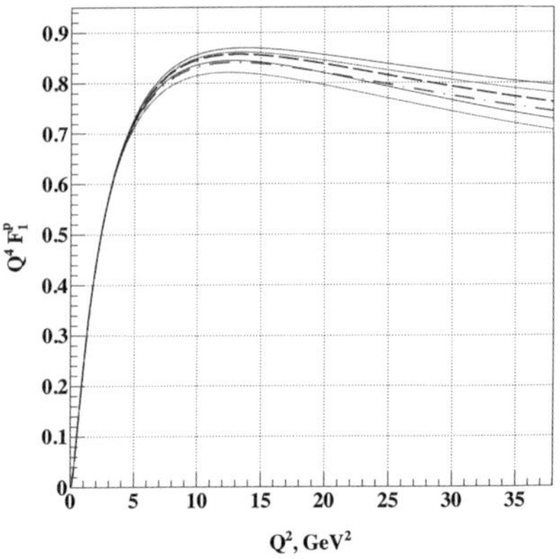

(b) $Q^4 F^p_1$

Fig. 2. Fit results for $G^p_M/(\mu_p G_D)$(a) and $Q^4 F^p_1$(b), before (dot-dashed) and after (dashed) including the results of Ref. [8]. Published values of G^p_M extracted from cross section (open circles) measurements.

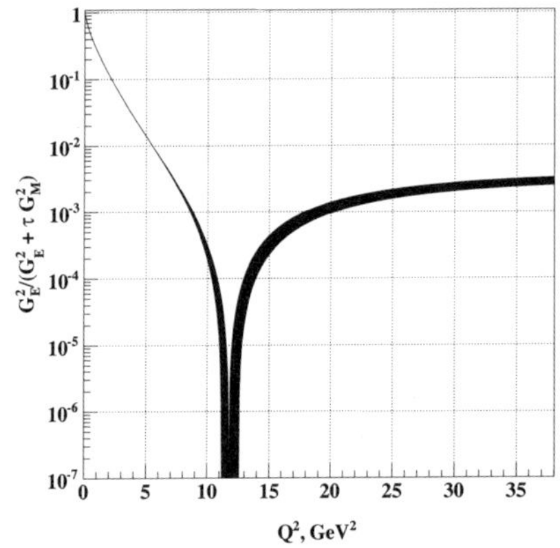

Fig. 3. $G^2_E/(G^2_E + \tau G^2_M)$, calculated from the results of the global fit, represents the maximum fraction of the reduced cross section carried by G_E.

G^p_E by roughly a factor of two in the measured Q^2 region and in the extrapolation to higher Q^2. Looking to the future, planned recoil polarization measurements of R_p to $Q^2 \approx 15$ GeV^2 and precision elastic ep differential cross section measurements to $Q^2 \approx 18$ GeV^2 using JLab's upgraded 11 GeV electron beam will complete the experimental knowledge of proton electromagnetic form factors in the spacelike region attainable with present day accelerators.

References

1. M. N. Rosenbluth, *Phys. Rev.* **79**, p. 615 (1950).
2. A. I. Akhiezer and M. P. Rekalo, *Sov. J. Part. Nucl.* **3**, p. 277 (1974).
3. R. G. Arnold, C. E. Carlson and F. Gross, *Phys. Rev. C* **23**, p. 363 (1981).
4. M. K. Jones *et al.*, *Phys. Rev. Lett.* **84**, p. 1398 (2000).
5. V. Punjabi *et al.*, *Phys. Rev. C* **71**, p. 055202 (2005).
6. O. Gayou *et al.*, *Phys. Rev. Lett.* **88**, p. 092301 (2002).
7. C. E. Carlson and M. Vanderhaeghen, *Annu. Rev. Nucl. Part. Sci.* **57**, p. 171 (2007).
8. A. J. R. Puckett *et al.*, *Phys. Rev. Lett.* **104**, p. 242301 (2010).
9. J. J. Kelly, *Phys. Rev. C* **70**, p. 068202 (2004).
10. S. J. Brodsky and G. R. Farrar, *Phys. Rev. D* **11**, p. 1309 (1975).
11. G. P. Lepage and S. J. Brodsky, *Phys. Rev. Lett.* **43**, p. 545 (1979).
12. T. Janssens *et al.*, *Phys. Rev.* **142**, p. 922 (1966).
13. C. Berger *et al.*, *Phys. Lett. B* **35**, p. 87 (1971).
14. L. E. Price *et al.*, *Phys. Rev. D* **4**, p. 45 (1971).
15. W. Bartel *et al.*, *Nucl. Phys. B* **58**, p. 429 (1973).
16. P. N. Kirk *et al.*, *Phys. Rev. D* **8**, p. 63 (1973).
17. F. Borkowski *et al.*, *Nucl. Phys. A* **222**, p. 269 (1974).
18. F. Borkowski *et al.*, *Nucl. Phys. B* **93**, p. 461 (1975).
19. G. G. Simon *et al.*, *Nucl. Phys. A* **333**, p. 381 (1980).
20. A. F. Sill *et al.*, *Phys. Rev. D* **48**, p. 29 (1993).
21. L. Andivahis *et al.*, *Phys. Rev. D* **50**, p. 5491 (1994).
22. M. E. Christy *et al.*, *Phys. Rev. C* **70**, p. 015206 (2004).
23. I. A. Qattan *et al.*, *Phys. Rev. Lett.* **94**, p. 142301 (2005).
24. M. J. Alguard *et al.*, *Phys. Rev. Lett.* **37**, p. 1258 (1976).
25. B. D. Milbrath *et al.*, *Phys. Rev. Lett.* **80**, p. 452 (1998).
26. T. Pospischil *et al.*, *Eur. Phys. J. A* **12**, p. 125 (2001).
27. O. Gayou *et al.*, *Phys. Rev. C* **64**, p. 038202 (2001).
28. C. Crawford *et al.*, *Phys. Rev. Lett.* **98**, p. 052301 (2007).
29. M. K. Jones *et al.*, *Phys. Rev. C* **74**, p. 035201 (2006).
30. G. Ron *et al.*, *Phys. Rev. Lett.* **99**, p. 202002 (2007).
31. R. Brun and F. Rademakers, *Nucl. Inst. and Meth. in Phys. Res. A* **389**, p. 81 (1997).
32. J. Arrington, W. Melnitchouk and J. A. Tjon, *Phys. Rev. C* **76**, p. 035205 (2007).

PION FORM FACTOR: STATUS AND OUTLOOK

G.M. HUBER
University of Regina, Regina, SK S4S 0A2, Canada

The pion elastic and transition form factors provide unique information on the interactions between light-quarks, and in particular on the interplay between perturbative and non-perturbative QCD. The recent experimental data on these observables from Jefferson Lab and BaBar are reviewed, and the future outlook discussed.

1. Introduction

There is much interest in trying to understand the structure of hadrons, both mesons and baryons, in terms of their constituents, the quarks and gluons. However, this structure is too complicated to be calculated rigorously in Quantum Chromodynamics (QCD) because perturbative QCD (pQCD) methods are not applicable in the confinement regime. For this reason, experimental data are absolutely required to help guide the development of theory and validate the approaches taken.

The pion holds a special role in this task, because it provides access to the simplest bound QCD system: a quark-antiquark pair. The hard contribution to meson form factors can be exactly calculated within the framework of pQCD. For example, in pQCD the pion elastic form factor (F_π) is given[1] as the sum of logarithms of Q^2 and powers of Q^2

$$F_\pi(Q^2) = \frac{4\pi C_F \alpha_s(Q^2)}{Q^2} \left| \Sigma_{n=0}^{\infty} a_n \left(log\left(\frac{Q^2}{\Lambda^2}\right)\right)^{-\gamma_n} \right|^2 [1 + O(\alpha_s(Q^2), m/Q^2)]. \quad (1)$$

At asymptotically high Q^2, only the hardest portion of the pion distribution amplitude (DA) contributes

$$\phi_\pi(x) \xrightarrow[Q^2\to\infty]{} \frac{3f_\pi}{\sqrt{n_c}} x(1-x), \quad (2)$$

and F_π takes the very simple form[1,2]

$$F_\pi(Q^2)\xrightarrow[Q^2\to\infty]{}\frac{16\pi\alpha_s(Q^2)f_\pi^2}{Q^2}, \tag{3}$$

where $f_\pi = 92.4$ MeV is the pion decay constant.[3]

However, in general the pion DA also contains soft contributions, governed by various types of quark-gluon correlations at lower Q^2. These are treated via QCD-based models, such as the constituent quark model, or models employing Light-Cone (LC) dynamics, the Bethe-Salpeter (plus Dyson-Schwinger) equation, or other approaches such as the use of dispersion relations or (QCD or LC) sum rules. The interplay of the hard and soft contributions to the pion DA is complex and poorly understood, leading to different theoretical viewpoints on whether higher-twist mechanisms dominate until very large momentum transfer, or not. Experimental measurements of the pion elastic and transition form factors, taken over a sufficiently wide kinematic range, offer the possibility to directly study the transition from constituent to current quark degrees of freedom.

2. $\gamma^*\gamma \to \pi^0$ Transition Form Factor

Recent data on the $\gamma^*\gamma \to \pi^0$ transition form factor from BaBar[4] have only deepened the mystery on how QCD transitions from the hard to the soft regime. In lowest order pQCD, the transition form factor is given[5] as

$$Q^2F_{\gamma\gamma\pi}(Q^2) = \frac{2f_\pi}{3}\int_0^1 \frac{dx}{x}\phi_\pi(x,Q^2) + O(\alpha_s) + O\left(\frac{\Lambda_{QCD}^2}{Q^2}\right), \tag{4}$$

which in the asymptotic limit becomes

$$F_{\gamma\gamma\pi}(Q^2)\xrightarrow[Q^2\to\infty]{}\frac{2f_\pi^2}{Q^2}. \tag{5}$$

While the previous data[6,7] were consistent with a leveling off around $Q^2 = 10$ GeV^2 of the transition form factor near the pQCD asymptotic limit, as expected if hard contributions dominate, the Babar data[4] indicate a continuing rise of the transition form factor, which must approach the asymptotic limit from above at much larger Q^2 (Fig. 1).

To obtain these data, the $e^+e^- \to e^+e^-\pi^0$ reaction in a storage ring is utilized. This is a very difficult measurement, as e^+e^- collider experiments are not designed to perform space-like measurements of this nature. One of the electrons is detected at large angles, yielding a virtual photon with large Q^2. The second electron is undetected, having presumably been

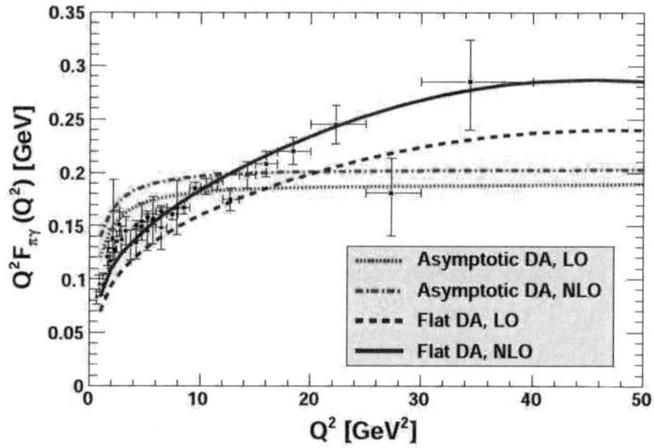

Fig. 1. Data for $\gamma^*\gamma \to \pi^0$ transition form factor from[4,6,7] multiplied by Q^2 in comparison with the calculations of Li and Mishima.[10] The asymptotic value $Q^2F_{\gamma\gamma\pi} \to \sqrt{2}f_\pi \approx 0.185$ GeV^2.

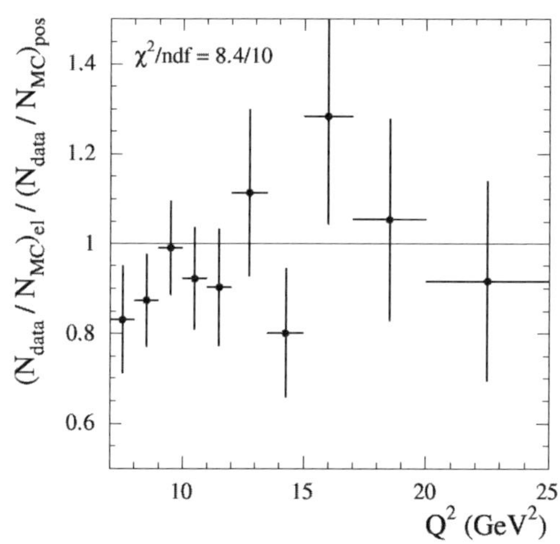

Fig. 2. Double-ratio for events with a detected e^- to those with a detected e^+ as a check of the reliability of the extracted cross sections.[4]

scattered at small angle and yielding a nearly real photon. One of the challenges of the BaBar analysis[4] is that the dominant background, Virtual Compton Scattering, exceeds the π^0 production cross section by a large factor, $> 10^3$. An additional challenge is posed by the high π^0 lab frame energy, which causes many of the $\pi^0 \to \gamma\gamma$ clusters in the BaBar calorimeter to overlap. The asymmetric beam energies (3.1 GeV positron, 9 GeV electron) produce very different energy and angle distributions and trigger corrections depending on whether an e^- or e^+ is detected. The double-ratio $(N_{data}/N_{MC})_{e^-}/(N_{data}/N_{MC})_{e^+}$ thus serves as a data quality check, and is found to be in reasonable agreement with unity, although for the higher statistics low Q^2 data the double-ratio is slightly low (Fig. 2).

The BaBar result has potentially significant ramifications on our understanding of the pion DA at moderate Q^2, and its evolution to the asymptotic limit. Various authors[8–10] argue that if these data are correct, they indicate a nearly flat shape of the pion DA with enhanced contributions at high and low x, in contradiction of most theoretical expectations (Fig. 1). The evolution of the DA with increasing Q^2, so that the asymptotic limit is eventually approached from above, is also an open theoretical question.

3. π^+ Elastic Form Factor

It is important to investigate also the pion elastic form factor, to see if the various "hard" exclusive processes involving the pion give a consistent picture. For example, Radyushkin[8] finds that if the "flat DA scenario" holds, the one gluon-exchange diagram should contribute negligibly and the pion elastic form factor should be calculated only non-perturbatively. High quality pion elastic form factor data are clearly required to delineate the role of hard versus soft contributions in the pion DA.

However, the experimental measurement of this observable poses its own challenges. At low Q^2, F_π can be determined in a model-independent manner from the scattering of high-energy pions by atomic electrons, yielding elastic form factor values governed by the charge radius of the pion.[11] To access the region above $Q^2 > 0.3$ GeV^2, one must employ high-energy electroproduction of pions on the proton. For selected kinematics near the pion pole at $t = m_\pi^2$, this process can be described as quasi-elastic scattering of the electron from a virtual π^+ in the proton. The physical region for t in pion electroproduction is negative, so measurements should be performed at the smallest attainable values of $-t$. To minimize background contributions, it is also necessary to separate out the longitudinal cross section σ_L, via a Rosenbluth L/T(/LT/TT) separation. The value of $F_\pi(Q^2)$ is then determined by comparing the measured longitudinal cross section at small values of $-t$ to the best available electroproduction model. The obtained F_π values are in principle dependent upon the model used, although one anticipates this dependence to be reduced at sufficiently small $-t$.

There are a number of approaches for the extraction of the pion elastic form factor from these cross sections, each with their associated uncertainties. Frazer[12] originally proposed that F_π be extracted from σ_L via a kinematic extrapolation to the pion pole, and that this be done in an analytical manner (Chew-Low method). However, this extrapolation procedure fails to produce a reliable answer, since different polynomial fits, each of which are equally likely in the physical region, differ considerably when continued to $t = m_\pi^2$.[13]

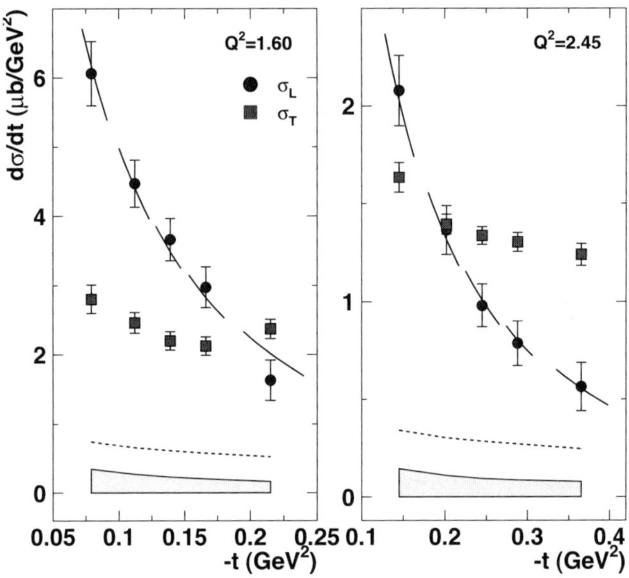

Fig. 3. Separated cross sections, σ_L and σ_T from the Jefferson Lab $F_\pi - II$ experiment[17] compared to the VGL Regge model[14] (full curve for L, dashed curve for T) with parameters $\Lambda_\pi^2 = 0.513(0.491)$ GeV^2 and $\Lambda_\rho^2 = 1.1$ GeV^2. Because of the various kinematic correlations, each t-bin has its own average $\overline{W}$ and $\overline{Q^2}$ which differ slightly from the nominal values. The error bars indicate the statistical and uncorrelated systematic uncertainty in both ϵ and t combined in quadrature; the error band denotes the correlated part of the systematic uncertainty by which all data points move collectively.

A more reliable approach is to use a model of the longitudinal electroproduction cross section incorporating the π^+ production mechanism and spectator neutron effects to extract $F_\pi(Q^2)$ from the σ_L data. Since the Regge model by Vanderhaeghen, Guidal and Laget (VGL)[14] is able, without fitted parameters, to provide a good description of both π^+ and π^- photoproduction data, and of σ_L electroproduction data over a range in W, t, and Q^2, it has been used to extract pion elastic form factor values from the Jefferson Lab σ_L data[15] up to a maximum Q^2 value of 2.45 GeV^2.[16] In this model, the exchange of high-spin, high-mass particles is taken into account by replacing the pole-like Feynman propagators of Born term models with Regge propagators. Many of the model's free parameters were fitted to pion photoproduction, with the result giving a good fit to existing data for $-t$ below 1 GeV^2. For electroproduction, the pion elastic form factor and the $\rho\pi\gamma$ form factor are added as adjustable parameters, parameterized with a monopole form

$$F_\pi(Q^2) = [1 + Q^2/\Lambda_\pi^2]^{-1}. \tag{6}$$

Over the range of $-t$ covered by the Jefferson Lab experiments, σ_L is completely determined by the π trajectory, whereas σ_T is also sensitive to the ρ exchange contribution. Λ_π^2 is determined by fitting the model to the $\sigma_L(t)$ data, as shown in Fig. 3, yielding the F_π values at each Q^2.

There has been considerable recent interest in $p(e, e'\pi^+)n$ model building.[18–21] At present, the respective authors have concentrated their efforts on the improved description of σ_T, since σ_L is already well described by the VGL model. We are in close contact with these authors and we are hopeful that alternative models of σ_L useful for the extraction of the pion elastic form factor will eventually be available. Access to more models would allow a better understanding of the model dependence of the F_π result to be made. The philosophy of the Jefferson Lab Fpi Collaboration remains to publish the experimentally measured σ_L, so that updated values of $F_\pi(Q^2)$ can be extracted as better models for the $p(e, e'\pi^+)n$ reaction become available.

3.1. *Prospects for higher Q^2 F_π data*

Unfortunately, the experimental knowledge of F_π above the $Q^2 = 2.45$ GeV^2 limit of the existing Jefferson Lab data is poor. The perception that F_π is difficult to extract at high Q^2 stems from the poor quality of the Cornell data. For example, Bebek, et al.[22,23] were unable to perform a L/T separation, and so were sensitive to the presence of isoscalar backgrounds. They were required to add an empirical isoscalar component to the theory of Berends[24] before performing a F_π extraction, amounting to an approximately 20% correction at $Q^2 = 4$ GeV^2.[22] Where they were able to perform a L/T separation,[25] the resulting uncertainties in σ_L were so large that for the actual determination of the pion elastic form factor, σ_T was simply assumed to be proportional to the total photon cross section and subtracted from the measured (differential) cross section.[23] Since no uncertainty was used in the assumed values of σ_T, the given uncertainties in their extracted values for F_π are (severely) underestimated. For the highest $Q^2 = 6.30$, 9.77 GeV^2 settings,[23] data were obtained only at low $\epsilon = 0.40$.

Table 1. The ratio M_{pQCD}/M_{pole} calculated by Carlson and Milana[26] for the kinematics of the Cornell $p(e, e'\pi^+)n$ data[23,25]. The non-pole backgrounds rise rapidly for $-t \gtrsim 0.20$ GeV^2, adversely affecting the reliability of F_π values extracted too far from the pion pole.

Q^2 (GeV^2)	W (GeV)	$-t$ (GeV^2)	M_{pQCD}/M_{pole}
		0.01	0.18
1.94	2.67	0.07	0.12
3.33	2.63	0.17	0.18
6.30	2.66	0.43	0.81
9.77	2.63	0.87	2.82

Carlson and Milana[26] point out that the existence of competing non-pole QCD processes complicate the extraction of F_π from these data, because of the large $|t|$ used, several settings of which have $-t_{min} > 20m_\pi^2$. The calculated backgrounds are reproduced in Table 1. It is seen that the background ratio M_{pQCD}/M_{pole} rises dramatically once $-t_{min} \gtrsim 0.20$ GeV^2. This, together with the already relatively large statistical and systematic uncertainties of those data, precludes a meaningful comparison with theoretical calculations, and led Carlson and Milana to conclude *"[we] question whether F_π has been truly determined for large Q^2"*.[26]

It is clear that *"more reliable measurements of F_π at high Q^2 require smaller $|t|$ and thus higher electron energy loss ν."*[26] The experimental requirements are a high quality, continuous electron beam of sufficiently high energy to allow data to be obtained over a wide $\Delta\epsilon$ range at high Q^2, low $-t$, and two spectrometers with sufficiently small forward angle capability, good angular reproducibility (to control systematic errors in the L/T separation) and sufficient missing mass resolution to cleanly separate $p(e, e'\pi^+)n$ events from $p(e, e'\pi^+)n\pi^0$. The Jefferson Lab upgrade, with a maximum electron beam energy of 11 GeV and SHMS+HMS spectrometers with good forward angle capabilities, make Hall C the only facility worldwide where these measurements can be pursued. This combination will allow F_π to be determined up to $Q^2 = 6$ GeV^2 in the best manner allowable by current models, and would provide a very significant advance

Table 1. The ratio M_{pQCD}/M_{pole} calculated by Carlson and Milana[26] for the kinematics of the Cornell $p(e,e'\pi^+)n$ data[23,25]. The non-pole backgrounds rise rapidly for $-t \gtrsim 0.20$ GeV^2, adversely affecting the reliability of F_π values extracted too far from the pion pole.

Q^2 (GeV^2)	W (GeV)	$-t$ (GeV^2)	M_{pQCD}/M_{pole}
		0.01	0.18
1.94	2.67	0.07	0.12
3.33	2.63	0.17	0.18
6.30	2.66	0.43	0.81
9.77	2.63	0.87	2.82

in our understanding of the pion elastic form factor (Fig. 4). Note that the $Q^2 = 6$ GeV^2 upper limit of our projected data is dictated primarily by the $-t_{min} \lesssim 0.20$ GeV^2 requirement to keep non-pole backgrounds small. If reliable theoretical calculations relating to the background processes were available (e.g. using updated GPD models), we could potentially extend the useful $-t$ range for F_π measurements, resulting in even higher Q^2 data from the Hall C facility.

It is equally important to check experimentally that the form factor extracted using the electroproduction technique corresponds to that from the physical pion. Our 11 GeV experiment plans a series of studies to give confidence in the extracted result. One straightforward test is to measure the elastic form factor using the electroproduction technique in kinematics which come close to the region that has been measured using elastic π^+ scattering. We plan to acquire low $Q^2 = 0.30$ GeV^2 data extremely close to the pion pole, $-t_{min} = 0.005$ GeV^2, to cross check the extracted F_π values using the electroproduction method against those obtained without approximation from elastic π^+e^- scattering at the CERN SPS.[11] Our data[27] will have 50% smaller $-t$ than any previous data,[33] and hence be a sensitive test of the electroproduction method.

Another test, of the model used to extract the form factor from the $p(e,e'\pi^+)n$ data, is to make measurements at different values of t_{min} but the same Q^2, i.e. various separations from the physical pion pole, $t = m_\pi^2$. While the longitudinal cross section, σ_L, depends on three Lorentz invariant quantities, t, W, and Q^2, the minimum accessible value of $-t$ is correlated with W at fixed Q^2. Hence, measurements at smaller values of $-t_{min}$ require larger values of W and are only practical at larger beam energies. We

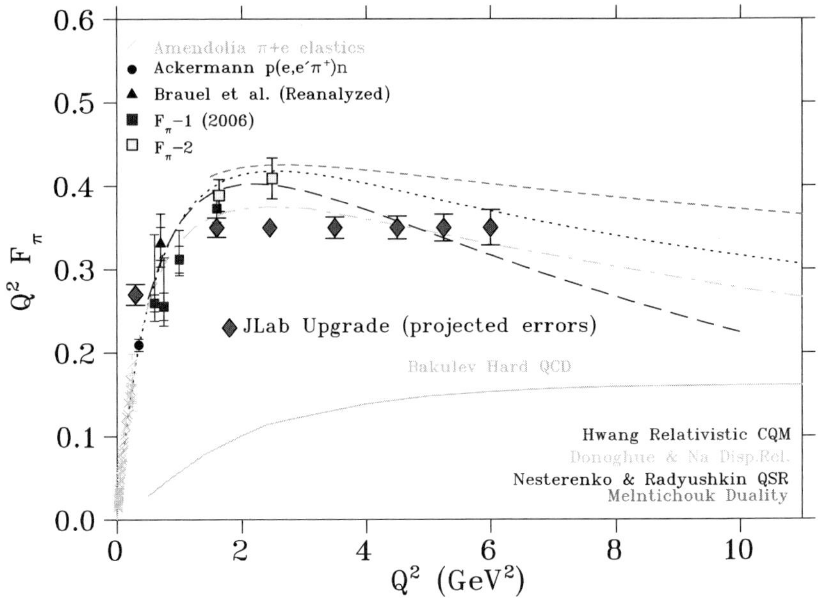

Fig. 4. Projected error bars for the Jefferson Lab 11 GeV pion elastic form factor experiment,[27] in comparison with a variety of theoretical models,[28–32] and existing precision data.[11,16,33,34] The error bars include all projected statistical and systematic uncertainties, as well as estimated model uncertainty in the elastic form factor extraction added in quadrature.

have already performed[16] one such test at Jefferson Lab, where the pion elastic form factor at $Q^2 = 1.6$ GeV^2 was extracted both at $W = 1.95$ GeV, $-t_{min} = 0.15$ GeV^2, and at $W = 2.22$ GeV, $-t_{min} = 0.093$ GeV^2. The second point was 30% closer to the physical pole to the first, and the extracted form factor from the second agreed with the first to ~4%, which is comparable to the experimental uncertainty. Additional measurements[27] at $Q^2 = 1.6$ GeV^2, W=3.00 GeV, even closer to the pion pole at $-t_{min} = 0.03$ GeV^2, as well as new measurements at $Q^2 = 2.45$ GeV^2, will allow the reliability of the model used to be further tested.

Finally, exclusive π^-/π^+ ratio measurements are sensitive to isoscalar backgrounds to the dominant (at low $-t$) isovector pion exchange amplitude. G-parity conservation restricts t-channel exchanges to $G = +1$ and -1 for the respective isoscalar and isovector photon amplitudes. Thus, if the photon possessed definite isospin, the two reactions $\gamma_L^* n \to \pi^- p$, $\gamma_L^* p \to \pi^+ n$ would be related to each other by simple isospin rotation and the cross sections would be equal.[35] Interference terms between the isoscalar and isovector photon amplitudes have opposite signs for these processes, which would lead to a difference in these two cross sections if significant isoscalar contributions were present. A departure from $R_L \equiv \sigma_L^{\pi^-}/\sigma_L^{\pi^+} = 1$ would indicate the presence of isoscalar backgrounds arising from mechanisms such as ρ meson[36] or pQCD contributions due to transverse quark momentum.[26] We plan[27] such tests of the t-channel dominance of the σ_L data at Q^2=1.6, 3.5 GeV^2, using a liquid deuterium target.

4. Summary

To summarize, experimental access to pion form factors in the space-like region is difficult. The $\gamma^*\gamma \to \pi^0$ measurements are the most direct, but the data analysis is challenging. The π^+ measurements require close attention to systematic errors and a model to extract the elastic form factor from electroproduction data. While the older $F_{\gamma\gamma\pi}(Q^2)$ data suggested the onset of perturbative behavior as early as $Q^2 \approx 1$ GeV^2, the recent BaBar data indicate that the perturbative region could be very far away. To be sure this picture is correct, it is essential to probe all exclusive π production channels to obtain a consistent, global understanding. The F_{π^+} results at $Q^2 = 2.5$ GeV^2 also seem to indicate that soft processes play an important role. The Jefferson Lab upgrade will allow us to dramatically improve the higher Q^2 π^+ data set, as well as to perform many checks to determine the reliability of the extracted form factors. Through careful measurements such as these, we hope to eventually obtain a better understanding of the transition to the perturbative regime.

Acknowledgements

The author would like to thank the F$_\pi$ Collaboration for their very productive and enjoyable scientific collaboration over the course of this work. This work is supported in part by the Natural Sciences and Engineering Research Council of Canada (NSERC).

References

1. G.P. Lepage, S.J. Brodsky, Phys. Lett. **87B** (1979) 359.
2. G.R.Farrar, D.R. Jackson, Phys. Rev. Lett. **43** (1979) 246.
3. O. Dumbrajs, et al., Nucl. Phys. **B 216** (1983) 277.
4. B. Aubert, et al., Phys. Rev. D **80** (2009) 052002.
5. G.P. Lepage, S.J. Brodsky, Phys. Rev. D **22** (1980) 2157.
6. H.J. Behrend, et al., Z. Phys. **C49** (1991) 401.

7. J. Gronberg, et al., Phys. Rev. D **57** (1998) 33.
8. A.V. Radyushkin, Phys. Rev. D **80** (2009) 094009.
9. M.V. Polyakov, arXiv:0906.0538 [hep-ph].
10. H. Li, S. Mishima, Phys. Rev. D **80** (2009) 074024.
11. S.R. Amendolia, et al., Nucl. Phys. **B277** (1986) 168; Phys. Lett. **146B** (1984) 116.
12. W.R. Frazer, Phys. Rev. **115** (1959) 1763.
13. R.C. Devenish, D.H. Lyth, Phys. Rev. D **5** (1972) 47; Phys.Rev.D **6** (1972) 2067.
14. M. Vanderhaeghen, M. Guidal and J.-M. Laget, Phys. Rev. C **57** (1998) 1454; Nucl. Phys. **A627** (1997) 645.
15. H.P. Blok, et al., Phys. Rev. C **78** (2008) 045202.
16. G.M. Huber, et al., Phys. Rev. C **78** (2008) 045203.
17. T. Horn, et al., Phys. Rev. Lett. **97** (2006) 192001.
18. M.M. Kaskulov, U. Mosel, Phys. Rev. D **81** (2010) 045202.
19. S.V Goloskokov, P. Kroll, Eur. Phys. J. **C65** (2010) 137.
20. C. Bechler, D. Mueller, arXiv:0906.2571 [hep-ph].
21. A. Faessler, T. Gutsche, V.E. Lyubovitskij, I.T. Obukhovsky, Phys. Rev. C **76** (2007) 025213.
22. C.J. Bebek, et al., Phys. Rev. D **13** (1976) 25.
23. C.J. Bebek, et al., Phys. Rev. D **17** (1978) 1693.
24. F.A. Berends, Phys. Rev. D **1** (1970) 2590.
25. C.J. Bebek, et al., Phys. Rev. Lett. **37** (1976) 1326.
26. C.E. Carlson, J. Milana, Phys. Rev. Lett. **65** (1990) 1717.
27. Experimental Proposal E12-06-101, "Measurement of the Charged Pion Form Factor to High Q^2", G.M. Huber, D. Gaskell, spokespersons. Approved by PAC30(2006), and reviewed again by PAC35(2010), where a scientific priority of 'A' was awarded.
28. A.P. Bakulev, K. Passek-Kumericki, W. Schroers, N.G. Stefanis, Phys. Rev. D **70** (2004) 033014, and erratum Phys. Rev. D **70** (2004)079906.
29. T. Huang, X.-G. Xu, Phys. Rev. D **70** (2004) 093013.
30. J.F. Donoghue, E.S. Na, Phys. Rev. D **56** (1997) 7073.
31. V.A. Nesterenko, A.V. Radyushkin, Phys. Lett. **115 B** (1982) 410.
32. W. Melnitchouk, Eur. Phys. J. A **17** (2003) 223 and private communication, 2006.
33. H. Ackermann, et al., Nucl. Phys.**B137** (1978) 294.
34. P. Brauel, et al., Z. Phys. **C3** (1979) 101.
35. A.M. Boyarski, et al, Phys. Rev. Lett. **21** (1968) 1767.
36. M. Vanderhaeghen, M. Guidal, and J.-M. Laget, Phys. Rev. C **57** (1998) 57.

HADRON FORM FACTORS AT LARGE MOMENTUM TRANSFER MOMENTUM FROM LATTICE QCD

HUEY-WEN LIN

Department of Physics, University of Washington
Seattle, WA 98195, USA
**E-mail: hwlin@phys.washington.edu*

We report recent progress in lattice QCD calculations of hadron form factors at larger transfer momentum. Conventional lattice form-factor calculations can only reach about 2.5 GeV^2, but in this work the transfer momentum is pushed as large as 4–7 GeV^2. We will demonstrate results for the nucleon and pion using 2+1-flavor anisotropic clover lattices. These measurements could give important theoretical input to experiments, such as those of Jefferson Lab's 12-GeV program, and provide insight into hadronic structure.

Keywords: Form Factors, Lattice QCD, Nucleon, Pion

Nucleons are building blocks of ordinary matter. In the Standard Model, they are composed from fundamental particles, quarks and gluons; their non-trivial interactions within the confined space of hadrons determines the properties of nucleons. Due to the confinement, we cannot directly observe the quarks and gluons of the nucleon; rather, experiments smash the nucleon into different final states (as in deep inelastic scattering) to provide us with understanding through observables like the parton distribution functions (PDFs). Processes such as electron elastic scattering with hadrons via photons give us electromagnetic form factors, which contain information in the momentum-space impact plane. Adding up information from various experiments, we can reconstruct these processes to gain a better understanding of how the quarks and gluons shape the nucleon.

Understanding form factors at high transfer momentum (Q^2) is particularly interesting. It not only helps to reveal the nature of hadrons at short distances but also challenges models whose parameters are calibrated in the low-Q^2 regime. Experiments such as those in the 12-GeV upgrade at Jefferson Lab will soon provide higher precision and higher momentum-transfer form-factor data; progress will be made not only in the nucleon sector but also for pions. Theoretically, perturbative QCD (pQCD) should perform better at larger Q^2, and we would like to get an understanding of when pQCD starts to become reliable. However, pQCD seems to fail to describe recent BaBar results for $\gamma^*\gamma \to \pi^0$ over a wide range of transfer momenta 4 $\text{GeV}^2 < Q^2 < 40\ \text{GeV}^2$,[1,2] suggesting that there may be a large gap before pQCD can describe the experimental data correctly. A nonperturbative QCD approach would be able to provide the QCD theory results to connect the missing dots if one could use such an approach for sufficiently large Q^2.

The techniques of lattice QCD have been applied to such varied phenomena as the spectroscopy of heavy-quark hadrons, the tower of excited baryon states, flavor physics involving the CKM matrix, hadron decay constants and baryon axial couplings. In many cases, it can provide higher precision data from the Standard Model than what can be measured experimentally. By discretizing spacetime into a four-dimensional grid with a fixed lattice spacing and finite volume, we are able to compute the path integral (in terms of discretized versions of the QCD Lagrangian and operators) directly via numerical integration, providing first-principles calculations of the consequences of QCD. The four-dimensional lattice size, lattice spacing, the quark masses in the sea sector and precision of the integration have been greatly improved due to the increasing computational resources and more efficient algorithms over the past decade. Preliminary calculations at physical pion masses are being explored. Lattice calculations with sub-percent precision have been achieved for many flavor-related quantities and used to guide tests of Standard Model and to probe possible new physics. Calculations of hadron structures and interactions require more computational resources but should soon enter an era of precision.

Hadron form factors have been calculated on the lattice by many groups, and calculations are still ongoing; see Ref. 3 and references within for a few examples of recent review articles. Recently, lattice calculations have also been used to calculate transition form factors involving excited nucleons[4] and experimentally challenge quantities, such as hyperon axial couplings.[5] However, the typical Q^2 range in lattice calculations of hadron form factors is less than 2.5 GeV^2. When one attempts higher-Q^2 calculations, they suffer from poor signal-to-noise ratio; for example, see the case study for the pion in Ref. 6.

There are two main restrictions in lattice QCD to achieve higher Q^2 form factors. One of the limitations is the available lattice spacing (a) used in the calculation, since the momenta allowed on the lattice (assuming periodic spatial boundary conditions) are $\frac{2\pi n}{L}a^{-1}$, where L is the length of the spatial volume while n ranges 0, 1, $\sqrt{2}$, $\sqrt{3}$, etc. Thus, the finer the lattice spacing, the higher momentum transfer one can achieve. However, these lattices may not be available in the next couple years, since one also needs to increase the number lattice points to keep the spatial volume sufficiently large to avoid finite-volume effects. In many existing calculations, one usually runs into high signal-to-noise in the form factors before reaching high n at any given lattice spacing. When we re-examine the typical steps in a form-factor calculation, we find the main causes of this noise and develop solutions. The traditional method simplifies the complicated form-factor analysis by working in a regime where only the ground-state signal remains. This greatly simplifies the analysis, since taking ratios of current-insertion matrix and the hadron two-point correlators will yield a constant in Euclidean time, which can be fit to a simple pleateau. To suppress excited states in the limited time extent of the lattice box, we usually introduce a technique called "smearing"; this gives the same ground-state signal at earlier Euclidean time. The parameters of the smearing are usually tuned when looking at the hadron masses at rest or checking the form factors at zero momentum transfer. However, as one increases the magnitude of the momentum transfer carried by the currents, the excessive momentum is carried by the initial or final hadron state (because of momentum conservation). The smearing parameters used to filter out the excited states will also filter out the boosted hadron signal. Thus, in a carefully tuned ground-state only system, one expects to see the signal die out after a few momentum points when fixing the smearing parameters throughout the calculation. The solution to this problem is simple: one can vary the smearing parameters as the current momentum changes but keep a close eye on the contamination from excited states for each momentum; or one can take not only the ground state but also the excited state into the analysis, extracting the ground state using multiple smearing parameters or operators.

We investigate the idea using a quenched lattice with pion mass set to 720 MeV, calculating the d-quark connected contribution to nucleon form factors. For this demonstration, we fix the final-state nucleon at rest so that the initial nucleon would carry the momentum transfer out at the inserted current. The left-hand side of Fig. 1 shows the results from both methods: the traditional method is shown as stars, and the new approach as circles. In the low-momentum region, we find the two approaches show consistent results with compatible statistical error bars. However, as soon as the momentum increases above the rest excited-state energy set by the smearing parameter, the initial nucleon operator contributes no further signals and presents a great deal of noise. At that point, some of our parameters that couple well to the rest excited state become dominant for the boosted nucleon ground state. These contribute to the form factors in the new method, and the signal-to-noise ratio remains relatively stable. Therefore, we can reach higher momentum transfer than the traditional fixed operator, ground-state only approach.

In this proceeding, we will demonstrate the results of this improved method, focusing on results from dynamical lattice. We will cover the nucleon electromagnetic form factors and preliminary results on the nucleon axial and pion form factors. In the end of this proceeding, we will mention an proposal for a future calculation of form factors at even higher momentum transfer and free of the systematics introduced in a calculation at fixed lattice spacing.

We primarily use the 2+1-flavor anisotropic lattices generated by the Hadron Spectrum Collaboration (HSC)[7,8] in this calculation. (Note that the same techniques will work equally well for isotropic lattices.) These lattices use Symanzik-improved gauge action with tree-level tadpole-improved coefficients and anisotropic clover action.[9] The the gauge links in the fermion action are 3-dimensionally stout-link smeared with a small smearing weight $\rho = 0.14$ and $n_\rho = 2$ iterations. The renormalized gauge and fermion anisotropies are around $\xi = 3.5$ (that is, $a_s = 3.5a_t$), and the inverse of the spatial lattice spacing is about 1.6 GeV. The quark propagators are calculated under antiperiodic boundary conditions in the time direction, while the spatial ones remain periodic. Quark propagators on the lattices are evaluated for source and sink operators with five Gaussian smearing parameters: $\sigma \in \{0.5, 1.5, 2.5, 3.5, 4.5\}$. We use four final momenta ($\vec{p_f} = \frac{2\pi}{L}\{0,0,0\}a^{-1}$, $\frac{2\pi}{L}\{-1,0,0\}a^{-1}$, $\frac{2\pi}{L}\{-1,-1,0\}a^{-1}$, $\frac{2\pi}{L}\{-2,0,0\}a^{-1}$) and vary the initial momentum over all $\vec{p_i} = \frac{2\pi}{L}\{n_x, n_y, n_z\}a^{-1}$ with integer $n_{x,y,z}$ and $n_x^2 + n_y^2 + n_z^2 \leq 10$ for current matrix elements. For more details on the parameters used in our work, please check out the Ref. 10.

From the HSC ensembles,[7,8] we use the $16^3 \times 128$ lattices with pion masses of 875, 580 and 450 MeV to calculate nucleon electromagnetic form factors. We extrapolate $F_{1,2}^{u,d}$ through a simple parametrization as used in interpolating experimental data:

$$F_1 = \frac{a_0 + \sum_{i=1}^{k-2} a_i \tau^i}{1 + \sum_{i=1}^{k} b_i \tau^i}; \quad \left(\frac{F_2}{\kappa}\right) = \frac{1 + \sum_{i=1}^{k-3} a_i \tau^i}{1 + \sum_{i=1}^{k} b_i \tau^i}. \tag{1}$$

We simultaneously fit the m_π and Q^2 (or τ) dependence of the lattice data, expanding each Q^2-fit parameter in terms of the pion mass: $a_i = a_i^{(0)} + a_i^{(1)} m_\pi^2$. The right-hand side of Fig. 1 shows an example of the extrapolations on the isovector Dirac form factor in the dynamical ensembles. The χ^2/dof for $F_{1,2}^v$ are 1.4 and 2.0, respectively. The lowest line/band represents the extrapolated form factors at the physical pion mass. We find good agreement between the Dirac form factor extrapolation at the physical pion mass and the experimental values; that is, the curvature (which is proportional to the product of squared nucleon mass and the mean-square charge radius) as a function of Q^2 fits nicely with interpolating forms derived from experimental data. However, the Pauli form factor at physical pion mass suffers slightly larger χ^2/dof, mostly due to the failure to reproduce the (anomalous) magnetic moment, which would require smaller-momentum data (< 500 MeV) that are not accessible with the current boundary conditions. At large momentum transfer, it becomes more consistent with the experimental form.

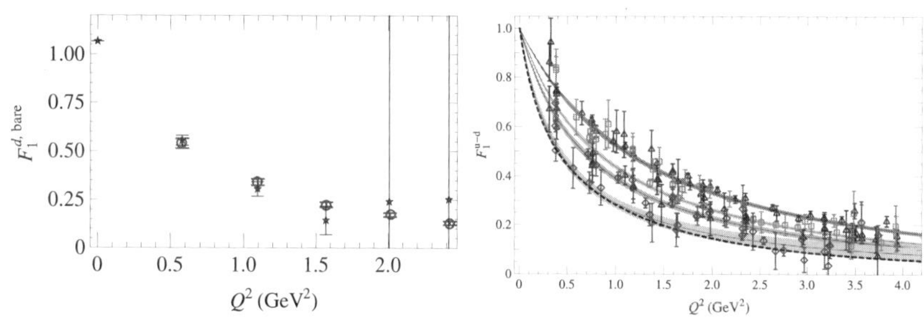

Fig. 1. Left: Unrenormalized d-quark contribution to the nucleon Dirac form factor F_1^d obtained from the ratio approach (stars) and the method adopted by this work (circles) on a quenched ensemble with $m_\pi = 720$ MeV. Right: Isovector Dirac form-factor extrapolation using $N_f = 2+1$ calculations in this work. The shaded bands show the simultaneous fit to the Q^2 dependence of each pion mass points (875, 580 and 450 MeV from top to bottom). The lowest band is the extrapolation to the physical pion mass.

Since the elastic form factors contain information about the spatial structure of the nucleon, we can convert our data into a description of its densities in the impact plane. Having form factors at larger momentum transfer can greatly influence our understanding of the short-distance hadron structure. Due to the relativistic effects of the transferred momentum on the wavefunction of the nucleon, we cannot use a simple three-dimensional Fourier transformation without some recourse to models. Instead, we use the model-independent formulation of Ref. 11 in terms of densities in a two-dimensional plane transverse to an infinite-momentum boost. The transverse charge density is defined as the Fourier transform of the form factor in such a plane:

$$\rho(\mathbf{b}) \equiv \int \frac{d^2\mathbf{q}}{(2\pi)^2} F_1(\mathbf{q}^2) e^{i\mathbf{q}\cdot\mathbf{b}}, \tag{2}$$

where bold vectors $\mathbf{b}$ and $\mathbf{q}$ lie in the transverse plane. Equivalently,

$$\rho(b) = \int_0^\infty \frac{Q\,dQ}{2\pi} J_0(bQ) F_1(Q^2), \tag{3}$$

for scalar b, where J_0 is a Bessel function. We can perform this integral numerically, using the $F_1(Q^2)$ obtained by extrapolating our fit form to the physical pion mass.

We demonstrate the importance of the high Q^2 form factors by restricting the data set to the region $Q < 2$ GeV^2 (which is the upper limit of transfer momentum for many lattice QCD calculations). We then compare this result to the density obtained from using all the available lattice Q^2. The left-hand side of Fig. 2 shows in red the effect of omitting the highest Q^2 data (above 2 GeV^2), compared to the blue which uses all Q^2 from the dynamical ensembles. The impact is significant in the central core, as we anticipated; omitting information about large transfer momenta results in a deviation in the density around 25%.

The right-hand side of Fig. 2 shows the neutron charge density; it is expected to have the largest systematics due to the neglect of the disconnected contributions in this work. Even so, we observe only about a 25% difference between the dynamical results and experiment, surprising considering that the disconnected value is on the same order as the neutron form factor itself. Once again, the difference becomes much more significant in the quenched case, and the even larger negative core charge density is consistent with suppression of mesonic contributions.

A similar approach can be applied to nucleon axial form factors, where experiments mostly derive results from neutron beta decay or pion form factors where various theoretical models predict a wide range of possibilities for the large-Q^2 region. The left-hand side of Fig. 3 shows the preliminary results for the isovector axial form factors using ensembles with 3 different pion masses. The data are simultaneously extrapolated in pion mass and Q^2 (as was the EM form factors). The lowest grey band is the result at physical pion mass, and the dashed line is the dipole form with best fit to the experimental data ($M_A = 1.026(21)$ GeV[14]). These high Q^2 form

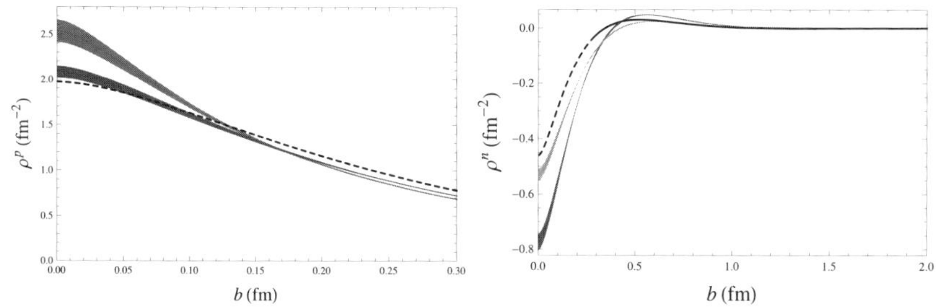

Fig. 2. Left: The transverse charge density $\rho(b)$ of the proton in the central region with different cuts applied to the F_1 data prior to fitting. The lower band uses all available Q^2 from the 2+1-flavor ensembles; the upper band uses only data with $Q^2 < 2$ GeV^2. The dashed line is parametrization of available experimental data.[12,13] Right: The transverse charge density $\rho(b)$ of the neutron. Results from the dynamical 2+1-flavor (quenched) ensembles are shown as upper (lower) bands. The dashed line is parametrization of available experimental data.[12,13]

factors can help us study the nucleon (and its quark component) helicity distribution.

Recent high-precision pion electroproduction data from Jefferson Lab using L/T-separated cross sections provide precise pion form factor measurements up to 2.45 GeV^2. Many theoretical models ranging from various quark models to holographic QCD provide diverse and diagreeing predictions for the intermediate and high-Q^2 dependence of the pion form factors. Future precision experiments will challenge these theoretical models even more, aiming at Q^2 through 6 GeV^2 and higher. The right-hand side of Fig. 3 shows the preliminary results for our dynamical ($N_f = 2+1$) pion form factors with 580, 875, 1350 MeV pion masses with Q^2 reaching nearly 7 GeV^2 for the highest-mass ensemble. The extrapolated form factor at the physical pion mass shows reasonable agreement with JLab precision measurements. Future attempts will focus on decreasing the pion masses and exploring Q^2-dependence of pion form factors for yet higher Q^2.

We demonstrate that one can extend form factor calculations to higher momentum transfer with reasonable statistical error at a fixed lattice spacing. However, systematic error could become significant when one reaches a large number of $|n|$. To reach even higher Q^2, we propose a numerical step-scaling calculation (for example, Ref. 15). On a small volume with very fine lattice spacing, we can easily reach high momentum. By calculating the step-scaling function at overlapping momentum points (or interpolating) we can reduce the systematic error due to finite-volume or lattice-discretization artifacts. However, generating several volumes and tuning for multiple lat-

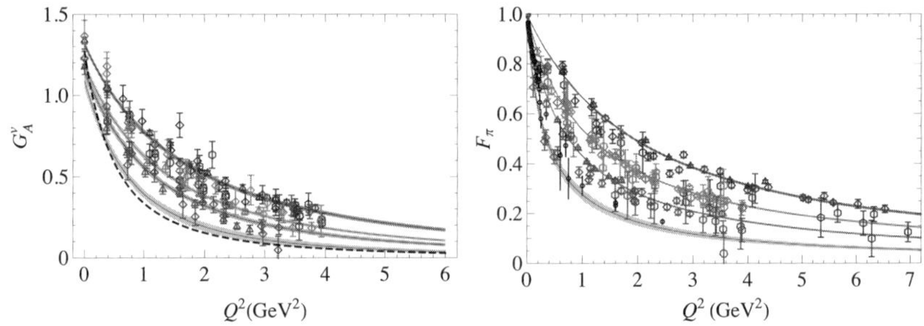

Fig. 3. Left: Preliminary results for nucleon isovector axial form factors using 3 pion masses at 875, 580 and 450 MeV. The lowest band is the extrapolation to the physical pion mass. The dashed line is a dipole form using $M_A = 1.03(2)$ GeV. Right: Preliminary pion form factor with multiple pion masses at 580, 875, 1350 MeV. The experimental points are shown in (black) circles while the lowest gray band is the extrapolation to the physical pion mass using our lattice points.

tice spacing of dynamical $N_f = 2+1$ lattices requires a large amount of computational resources; we hope such a proposal will become feasible as petascale computing facilities become available in the near future.

References

1. B. Aubert *et al.*, *Phys. Rev.* **D80**, p. 052002 (2009).
2. A. V. Radyushkin, *Phys. Rev.* **D80**, p. 094009 (2009).
3. P. Hagler, *Phys. Rept.* **490**, 49 (2010).
4. H.-W. Lin, S. D. Cohen, R. G. Edwards and D. G. Richards, *Phys. Rev.* **D78**, p. 114508 (2008).
5. H.-W. Lin and K. Orginos, *Phys. Rev.* **D79**, p. 034507 (2009).
6. P.-h. J. Hsu and G. T. Fleming, *PoS* **LAT2007**, p. 145 (2007).
7. R. G. Edwards, B. Joo and H.-W. Lin, *Phys. Rev.* **D78**, p. 054501 (2008).
8. H.-W. Lin *et al.*, *Phys. Rev.* **D79**, p. 034502 (2009).
9. P. Chen, *Phys. Rev.* **D64**, p. 034509 (2001).
10. H.-W. Lin, S. D. Cohen, R. G. Edwards, K. Orginos and D. G. Richards (2010).
11. G. A. Miller, *Phys. Rev. Lett.* **99**, p. 112001 (2007).
12. J. Arrington, W. Melnitchouk and J. A. Tjon, *Phys. Rev.* **C76**, p. 035205 (2007).
13. J. J. Kelly, *Phys. Rev.* **C70**, p. 068202 (2004).
14. V. Bernard, L. Elouadrhiri and U. G. Meissner, *J. Phys.* **G28**, R1 (2002).
15. H.-W. Lin and N. Christ, *Phys. Rev.* **D76**, p. 074506 (2007).

PROTON FORM FACTOR RATIO MEASUREMENTS AT HIGH Q^2 WITH SUPER BIGBITE

M. KHANDAKER
for the JLab Hall A EMFF Collaboration
Norfolk State University, Norfolk, VA 23504, USA
and
Jefferson Lab, Newport News, VA 23606, USA
E-mail: mahbub@jlab.org

A proposed measurement of the proton electromagnetic form factor ratio, $\mu_p G_E^p/G_M^p$ using the Super Bigbite apparatus is presented. The proposed instrumentation will allow measurements of three nucleon electromagnetic form factors G_E^p, G_E^n, and G_M^n with unprecedented precision to Q^2-values up to three times higher than existing data.

1. Introduction

The ground-state electromagnetic nucleon form factors (FFs) are among the most fundamental quantities that describe the non-perturbative structure of the nucleons. Their study provides a powerful test of our understanding of non-perturbative QCD and confinement. The challenge of understanding the nucleon's structure and dynamics has occupied a central place in nuclear physics, and experimental and theoretical physicists have spent a considerable amount of effort in the last five decades to understand it. A break-through was made towards this goal in the last decade, when the two Jefferson Lab (JLab) Hall A experiments obtained the elastic electromagnetic FF ratio of the proton, G_E^p/G_M^p, from measured recoil proton polarization components, using the recoil polarization method.[1]

The results of the GEp(1)[2,3] and GEp(2)[4] experiments in Hall A have stimulated a huge amount of theoretical activity, as evidenced by the nearly 1000 combined citations of these papers, and by a revival of this field. An entirely new picture of the structure of the proton has emerged after the GEp(1) and GEp(2) experiments showed that the ratio G_E^p/G_M^p was in fact not constant, $\mu_p G_E^p/G_M^p \approx 1$, as suggested by the Rosenbluth separation method,[5] but decreased almost linearly by a factor of ~ 3.7 over the Q^2

range of 1 to 5.6 GeV^2. It is widely accepted that these results have forced a significant rethinking of nucleon structure. Recently, a third experiment, GEp(3),[6] measured the ratio to Q^2 of 8.5 GeV^2 in a Hall C experiment using the same recoil polarization method.[1] Figure 1 shows the existing data for G_E^p/G_M^p, the projected errors for a new approved JLab experiment E12-07-109, GEp(5),[7] and the results of several theoretical predictions.

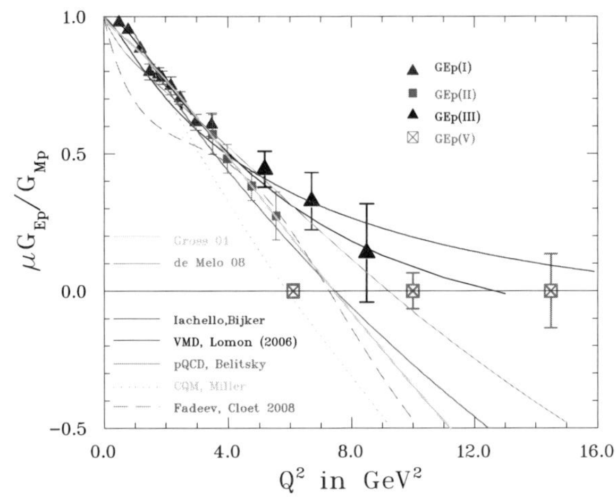

Fig. 1. The published results of GEp(1),[2,3] GEp(2),[4] preliminary results from GEp(3)[6] and the projected results of GEp(5)[7] in a 60-day run are shown here. Theoretical predictions are from Refs. 8–14

A common thread in virtually all of the theoretical explanations of the Q^2-dependence of G_E^p/G_M^p is an important role for quark orbital angular momentum (quark OAM) in the nucleon's dynamics. One approach to explaining the proton FF ratio data involves refined perturbative QCD calculations that include an $L = 1$ component in the quark light-cone wave function.[12] Also notable are relativistic constituent-quark calculations,[15] some of which even preceded the discovery by GEp(1).[2] Perhaps the most realistic model is a calculation out of Argonne by Cloët, Roberts and co-workers that uses an approach based on the Dyson Schwinger Equations (DSEs) together with the Faddeev equations.[14]

2. The GEp(5) Experiment (E12-07-109)

The measurement of nucleon FFs to the highest possible Q^2 is one of the prime tenets of the JLab 12-GeV upgrade. Figure 1 makes it clear that the only way to achieve clarity in discriminating between theoretical explanations of the form-factor data is to measure the proton FF ratio with considerable precision to high values of Q^2.

For example, three of the predictions shown, the covariant spectator model (Gross 04),[8] the refined pQCD calculation including quark OAM ($F_2/F_1 \propto \ln^2(Q^2/\Lambda^2)/Q^2$),[12] and the DSE/Faddeev calculation,[14] all predict G_E^p/G_M^p to cross zero somewhere in the neighborhood of $Q^2 \approx 7$ GeV^2. At the same time, the two vector-meson dominance (VMD) models show G_E^p/G_M^p approaching zero much more gradually.[10,11] GEp(1) and GEp(2) agree reasonably well with all of these predictions. In contrast, the GEp(3) data appear to favor the VMD models. A definitive resolution to this issue has profound physical implications, and cannot occur without high-precision data at high Q^2.

The JLab PAC32 in August of 2007 approved experiment E12-07-109 (GEp(5))[7] which will measure the proton's electromagnetic FF ratio, G_E^p/G_M^p, up to 14.5 GeV^2 using the recoil polarization technique. The experiment will use a large-acceptance spectrometer, to detect the recoil protons. A schematic representation of the experiment is shown in Fig. 2.

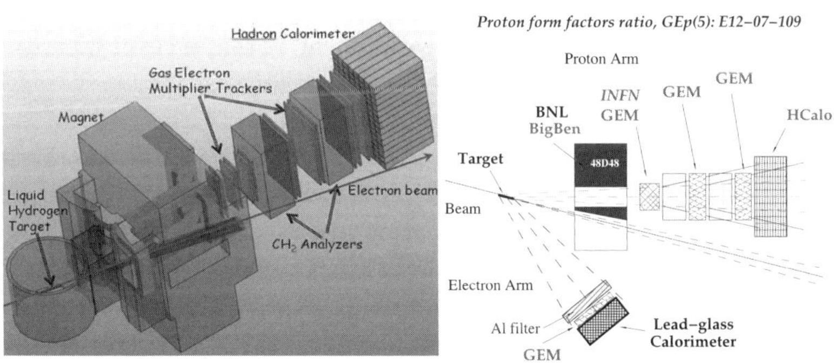

Fig. 2. The left panel shows a SBS configuration for the GEp(5) experiment (artist's view). The right panel shows a top view of the experiment layout.

The measurement of the ground-state electromagnetic FFs of the proton becomes quite challenging at high momentum transfers, because the effective rate drops as roughly E_{beam}^2/Q^{16}. The polarization-transfer method also requires very large statistics because of the relatively low analyzing power of the recoil polarimeter. The GEp(5) experiment meets these challenges through an innovative approach. The spectrometer in the proton arm is based on a large open-geometry dipole magnet that is placed quite close to the target, and at a small angle, a configuration that provides large solid angle. The placement of the dipole at small angle is made possible by incorporating a cut in the magnet yoke for the beam line. The spectrometer will also be equipped with a double focal plane polarimeter (FPP), instrumented with gas electron multiplier (GEM) trackers, that provides improved efficiency and very high rate capability, and a highly-segmented hadron calorimeter which will be part of the trigger and will help reduce the event rate. The electron will be detected in coincidence by an existing electromagnetic calorimeter, "BigCal". It will also be equipped with a GEM tracker to further improve its position resolution. The GEp(5) experiment will use the 11 GeV polarized electron beam and a 40-cm long liquid hydrogen target.[16] GEp(5) will have excellent statistical power. Measurements will be made at three values of Q^2: 5, 10, and 14.5 GeV^2 thus nearly doubling the maximum value of Q^2 achieved by GEp(3), while achieving an error in the ratio G_E^p/G_M^p of 0.10.

3. Concept of the GEp(5) experiment

In GEp(5) elastic scattering of longitudinally polarized 11 GeV electrons from a liquid hydrogen target will be used, $H(\vec{e}, e'\vec{p})$. Polarization transfer to the recoiling proton provides a measure of the G_E^p/G_M^p as proposed in the double-polarization method.[1] In measurement of the nucleon FF ratio this method mitigates the difficulties of the Rosenbluth separation method at high momentum transfer and is almost insensitive to the two-photon effects. The most effective realization of the double-polarization approach for the large momentum transfer G_E^p/G_M^p-ratio experiments was proposed in Ref. 17, which is based on a combination of the spin precession in the magnetic spectrometer and the polarimeter. This provides directly the ratio of the polarization components of the proton and has a high figure-of-merit as well as a cancellation of most systematic uncertainties related to the polarimeter analyzing power and the efficiency.

In the one-photon exchange approximation, the scattering of longitudinally polarized electrons from unpolarized hydrogen results in a transfer of polarization to the recoiling proton with two components, P_t perpendicular to, and P_ℓ parallel to the proton momentum in the scattering plane,

$$P_t = -\sqrt{\frac{2\varepsilon(1-\varepsilon)}{\tau}} \frac{G_M G_E}{\sigma_{red}}, \quad P_l = \sqrt{(1-\varepsilon^2)} \frac{G_M^2}{\sigma_{red}} \tag{1}$$

where $\tau = Q^2/4M_p^2$, $\varepsilon = [1 + 2(1+\tau)\tan^2(\theta_e/2)]^{-1}$, θ_e is the lab electron angle, and σ_{red} is the reduced cross-section $G_M^2 + \varepsilon G_E^2/\tau$. Measuring these two components simultaneously and taking their ratio gives the ratio of the form factors

$$\frac{G_E}{G_M} = -\frac{P_t}{P_l}\sqrt{\frac{\tau(1+\varepsilon)}{2\varepsilon}} \tag{2}$$

Using the double-polarization technique, the FF ratio G_E/G_M at a given Q^2 can be obtained without measuring the absolute cross sections and without change of beam energy or detector angle, thus eliminating important sources of systematic uncertainties. However, the analyzing power of the polarimeter and the beam polarization that come into an actual measurement of P_t and P_ℓ cancel out in Eq. 2, and thus do not need to be known particularly well. They do affect the size of the measured asymmetries, however, so they still need to be maximized to increase the experimental figure-of-merit.

4. The Super Bigbite apparatus

The Super Bigbite apparatus will be used in GEp(5) as a large-acceptance spectrometer, a Super Bigbite spectrometer (SBS). The plan is based on an existing, resistive coil dipole magnet, 48D48, available from Brookhaven National Laboratory. It will provide angular coverage up to ≈70 msr, with a detector package that will be capable of operating at the highest possible luminosity, almost 10^{39} electron/s×nucleon/cm^2. Table 1 lists the key features of the SBS.

Table 1. Key features of the Super Bigbite spectrometer.

Large solid angle: $\Omega \sim$ 5–76 msr
Large momentum acceptance: $p \sim$ 2–10 GeV/c
High luminosity capability: $\mathcal{L} \sim 8 \times 10^{38}$ [electron/s×nucleon/cm^2]
Small scattering angle capability: $\theta_{\min} \sim 3.5°$
Very good angular resolution: σ_θ [mrad] $= 0.14 + 1.3/p\,[\mathrm{GeV}/c]$
Good vertex resolution: $\sigma_y \approx$ 1–2 mm
Good momentum resolution: $\sigma_p/p\,[\%] = 0.29 + 0.03 \times p\,[\mathrm{GeV}/c]$
High energy trigger threshold: $E_{trig} \sim$ 3–4 GeV (for proton)

The use of a simple dipole placed close to the target allows one to achieve the high acceptance in the spectrometer. Such an installation is made possible by a deep cut through the iron yoke of the magnet for the beam line. The magnet deflects charged particles vertically and will be used with a field integral up to 2.5 T·m. The field in the beam line will be reduced to an acceptable level by specially developed magnetic shields. The relatively small bend angle is compensated for by the high coordinate resolution (70 μm) of the front GEM-based chambers resulting in a momentum resolution of $\sim 0.5\%$ at 8 GeV/c in GEp(5) with the 40-cm long LH_2 target. The GEM technology solves the main challenge of the SBS, the very high counting rates, allowing tracking at very high background rates. *These features combined will give SBS at least a factor of 10 advantage compared with any existing or proposed spectrometer at JLab for nucleon form-factor measurements.*

The proposed instrumentation, that is collectively referred as the Super Bigbite Project, includes a set of components that will be used in three different configurations for each of the nucleon FF measurements of G_E^p (GEp(5)[7]), G_E^n (GEn(2)[18]), and G_M^n (GMn[19]), respectively. In all cases, the design philosophy incorporates the use of large open-geometry detection, high-rate-handling capability through the use of GEM technology, and the ability to measure at relatively forward angles.

4.1. *Trackers of the Super Bigbite apparatus*

Three trackers are required in the SBS for the GEp(5) experiment. The first tracker (FT) will be used to measure the proton momentum and its direction before interaction with the first CH_2 analyzer. This tracker with an active area of 40×150 cm^2 will include six chambers, each with two-dimensional read-out and three GEM amplification foil planes. The front tracker will be followed by a double polarimeter consisting of two trackers each preceded by a CH_2 analyzer. Adding this second polarimeter increases the figure-of-merit by a factor of 1.7, equivalent to a 30% reduction of the experimental statistical errors. The second tracker (ST) will measure the proton track after the proton passes through the first CH_2 analyzer. The third tracker (TT) will measure the direction of the track after the proton passes through the second CH_2 analyzer. The dimensions of the ST and the TT were chosen to be 50×200 cm^2. Each GEM chamber will consist of 50×40 cm^2 sub-sections.

At the FT we expect background hit rates of about 400 kHz/cm^2 (based on GEANT simulations) due to the direct view of the target.[20] This background is dominated by soft photons originating from the target. Low-momentum charged particles are swept away by the magnet. The rates on the second and third trackers are expected to be 130 kHz/cm^2 and

64 kHz/cm^2, respectively, dominated by soft electrons/positrons produced by photons converting in the analyzers.

4.2. *Hadron and electromagnetic calorimeters*

The hadron calorimeter (HCalo), positioned at the end of the SBS detector package will be used to detect the recoiling protons in GEp(5), the neutrons in GEn(2), and the neutrons and protons in GMn experiments in the momentum range $\sim$ 2.5 to 10.5 GeV/c.

The total active area of HCalo will be 5.5 m^2. It has a good time resolution of 1.5 ns, high granularity (15$\times$15 cm^2), a very good coordinate resolution of 2 cm, and, in addition, a high energy threshold.

In GEp(5), the large energy of the proton will allow the use of a 3–4 GeV energy threshold in the calorimeter without a significant loss of detection efficiency. Such a high energy threshold leads to the suppression of pion triggering and opens the possibility of using a coincidence for the DAQ trigger. HCalo will be used to trigger the DAQ, in coincidence with the signals from the elastic electrons detected in the electromagnetic calorimeter (BigCal). In addition, these two calorimeters will provide coordinate information that will be used to locate the proton track in the front tracker and in the FPP trackers by applying kinematical constraints. The energy and coordinate resolutions of BigCal of about 9% and 6 mm, respectively, for 3.5 GeV electrons satisfy the trigger and tracking requirements. The expected energy resolution for 8 GeV protons in HCalo is 20%.[20] The active size of HCalo in GEp(5) will be 150$\times$300 cm^2, which is a little smaller than its full size and the total number of active modules will be 200.

The lead-glass blocks of the existing BigCal calorimeter will be rearranged for GEp(5) in a matrix 20$\times$75, a shape optimized for the largest acceptance at Q^2=14.5 GeV^2. BigCal will be installed 3 m from the target at a central angle of 39° (at the largest Q^2). It will be shielded from the target by a 20 cm Al plate to reduce the radiation dose on the lead glass. This also reduces the hit rate in the coordinate detector, a two-plane GEM-based chamber, placed in front of the BigCal.

The angular correlation between the scattered electron and the recoil proton will be measured very accurately, especially the angle between the electron scattering plane and the proton recoil plane. Because of the small size of the electron beam, the angles of the electron and of the recoil proton can be determined with a good accuracy of $\sim$ 0.5 mrad. This corresponds to a 5 MeV/c uncertainty in the perpendicular momentum balance. Such an uncertainty is much smaller than the typical out-of-plane momentum for inelastic events, 150 MeV/c, which results in a powerful suppression of the background. In order to achieve this, a 1 mm coordinate accuracy is required for the scattered electron which will be achieved by using the coordinate detector.

4.3. *Experimental trigger in GEp(5)*

The trigger is a critical part of the GEp(5) experimental design. Achieving a reasonable data rate in this measurement is particularly challenging given the high luminosity of 8×10^{38} cm^{-2}s^{-1} with open detectors. To address this challenge, the trigger for GEp(5) will have two main features: (1) a high energy threshold in the calorimeters in both arms (3–4 GeV for the proton and 2.5 GeV for the electron); and (2) "smart" FPGA-based coincidence electronics that allow second-level triggering on spatial correlations of hits in the two detector arms. The high energy thresholds will reduce the rate in the electron arm to 60 kHz and in the proton arm to about 1.5 MHz. With a coincidence time window of 50 ns, forming a simple coincidence between the two arms would result in a trigger rate of at least 5 kHz, which is too high (or at least close to the limit) for the expected total event size of $\sim$ 20 kB. However, the small deflection of the proton trajectory in the SBS makes it possible to use the angular correlation between the elastic electron and the proton trajectories, at the trigger level, to reduce the coincidence trigger rate further by a factor of five.

5. Summary

The study of the ground-state electromagnetic form factors to the highest possible Q^2 is one of the top priorities of the JLab 12-GeV upgrade. The Super Bigbite Project will make possible three ground-breaking precision measurements of the nucleon's elastic form factors and it provides great potential for other experiments in DIS and SIDIS, where the kinematics will be well into the scaling regime. The project has already been through two JLab technical reviews. Funding for part of the instrumentation (the first GEM chambers FT) has already been approved by INFN-Italy and JLab has incorporated the project proposal in its long-term capital funding request to DOE.

This work is supported by DOE contract DE-AC05-06OR23177, under which Jefferson Science Associates, LLC, operates the Thomas Jefferson National Accelerator Facility.

References

1. A.I. Akhiezer *et al.*, *Sov. Phys. JETP* **33**, 765 (1957); *Sov. J. Part. Nucl.* **3**, 277 (1974); R. Arnold *et al.*, *Phys. Rev.* C **23**, 363 (1981).
2. M.K. Jones *et al.*, *Phys. Rev. Lett.* **84**, 1398 (2000).
3. V. Punjabi *et al.*, *Phys. Rev.* C **71**, 055202 (2005).
4. O. Gayou *et al.*, *Phys. Rev. Lett.* **88**, 092301 (2002).
5. M.N. Rosenbluth, *Phys. Rev.* **79**, 615 (1950).
6. C.F. Perdrisat, V. Punjabi, M.K. Jones, E. Brash (spokespersons), Jefferson Lab experiment E04-108 (Gep(3)) (2004).
7. E. Brash, E. Cisbani, M. Jones, M. Khandaker, L. Pentchev, Ch. Perdrisat, V. Punjabi, B. Wojtsekhowski (spokespersons), Jefferson Lab experiment E12-07-109 (Gep(5)) (2007).
8. F. Gross and P. Agbakpe, *Phys. Rev.* C **73**, 015203 (2006).
9. J.P.B. de Melo *et al.*, *Phys. Lett.* B **671**, 153 (2009).
10. R. Bijker and F. Iachello, *Phys. Rev.* C **69**, 068201 (2004); F. Iachello and Q. Wan, *Phys. Rev.* C **69**, 055204 (2004).
11. E.L. Lomon, *Phys. Rev.* C **64**, 035204 (2001), *Phys. Rev.* C **66**, 045501 (2002), and arXiv:nucl-th/0609020 (2006).
12. A.V. Belitsky, X. Ji, and F. Yuan, *Phys. Rev. Lett.* **91**, 092003 (2003).
13. M.R. Frank, B.K. Jennings, and G.A. Miller, *Phys. Rev.* C **54**, 920 (1996).
14. I.C. Cloët and C.D. Roberts, arXiv:0811.2018 [nucl-th] (2008); I.C. Cloët *et al.*, *Few Body Syst.* **46**, 1 (2009).
15. G.A. Miller and M.R. Frank, *Phys. Rev.* C **65**, 065205 (2002); G.A. Miller, *Phys. Rev.* C **66**, 032201 R (2002).
16. "*Electromagnetic Form Factors of the Nucleon using the 12-GeV CEBAF*", C.W. de Jager *et al.*, Funding Proposal submitted to DOE, November, 2009, `http://www.jlab.org/~bogdanw/EMFF.pdf`.
17. C.F. Perdrisat, V. Punjabi and M. Jones (spokespersons), Jefferson Lab experiment E89-014 and E93-027 (1993).
18. G. Cates, S. Riordan, and B. Wojtsekhowski (spokespersons), Jefferson Lab experiment E12-09-016 (GEn(2)) (2009).
19. R. Gilman, B. Quinn and B. Wojtsekhowski (spokespersons), Jefferson Lab experiment E12-09-019 (GMn) (2009).
20. The Super Bigbite Spectrometer (SBS) Conceptual Design Report and SBS Technical Review Report can be found on the web-page `http://hallaweb.jlab.org/12GeV/SuperBigBite/SBS-CDR/`.

THE ELECTRIC AND MAGNETIC FORM FACTORS OF THE NEUTRON WITH SUPER BIGBITE

S. RIOIRDAN* and G.D. CATES

Department of Physics, University of Virginia,
Charlottesville, VA 22904, USA
**E-mail: spr4y@virginia.edu*

B. WOJTSEKHOWSKI

Thomas Jefferson National Accelerator Facility
Newport News, VA 23606, USA

Electromagnetic form factors provide experimental access to information about the underlying quark charge and magnetic moment distributions of the nucleon. By performing measurements of these form factors and comparing them to theories containing important aspects of QCD, we gain insight into the generation nucleon structure. Measurements of the form factors of the neutron are generally over a smaller range of Q^2 compared to that of the proton, but are are equally as important to complete our picture of the nucleon. Presented are two approved experiments using the proposed Super Bigbite spectrometer which will extend these measurements to a range of Q^2 comparable to that of the present proton form factors as well as with comparable precision.

Keywords: Neutron, Form factors, Electric form factor, Sachs form factors

1. Introduction

The electromagnetic current of a spin-$\frac{1}{2}$ target with structure interacting with a single virtual photon can be described using two independent functions of invariant four-momentum transfer, the Sachs electric and magnetic form factors, G_E and G_M,

$$J^\mu = e\bar{u}(p')\left[\gamma^\mu \frac{G_E(Q^2)+\tau G_M(Q^2)}{1+\tau} + \frac{i\sigma^{\mu\nu}}{2M} q_\nu \frac{G_M(Q^2)-G_E(Q^2)}{1+\tau}\right] u(p)$$

where p and p' are the final and initial momenta of the nucleon, respectively, q^μ is the four momentum of the virtual photon, Q^2 is the four-momentum transfer, M is the mass of the target, and τ is $Q^2/4M^2$. These form factors carry the physical interpretation of being related to the Fourier transforms in the Briet frame of the electric charge distribution for G_E, and magnetic moment distribution for G_M.

Form factors are important to our understanding of QCD as they provide experimental access to the underlying structure of the valence and sea quarks in the nucleon. Measurements of these form factors are important tests for QCD and QCD-inspired models and can provide insight into the dominant mechanisms of nucleon structure.

While the Sachs form factors of the proton have presently been measured up to at least about $Q^2 = 10$ GeV^2, precise data on the form factors of the neutron only range up to 3–4 GeV^2 [1,2]. One of the challenges presented by neutron form factor measurements is the fact that there are no sufficient free neutron targets to perform these measurements to an adequate precision with present experimental techniques. Because of this, one must study it bound in a nucleus. This introduces inherent theoretical complications in the interpretation of the measurements and how they would relate to free neutron measurements.

Additionally, the electric form factor of the neutron, G_E^n, is intrinsicly small compared to G_M^n as the neutron is an overall neutron particle. This adds an additional experimental challenge, and accordingly, G_E^n has the smallest Q^2 coverage of all the form factors, with the highest measurements up to 3.4 GeV^2 coming from the soon-to-be-published E02-013 experiment at Jefferson Lab.

2. G_E^n to $Q^2 = 10$ GeV^2: E12-09-016

Measuring G_E^n to a Q^2 range comparable to the other form factors offers unique opportunities to gain information on nucleon structure. If all four form factors have been measured, a Dirac/Pauli form factor deconvolution can be performed, as well as quark flavor separation assuming charge symmetry. Additionally, comparisons to models which were successful for the proton form factors can be done. G_E^n is unique in that there are a wide range of predictions between several successful models in the 8 – 10 GeV^2 range.

To access information on electric form factors at high Q^2, polarization measurements are frequently employed which measure the ratio G_E/G_M. They provide a method which generally has fewer systematic errors from two-photon effects than Rosenbluth separation, and in the case of the neutron, fewer systematics in the case of nuclear effects.

One such method involves scattering a longitudinally polarized electron beam from a polarized target. The helicity-dependent cross section asymmetry can be written

$$
\begin{aligned}
A_{\text{phys}} &= \frac{\sigma_+ - \sigma_-}{\sigma_+ + \sigma_-} \\
&= -\frac{2\sqrt{\tau(\tau+1)}\tan(\theta_e/2)G_E/G_M \sin\theta^* \cos\phi^*}{(G_E/G_M)^2 + (\tau + 2\tau(1+\tau)\tan^2(\theta_e/2))} \\
&\quad -\frac{2\tau\sqrt{1+\tau+(1+\tau)^2\tan^2(\theta_e/2)}\tan(\theta_e/2)\cos\theta^*}{(G_E/G_M)^2 + (\tau + 2\tau(1+\tau)\tan^2(\theta_e/2))},
\end{aligned}
$$

where $\sigma_{+/-}$ is the measured cross section for two electron beam helicities, θ_e is the scattering angle of the electron, and θ^* and $\phi*$ are the polar and azimuthal angles of the momentum transfer with respect to the target polarization and electron scattering plane.[3] When the target is polarized transverse to the three-momentum transfer in the electron scattering plane ($\theta^* = 90°$, $\phi^* = 0°$) the asymmetry is approximately proportional to the ratio G_E/G_M when G_E/G_M is small.

Experiment E12-09-016,[4] measurements of G_E^n/G_M^n up to Q^2 of 10 GeV^2 was approved for 50 days of running in Hall A at Thomas Jefferson National Accelerator Facility through the quasielastic coincidence reaction $^3\overrightarrow{\text{He}}(e, e'n)pp$. This experiment utilizes the Super Bigbite apparatus, which employs the concept of large angular and momentum acceptance spectrometers with moderate momentum resolution. Additionally, it will employ an upgraded version of the Hall A polarized ^{3}He target, which will allow for unprecedentedly high in-beam nucleus polarization at beam currents up to 60 μA. A diagram of the experiment can be seen in Fig. 1. This experiment will provide data on the ratio G_E^n/G_M^n at three Q^2 points with precision comparable to previous measurements, Fig. 2.

2.1. *Super Bigbite*

The Super Bigbite apparatus is a general purpose set of detectors and magnets which can be configured to meet a wide range of experimental needs. For this experiment, the electron arm will be composed of the existing Bigbite magnet followed by a set of GEM chambers, a gas Čerenkov detector, a segmented plane of scintillators, and a segmented lead-glass calorimeter, Fig. 3.

The set of four 1.4×4.0 m^2 GEM chambers, each consisting of an X-Y readout, will provide high rate hit-based tracking in a high luminosity environment with an expected momentum resolution of $\sigma_p/p \sim 1\%$. The highest rates for this experiment are predicted to be about 60 kHz/cm^2, well below

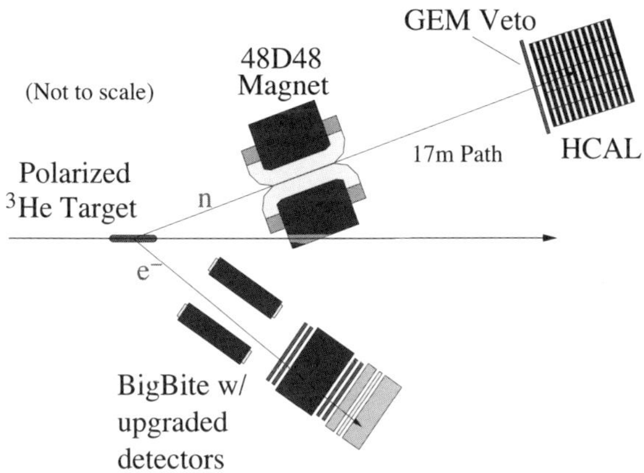

Fig. 1. Schematic experimental setup for E12-09-016.

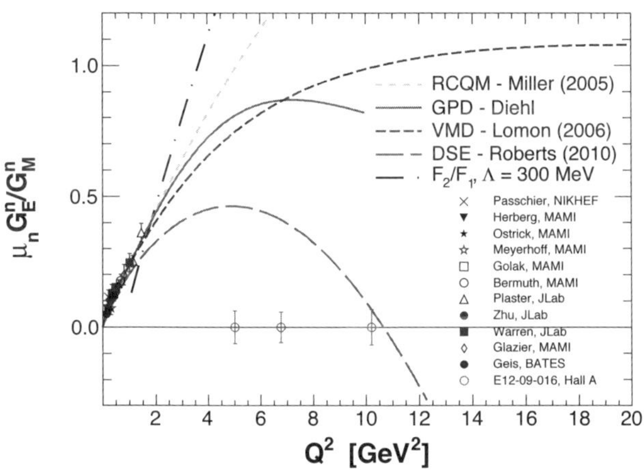

Fig. 2. Anticipated Q^2 and uncertainties for E12-09-016. The references for the data points can be found in the review Ref. 1. The theory curves are Refs. [5–10].

the demonstrated capability of the GEM chambers.[11] The combination of the gas Čerenkov and lead-glass calorimeter will provide a pion rejection factor of about 10000:1, which leads to an overall pion contamination of $< 1\%$.

The neutron arm of the setup is composed of the hadronic calorimeter, HCAL, behind the 48D48 magnet about 17 m from the target. Each

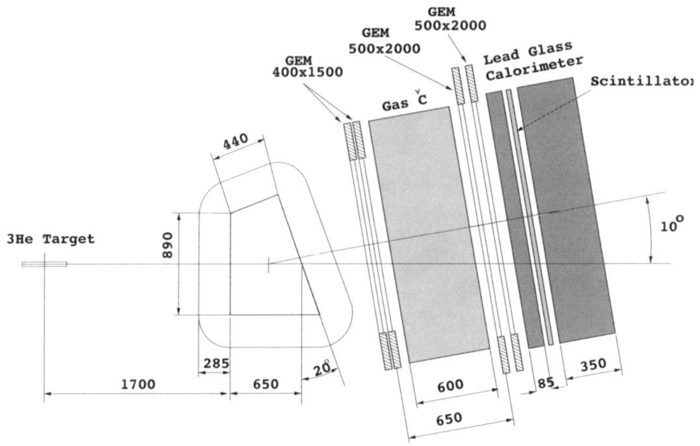

Fig. 3. The electron arm configuration.

calorimeter block consists of 40 interleaved iron and scintillator plates, each of which covers and area of 15×15 cm^2. These blocks will be approximately arranged in a 9 column $\times$ 27 row arrangement to match the electron arm acceptance. An important task of this arm is to provide enough information to separate of neutron quasielastic scattering events from all other processes. This is done by measuring the nucleon three-momentum through position reconstruction and a momentum measurement through time-of-flight. The 48D48 magnet provides a field integral of approximately 1 T $\cdot$ m which provides clean separation of quasielasticly scattered protons and neutrons from ^{3}He.

The use of a hadronic calorimeter allows for a high effective threshold, virtually eliminating accidental background with a detection efficiency of more than 95%, as demonstrated by Geant4 simulations. Two important characteristics of this detector are the time-of-flight resolution and the spatial resolution, both of which are used in separating quasielastic and inelastic events in the data. Time-of-flight is anticipated to be close to 500 ps using an optimized scintillator/WLS bar readout-scheme. The spatial resolution has been simulated to be approximately $\sigma_x = 5$ cm.

2.2. *Polarized* ^{3}He *Target*

The polarized ^{3}He target used will be similar to that used in previous Hall A experiments, with several important modifications. This target is polarized through spin-exchange optical pumping with two alkali metal vapors, rubidium and potassium. This technique, in additional with spectrally narrow diode lasers, has produced sustained polarizations has high as 60% in previous Hall A experiments with beam currents of approximately $10 - 15\ \mu$A.

To allow the collection of high statistics, this target will have two important modifications to allow it to operate at beam currents of 60 μA. First, the target chamber of the cell will be constructed of metal or be metal plated instead of being constructed only of glass, as with the previous cells. Additionally, it will employ a double-barrel convection design instead of a single-tube diffusion design, which allows for rapid polarization of the gas and higher sustained in-beam polarization. With such a design, polarizations in excess of 60% are predicted, providing a large effective luminosity gain.

2.3. *Analysis and Nuclear Corrections*

Analysis of the coincident electron-nucleon events heavily involves the separation of quasielastic neutron events from inelastic background which overlap due to Fermi-smearing and limited detector resolution. By placing tight cuts on missing momentum and invariant mass, a good selection of quasielastic events can be made. Additionally, these cuts suppress final state interactions of the outgoing nucleon with the residual ^{3}He nucleus and select more heavily on the ^{3}He S-wave state, which is of higher effective polarization.

Some fraction of this sample inevitably contains inelastic events, which require calculations for a correction to be applied. It is anticipated a Geant4 Monte Carlo using a realistic ^{3}He model and side-band subtraction can be employed to determined the inelastic yield as well as the asymmetry in the sample.

Nuclear effects were successfully calculated for a similar, but lower Q^2, measurement by Misak Sargsian within the generalized eikonal approximation[12] (GEA) using a ^{3}He wave-function constructed from a variety of potentials. This GEA model includes contributions from nucleon-nucleon rescattering, meson exchange currents, delta-isobar currents, and charge exchange effects. Such an approach has already been applied to previous measurements in E02-013, which produced corrections on the order of 10% to the measured neutron asymmetry.

The acceptance-averaged ratio G_E^n/G_M^n can then be extracted from the corrected asymmetry. Approved measurements of G_M^n (discussed in the following section) will allow for the final extraction of G_E^n.

2.4. *Anticipated Results*

Shown in Fig. 2 are the three approved points and error bars. These three points will bring measurements of G_E^n to a comparable range as other form factors allowing for a comparison to many different models as well as the extraction of quark-flavor distributions. It is interesting that many of the models shown in Fig. 2 have strong deviations from each other at higher Q^2. In particular, the DSE model shows a zero-crossing for G_E^n at Q^2 near 10 GeV^2, which could be observed from these measurements.

3. G_M^n to $Q^2 = 13.5$ GeV^2: E12-09-019

Experiment E12-09-019,[13] also in Hall A at Jefferson Lab, will provide measurements of G_M^n to $Q^2 = 13.5$ GeV^2. This is done by measuring the ratio of the cross section from reactions $D(e,e')n$ and $D(e,e')p$. This ratio can then be related to the free nucleon cross sections, which contain information on G_E^n and G_M^n,

$$\frac{\frac{d\sigma}{d\Omega}\big|_{\mathrm{n(e,e')}}}{\frac{d\sigma}{d\Omega}\big|_{\mathrm{p(e,e')}}} = \frac{\eta\frac{\sigma_{\mathrm{Mott}}}{1+\tau}\left(\left(G_E^n\right)^2 + \frac{\tau}{\epsilon}\left(G_M^n\right)^2\right)}{\frac{d\sigma}{d\Omega}\big|_{\mathrm{p(e,e')}}}$$

where $\eta = E'/E$, E and E' are the energies of the initial and scattered electron, respectively, $\epsilon = \left(1 + \vec{q}^2/Q^2\tan^2(\theta_e/2)\right)^{-1}$, and $\vec{q}$ is the three-momentum transfer. This technique exploits that fact that the proton cross sections are well known in this range and that many effects due to nuclear binding cancel. Calculations have shown that such effects are a less than 1% correction to the measured ratio.[14]

To perform this measurement, the Super Bigbite apparatus in a configuration similar to the one used for the previously discussed G_E^n measurement. In this case, a liquid deuterium target is used in place of the polarized ^{3}He. To calibrate the proton and neutron detection efficiencies, the reactions $p(e,e')p$ and $p(\gamma,\pi^+)n$ can be studied using a liquid hydrogen target.

Anticipated results for the awarded 25 days of beam-time are shown in Fig. 4.

4. Summary

By measuring the electric and magnetic form factors of the neutron, we help complete our picture of nucleon form factors and gain new information to compare to models that are favored in our descriptions of nucleons.

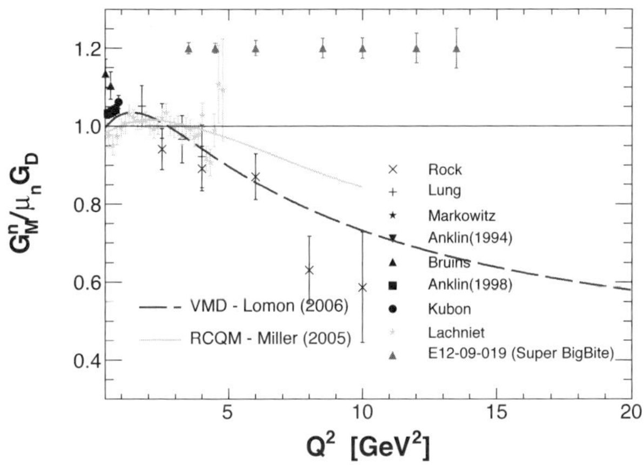

Fig. 4. Anticipated Q^2 and uncertainties for E12-09-019. The references for the data points can be found in the review [1] and the recent G^n_M data in Ref. [2].

The Super Bigbite apparatus provides opportunities to push these measurements to unprecedentedly high Q^2 by allowing the collection of data at high luminosities.

References

1. C.F. Perdrisat, V. Punjabi, and M. Vanderhaeghen, Prog. in Part. and Nucl. Phys. **59**, 694 (2007).
2. J. Lachniet *et al.*, Phys. Rev. Lett. **102**, 192001 (2009).
3. T.W. Donnelly and A.S. Raskin, Ann. Phys. (N.Y.) **169** 247 (1986).
4. E12-09-016 proposal, http://hallaweb.jlab.org/collab/PAC/PAC34/PR-09-016-gen.pdf
5. G.A. Miller, Phys. Rev. C **66**, 032201(R) (2002); H.H. Matevosyan, A.W. Thomas, G.A. Miller, Phys. Rev. C **72**, 065204 (2005); private communication (2009).
6. E.L. Lomon, Phys. Rev. **C** 66, 045501 (2002); arXiv:nucl-th/0609020.
7. M. Diehl, T. Feldmann, R. Jakob, and P. Kroll, Eur. Phys. J. C **39**, 1 (2005).
8. C.D. Roberts *et al.*, Eur. Phys. J. **ST 140**, 53 (2007).
9. I.C. Cloet *et al.*, Few-Body Systems **46**, 1(2009).
10. A.V. Belitsky, X. D. Ji, and F. Yuan, Phys. Rev. Lett. **91**, 092003 (2003).
11. B. Ketzer *et al.*, Nucl. Instr. and Meth. **A. 535**, 314 (2004).
12. M.M. Sargsian, T.V. Abrahamyan, M.I. Strikman, and L.L Frankfurt, Phys. Rev. C **71**, 044614 (2005).
13. E12-09-019 proposal, http://hallaweb.jlab.org/collab/PAC/PAC34/PR-09-019-gmn.pdf
14. S. Jeschonnek and J.W. Van Orden, Phys. Rev. C **62** (2000) 044613

MEASUREMENT OF THE NEUTRON MAGNETIC FORM FACTOR AT HIGH Q^2 USING THE RATIO METHOD ON DEUTERIUM

G.P. GILFOYLE*, W.K. BROOKS, K. HAFIDI *et al.*
for the CLAS Collaboration
* *E-mail: ggilfoyl@richmond.edu*

The 12-GeV Upgrade at Jefferson Lab will create an opportunity to dramatically extend our knowledge of the magnetic form factor of the neutron (G_M^n) and the other elastic, electromagnetic form factors. We describe here an approved experiment that will cover a Q^2 range ($3.5 - 14$ GeV^2) with significant discovery potential. Different theoretical approaches (generalized parton distributions, Dyson-Schwinger equations, *etc*) diverge is this region and existing data cannot distinguish among them. The proposed measurement will be performed in Hall B with the CLAS12 detector and will have statistical and systematic uncertainties below 3%. It is based on the ratio of electron-neutron to electron-proton scattering that was successfully applied in the CLAS detector at Jefferson Lab.

Keywords: Form factors

1. Introduction

The 12 GeV Upgrade at the Thomas Jefferson National Accelerator Facility (JLab) has begun and will double the beam energy and open new physics opportunities. The new physics program at JLab will dramatically extend the reach of our understanding of a fundamental feature of the neutron; its magnetic form factor G_M^n. The elastic electromagnetic form factors(EEFFs) describe the distribution of charge and magnetization inside the nucleon at low Q^2 and probe the quark structure at higher Q^2. A broad program is planned at JLab to measure the EEFFs, map the internal landscape of the nucleon, and test non-perturbative Quantum Chromodynamics (QCD) and QCD-inspired models of the nucleon (see NSAC Long-Range Plan and others [1–7]). The measurement we discuss here will be performed with the new CLAS12 detector in Hall B (Experiment E12-07-104 [2]) at JLab which will replace the current and similar CLAS6 detector. CLAS12 is a large acceptance spectrometer consisting of layers of silicon counter, drift chambers, time-of-flight scintillators, electromagnetic calorimeters, and Ćerenkov counters. The forward part is based on a toroidal magnet and there is a central detector for particles emitted at large angles based on a solenid magnet. The proposed measurement we describe here will cover the range $Q^2 = 3.5 - 14.0$ GeV^2 with systematic uncertainties less than 3%. Statistical uncertainties will be about 3% in the highest Q^2 bin in this range and significantly less at lower Q^2. The anticipated range and uncertainties of the experiment are shown in Figure 1. The reduced magnetic form factor $G_M^n/(\mu_n G_D)$ is plotted versus Q^2 where μ_n is the neutron magnetic moment and $G_D = 1/(1 + Q^2/\Lambda^2)^2$ is the dipole form factor with $\Lambda^2 = 0.71$ GeV^2. We used the recent parameterization of the world's data on G_M^n in Ref [16] to predict the reduced form factor. Also shown are selected world's data for G_M^n including the recent CLAS6 results (red, open circles)[8]. The proposed CLAS12 experiment (black, closed squares) will nearly triple the upper limit of the previous CLAS6 measurement and provide precise data well beyond any existing measurement. In this paper we discuss the current experimental and theoretical situation and our method for measuring G_M^n with CLAS12.

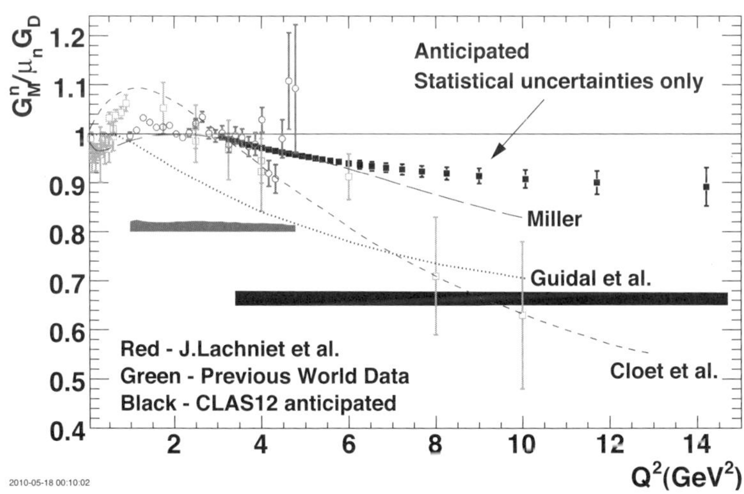

Fig. 1. Selected data [8–15] and anticipated results for G_M^n for 30 days of running with CLAS12 (black, filled squares) in units of $\mu_n G_D$ as a function of Q^2. The anticipated CLAS12 results follow a fit to the world data on G_M^n that includes the CLAS6 G_M^n results [16] (red, open circles). Curves are described in the text.

2. Experimental Status

Figure 1 shows the world data for G_M^n up to $Q^2 = 14$ GeV^2. For $Q^2 > 4.5$ GeV^2 the data are sparse and have large uncertainties. In the range $Q^2 < 4.5$ GeV^2, the situation is much better especially with the addition of the CLAS6 measurement [8] (red, open circles). The CLAS6 measurement was performed using largely the same method we propose here for CLAS12. The CLAS6 data are surprisingly consistent with the dipole parameterization. This was a unexpected because previous measurements show G_M^n decreasing at large Q^2 although with large uncertainties (green points in Figure 1 for $Q^2 \geq 4.5$ GeV^2.

The broad effort at JLab to measure the EEFFS is in a Q^2 region with significant discovery potential. All these form factors are needed to untangle nucleon structure [17]. Measuring the ratio of the proton electric to magnetic form factor G_E^p/G_M^p using polarization observables revealed a striking dropoff contradicting previous measurements [18]. On the neutron side the ratio G_E^n/G_M^n was recently measured with greater precision and at higher Q^2 than ever before [19]. Those researchers used the recent CLAS6 measurement of G_M^n to extract G_E^n. The preliminary results for the points at Q^2 of 2.5 GeV^2 and 3.5 GeV^2 are 2-4 standard deviations away from the Galster parameterization; suggesting the onset of changes from the lower Q^2 behavior. All of these new, intriguing results are in the same Q^2 range as the proposed CLAS12 G_M^n measurement.

3. Theoretical Status

Progress has been made on the theory side of our understanding of the EEFFs. The interpretation of the form factors in relativistic kinematics has proven more challenging than expected and has required a new picture based on generalized parton distributions (GPDs) [20,21]. To understand non-perturbative QCD Cloët *et al.* employ a Dyson-Schwinger Equation approach and assume two of the valence quarks form a di-quark. Figure 2 shows the ratio of the neutron electric to magnetic form factors for two different values of the diquark radius (solid curve and long-dashed curve), data from Madey *et al.* [22], and the Kelly parameterization (dashed curve) [23]. The Cloët *et al.* prediction diverges dramatically from the data parameterization at $Q^2 \approx 5$ GeV^2 and crosses zero at $Q^2 \approx 11$ GeV^2. This behavior marks this region of Q^2 as one of potential discovery value and lies well within the Q^2 range of our proposed experiment. The prediction of G_M^n by Cloët *et al.* for G_M^n is shown in Figure 1 along with two other

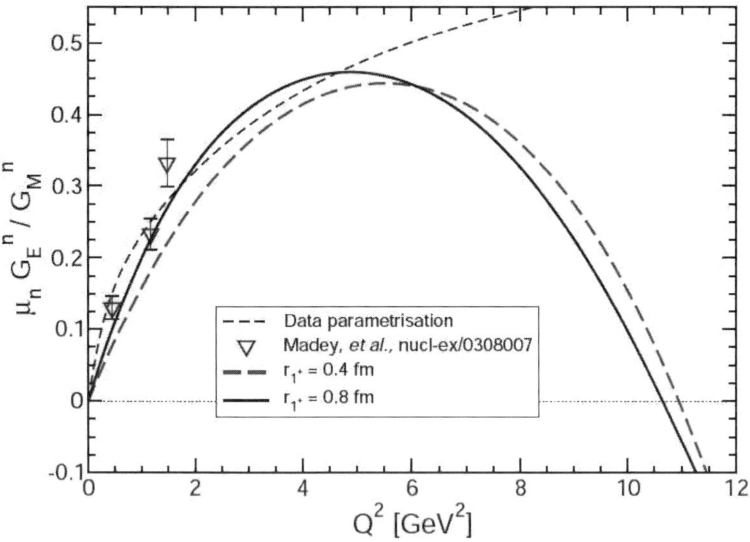

Fig. 2. Result for the normalized ratio of Sachs electric and magnetic form factors for the neutron computed with two different diquark radii. Short-dashed curve: parameterization of Ref. [23]. Down triangles: data from Ref. [22].

calculations. In Miller's calculation the nucleon is treated using light-front dynamics as a relativistic system of three bound quarks and a surrounding pion cloud [24]. The model gives a good description of much of the previous data (including the other three EEFFs) even at high Q^2 and is consistent with the CLAS6 results. The curve from Guidal *et al* [25] uses GPDs to characterize the EEFFs at low Q^2 and then extend their calculations to higher Q^2, but fails to reproduce the data at low Q^2. All three curves differ measurably in magnitude and/or slope in the range $Q^2 = 6 - 14$ GeV^2. There is an opportunity here to distinguish among competing pictures.

We also want to touch on truly *ab initio* calculations performed using lattice QCD. These calculations are still limited in reach to $Q^2 \approx 1$ GeV^2, but we expect that significant progress will be made by the time the proposed experiment is complete. This is where the broad assault on the EEFFs at JLab is essential. The EEFFs can be formed into isovector and isoscalar combinations that are sensitive to different physical effects. The isovector combination is free of disconnected contributions which are notoriously difficult to compute on the lattice. This freedom will make the isovector form factor an early test of lattice QCD as the calculations reach higher Q^2.

4. Experimental Method

We now outline the experimental technique (more details are in Ref. [2]). We will use the ratio of quasielastic $e-n$ to $e-p$ scattering from a deuterium target to measure G_M^n. This technique reduces uncertainties associated with other methods and has been used by us [8] and others [12,26–29] (see Figure 1). The method is based on the ratio

$$R = \frac{\frac{d\sigma}{d\Omega}(^2\mathrm{H}(e,e'n)_{QE})}{\frac{d\sigma}{d\Omega}(^2\mathrm{H}(e,e'p)_{QE})} = a(Q^2)\frac{\sigma^n_{mott}({G_E^n}^2 + \frac{\tau_n}{\varepsilon_n}{G_M^n}^2)\left(\frac{1}{1+\tau_n}\right)}{\frac{d\sigma}{d\Omega}(^1\mathrm{H}(e,e')p)} \tag{1}$$

for quasielastic (QE) kinematics. The right-hand side is written in terms of the free nucleon form factors and where $\tau = Q^2/4M^2$ and $\epsilon = 1/[1 + 2(1+\tau)\tan^2(\theta/2)]$. Deviations from this 'free ratio' assumption are parameterized by the factor $a(Q^2)$ which can be calculated from deuteron models and is close to unity at large Q^2. The results of other measurements of the proton cross section and the neutron electric form factor are used to extract G_M^n. The ratio R is insensitive to the luminosity, electron acceptance, electron reconstruction efficiency, trigger efficiency, the deuteron wave function used in $a(Q^2)$, and radiative corrections [8].

We will first select quasielastic (QE) events by applying a cut on θ_{pq}, the angle between the nucleon 3-momentum and the momentum transfer $\vec{q}$. Nucleons from inelastic events nucleons tend to be emitted at large θ_{pq} while QE nucleons are emitted along $\vec{q}$. Analyzing protons and neutrons the same way avoids biasing the results. Next, we select events with $W^2 = M_N^2$, but at high Q^2 the width of the residual mass spectrum W^2 and contaminates the the QE peak. The top panel of Figure 3 shows the results of a simulation for such high-Q^2 $e-n$ events. The red histogram shows the QE events, the green one shows the inelastic events, and the black one is the total. The QE events are overwhelmed by the inelastic background. Requiring $\theta_{pq} < 1.5°$ for the neutron (lower-left-hand panel) considerably reduces the inelastic background, but the QE events are still a shoulder on a larger inelastic peak. We now take advantage of the large acceptance of CLAS12. The inelastic events emit additional particles (pions, photons, *etc.*). and CLAS12 will detect many of those particles. In the lower-right-hand panel we have applied a hermiticity cut; rejecting events with a third particle. The quasielastic peak now clearly rises out of the inelastic noise. We emphasize here that Figure 3 is a worst case scenario. We simulated the reaction for the highest Q^2 bin (12.5 – 14 GeV^2) where the inelastic contamination is largest. The data at lower Q^2 will have less kinematic spreading and inelastic contamination.

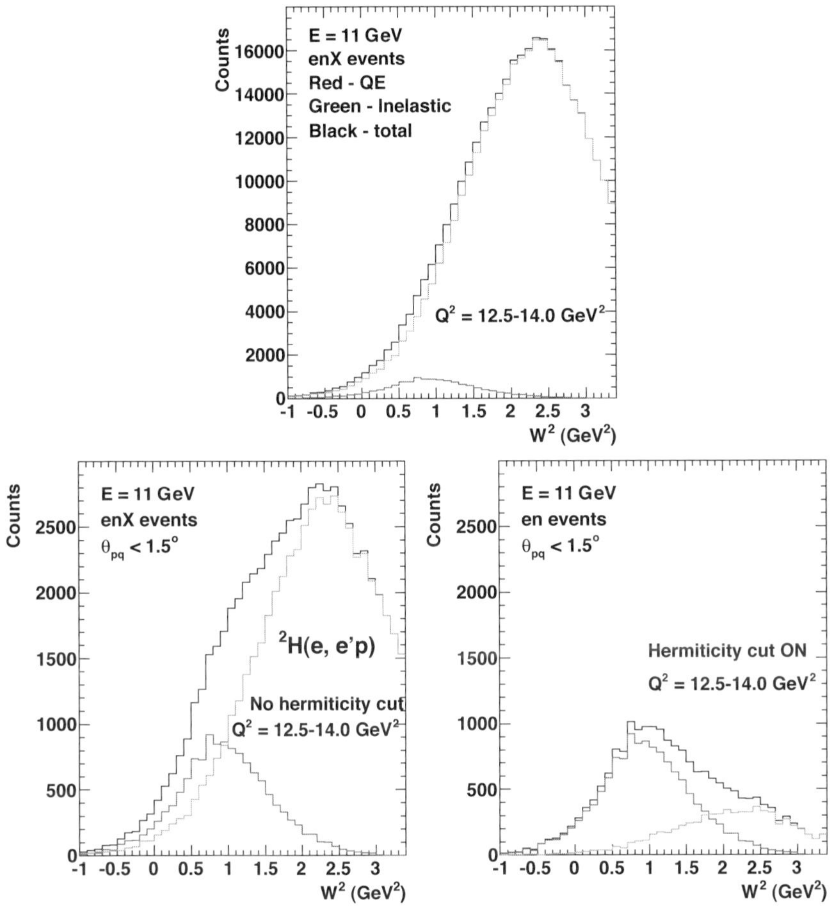

Fig. 3. The impact of the hermiticity cut is shown. The top panel displays the W^2 spectra for simulated $e-n$ events. In the lower-left-hand panel, the neutrons are required to have $\theta_{pq} < 1.5°$. In the lower-right-hand panel, we add the hermiticity cut. The number of events shown is not representative of the anticipated value.

The CLAS12 G_M^n measurement has important consistency checks. Neutrons will be measured in two subsystems of CLAS12: the forward electromagnetic calorimeter (FC) and the forward time-of-flight (TOF). These two subsystems will enable us to make semi-independent measurements of the $e-n$ production and provide a cross check. The CLAS12 measurement will also have a large overlap ($Q^2 = 3.5 - 4.8$ GeV^2) with the previous CLAS6 G_M^n one (see Figure 1).

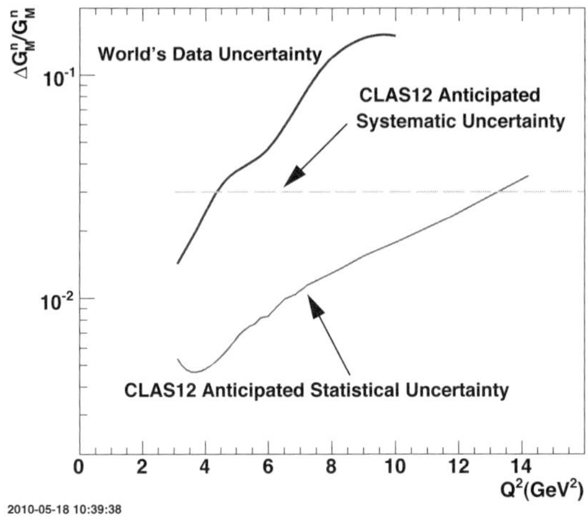

Fig. 4. Run statistics for the CLAS12 G^n_M experiment (E12-07-104) for 30 days of running. The blue, solid curve is the current uncertainty on G^n_M for the world data, the green, dashed line is our anticipated statistical uncertainty, and the red, solid curve is our goal of 3% systematic uncertainty.

An essential aspect of the neutron measurement in the TOF and FC systems is measuring the neutron detection efficiency. We will use the $p(e, e'\pi^+n)$ reaction as a source of tagged neutrons. Electrons and π^+'s will be detected in CLAS12 and missing mass used to select candidate neutrons (found events). We then predict the position of the neutron in CLAS12 and search for it (if a neutron is observed we call these reconstructed events). The ratio of reconstructed to found events is the detection efficiency. This will be done in CLAS12 with a unique, dual target. Co-linear, liquid hydrogen and deuterium cells will provide production and calibration events simultaneously and under the same conditions (*in situ*). This reduces our vulnerability to variations in detector gains, beam properties, *etc.* In the CLAS6 measurement we used the same techniques and found the different data sets were consistent within the statistical uncertainties [8].

In Figure 4 we show the anticipated uncertainty on G^n_M (red, solid line). Proton cross sections came from Ref [23] and neutron cross sections from Galster (G^n_E) and Alberico *et al.* (G^n_M) [16]. We included the effect of a cut requiring $W^2 < 1.2$ GeV^2 that reduces residual inelastic contamination. Over the range $Q^2 = 3.5 - 14.0$ GeV^2 we have statistical uncertainties at or below 3% (green, dashed line) and typically much lower. These uncertainties are far better than the current precision on G^n_M (blue, solid line).

5. Summary

To summarize, the scientific motivation for measuring the neutron magnetic form factor is compelling. At higher Q^2 ($5 - 13$ GeV^2) we have observed new, surprising behavior in the other form factors and developed QCD-inspired models that diverge widely. To explore this new territory JLab will measure all of the EEFFs. The CLAS12 experiment will use a tested method for measuring G_M^n and push the frontier of our understanding of the neutron magnetic form factor up to $Q^2 = 14$ GeV^2 and with high precision. The ability to veto multiparticle final states dramatically reduces the background from inelastic events that contaminate the QE peak. We will use a dual-cell target for precise, *in situ* measurements of the neutron detection efficiency as demonstrated in the CLAS6 experiment. Finally, with CLAS12 we have several important consistency checks using different detector subsystems of CLAS12 (TOF and EC) to validate our results.

References

1. Nuclear Science Advisory Committee, *The Frontiers of Nuclear Science* (US Department of Energy, 2007).
2. G. Gilfoyle *et al.*, Experiment E12-07-104, Jefferson Lab (Newport News, VA, 2007).
3. E. Brash *et al.*, Experiment E12-09-001, Jefferson Lab (Newport News, VA, 2009).
4. S. Gilad *et al.*, Experiment E12-07-108, Jefferson Lab (Newport News, VA, 2007).
5. B. Quinn *et al.*, Experiment E12-09-119, Jefferson Lab (Newport News, VA, 2009).
6. B. Anderson *et al.*, Experiment E12-09-006, Jefferson Lab (Newport News, VA, 2009).
7. B. Wojtsekhowski *et al.*, Experiment E12-10-005, Jefferson Lab (Newport News, VA, 2010).
8. J. Lachniet, A. Afanasev, H. Arenhövel, W. K. Brooks, G. P. Gilfoyle, D. Higinbotham, S. Jeschonnek, B. Quinn, M. F. Vineyard *et al.*, *Phys. Rev. Lett.* **102**, p. 192001 (2009).
9. W. Bartel *et al.*, *Nucl. Phys. B* **58**, p. 429 (1973).
10. G. Kubon *et al.*, *Phys. Lett. B* **524**, 26 (2002).
11. A. Lung *et al.*, *Phys. Rev. Lett.* **70**, 718(Feb 1993).
12. H. Anklin *et al.*, *Phys. Lett. B* **428**, 248 (1998).
13. B. Anderson *et al.*, *Phys. Rev. C* **75**, p. 034003 (2007).
14. R. G. Arnold *et al.*, *Phys. Rev. Lett.* **61**, 806(Aug 1988).
15. S. Rock *et al.*, *Phys. Rev. Lett.* **49**, 1139(Oct 1982).
16. W. M. Alberico, S. M. Bilenky, C. Giunti and K. M. Graczyk, *Phys. Rev. C (Nuclear Physics)* **79**, p. 065204 (2009).

17. M. Diehl, T. Feldmann, R. Jakob and P. Kroll, *Eur. Phys. J. C* **39**, p. 1 (2005).
18. M. K. Jones *et al.*, *Phys. Rev. Lett.* **84**, 1398(Feb 2000).
19. K. de Jager (2009).
20. G. A. Miller, *Phys. Rev. Lett.* **99**, p. 112001 (2007).
21. G. A. Miller and J. Arrington, *Phys. Rev.* **C78**, p. 032201 (2008).
22. R. Madey *et al.*, *Phys. Rev. Lett.* **91**, p. 122002 (2003).
23. J. J. Kelly, *Phys. Rev. C* **70**, p. 068202(Dec 2004).
24. G. Miller, *Phys. Rev. C* **66**, p. 032201 (2002).
25. M. Guidal *et al.*, *Phys. Rev. D* **72**, p. 054013 (2005).
26. C. Hyde-Wright and K.deJager, *Ann. Rev. Nucl. Part. Sci.* **54**.
27. H. Anklin *et al.*, *Phys. Lett. B* **336**, 313 (1994).
28. G. Kubon *et al.*, *Phys. Lett. B* **524**, 26 (2002).
29. E. E. W. Bruins *et al.*, *Phys. Rev. Lett.* **75**, 21(Jul 1995).

CLAS DATA AND PROGRESS IN THE INVESTIGATION OF ELECTRO-EXCITATION OF NUCLEON RESONANCES

M. UNGARO

Physics Department, University of Connecticut,
Storrs, CT, 06269, USA

Jefferson Laboratory
Newport News, VA, 23606

E-mail: ungaro@jlab.org
www.jlab.org/~ungaro

We present an overview of selected CLAS nucleon excitations results: electro-production of single pion from the $\Delta(1232)$, Roper, $D_{13}(1520)$ resonances and electro-production of double pion $N\pi^+\pi^-$. These and other CLAS measurements, coupled with recent theoretical developments, will allow non-perturbative approaches to shed lights on the role of quarks and gluons in nuclei and address issues such as confinement and the non-zero quark mass in the chiral limit.

Keywords: CLAS; barion resonances; transition form factors

1. Introduction: N* Beyond CQM and pQCD

A major goal of hadron physics is to probe the structure of the nucleon. Measurements of N* play a crucial role in this investigation in that:

- The N* spectrum is a direct reflection of the underlying degrees of freedom of the nucleon.
- Electromagnetic transition form factors probe the underlying spatial and spin structure versus distance scale.

Constituent Quark Models (CQM) provide simple and effective interpretations of many phenomena [1–6], especially near the photon point. On the other hand, an investigation of the fundamental structure of the nucleon and its excited states has to involve partonic degrees of freedom.

At very large Q^2, perturbative Quantum Chromo Dynamics (pQCD) prescribe scaling of the N* helicity amplitudes [7,8]:

$$A_{1/2} \propto 1/Q^3, \;\; A_{3/2} \propto 1/Q^3, \;\; G_M^* \propto 1/Q^4,$$

as seen in Fig. 1. However pQCD cannot answer fundamental questions such as:

- What is the nature of quark confinement?
- What is the dynamic of quark-gluon interactions at low energy?

To address these challenges, we need to step in the domain of relativistic quantum field theory where the key phenomena can only be understood with **non-perturbative** methods.

Thanks to recent theoretical developments, understanding the microscopic nature of the nucleon using non-perturbative approaches is beginning to become a reality... and it's becoming increasingly clear that measurements of nucleon resonance play a crucial role. I will present three examples of such approaches:

- Lattice QCD
- Dyson-Schwinger Equations and Dynamical Chiral Symmetry Breaking
- Light Cone Sum Rules

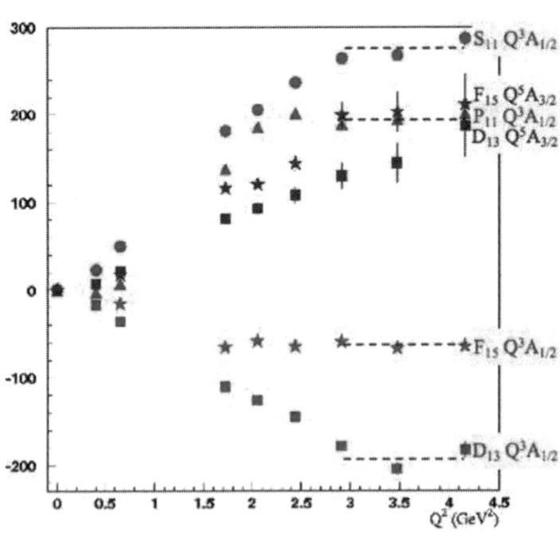

Fig. 1. Constituent Counting Rule Scaling. These SLAC data show the beginning of the scaling behavior of the S_{11}, P_{11}, D_{13}, F_{15} helicity amplitudes. The Delta form factor begin scaling at around 10 GeV^2.

1.1. *Lattice QCD*

The Hadron Spectrum Collaboration involving the Lattice Group at Jefferson Lab, Carnegie Mellon University, Univ. of Maryland, and Trinity College (Dublin) has embarked on an ambitious program to compute the high lying excited state spectrum of baryons and mesons, as well as their (excited state) electromagnetic transition form-factors up to Q^2 of 10 GeV^2 [9,10].

New techniques on the lattice allow the calculation of baryon spectra not only in the ground state but also of the excited states [11–13], see Fig. 2 (left), and first results have been obtained for the N→P_{11} transition form factor [14], shown in Fig. 2 (right).

Although still exploratory, these results are very promising and underline the importance of experimental N* measurements.

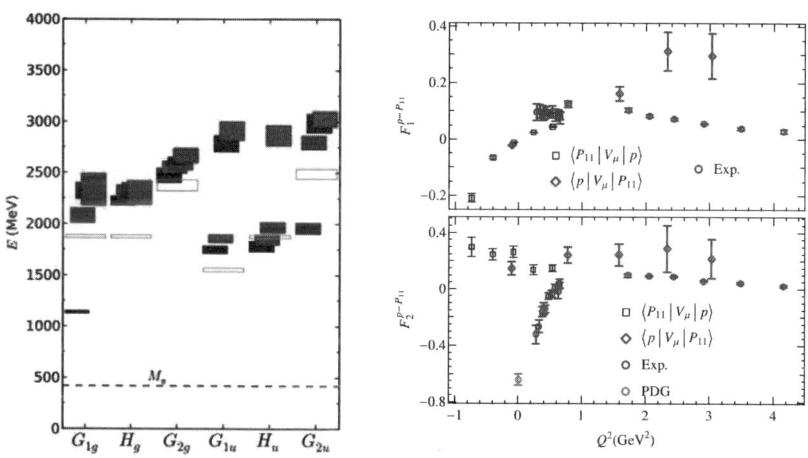

Fig. 2. Lattice QCD calculations. Left: The energies obtained for each symmetry channel of isospin 1/2 baryons. The pion mass was 400 Mev and the lattice spacing was 2.64 fm^3. The errors are indicated by the vertical size of the boxes. Right: proton-Roper form factors $F*_{1,2}$ obtained from CLAS experiments and PDG number (circles) and lattice methods (squares, diamonds). The calculations[14] involve for the first time radial excitations in the calculation of the form factors, and a way to extract signal to noise that help the extraction at high Q^2. Typical previous lattice calculation are up to 3 GeV^2.

1.2. *Dynamical Chiral Symmetry Breaking and Dyson-Schwinger Equations*

Dynamical Chiral Symmetry Breaking (DCSB) explains the origin of constituent-quark masses [15,16]: the dressed-quark mass function is depicted in Fig. 3. Notice that in perturbation theory it is impossible in the chiral limit to obtain a quark mass different than zero.

The complex of Dyson-Schwinger equations (DSEs) is a powerful tool that has been employed with marked success to study confinement and DCSB, and their impact on hadron observables [17–21].

International studies of nucleon electromagnetic form factors lead by Cloet, Bentz, Thomas are underway. In their calculations dressed quarks are the degrees of freedom and correlations are expressed with di-quarks. The diquark current depends on their charge radius. A preliminary result is the N→ Δ transition form factor shown in Fig. 3.

Heres the key: we can use high precision measurements of nucleon-resonance transition form factors, DSEs and DCSB to chart the momentum evolution of the dressed-quark mass. In particular, the transition between current and constituent quark domains, believed to happen between 5-10 GeV^2.

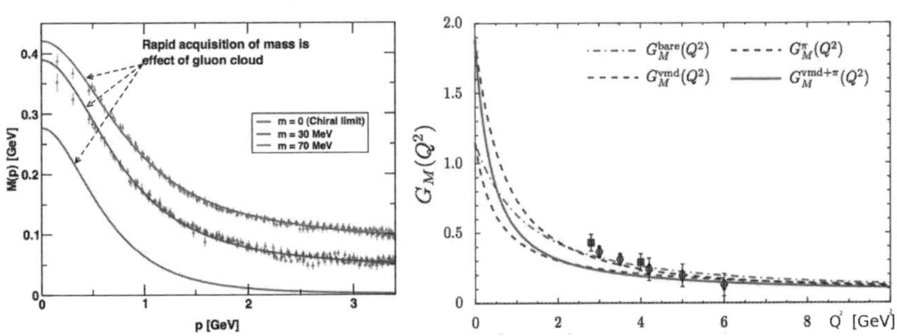

Fig. 3. Left: Dressed-quark mass function. solid curves: DSE results[15,16] . Data: numerical simulations of unquenched lattice QCD[22] . In this figure one observes the current-quark of perturbative QCD evolving into a constituent-quark as its momentum becomes smaller. The constituent-quark mass arises from a cloud of low-momentum gluons attaching themselves to the current-quark. Right: preliminary results for the N→ Δ transition form factor using the DSEs framework compared with data.[23,24] The form factor lacks significant strength at the photon point (its value should be 3) but qualitatively the calculations are very promising.

1.3. *Light Cone Sum Rule and Form Factors*

One of the prediction of QCD is that the, at large momentum transferred, the form factor become dominated by the contribution of valence states with small transverse separation between them. Such distributions are dubbed Distribution Amplitudes (DA).

A non perturbative approach that allows quantitative descriptions of form factor using is the Light-Cone Sum Rule (LCSR): within this framework, the the DA are directly connected to the transition form factors.

The DA can be calculated on the lattice. Recent effort from Warkentin, Braun, of Theoretical Groups in Regensburg and others from the QCDSF collaboration calculated the DA for both the proton and the S11 [25]. The results are shown in Fig. 4. One can notice that the momentum fraction of the u quark for example is significantly higher in the S11 than in the proton. The pion mass used in this calculation was of the order of the GeV and the calculation was done at the leading order of the DA.

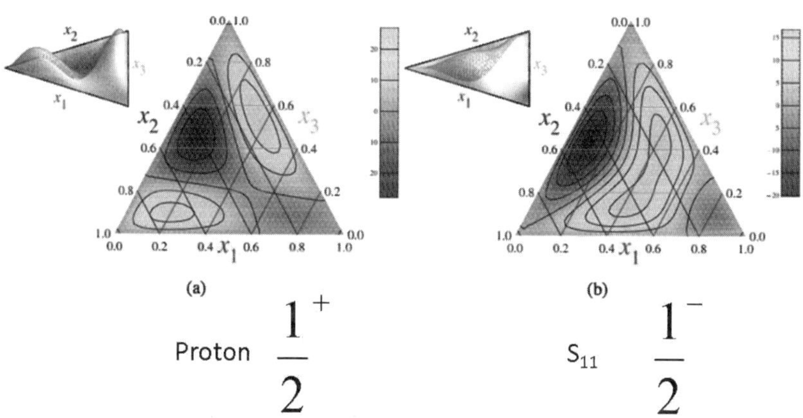

Fig. 4. Barycentric plot of the distribution amplitudes for proton and the $S_{11}(1535)$ using a pion mass of 1 GeV[25] . The lines of constant x_1, x_2 and x_3 are parallel to the sides of the triangle labeled by x_2, x_3 and x_1, respectively.

The results of the calculation of the form factors within the LCSR framework using these DA [26] are shown in the Fig. 5.The band represents uncertainties in the DA (that will be improved with future calculations) and assumptions in the LCSR. The LCSR expect to converge at $Q^2 > 2$. Here the prediction seems to describe the data reasonably well.

The group has plan in the next couple of years to a) decrease the pion mass for more precise measurement and b) to repeat this calculation for all the baryons in the octect and the decuplet.

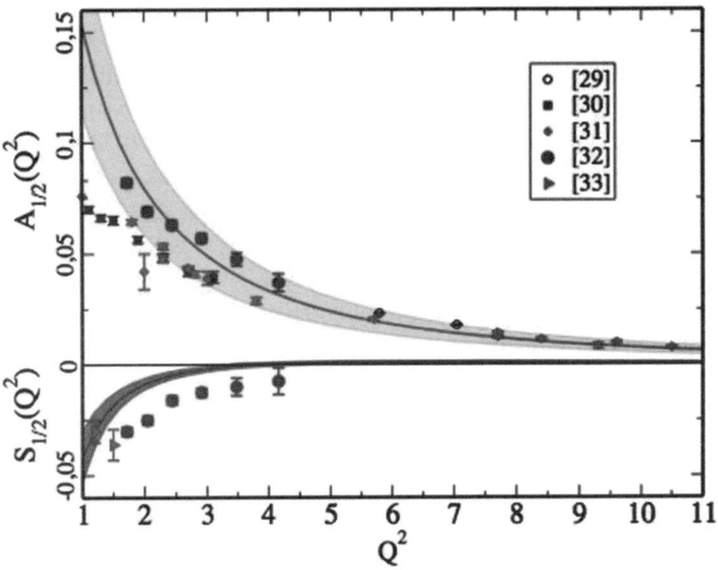

Fig. 5. The LCSR calculation for the helicity amplitudes $A_{1/2}$ and $S_{1/2}$ for the electroproduction of $S_{11}(1535)$ resonance using lattice results compared to the available experimental data.[27,28]

2. The N* Program in CLAS

The N* Program in CLAS is aimed at mapping the the $\gamma_v NN^*$ electrocouplings as a function of photon virtuality with the combined analysis of the major electroproduction channels. Among the channels analyzed: $p\pi^0$, $n\pi^+$, $p\eta$, $K\Lambda$ and double charged pion electro-production. The CLAS data for the first time provided information on many observables in these exclusive channels, including fully integrated cross sections and a variety of 1-fold differential cross sections complemented by single and double polarization asymmetries in a range of Q^2 from 0.2 to 5 GeV^2 .

In Fig. 6 the branching ratios of several meson channels in the resonance region are shown. All these channels are detected in CLAS. The main advantages of the combined analysis of single and double meson production channels are:

a) independent calculation of electro-coupling amplitudes
b) reliable determination of non-resonant amplitudes

In the following sections a selection of CLAS results will be shown. But before that, it's important to present the methodology of extraction of physics observables from the data.

State	$\beta_{N\pi}$	$\beta_{N\eta}$	$\beta_{N\pi\pi}$
$\Delta(1232)P_{33}$	0.995		
$N(1440)P_{11}$	0.55-0.75		0.3-0.4
$N(1520)D_{13}$	0.55-0.65		0.4-0.5
$N(1535)S_{11}$	0.35-0.55	0.45-0.60	<0.1
$N(1620)S_{31}$	0.20-0.30		0.7-0.8
$N(1650)S_{11}$	0.60-0.95	0.03-0.10	0.1-0.2
$N(1685)F_{15}$	0.65-0.70		0.3-0.4
$\Delta(1700)D_{33}$	0.1-0.2		0.8-0.9
$N(1720)P_{13}$	0.1-0.2	0.01-0.15	> 0.7

Fig. 6. Branching Ratios of several meson channels in the resonance region. Most single and double pion channels overlap, with double pion production becoming more relevant with higher state mass. CLAS ability of detecting multiple final states, couple with the combined analysis of these channels, allow a reliable extraction of resonant and non-resonant amplitudes.

2.1. *Single and double Pion Analysis Tools*

In order to access the helicity amplitudes, several effective Lagrangian approaches have been developed at JLAB:

Unitary Isobar Model(UIM) [29] :

- non-resonant amplitudes incorporate nucleon and meson Born terms, vector meson exchanges, Regge terms at high energy.
- resonant terms incorporate relativistic Breit-Wigner amplitudes with energy-dependent widths.
- full amplitude unitarized in K-matrix approximation.

Fixed-t dispersion relations (DR) [29] :

- resonance amplitudes constructed in same way as in UIM
- non-resonant amplitudes from Born terms and dispersion relation.

Isobar Model JLab-Moscow State University (JM) [30] :

- In this approach the resonant part of the amplitudes is isolated and the Q^2 evolution of the individual electro-coupling parameters of the contributing nucleon resonances are determined from a simultaneous fit to all measured observable.

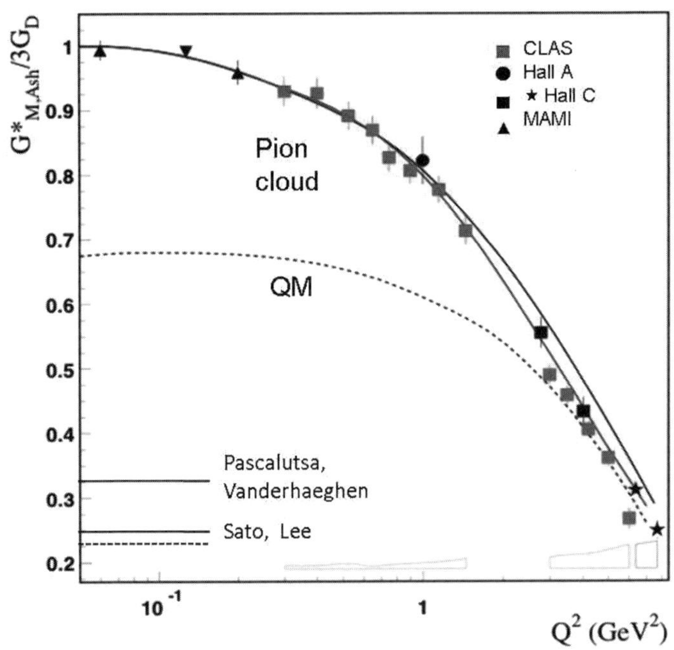

Fig. 7. The $N \to \Delta$ magnetic form factor G^*_M. The meson cloud contributes more than 30% to the "bare" quark model (QM).[31,32] The middle line is the Pascalutsa-Vanderhaeghen calculation:[33] at the large N_c limit, the $N \to \Delta$ GPD H_M is related to the nucleon (isovector) Pauli form factor.[33]

2.2. *The Delta Resonance*

The $N \to \Delta$ transition amplitudes have been measured for 3 decades up to $Q^2 = 7$ GeV^2. In Fig. 7 the magnetic form factor G^*_M normalized to the dipole form G_D is shown. In the Constituent counting rule G^*_M should scale as $1/Q^4$, but instead it continues to to drop faster with Q^2 than the G_D. We expect G^*_M to start scaling at around 10 GeV^2. In the plot is also highlighted the importance of the meson cloud contribution [31,32] and the GPD link of the $N \to \Delta$ form factor to the electromagnetic properties of the nucleon [33].

2.3. $P\pi^+\pi^-$ *Results*

With the analysis of two pions data our knowledge of the excitations in the second resonance region improves dramatically. Many of these resonances decay predominantly to $N\pi\pi$ final states, as shown in Fig. 6. The analysis of the CLAS two pions data allowed us for the first time to map out the Q^2 evolution of electro-coupling parameters for resonances with masses above 1.6 GeV that decay preferably by two pion emission such as $S_{31}(1620)$, $D_{33}(1700)$ and $P_{13}(1720)$ [9].

The CLAS data was analyzed within the JM framework [30]. Various production mechanisms have different shapes in different observables. On the other hand, any given mechanism manifest itself characteristically in every observable. The combined analysis of all observables gives us access to all relevant mechanisms, and allow a reliable and unambiguous separation of resonant and non-resonant amplitudes.

As a typical example, the model description of the nine 1-fold differential cross sections at W = 1.51 GeV and $Q^2 = 0.375$ GeV^2 are presented in Fig. 8 with the extracted contributing mechanisms.

2.4. *The Roper and the* $D_{13}(1520)$

For the first time transition amplitudes for the Roper resonance have been extracted, see Fig. 9 (top). The transverse amplitude become positive and large at Q^2 of around 2 GeV^2. This sign has been observed on both the single and the double pion channels. The longitudinal amplitude drops smoothly with Q^2. Both amplitudes are qualitatively consistent with Light Cone Quark Models. However, meson contributions might be required at low Q^2 to explain the behavior quantitatively. The interpretation of the Roper as a hybrid baryon is ruled out, since the color component couples only with the transverse amplitude [34].

CLAS also measured for the first time the $S_{1/2}$ amplitude of the $D_{13}(1520)$ state. All three amplitudes are shown in Fig. 9 (middle and bottom rows). There is a clear evidence of a switch from a helicity 3/2 dominance at the photon point to a helicity 1/2 dominance at high Q^2. Worth to mention is the very accurate results for these amplitudes.

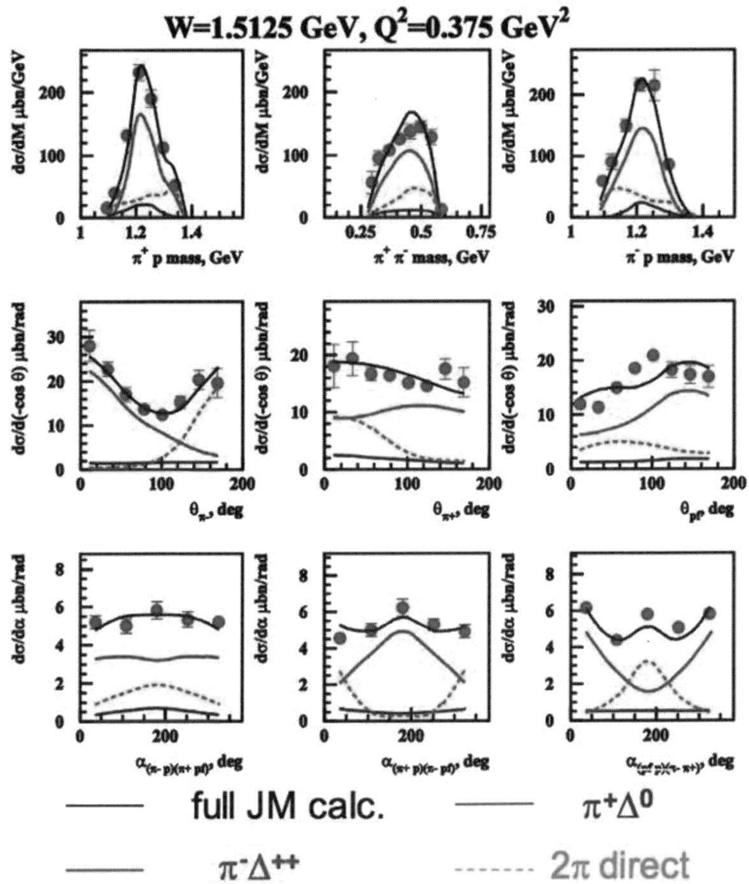

Fig. 8. Description of the CLAS charged double pion differential cross sections at W = 1.51 GeV and Q^2 = 0.375 GeV^2 within the framework of the JM model.[30] Full calculations are shown by the solid lines. Contributions from $\pi^-\Delta^{++}$ and $\pi^+\Delta^0$ isobar channels are shown by the red and blue lines respectively, and contributions from the direct charged double pion production processes are shown by the dot-dashed purple lines. The combined analysis of all observables gives us access to all relevant mechanisms, and allow a reliable separation of resonant and non-resonant electro-couplings amplitudes.

3. Conclusions

Thanks to recent theoretical developments, one could argue that the N* resonances are "back in fashion": high precision measurements of N* amplitudes will be indispensable to non-perturbative approaches such as lattice-QCD, DSEs and LCSR as they are refined. CLAS results show impressive

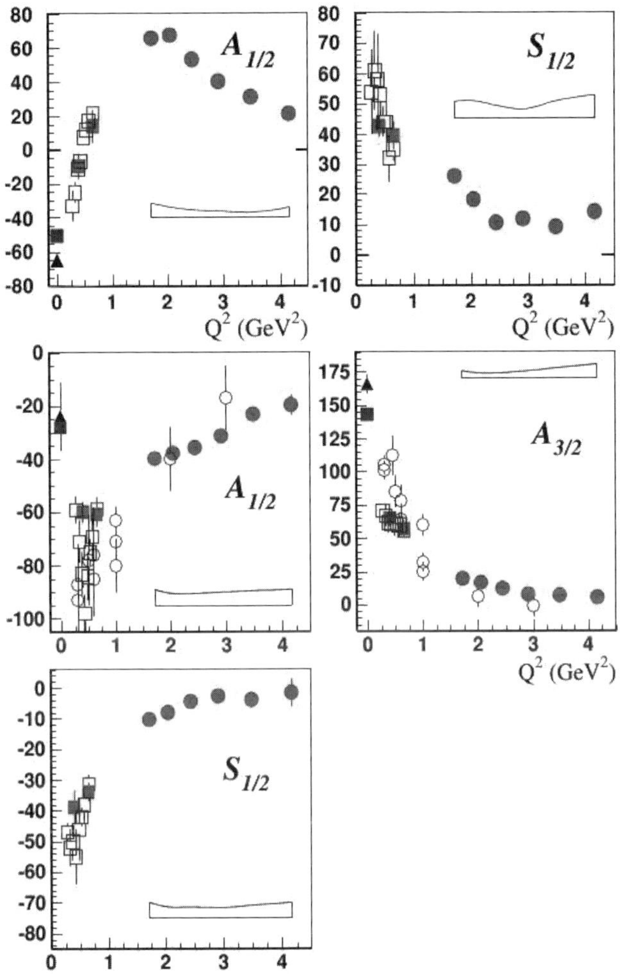

Fig. 9. Electro-coupling parameters of the $P_{11}(1440)$ (top) and $D_{13}(1520)$ (middle and bottom rows) on the proton in units of 10^3 $GeV^{1/2}$. CLAS results[35] of the single pion production data are represented by the red circles and squares. Open squares are for the combined analysis of the single and double pion channel.[36] Shown also are the results of the double pion data.[37] World data from single pion electro-production, available before CLAS, are represented by open circles.

accuracy in determining the electro-coupling amplitudes in existing data up to Q^2 of ~ 5 GeV^2, and the 12 GeV upgrade of JLAB will extend these measurements up to 11 GeV^2.

References

1. M. D. Sanctis, M. M. Giannini, L. Repetto and E. Santopinto, *Phys. Rev.* **C62**, p. 025208 (2000).
2. B. Julia-Diaz, D. O. Riska and F. Coester, *Phys. Rev.* **C69**, p. 035212 (2004).
3. P. L. Chung and F. Coester, *Phys. Rev.* **D44**, 229 (1991).
4. M. M. Giannini, *Rept. Prog. Phys.* **54**, 453 (1990).
5. N. Isgur and G. Karl, *Phys. Rev.* **D18**, p. 4187 (1978).
6. F. Cardarelli, E. Pace, G. Salme and S. Simula, *Phys. Lett.* **B357**, 267 (1995).
7. G. P. Lepage and S. J. Brodsky, *Phys. Rev.* **D22**, p. 2157 (1980).
8. J. J. Kelly, *Phys. Rev.* **C70**, p. 068202 (2004).
9. I. Aznauryan *et al.*, **D79** (2009).
10. H.-W. Lin *et al.*, *Phys. Rev.* **D79**, p. 034502 (2009).
11. S. Basak *et al.*, *Phys. Rev.* **D72**, p. 074501 (2005).
12. J. M. Bulava *et al.*, *Phys. Rev.* **D79**, p. 034505 (2009).
13. J. J. Dudek, R. Edwards and C. E. Thomas, *Phys. Rev.* **D79**, p. 094504 (2009).
14. H.-W. Lin, S. D. Cohen, R. G. Edwards and D. G. Richards, *Phys. Rev.* **D78**, p. 114508 (2008).
15. M. S. Bhagwat, M. A. Pichowsky, C. D. Roberts and P. C. Tandy, *Phys. Rev.* **C68**, p. 015203 (2003).
16. M. S. Bhagwat and P. C. Tandy, *AIP Conf. Proc.* **842**, 225 (2006).
17. C. D. Roberts and A. G. Williams, *Prog. Part. Nucl. Phys.* **33**, 477 (1994).
18. C. D. Roberts and S. M. Schmidt, *Prog. Part. Nucl. Phys.* **45**, S1 (2000).
19. R. Alkofer and L. von Smekal, *Phys. Rept.* **353**, p. 281 (2001).
20. P. Maris and C. D. Roberts, *Int. J. Mod. Phys.* **E12**, 297 (2003).
21. C. D. Roberts, M. S. Bhagwat, A. Holl and S. V. Wright, *Eur. Phys. J. ST* **140**, 53 (2007).
22. P. O. Bowman *et al.*, *Phys. Rev.* **D71**, p. 054507 (2005).
23. M. Ungaro *et al.*, *Phys. Rev. Lett.* **97**, p. 112003 (2006).
24. V. V. Frolov *et al.*, *Phys. Rev. Lett.* **82**, 45 (1999).
25. N. Warkentin *et al.*, *PoS* **LATTICE2008**, p. 154 (2008).
26. V. M. Braun *et al.*, *Phys. Rev. Lett.* **103**, p. 072001 (2009).
27. V. D. Burkert, *AIP Conf. Proc.* **1056**, 348 (2008).
28. M. M. Dalton *et al.*, *Phys. Rev.* **C80**, p. 015205 (2009).
29. I. G. Aznauryan, *Phys. Rev.* **C67**, p. 015209 (2003).
30. V. I. Mokeev *et al.*, *Phys. Rev.* **C80**, p. 045212 (2009).
31. T. Sato and T. S. H. Lee, *Phys. Rev.* **C63**, p. 055201 (2001).
32. S. S. Kamalov and S. N. Yang, *Phys. Rev. Lett.* **83**, 4494 (1999).
33. V. Pascalutsa, M. Vanderhaeghen and S. N. Yang, *Phys. Rept.* **437**, 125 (2007).
34. Z.-p. Li, V. Burkert and Z.-j. Li, *Phys. Rev.* **D46**, 70 (1992).
35. I. G. Aznauryan *et al.*, *Phys. Rev.* **C78**, p. 045209 (2008).
36. I. G. Aznauryan, V. D. Burkert, G. V. Fedotov, B. S. Ishkhanov and V. I. Mokeev, *Phys. Rev.* **C72**, p. 045201 (2005).
37. V. I. Mokeev *et al.* (2009).

A COVARIANT FORMALISM FOR THE N^* ELECTROPRODUCTION AT HIGH MOMENTUM TRANSFER

G. RAMALHO[†]

Centro de Física Teórica de Partículas, Av. Rovisco Pais, 1049-001 Lisboa, Portugal
[†] *E-mail: gilberto.ramalho@cftp.ist.utl.pt*

F. GROSS

Thomas Jefferson National Accelerator Facility, Newport News, VA 23606, USA
College of William and Mary, Williamsburg, Virginia 23185, USA

M. T. PEÑA

Centro de Física Teórica de Partículas, Av. Rovisco Pais, 1049-001 Lisboa, Portugal
Departamento de Física, Instituto Superior Técnico, 1049-001 Lisboa, Portugal

K. TSUSHIMA

EBAC in Theory Center, Thomas Jefferson National Accelerator Facility, Newport News, Virginia 23606, USA

A constituent quark model based on the spectator formalism is applied to the $\gamma N \to N^*$ transition for the three cases, where N^* is the nucleon, the Δ and the Roper resonance. The model is covariant, and therefore can be used for the predictions at higher four-momentum transfer squared, Q^2. The baryons are described as an off-mass-shell quark and a spectator on-mass-shell diquark systems. The quark electromagnetic current is described by quark form factors, which have a form inspired by the vector meson dominance. The valence quark contributions of the model are calibrated by lattice QCD simulations and experimental data. Contributions of the meson cloud to the inelastic processes are explicitly included.

Keywords: Covariant quark model; Nucleon resonances; Meson cloud

1. Introduction

Study of the nucleon structure and its electromagnetic excitation is one of the important topics associated with the missions and activities of modern accelerator facilities. At Jefferson lab very accurate data have been

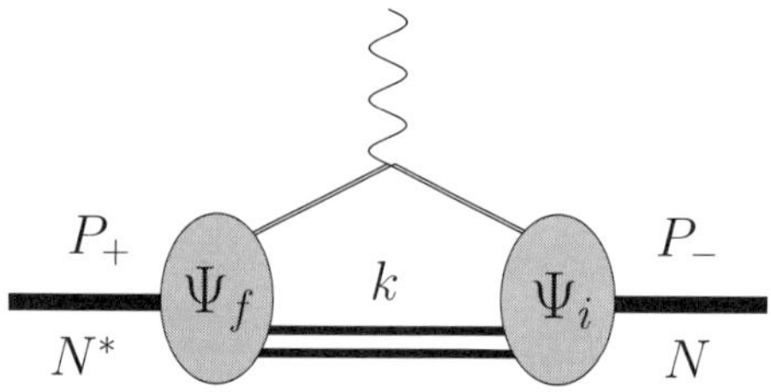

Fig. 1. $\gamma N \to N^*$ transition in the covariant spectator quark model (diquark on-shell) in relativistic impulse approximation. P_+ (P_-) represents the final (initial) baryon momentum and k the intermediate diquark momentum. The baryon wave functions are represented by Ψ_f and Ψ_i for the final and initial states, respectively.

extracted for the $\gamma N \to N^*$ reactions, for several N^* resonances at low and high Q^2 [1,2], defining new challenges for the theoretical models. Although one believes that the nucleon excitations are governed by QCD with quarks and gluons in a non-perturbative regime, it is at present nearly impossible to solve QCD exactly in the region $Q^2 = 0 - 10$ GeV^2. Thus, one has to rely on some effective and phenomenological approaches. One of popular approaches is the dynamical coupled channel reaction models [3–6], where the effective degrees of freedom are mesons and baryons. In these models a baryon core structure is assumed, and it is modified by the meson cloud dressing resulting from the meson-baryon interactions. Effective field theories based on chiral symmetry, with pions and baryons alone as degrees of freedom, are applicable only in the very low Q^2 region. On the other hand, perturbative QCD works only in the very large Q^2 region [7,8]. Alternative descriptions are constituent quark models [9]. A constituent quark has an internal structure resulting from the quark-antiquark dressing, and from the short range interaction with gluons. The quark structure of a baryon can be represented by electromagnetic valence quark form factors. In this work we present the covariant spectator quark model [7,8,10], and show several applications of the model. Covariance is important in the applications in the higher Q^2 region.

2. Spectator quark model

In the covariant spectator quark model baryons are described as a three-valence quark systems with an on-shell quark-pair, or diquark, while the remaining quark is off-shell and free to interact with electromagnetic fields. The quark-diquark vertex is then represented by a baryon B wave function Ψ_B that effectively describes quark confinement [10]. See Fig. 1.

The quark electromagnetic current j_I^μ is given by the Dirac and Pauli structures:

$$j_I^\mu = \left(\frac{1}{6}f_{1+} + \frac{1}{2}f_{1-}\tau_3\right)\left(\gamma^\mu - \frac{\not{q}q^\mu}{q^2}\right) + \left(\frac{1}{6}f_{2+} + \frac{1}{2}f_{2-}\tau_3\right)\frac{i\sigma^{\mu\nu}q_\nu}{2M_N}, \quad (1)$$

where M_N is the nucleon mass, $f_{1\pm}$ and $f_{2\pm}$ are the quark form factors as functions of Q^2, and τ_3 the isospin operator. To represent the quark structure we adopt a vector meson dominance motivated parametrization, where the form factors are written in terms of two vector meson poles:

$$f_{1\pm}(Q^2) = \lambda_q + (1-\lambda_q)\frac{m_v^2}{m_v^2+Q^2} + c_\pm \frac{Q^2 M_h^2}{(M_h^2+Q^2)^2} \quad (2)$$

$$f_{2\pm}(Q^2) = \kappa_\pm \left\{ d_\pm \frac{m_v^2}{m_v^2+Q^2} + (1-d_\pm)\frac{Q^2}{M_h^2+Q^2}\right\}. \quad (3)$$

In the above $m_v = m_\rho$ is a light vector meson mass that effectively represents the ρ and ω poles and M_h is the an effective heavy vector meson mass, that takes into account the short range phenomenology. We chose $M_h = 2M_N$ in the present study. The isoscalar κ_+ and isovector κ_- quark anomalous moments are fixed by the nucleon magnetic moments. The adjustable parameters are λ_q and the mixture coefficients $c_\pm$ and $d_\pm$. In the study of the nucleon properties, it turned out that $d_+ = d_-$ gives a very good description of the nucleon electromagnetic form factor [10]. This reduces the number of adjustable parameters to 4. The quality of the model description for the nucleon form factors is illustrated in Fig. 2. The quark current fixed by the nucleon form factors will be used for all other applications discussed below.

To write the baryon B wave function Ψ_B, we start from the baryon rest frame, $P = (M_B, 0, 0, 0)$, with M_B the baryon mass. We represent the baryon wave function as the direct product of the diquark and quark states of flavor, spin, orbital angular momentum and radial excitation, consistent with the baryon quantum numbers. The flavor states are written using the $\mathrm{SU}_F(3)$ quark states $\Phi_I^{0,1}$, with the diquark of total isospin $I = 0, 1$. Similarly, the diquark spin states associated with spin $S = 0, 1$, $\Phi_S^{0,1}$, can be written in terms of the polarization vectors $\varepsilon^\mu(0) = (1, 0, 0, 0)$ and $\varepsilon^\mu(\pm 1) = \mp\frac{1}{\sqrt{2}}(0, 1, \pm i, 0)$, where $\lambda = 0, \pm 1$ is the diquark polarization [7,10,11]. Once the wave functions are written explicitly in terms of the baryon properties in the rest frame, the relativistic generalization is performed with a boost to the moving frame. The diquark polarization vectors will be represented by a function $\varepsilon_P^\mu(\lambda)$ of the center-of-mass momentum P in the fixed-axis

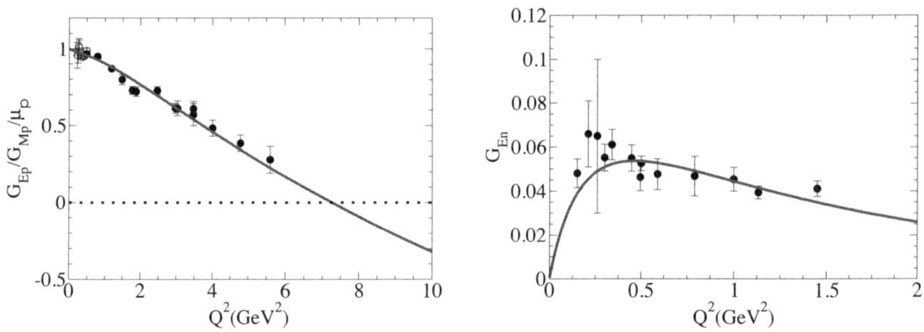

Fig. 2. Ratio for the electric and magnetic form factors for the proton, and neutron electric form factor for model II in Ref. [10]. Data are as presented in Ref. [10].

representation, as described in Ref. [11]. The explicit covariant form for the nucleon, Δ and Roper wave functions can be found in Refs. [7,8,10,12].

The electromagnetic current associated with the final state N^* in the covariant spectator quark model (see Fig. 1) is determined by

$$J^\mu = 3 \sum_\lambda \int_k \overline{\Psi}_f(P_+, k) j_I^\mu \Psi_i(P_-, k). \tag{4}$$

In the above, $\int_k$ represents the covariant integral with respect to the on-mass-shell diquark momentum and λ the diquark polarization. For simplicity, diquark and baryon polarization indices are suppressed.

3. Applications

In Eq. (4) we can write the electromagnetic transition current in terms of $q = P_+ - P_-$ and $P = \frac{1}{2}(P_+ + P_-)$. The corresponding form factors, invariant functions of Q^2, are G_E and G_M for the nucleon, G_M^*, G_E^* and G_C^* for the Δ, and F_1^* and F_2^* for the Roper.

3.1. *Nucleon*

For the nucleon, the simplest wave function has a quark-diquark S-wave configuration [10]:

$$\Psi_N = \frac{1}{\sqrt{2}} \left[\Phi_I^0 \Phi_S^0 + \Phi_I^1 \Phi_S^1 \right] \psi_N(P, k), \tag{5}$$

with $\Phi_I^{0,1}$ and $\Phi_S^{0,1}$, the diquark spin and isospin states of 0 and 1, and ψ_N a scalar wave function. Results for the nucleon form factors [10] are shown in Fig. 2. No explicit pion cloud is included for the results.

3.2. $\gamma N \to \Delta$ *transition*

The $\gamma N \to \Delta$ transition is more complex than the nucleon elastic reaction. The transition current (4), with Ψ_Δ, associated exclusively with the quark valence degrees of freedom, is insufficient to explain the data [7,8]. As near the Δ region the nucleon has enough energy to create a pion, the electromagnetic interaction with intermediate pion-baryon states should also be considered. Then, the transition form factors can be decomposed as

$$G_X^* = G_X^b + G_X^\pi, \tag{6}$$

where G_X^b stands for the contribution of the quark core (bare) and G_X^π for the contribution due to the pion cloud. The label X holds for M (magnetic dipole), E (electric quadrupole) and C (Coulomb quadrupole) form factors. This decomposition is justified when the pion is created by the overall baryon three-quark system and not from a single quark.

As a first application we describe the Δ as a quark-diquark S-state coupled to a spin 3/2 to form a total $J = 3/2$ state [7]. The transition proceeds then only through the magnetic form factor [7,8]:

$$G_M^b(Q^2) = 4\eta f_v \mathcal{I}, \tag{7}$$

where $\eta = \frac{2}{3\sqrt{3}} \frac{M_N}{M_N + M_\Delta}$, $f_v = f_{1-} + \frac{M_N + M_\Delta}{2M_N} f_{2-}$ and $\mathcal{I}$ is the overlap integral between the nucleon and Δ S-state scalar wave functions. This result allows us to understand why the pion cloud is essential to describe the data, and necessary to be added. In a pure constituent quark model the overlap integral is limited by the wave function normalization [7]. At $Q^2 = 0, \mathcal{I} \leq 1$, and for the spectator quark model this implies an upper value for $G_M^b(0)$ of 2.07, to be compared with the experimental result 3.02 [7]. Higher angular momentum partial waves for the relative quark-diquark motion are only possible to contribute to the quadrupole form factors. Since these are small compared to G_M^*, they have a reduced weight in the wave function, and consequently in G_M^*. Therefore, the discrepancy found in the leading form factor, between constituent quark models and experimental data, is mainly to be compensated by the pion cloud contributions. To adjust the valence quark contributions we use the results of the Sato-Lee model obtained from the data [3], subtracted by the pion cloud contributions. The result of the fit is presented in the left panel of Fig. 3. The experimental data points are reached when $G_M^\pi = \lambda_\pi \left(\frac{\lambda_\pi^2}{\Lambda_\pi^2 + Q^2} \right)^2 (3G_D)$ is added to G_M^b (G_D the nucleon dipole factor). See the right panel in Fig. 3.

The next step is to include D-state admixtures in the wave function

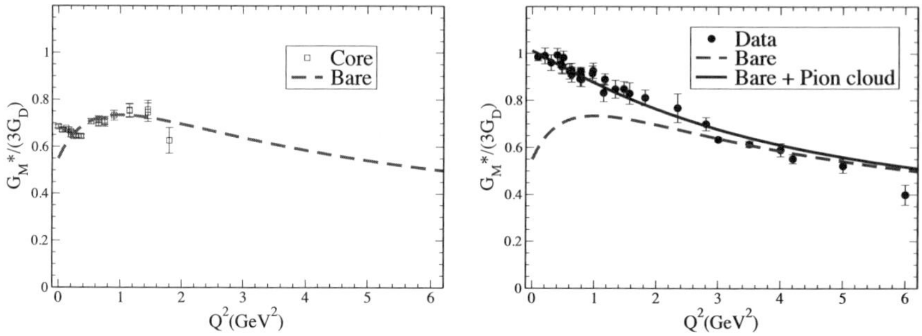

Fig. 3. G_M^* normalized by $3G_D$ in the $\gamma N \to \Delta$ transition. Left: Valence quark contributions from our model compared with the bare contributions of Sato-Lee model [3]. Right: Bare plus phenomenological pion cloud contributions, compared with the data [7].

as [8],

$$\Psi_\Delta = N\left[\Psi_S + a\Psi_{D3} + b\Psi_{D1}\right], \tag{8}$$

where Ψ_{D3} represents a D-state with a core spin 1/2 and Ψ_{D1} a D-state with a core spin 3/2. The D-state generates contributions for G_E^* and G_C^* form factors, which, otherwise for a pure S-wave function would vanish identically. To separate the valence quark contributions, we have also extended the model to the lattice QCD regime [13–15] and adjusted the D-state parameters to the quenched lattice QCD data [16] for a pion mass region where pion cloud effects are expected to be small [14]. Once the valence quark contributions are fixed from the lattice regime, the results are extrapolated back to the physical region. Finally, by adding the pion cloud contributions derived from the large-N_c limit [8,14] to the valence quark contributions G_X^b, we obtain the final result shown in Fig. 4. The results agree well with the physical data. See Refs. [8,14] for details.

3.3. $\gamma N \to$ *Roper transition*

Within the covariant spectator quark model, we can also describe the Roper system, as the first radial excitations of the nucleon [12]. Thus, the Roper wave function has the same structure as that for the nucleon Eq. (5), except for the scalar wave function, which is replaced by ψ_R. Under this assumption, the orthogonality between the Roper and nucleon wave functions is reduced to the orthogonality between the corresponding scalar wave functions: $\int_k \psi_R \psi_N = 0$ at $Q^2 = 0$. This fixes the free parameters in ψ_R completely, assuming that the nucleon and the Roper have the same short

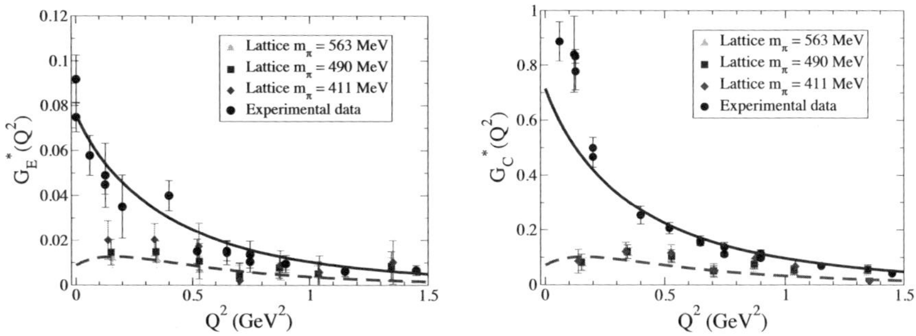

Fig. 4. Electric and Coulomb quadrupole form factor [14] for the $\gamma N \to \Delta$ transition. Lattice data are taken from [16].

range behavior, but differ in the long range structure. No extra parameter is needed additionally to the ones already fixed in the nucleon wave function [12]. Once ψ_R is defined, we can calculate and predict the nucleon to Roper transition form factors F_1^* and F_2^*. The results are shown in Fig. 5, and are consistent with the CLAS data [2] for $Q^2 > 2$ GeV^2. These facts support the idea that the valence quark degrees of freedom are well described in the covariant spectator quark model. Once the valence quark contributions are determined, we can then estimate the meson cloud contributions using the decomposition $F_i^* = F_i^b + F_i^{mc} (i = 1, 2)$, where F_i^b is the bare contribution and F_i^{mc} is the meson cloud contribution [12]. The results are also in Fig. 5.

4. Conclusions

We have developed a formalism which is successful in describing the valence quark contributions to the nucleon form factors, without the inclusion of pion cloud. The present approach also describes very well the $\gamma N \to \Delta$ data, both in the physical regime and the lattice regime, where the pion cloud effects are suppressed in the lattice regime. Furthermore, the results are consistent with the estimate of the core contributions of the Sato-Lee model. As for the $\gamma N \to$ Roper transition, we have obtained a very good description for the high Q^2 data, where valence quark degrees of freedom are expected to be dominant.

Other applications of the present approach have been made also for the determination of the Δ [17–19] and decuplet [15] electromagnetic form factors, and the octet magnetic moments (in this case including pion cloud effects).

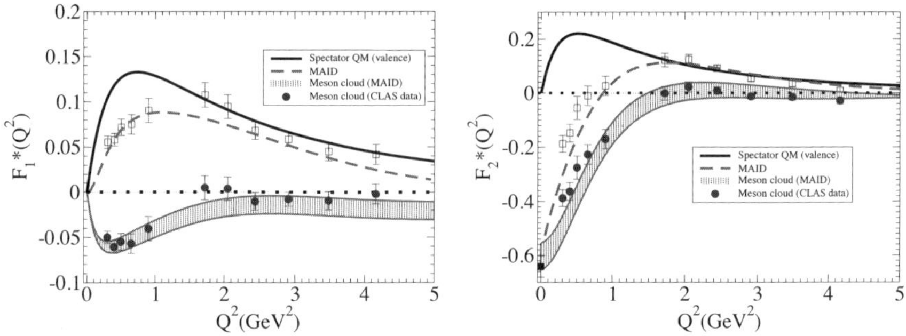

Fig. 5. $\gamma N \to$ Roper transition form factors, F_1^* and F_2^*. Data are from [2]. Meson cloud contributions based on the MAID fit [20] are represented by the band [12]. The meson cloud contributions from CLAS are also shown.

Acknowledgments:

G. R. would like to thank Viktor Mokeev for the invitation to the workshop. G. R. was supported by the Portuguese Fundação para a Ciência e Tecnologia (FCT) under the grant SFRH/BPD/26886/2006.

References

1. V. D. Burkert and T. S. H. Lee, *Int. J. Mod. Phys.* **E13**, 1035 (2004).
2. I. G. Aznauryan *et al.*, *Phys. Rev.* **C80**, p. 055203 (2009).
3. B. Julia-Diaz, T. S. H. Lee, T. Sato and L. C. Smith, *Phys. Rev.* **C75**, p. 015205 (2007).
4. S. S. Kamalov, S. N. Yang, D. Drechsel, O. Hanstein and L. Tiator, *Phys. Rev.* **C64**, p. 032201 (2001).
5. S. Schneider, S. Krewald and U.-G. Meissner, *Eur. Phys. J.* **A28**, 107 (2006).
6. A. V. Anisovich *et al.*, *Eur. Phys. J.* **A44**, 203 (2010).
7. G. Ramalho, M. T. Peña and F. Gross, *Eur. Phys. J.* **A36**, 329 (2008).
8. G. Ramalho, M. T. Peña and F. Gross, *Phys. Rev.* **D78**, p. 114017 (2008).
9. M. M. Giannini, *Rept. Prog. Phys.* **54**, 453 (1991).
10. F. Gross, G. Ramalho and M. T. Peña, *Phys. Rev.* **C77**, p. 015202 (2008).
11. F. Gross, G. Ramalho and M. T. Peña, *Phys. Rev.* **C77**, p. 035203 (2008).
12. G. Ramalho and K. Tsushima, *Phys. Rev.* **D81**, p. 074020 (2010).
13. G. Ramalho and M. T. Peña, *J. Phys.* **G36**, p. 115011 (2009).
14. G. Ramalho and M. T. Peña, *Phys. Rev.* **D80**, p. 013008 (2009).
15. G. Ramalho, K. Tsushima and F. Gross, *Phys. Rev.* **D80**, p. 033004 (2009).
16. C. Alexandrou *et al.*, *Phys. Rev.* **D77**, p. 085012 (2008).
17. G. Ramalho and M. T. Peña, *J. Phys.* **G36**, p. 085004 (2009).
18. G. Ramalho, M. T. Peña and F. Gross, *Phys. Lett.* **B678**, 355 (2009).
19. G. Ramalho, M. T. Peña and F. Gross, *Phys. Rev.* **D81**, p. 113011 (2010).
20. D. Drechsel, S. S. Kamalov and L. Tiator, *Eur. Phys. J.* **A34**, 69 (2007).

TRANSVERSE MOMENTUM DEPENDENT DISTRIBUTIONS IN HARD SCATTERING

A. PROKUDIN*

Jefferson Lab, 12000 Jefferson Avenue, Newport News, VA 23606
** E-mail: prokudin@jlab.org*

Transverse Momentum Dependent Distributions (TMDs) describe the spin structure of the proton. At leading twist spin structure of spin-1/2 hadron can be described by 8 TMDs. TMDs reveal three-dimensional distribution of partons inside polarised nucleon. Experimentally these functions can be studied in polarised experiments using Spin Asymmetries in particular Single Spin Asymmetries (SSAs). We discuss transversity that measures distribution of transversely polarised quarks in a transversely polarised nucleon and Sivers distribution function that describes distribution of unpolarised quarks in a transversely polarised nucleon.

1. Introduction

Transverse Momentum Dependent distribution functions (TMDs) depend on two independent variables: fraction of hadron momentum carried by parton, x, and intrinsic transverse momentum of the parton, $\mathbf{p}_T$.

The quark-quark distribution correlation function is defined as[1]

$$\Phi_{ij}(x, \mathbf{p}_T) = \int \frac{d\xi^- d^2\xi_T}{(2\pi)^3} \, e^{ip\cdot\xi} \langle P, S_P | \bar{\psi}_j(0)\, \mathcal{U}^{n_-}_{(0,+\infty)}\, \mathcal{U}^{n_-}_{(+\infty,\xi)}\, \psi_i(\xi) | P, S_P \rangle \bigg|_{\xi^+=0} \qquad (1)$$

with $p^+ = xP^+$, $P^+ = (P^0 + P^3)/\sqrt{2}$ is the big component of proton's momentum. $\mathcal{U}$ is the gauge link (Wilson line) that assures color gauge invariance of the correlator Φ_{ij}.

Corresponding distribution functions can be obtained by projecting the correlator onto the full basis of γ matrices $\Phi^{[\Gamma]} = \frac{1}{2}\mathrm{Tr}[\Phi\,\Gamma]$. At leading twist (expansion in P^+) the spin structure of the proton can be described by 8

TMDs (see[1] and references therein):

$$\Phi^{[\gamma^+]}(x,p_T^2) = f_1(x,p_T^2) - \frac{\epsilon_T^{\rho\sigma}\, p_{T\rho} S_{T\sigma}}{M}\, f_{1T}^{\perp}(x,p_T^2)\,, \tag{2}$$

$$\Phi^{[\gamma^+\gamma_5]}(x,p_T^2) = S_L\, g_{1L}(x,p_T^2) - \frac{p_T\cdot S_T}{M}\, g_{1T}(x,p_T^2)\,, \tag{3}$$

$$\Phi^{[i\sigma^{\alpha+}\gamma_5]}(x,p_T^2) = S_T^{\alpha}\, h_1(x,p_T^2) + S_L\, \frac{p_T^{\alpha}}{M}\, h_{1L}^{\perp}(x,p_T^2) \tag{4}$$
$$- \frac{p_T^{\alpha}p_T^{\rho} - \frac{1}{2}\, p_T^2\, g_T^{\alpha\rho}}{M^2}\, S_{T\rho} h_{1T}^{\perp}(x,p_T^2) - \frac{\epsilon_T^{\alpha\rho}\, p_{T\rho}}{M} h_1^{\perp}(x,p_T^2)\,.$$

Here the projections are done with respect to the quark's polarisation, $\Phi^{[\gamma^+]}$ is distribution of unpolarised quarks, $\Phi^{[\gamma^+\gamma_5]}$ is distribution of longitudinally polarised quarks and $\Phi^{[i\sigma^{\alpha+}\gamma_5]}(x,p_T^2)$ is the distribution of transversely polarised quarks.

A powerful tool to study the spin structure is measurement of spin asymmetries in the SIDIS process $\ell P(S_P) \to \ell' h X$ or in Jet-SIDIS $\ell P(S_P) \to \ell' Jet X$. A number of structures can be measured in DIS process $\ell P(S_P) \to \ell' X$ or in SIDIS with two detected hardons $\ell P(S_P) \to \ell' h_1 h_2 X$.

Applicability of QCD factorization theorems[2] at high values of photon's virtuality $Q^2 \gg \Lambda_{QCD}^2$ allows to decompose unpolarised and polarised cross-sections of the process under consideration as convolution of parton distribution functions f_q and fragmentation functions $D_{q/h}$:

$$d\sigma^{\ell P(S_P)\to \ell' h X} \propto \sum_q e_q^2\, f_q \otimes d\sigma^{\ell q(s_q)\to \ell' q'(s_{q'})} \otimes D_{q/h}\,, \tag{5}$$

where $\otimes$ stands for convolution of the transverse momenta.

In case of SIDIS process $\ell P(S_P) \to \ell' h X$ two scales exist: virtuality of the photon Q^2 and transverse momentum of the produced hadron P_{hT} with respect to the lepton scattering plane. Consequently two distinct factorization theorems exist, if $Q^2 \simeq P_{hT}^2 \gg \Lambda_{QCD}^2$ collinear factorization theorem for unpolarised cross section[3] and Single Spin Asymmetries (SSA)[4] is valid, SSA is described as twist-3 quantum interference[5] of scattering amplitudes with different number of active partons. Partons are considered to be parallel to the parent hadron and small transverse intrinsic momentum of partons is neglected. If the produced hadron has low transverse momentum $Q^2 \gg P_{hT}^2 \geqslant \Lambda_{QCD}^2$ then so-called Transverse Momentum Dependent factorisation theorem[6] is valid. The cross section is written as convolution of TMDs.

Let us describe some of TMDs.

2. Transversity

Transversity distribution $h_1(x, p_T^2)$ [7] describes distribution of transversely polarised quarks inside a transversely polarised nucleon. Tensor charge measures net transverse polarisation of quarks

$$\delta_T q = \int_0^1 dx(h_{1q}(x) - h_{1\bar{q}}(x)). \tag{6}$$

Lorentz boost and rotation do not commute thus transversity and helicity distribution (g_1 distribution of longitudinally polarised quarks inside a longitudinally polarised nucleon) are different, as soon as interaction with virtual photon takes place the preferred direction appear to exist and longitudinal and transverse polarisation of the nucleon can be clearly defined. We may write transversity in so called helicity basis and obtain that $h_1 \propto \mathrm{Im} A_{+-,-+}$, here A is one of 16 helicity amplitudes describing spin structure of the spin-1/2 nucleon. $A_{\Lambda\lambda,\Lambda'\lambda'}$where $\lambda\lambda'$, $(\Lambda\Lambda')$ are quark (nucleon) helicities. Transversity is a chiral-odd quantity and thus cannot be measured in DIS (observables should be chiral-even) and it requires helicity flip of 1 unit of angular momentum thus gluon helicity distribution does not exist. Experimentally spin distributions can be measured using SSA on transversely polarised target defined as

$$A_{UT} \propto \frac{\sigma^{\uparrow} - \sigma^{\downarrow}}{\sigma^{\uparrow} + \sigma^{\downarrow}} \,. \tag{7}$$

The "golden" channel to measure transversity is double spin asymmetry in Drell-Yan experiment with polarised proton and anti-proton beams, product of transversities can be measured[8]

$$A_{NN} \propto \sum_q e_q^2 h_{1q/P} h_{1\bar{q}/\bar{P}} \,. \tag{8}$$

In SIDIS transversity can be measured together with chiral-odd fragmentation function $H_{1q}^{\perp}(z, k_T^2)$ so-called Collins fragmentation function[9]

$$A_{UT}^{\sin(\Phi_h+\Phi_S)} \propto \frac{\sum_q e_q^2 h_{1q} \otimes H_{1q}^{\perp}}{\sum_q e_q^2 f_{1q} \otimes D_{1q}} \,, \tag{9}$$

where Φ_h and Φ_S are azimuthal angles of produced hadron and target polarisation vector with respect to lepton scattering plane.

Experimental data on SIDIS asymmetries $A_{UT}^{\sin(\phi_h+\phi_S)}$ from HERMES, COMPASS collaborations[10,11] and $e^+e^- \to h_1 h_2 X$ data from BELLE collaboration[12] allow extraction[13,14] of transversity and Collins fragmentation function.

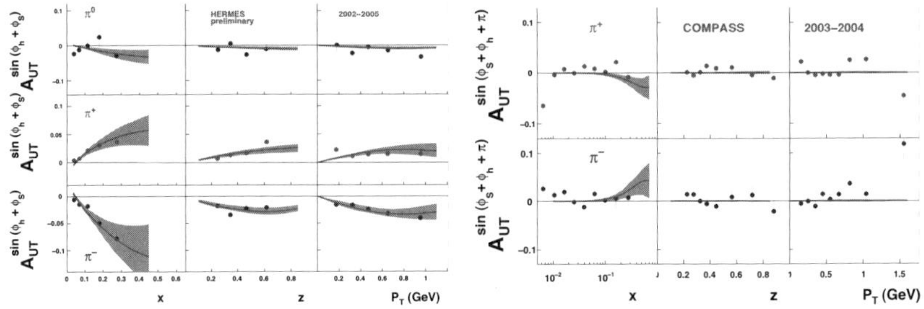

Fig. 1. Fit[14] of HERMES[10] (left panel) and COMPASS[11] data (right panel).

In Fig. 1 we show the best fit to the HERMES[10] and COMPASS[11] data, respectively.

The two sets of Belle[12] data, coming from two analyses of the same experimental events, are not independent. Therefore we include only one set of data in the fit, either A_{12} or A_0 data, see for details Refs.[13,14] In this analysis we report the results obtained by using A_{12} data, the $\cos(\phi_1 + \phi_2)$ method.

$$A_{12}(z_1, z_2, \theta, \varphi_1 + \varphi_2) \propto \cos(\phi_1 + \phi_2) \frac{\sum_q e_q^2 \, H_{1q}^{\perp}(z_1) \, H_{1\bar{q}}^{\perp}(z_2)}{\sum_q e_q^2 \, D_{1q}(z_1) D_{1\bar{q}}(z_2)}, \tag{10}$$

here ϕ_1 and ϕ_2 are azimuthal angles of produced hadrons h_1 and h_2 with respect to the plane formed by the quark thrust axis and electron-positron pair, in a so-called Collins-Soper frame.

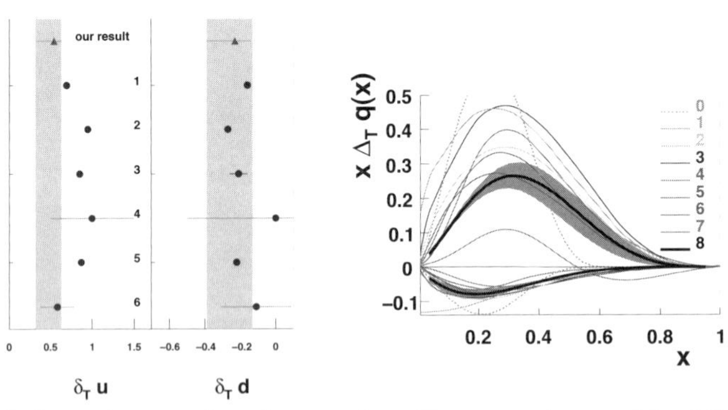

Fig. 2. Left panel: extraction of tensor charge[14] compared to models of Refs. [15–20]. Right panel: transversity for u and d quarks[14] compared to models of Refs. [15–17,21–24].

The extracted values[14] of tensor charge are $\delta_T u = 0.54^{+0.09}_{-0.22}$, $\delta_T d = -(0.23^{+0.09}_{-0.16})$ at $Q^2 = 0.8$ GeV^2.

In Fig. 2 we plot tensor charge and transversity distributions compared to models. As can be seen from Fig. 2 the experimental precision is still not good enough to discriminate among models. Apart from that there is no information on anti-quark transversity distributions and Q^2 dependence of asymmetry is not yet known experimentally. Electron Ion Collider will be able to shed light on both anti-quark distributions and Q^2 dependence. JLab data will allow to expand x range of transversity extraction. Predictions for JLab are shown in Fig.3.

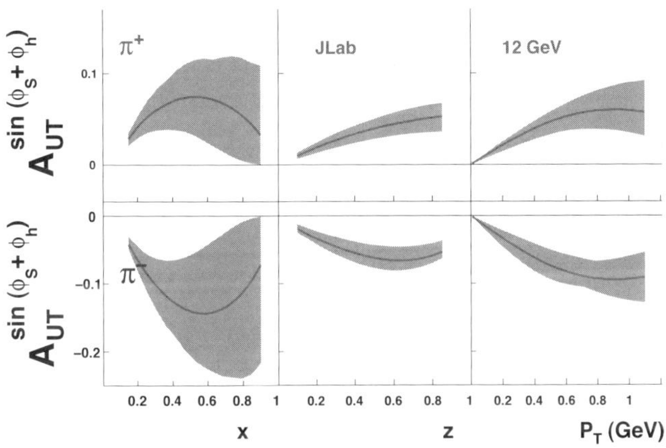

Fig. 3. Predictions[14] of Collins asymmetry for JLab at 12 GeV.

3. Sivers distribution function

$f_{1T}^{\perp}(x, p_T^2)$ is so-called Sivers function,[25] it describes correlation between orbital angular motion of quarks and the spin of the proton $\epsilon_T^{\rho\sigma} p_{T\rho} S_{T\sigma}$. This function exists due to the presence of Final State Interactions of the struck quark and the remnant of the nucleon after the interaction.

Distribution of unpolarised quarks inside a transversely polarised nu cleon can be written as

$$f_{q/p^\uparrow}(x, \mathbf{p_T}) = f_1(x, p_T^2) - f_{1T}^{\perp}(x, p_\perp^2) \frac{\mathbf{S} \cdot (\hat{\mathbf{P}} \times \mathbf{p_T})}{M} , \quad (11)$$

here $\mathbf{S} \cdot (\hat{\mathbf{P}} \times \mathbf{p_T})$ is correlation between quark motion and the spin of the nucleon. The function that measures this correlation is Sivers function $f_{1T}^{\perp}$.

This function obeys a modified universality, it changes sign from SIDIS to Drell-Yan[26] process

$$f_{1T}^{\perp}(x,p_T^2)^{\rm SIDIS} = -f_{1T}^{\perp}(x,p_T^2)^{\rm DY}, \tag{12}$$

the prediction of change of sign based on color gauge symmetry and parity and time reversal invariance $\mathcal{P},\mathcal{T}$ of strong interactions. Experimental test of this relation is very important for our understanding of QCD.

The asymmetry associated with Sivers effect is

$$A_{UT}^{\sin(\Phi_h-\Phi_S)} \propto \frac{\sum_q e_q^2 f_{1T}^{\perp} \otimes D_{1q}}{\sum_q e_q^2 f_{1q} \otimes D_{1q}} . \tag{13}$$

The first moment of Sivers function $f_{1T}^{\perp(1)}(x) = \int d^2k_\perp \frac{k_\perp^2}{2M^2} f_{1T}^{\perp}(x,p_T^2)$ is related[27] to twist three Qiu-Sterman matrix element $T_F(x,x)$ [5]

$$T_F(x,x) = f_{1T}^{\perp(1)}(x) . \tag{14}$$

Twist three matrix elements are ingredients of collinear QCD to describe spin asymmetries and the formalism is valid at high $P_{hT} \gg \Lambda_{QCD}$. The Sivers function describes the asymmetry at low values of $P_{hT} \sim \Lambda_{QCD}$. Interplay between two formalisms is very important check for EIC, the collider will be able to cover P_{hT} region up to 4 GeV.

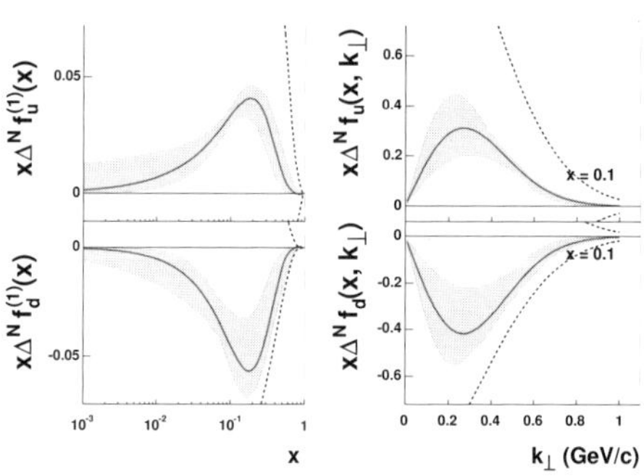

Fig. 4. Extraction of Sivers function.[28]

In Fig. 4 we plot Sivers function extracted from the experimental data from HERMES[29] and COMPASS[11] collaborations. In Fig. 5 we show the three-dimensional parton distribution at $x = 0.01$. As can be seen from Fig. 5 the distribution of partons in a transversely polarised hadron is not rotational symmetric, the distributions have dipole deformation with respect to the "center" of the hadron.

Fit[28] of HERMES[29] and COMPASS[11] data for $\pi^\pm$ and $K^\pm$ production is presented in Fig. 6 and Fig. 7.

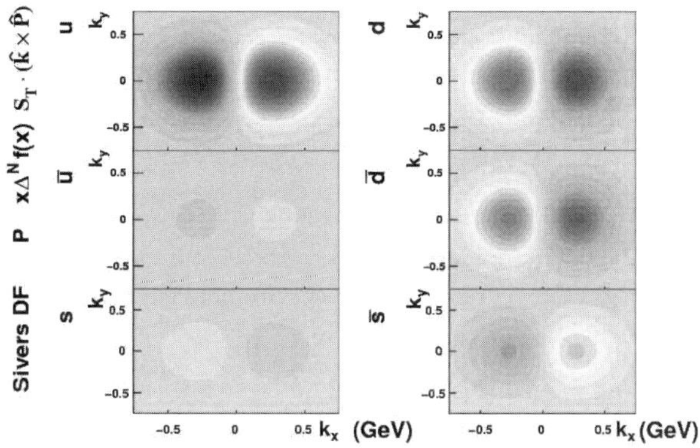

Fig. 5. Three-dimensional parton distribution.

4. Conclusions

Transverse Momentum Dependent distributions describe spin structure of the proton. Experimental data from HERMES, COMPASS and BELLE collaborations allow extraction of transversity and Collins fragmentation functions. Sivers functions for u and d quarks were extracted from HERMES and COMPASS data. Future experimental data from JLab will widen explored x region and refine precision of extraction of TMDs. Q^2 range at Electron Ion Collider will allow to check evolution of asymmetries, wide P_{hT} range will allow to measure asymmetries in both regions where TMD and collinear QCD factorizations are valid.

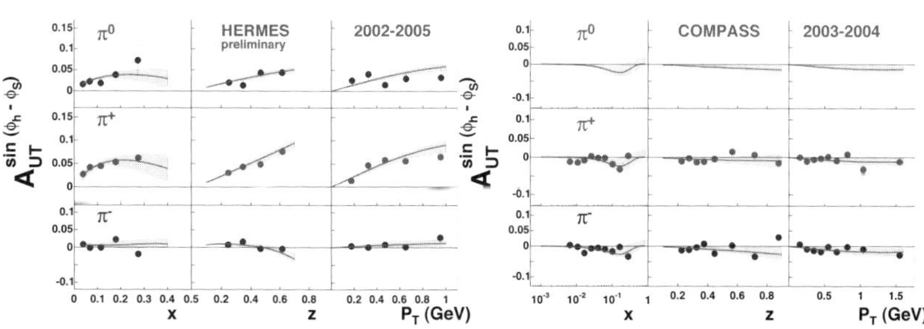

Fig. 6. Fit[28] of pion HERMES[29] (left panel) and COMPASS[11] data (right panel).

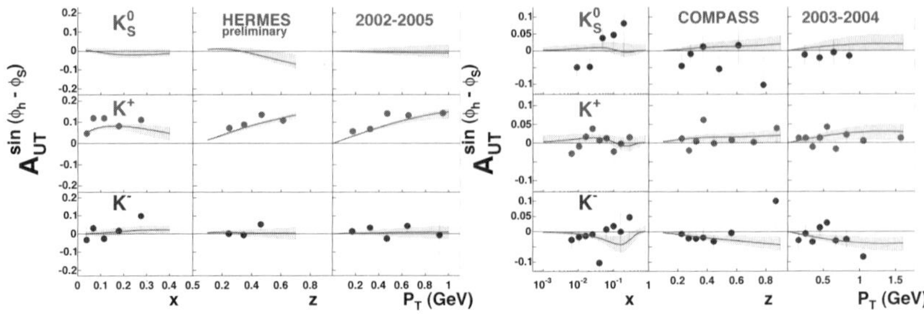

Fig. 7. Fit[28] of kaon HERMES[29] (left panel) and COMPASS[11] data (right panel).

References

1. A. Bacchetta *et al.*, *JHEP* **02**, p. 093 (2007).
2. J. C. Collins, D. E. Soper and G. F. Sterman, *Adv. Ser. Direct. High Energy Phys.* **5**, 1 (1988).
3. R. Meng, F. I. Olness and D. E. Soper, *Phys. Rev.* **D54**, 1919 (1996).
4. J.-w. Qiu and G. Sterman, *Phys. Rev.* **D59**, p. 014004 (1999).
5. J.-w. Qiu and G. Sterman, *Phys. Rev. Lett.* **67**, 2264 (1991).
6. X.-d. Ji, J.-p. Ma and F. Yuan, *Phys. Rev.* **D71**, p. 034005 (2005).
7. J. P. Ralston and D. E. Soper, *Nucl. Phys.* **B152**, p. 109 (1979).
8. V. Barone *et al.* (2005).
9. J. C. Collins, *Nucl. Phys.* **B396**, 161 (1993).
10. M. Diefenthaler (2007).
11. M. Alekseev *et al.* (2008).
12. R. Seidl *et al.*, *Phys. Rev.* **D78**, p. 032011 (2008).
13. M. Anselmino *et al.*, *Phys. Rev.* **D75**, p. 054032 (2007).
14. M. Anselmino *et al.*, *Nucl. Phys. Proc. Suppl.* **191**, 98 (2009).
15. I. C. Cloet, W. Bentz and A. W. Thomas, *Phys. Lett.* **B659**, 214 (2008).
16. M. Wakamatsu, *Phys. Lett.* **B653**, 398 (2007).
17. B. Pasquini, M. Pincetti and S. Boffi, *Phys. Rev.* **D72**, p. 094029 (2005).
18. M. Gockeler *et al.*, *Phys. Lett.* **B627**, 113 (2005).
19. H.-x. He and X.-D. Ji, *Phys. Rev.* **D52**, 2960 (1995).
20. L. P. Gamberg and G. R. Goldstein, *Phys. Rev. Lett.* **87**, p. 242001 (2001).
21. V. Barone, T. Calarco and A. Drago, *Phys. Lett.* **B390**, 287 (1997).
22. B.-Q. Ma, I. Schmidt, J. Soffer and J.-J. Yang, *Phys. Rev.* **D64**, p. 014017 (2001).
23. P. Schweitzer *et al.*, *Phys. Rev.* **D64**, p. 034013 (2001).
24. A. Bacchetta, F. Conti and M. Radici, *Phys. Rev.* **D78**, p. 074010 (2008).
25. D. W. Sivers, *Phys. Rev.* **D41**, p. 83 (1990).
26. J. C. Collins, *Phys. Lett.* **B536**, 43 (2002).
27. D. Boer, P. J. Mulders and F. Pijlman, *Nucl. Phys.* **B667**, 201 (2003).
28. M. Anselmino *et al.*, *Eur. Phys. J.* **A39**, 89 (2009).
29. A. Airapetian *et al.*, *Phys. Rev. Lett.* **94**, p. 012002 (2005).

STUDIES OF TMDs AT COMPASS

H. WOLLNY on behalf of the COMPASS Collaboration
Physikalisches Institut, Albert–Ludwigs University of Freiburg,
79104 Freiburg, Germany
** Heiner.Wollny@cern.ch*

Transverse spin and transverse momentum distribution functions of the constituents of the nucleon are a crucial input for a complete description of the nucleon. COMPASS measured such for longitudinally and transversely polarized deuterons and protons. In the following we will focus on recent results from the 2007 transverse proton data and on the results for unpolarized deuterons.

Keywords: SIDIS, polarized target, Transversity, TMD, Collins, Sivers

1. Introduction

At leading order and integrating over transverse quark momenta, three parton distribution functions (PDFs) are needed for a complete description of the nucleon. The quark density $q(x)$, the helicity distribution $\Delta q(x)$ and the transversity distribution $\Delta_T q(x)$. The first PDF, $q(x)$, describes the probability to scatter off a quark inside the nucleon carrying the momentum fraction x. The helicity PDF, $\Delta q(x)$, is defined for a longitudinal polarized proton as the difference of the probabilities that the struck quark carries momentum fraction x and its spin is parallel or anti-parallel to the spin of the parent nucleon. The transversity PDF, $\Delta_T q(x)$, is analogously defined to $\Delta q(x)$, however for a transversely polarized nucleon. Because $\Delta_T q(x)$ is chiral-odd it cannot be measured in inclusive DIS. However it can be accessed in semi-inclusive deep-inelastic scattering in combination with a chiral-odd fragmentation function. Three different channels to address the transversity distribution, have been analysed, both for deuteron and proton targets. They will be discussed in Sec. 3.

If one considers intrinsic transverse momenta of the quarks inside the nucleon, several further PDFs, so called transverse momentum dependent distribution functions (TMDs), contribute to the SIDIS cross-section, which will be discussed in Sec. 4.

2. The COMPASS Experiment

COMPASS is a fixed target experiment situated at the M2 beam line of the SPS accelerator at CERN. It has a wide physics program dedicated to the study of the nucleon spin structure and of hadron spectroscopy. In the years 2002, 2003, 2004 and 2006 data with a polarized ^{6}LiD target were taken. In the years 2002-2004 approximately 20 % of the data taking was dedicated to transverse target polarization. In the year 2007 COMPASS took data with a NH_3 target, equally shared between longitudinal and transverse target polarization. After the years 2008 and 2009, dedicated to baryon and meson spectroscopy measurements with hadron beams, in 2010 COMPASS continues with muon beam and transversely polarized protons.
The target consists of several cells along the beam direction, which are oppositely polarized. The polarization is periodically reversed to reduce systematic effects due to the different acceptances of the cells. The detector consists of two open field dipole magnets, allowing the detection of particles with ± 180 mrad. Particle identification is done with a Ring Imaging Cerenkov detector, two hadron calorimeters and muon filters. For a detailed description we refer to reference [1].

3. Transversity

Collins Asymmetry: The chiral-odd transversity distribution can be measured in SIDIS involving the chiral-odd Collins fragmentation function $\Delta_T^0 D_q^h$ [2], which describes the fragmentation of transversely polarized quarks into unpolarized hadrons. According to Collins the fragmentation leads to an asymmetry A_{Coll} of the number of produced hadrons in $\sin(\phi_h + \phi_S - \pi)$. Here ϕ_h is the azimuthal angle, measured around the direction of the virtual photon, of the hadron with respect to the scattering plane and ϕ_S is the azimuthal angle between the spin of the initial quark and the scattering plane [3–5]. In first order the measured asymmetry A_{Coll} is proportional to a convolution over intrinsic quark transverse momenta of the transversity distribution and the Collins fragmentation function measured at Belle [6]. Because of the convolution, assumptions about the transverse momentum dependence of the distribution and the fragmentation functions have to be made to extract transversity [7]. In addition the Q^2 evolution of the Collins fragmentation function from Belle to COMPASS energies is not known, introducing a further uncertainty.

Dihadron Interference: Another possibility to access the transversity distribution is to measure it in SIDIS in combination with the polarized dihadron interference fragmentation function (DiFF) [8–11]. Here, the fragmentation of a transversely polarized quark into two unpolarized hadrons leads to an azimuthal modulation in $\sin(\phi_R+\phi_S-\pi)$ in the number of produced hadron pairs. Here ϕ_S is defined as above and ϕ_R is the azimuthal angle, measured around the direction of the virtual photon, between the vector $\vec{R} = (z_2\vec{P}_1 - z_1\vec{P}_2)/(z_1 + z_2)$ and the scattering plane. Where P_1 and P_2 are the momenta of the two hadrons and z_1 and z_2 are their fractional energies. For oppositely charged hadron pairs P_1 is per definition the momentum of the hadron with positive charge, otherwise P_1 is the momentum of the hadron with the largest energy fraction. Here, the measured asymmetry A_{RS} is proportional to a product of the transversity distribution and the polarized DiFF, recently measured at Belle [12]. Therefore, no assumptions about transverse momentum dependences of distribution and fragmentation functions are needed, to extract transversity. In addition, the Q^2 evolution of the polarized DiFF from Belle energies to COMPASS energies is known [13]. Hence, the extraction of transversity with this channel is from theoretical point of view cleaner than the extraction from the Collins asymmetries.
Lambda Polarization: A third channel to measure transversity is to study Λ and $\bar{\Lambda}$ polarization in SIDIS with a transversely polarized target [14]. The polarization P_T^Λ of the produced $\Lambda/\bar{\Lambda}$-hyperons, measured via the parity violating decay into $\Lambda \to p\pi^-$ and $\bar{\Lambda} \to \bar{p}\pi^+$, respectively, is used as a polarimeter of the initial transverse spin of the fragmenting quark. It is in first order proportional to a product of transversity and the fragmentation function $\Delta_T D_q^\Lambda$, which describes the fragmentation of a transversely polarized quark into a transversely polarized Λ-hyperon.

4. Transverse Momentum Dependent Distribution Functions, TMDs

Taking into account intrinsic transverse momenta of the quarks in total eight TMDs are needed for a complete description at leading-twist [15]. For unpolarized nucleons the Cahn effect, the Boer-Mulders TMD and perturbative QCD effects, like gluon radiation, contribute to $\cos\phi_h$ and $\cos 2\phi_h$ modulations in the SIDIS cross section of single hadrons [16–18]. For a transversely polarized nucleon the Sivers TMD $\Delta_0^T q$ [19] is related to the quark angular orbital momentum inside a transversely polarized nucleon and is therefore of special interest since this could be a crucial piece to solve the nucleon spin puzzle. The Sivers effect leads to an azimuthal modulation of the number of produced hadrons in $\sin(\phi_h - \phi_S)$, where ϕ_h and ϕ_S are defined as for the Collins asymmetry discussed in Sec. 3. The asymmetry A_{Siv} is proportional to a convolution over intrinsic quark transverse momenta of the Sivers function and the well-known unpolarized fragmentation function D_q^h.

Recent results of the Sivers function measured with protons and results of $\cos\phi_h$ and $\cos 2\phi_h$ modulations for unpolarized deuterons will be reviewed in this article. For the six remaining azimuthal asymmetries measured with transversely polarized deuterons the reader is referred to [20]. Detailed results of azimuthal asymmetries measured with longitudinally polarized deuterons can be found in [21].

5. Event selection

Kinematic cuts on the squared four momentum transfer $Q^2 > 1\,(\mathrm{GeV}/c)^2$, on the fractional energy transfer of the muon $0.1 < y < 0.9$ and the hadronic invariant mass $W > 5\,\mathrm{GeV}/c^2$ are applied to select DIS events. The selection of charged hadrons differs slightly for the various analyses. However, for the single hadron analyses in general a fractional energy $z > 0.2$ of the observed final hadron is required to select hadrons in the current fragmentation region. In addition a transverse momentum $p_T^h > 0.1\,\mathrm{GeV}/c$ of the hadron with respect to the virtual photon is required to ensure a good definition of ϕ_h. Muons are rejected in demanding a minimal energy deposit in the hadronic calorimeters.

For the hadron pair analysis $z_{1,2} > 0.1$ is required for each hadron and for the sum $z_1 + z_2 < 0.9$ to reject exclusively produced ρ^0-mesons. In order to have a good definition of ϕ_R a cut on $R_T > 70\,\mathrm{MeV}/c$, the transverse component of $\vec{R}$ with respect to the virtual photon direction, is applied.

6. Results

6.1. *Transversity*

The Collins asymmetries A_{Coll}^p have been evaluated for charged hadrons in bins of Bjorken x, energy fraction z and transverse momentum p_T^h. The results for the full 2007 proton statistics [22] are shown in Fig. 1. For positive hadrons the asymmetry is negative and for negative hadrons it is positive. For both charges the size of the asymmetry increases with x and is compatible in their strengths. The values agree both in magnitude and in sign with the previous measurements of HERMES [23], which have been performed at a considerably lower electron beam momentum of 27.5 GeV/c. From the transversely polarized deuteron data Collins asymmetries for unidentified charged hadrons, charged pions and charged kaons have been extracted [3–5]. All asymmetries found to be small and compatible with zero within the statistical errors. Hence for deuteron the transversity distribution must be small or even vanishing due to isospin symmetry, because the proton results and the Belle results [6] confirmed the existence of a non-zero Collins fragmentation function. The results for the transverse 2007 proton data of the dihadron asymmetries A^p_{RS} for oppositely charged hadron pairs [24] are shown as a function of x, z and M_{inv} in Fig. 2. A strong asymmetry is observed in the valence x-region. In the invariant mass the asymmetry is negative over the whole range. Compared to the Collins asymmetries, the size of A^p_{RS} is approximately a factor of two larger, emphasizing the good analyzing power of the dihadron interference fragmentation function to measure transversity. From the transversely polarized

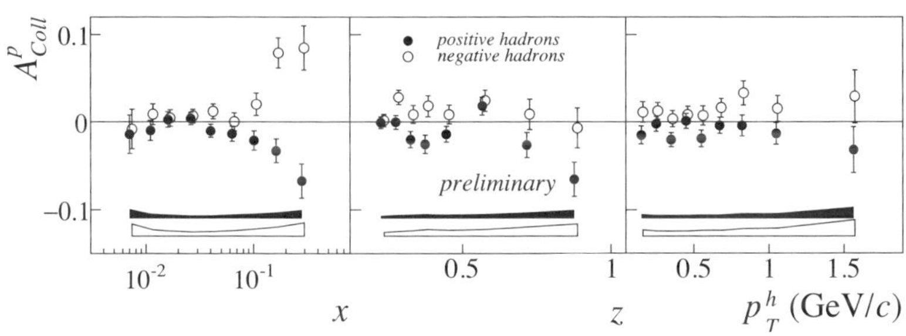

Fig. 1. Collins asymmetries for positive and negative hadrons as a function of x, z and p_T. The horizontal bands indicate the systematic uncertainties.

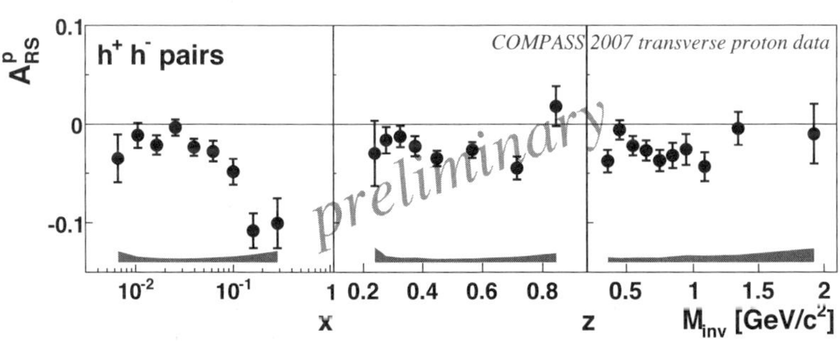

Fig. 2. Dihadron asymmetries A^p_{RS} for oppositely charged hadron pairs as a function of x, z and M_{inv}. The horizontal band indicates the systematic uncertainties.

deuteron data COMPASS extracted dihadron asymmetries for unidentified

h^+h^--pairs and identified $\pi^+\pi^-$, K^+K^-, π^+K^- and $K^+\pi^-$-pairs [25–27]. In addition also different charge combinations have been analysed, considering the two most energetic hadrons in each event. All asymmetries found to be small and compatible with zero within the statistical errors.
The polarization of Λ and $\overline{\Lambda}$-hyperons measured with transversely polarized ^{6}LiD and NH_3 targets, respectively, have been evaluated in bins of x and z [28,29]. For both targets the evaluated polarizations of Λ and $\overline{\Lambda}$-hyperons are small and compatible with zero within the errors. This indicates, taking into account the sizeable Collins and dihadron asymmetries measured for protons, that the analyzing power of this channel seems to be small.

6.2. *Sivers-Function*

The Sivers asymmetries for charged hadrons from the transverse 2007 proton data [22] are shown, as a function of x, z and p_T^h, in Fig. 3. In addition to the indicated systematic uncertainty, for positive hadrons an additional systematic uncertainty of ± 0.01 is assigned reflecting a time variation of the spectrometer acceptance between the first and the second half of data taking used for this analysis. The asymmetry for positive hadrons is significantly positive over almost the complete x-range. The asymmetry for negative hadrons is small and compatible with zero within the statistical errors. Both results show the same trend as the previous measurement of HERMES [30]. However, the magnitude of the asymmetry for positive hadrons is smaller than the one of HERMES. This might be explained by a dependence on W, the invariant mass of the photon-nucleon system. Studies indicate that the asymmetry seems to be only sizeable at small W, where HERMES measures and approach zero at high W. The Sivers asymmetries for unidentified charged hadrons, charged pions and kaons extracted from transversely polarized ^{6}LiD [3–5] were all found to be small and compatible with zero within the statistical errors.

6.3. *Azimuthal Asymmetries from Unpolarized Deuterons*

An unpolarized deuteron sample is obtained by combining data samples with opposite target polarization taken in the year 2004. The magnitudes of $\cos\phi_h$ and $\cos 2\phi_h$ modulations for charged hadrons [31] as a function of x, z and p_T are shown in Fig. 4. Large asymmetries up to 20 % are found for the $\cos\phi_h$ modulation and asymmetries in the order of 5 % for the $\cos 2\phi_h$ modulation. The trend of the asymmetries for positive and negative hadrons are similar, however, the magnitudes of the two differs significantly. The

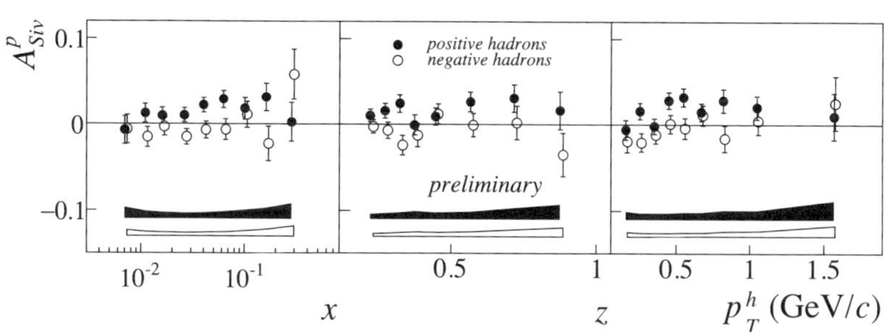

Fig. 3. Sivers asymmetries for positive and negative hadrons as a function of x, z and p_T. The horizontal bands indicate the systematic uncertainties.

systematic uncertainties have been evaluated to be sizeable. The dominant contribution is given by the acceptance corrections determined with Monte Carlo simulations.

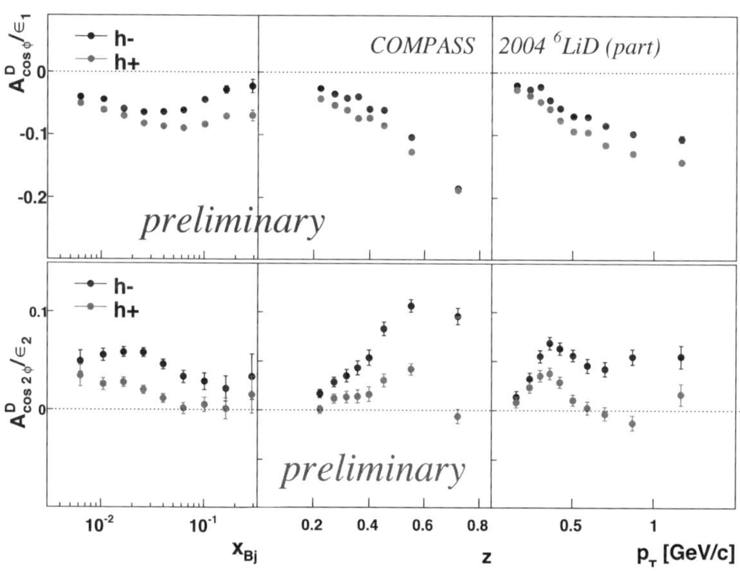

Fig. 4. Unpolarized asymmetries for charged hadrons as a function of x, z and p_T. The kinematical factors are: $\epsilon_1 = (2-y)\sqrt{1-y}/(1+(1-y)^2)$ and $\epsilon_2 = (1-y)/(1+(1-y)^2)$.

7. Summary and Outlook

Recent results of COMPASS measurements related to transversity and transverse momentum dependent distribution functions have been presented. In the year 2010 and 2011 COMPASS will continue its measurements with transversely and longitudinally polarized protons, which will significantly reduce the statistical errors.

References

1. P. Abbon *et al.*, *Nucl. Instrum. Meth.* **A577**, 455 (2007).
2. J. C. Collins, *Nucl. Phys.* **B396**, 161 (1993).
3. V. Y. Alexakhin *et al.*, *Phys. Rev. Lett.* **94**, p. 202002 (2005).
4. E. S. Ageev *et al.*, *Nucl. Phys.* **B765**, 31 (2007).
5. M. Alekseev *et al.*, *Phys. Lett.* **B673**, 127 (2009).
6. R. Seidl *et al.*, *Phys. Rev.* **D78**, p. 032011 (2008).
7. M. Anselmino *et al.*, *Nucl. Phys. Proc. Suppl.* **191**, 98 (2009).
8. A. V. Efremov *et al.*, *Phys. Lett.* **B284**, 394 (1992).
9. X. Artru and J. C. Collins, *Z. Phys.* **C69**, 277 (1996).
10. M. Radici, R. Jakob and A. Bianconi, *Phys. Rev.* **D65**, p. 074031 (2002).
11. A. Bacchetta and M. Radici, *Phys. Rev.* **D69**, p. 074026 (2004).
12. A. Vossen *et al.*, First Measurement of the Interference Fragmentation Function in e^+e^- at Belle, in *Proceedings of XIII Advanced Research Workshop on High Energy Spin Physics, Dubna, Russia*, Sept. 2009.
13. F. A. Ceccopieri, M. Radici and A. Bacchetta, *Phys. Lett.* **B650**, 81 (2007).
14. X. Artru and M. Mekhfi, *Nucl. Phys.* **A532**, 351 (1991).
15. P. J. Mulders and R. D. Tangerman, *Nucl. Phys.* **B461**, 197 (1996).
16. R. N. Cahn, *Phys. Lett.* **B78**, p. 269 (1978).
17. D. Boer and P. J. Mulders, *Phys. Rev.* **D57**, 5780 (1998).
18. H. Georgi and H. D. Politzer, *Phys. Rev. Lett.* **40**, p. 3 (1978).
19. D. W. Sivers, *Phys. Rev.* **D41**, p. 83 (1990).
20. A. Kotzinian, Beyond Collins and Sivers: further measurements of the target transverse spin-dependent azimuthal asymmetries in semi-inclusive DIS from COMPASS, in *Proceedings of DIS 2007, Munich, Germany*, April 2007.
21. I. A. Savin, Azimuthal asymmetries in production of charged hadrons by high energy muons on polarized deuterium targets, in *Proceedings of XIII Advanced Research Workshop on High Energy Spin Physics, Dubna, Russia*, Sept. 2009.
22. H. Fischer, Review of Spin Physics in DIS, a personal selection, in *Proceedings of DIS 2010, Florence, Italy*, April 2010.
23. A. Airapetian *et al.*, *arXiv:1006.4221* (2010).
24. H. Wollny, Transversity Signal in two Hadron Pair Production in COMPASS, in *Proceedings of DIS 2009, Madrid, Spain*, April 2009.
25. R. Joosten, *AIP Conf. Proc.* **792**, 957 (2005).
26. R. Joosten, *AIP Conf. Proc.* **915**, 646 (2007).
27. A. Vossen, Measurement of Transverse Spin Effects at COMPASS, in *Proceedings of 42nd Rencontres de Moriond on QCD and Hadronic Interactions, La Thuile, Italy*, March 2007.
28. A. Ferrero, *AIP Conf. Proc.* **915**, 436 (2007).

29. D. Kang, Longitudinal and transverse Lambda polarization at COMPASS, in *Proceedings of DIS 2010, Florence, Italy*, April 2010.
30. A. Airapetian *et al.*, *Phys. Rev. Lett.* **103**, p. 152002 (2009).
31. W. Kafer, Measurements of Unpolarized Azimuthal Asymmetries at COMPASS, in *Proceedings of Transversity 2008, Ferrara, Italy*, May 2008.

STUDIES OF TMDs AT HERMES

L. L. PAPPALARDO* (on behalf of the HERMES Collaboration)

Dipartimento di Fisica, Universita' di Ferrara and INFN Sezione di Ferrara, Ferrara, 44100, Italy
**E-mail: pappalardo@fe.infn.it*

Single-spin asymmetries measured at the HERMES experiment for pions and charged kaons in semi-inclusive deep-inelastic scattering of electrons and positrons off a transversely polarized hydrogen target, are reported. Two of them reveal clear evidences of non-zero leading-twist parton distribution functions, such as the poorly known transversity and Sivers functions, describing crucial aspects of the dynamics of quarks in a transversely polarized nucleon. Evidences of a non-zero favored and disfavored Collins fragmentation function with opposite sign are also observed.

Keywords: Semi-inclusive DIS; Single-spin asymmetries; Transversity, TMDs.

1. Introduction

For decades, inclusive deep-inelastic scattering (DIS) experiments have crucially contributed to the present knowledge on the internal structure of nucleons. In particular, these experiments have provided precise measurements of two fundamental parton distribution functions (PDFs)[1] : the momentum distribution, $f_1(x)$, and the helicity distribution, $g_1(x)$, where x denotes the longitudinal momentum fraction carried by the partons.

A complete description of the nucleon structure at leading order in an expansion in M/Q (twist expansion) requires, after integrating over the quark transverse momentum $\mathbf{p}_T$, the knowledge of a third fundamental PDF, named transversity, which is presently poorly known. It is denoted as $h_1(x)$ and reflects the difference between the number densities of quarks with spin parallel or antiparallel to the spin of the parent nucleon when the latter is transversely polarized (see Ref. [2] for a review on the subject). Differently from $f_1(x)$ and $g_1(x)$, $h_1(x)$ does not couple with gluons for spin-$\frac{1}{2}$ targets due to helicity conservation. As a consequence, although $g_1(x)$ and $h_1(x)$ might be similar at very low scales,[2] they are expected to have a different QCD evolution, and can thus differ significantly at higher scales. Furthermore, it has been shown[3] that any departure from the identity $h_1(x) = g_1(x)$ is a signature of relativistic effects in the nucleon.

Transversity has long remained unmeasured due to its chiral-odd nature, which prevents its measurement in inclusive deep-inelastic scattering. In order to preserve chirality conservation in hard electromagnetic and strong interactions, the transversity distribution can only be measured in conjunction with another chiral-odd object. One possibility is represented by the semi-inclusive deep-inelastic scattering (SIDIS), in which at least one final state hadron is detected in coincidence with the scattered lepton. In this process the transversity distribution can be probed, e.g., in conjunction with the chiral-odd Collins fragmentation function, $H_1^{\perp}(z)$, which describes the correlation between the transverse polarization of the quarks and the transverse momentum of the produced hadrons $P_{h\perp}$. Here z denotes the fraction of the original virtual photon energy carried by the produced hadron. This correlation, commonly denoted as the Collins effect, produces left-right asymmetries in the direction of the final-state hadrons.

Besides allowing to access transversity, semi-inclusive DIS experiments also opened the way to the extraction of a variety of new PDFs which arise when the transverse momentum $\mathbf{p}_T$ of the quarks is not integrated out. These poorly known PDFs, typically denoted as transverse-momentum-dependent (TMD) PDFs, are increasingly gaining theoretical and experimental interest. Describing correlations between the quark or the nucleon polarization and the quark transverse momentum $\mathbf{p}_T$, i.e. spin-orbit correlations, the TMD distribution functions encode information on the 3-dimensional structure of nucleons. At leading-twist, eight TMD distribution functions, each describing a peculiar aspect of the nucleon structure, enter the SIDIS cross section in conjunction with a fragmentation function.[4,5] When the polarization of the final hadrons is not detected, this can be either the Collins function $H_1^{\perp}(z)$ or the relatively well known spin-independent fragmentation function $D_1(z)$. Among these PDFs, particularly interesting is the Sivers function[6] , denoted as $f_{1T}^{\perp}(x)$. It describes the correlation between the transverse momentum $\mathbf{p}_T$ of the quarks and the transverse spin of the parent nucleon, also known as the Sivers effect. Similarly to the Collins effect, the Sivers effect causes left-right (azimuthal) asymmetries in the direction of the final-state hadrons. The Sivers function is naïve T-odd, i.e. odd under a special (naïve) time-reversal operator that does not interchange initial and final states. The interest on this TMD distribution function suddenly increased after it was demonstrated[7] that non-vanishing orbital angular momentum of quarks, the main still unmeasured contribution the nucleon spin, is needed for a non-vanishing Sivers effect, although no direct simple relations can be established between the two, yet.

The lepton-hadron SIDIS cross section can be written as a superposition of spin-averaged and spin-dependent terms with various polarization states of target and beam.[4,5] Each of these terms is modulated by a specific combination of the azimuthal angle of the target polarization, ϕ_S, and of the produced hadron, ϕ, both referred to the lepton scattering plane. In the present work, a selection of Fourier amplitudes $2\langle\sin(m\phi \pm n\phi_S)\rangle^h_{\mathrm{UT}}$, related to some of these terms, are presented for various hadron types h. Here m and n are integers, and the subscript UT denotes unpolarized beam and transversely polarized target. These Fourier amplitudes were extracted through a maximum-likelihood fit of the SIDIS events, alternately binned in x, z and $P_{h\perp}$, but unbinned in ϕ and ϕ_S.

2. The Collins and Sivers amplitudes

Preliminary HERMES results for the Collins amplitudes for pions are shown in Fig. 1 as a function of x, z, or $P_{h\perp}$ (see Ref. [8] for the final results). They are positive for π^+ and negative for π^-. The amplitudes of π^+ and π^- are observed to increase in magnitude with x. This is consistent with the expectation that transversity mainly receives contributions from the valence quarks. A large contribution from the sea quarks, which would dominate in the low-x region, is indeed not expected due to the fact that transversity can not be generated in gluon splitting in a proton target. A possible explanation for the π^- amplitude, observed to be of opposite sign to that of π^+ and of similar magnitude, can be the dominance of the u-quark flavour among struck quarks in conjunction with a substantial magnitude with opposite sign of the unfavored Collins fragmentation function, describing, *e.g.* the fragmentation of u quarks into π^- mesons: $H^{\perp,u\to\pi^-}_{1,\mathrm{unfav}} \approx -H^{\perp,u\to\pi^+}_{1,\mathrm{fav}}$. This observation is supported by the BELLE results[9] and by the combined fits reported in Refs. [10,11]. The Collins amplitudes for charged kaons are also shown in Fig. 1. Due to the common u-quark dominance, *i.e.* the dominant contribution to π^+ and K^+ production from scattering off u-quarks, one would naïvely expect similar amplitudes for these two mesons. The observed differences can be ascribed, for instance, to the the different Collins fragmentation functions involved. On the other hand, the amplitudes for π^- and K^-, which also exhibit a very different behavior, are not expected to be similar not only because of the different fragmentation functions involved, but also because, in contrast to π^-, a K^- has no valence quarks in common with the target proton.

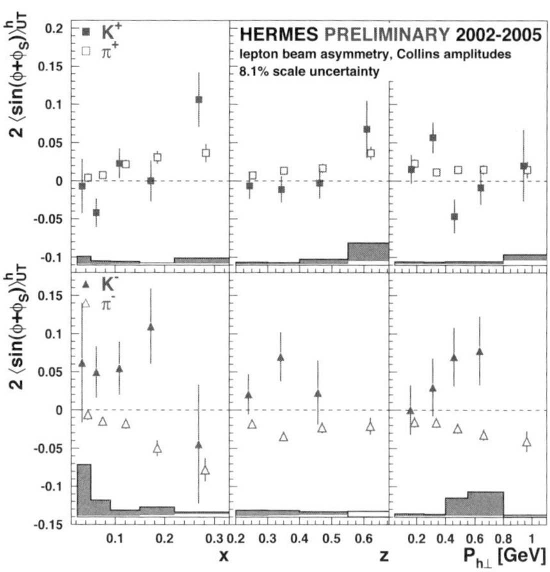

Fig. 1. Collins amplitudes for charged pions (open dots) and charged kaons (full dots) as a function of x, z, or $P_{h\perp}$. A 7.3% scale uncertainty, not shown, arises from the accuracy in the measurement of the target polarization.

The HERMES results for the Sivers amplitudes,[12] are shown for pions and charged kaons in Fig. 2 as a function of x, z, or $P_{h\perp}$. They are positive for all hadrons except for π^-, for which they are consistent with zero, and in general are found to increase with increasing z. In the cases of π^+ and K^+, the amplitudes show a saturation for $P_{h\perp} \geq 0.4$ GeV and are consistent with the predicted linear decrease in the limit of $P_{h\perp}$ going to zero. Despite the expectation of similar magnitudes for the π^+ and K^+ amplitudes, based on the common u-quark dominance, the amplitudes for K^+ are found to be significantly larger than those for π^+. This difference may be due to a significant role of other quark flavors, e.g. the sea quarks, in conjunction with possible higher-twist contributions in kaon production. Assuming that scattering from u-quarks is the dominant process, the positive Sivers amplitudes for π^+ and $K^\pm$ are compatible with a large negative Sivers function for u-quarks. The vanishing amplitudes for π^- then require cancellation effects, e.g. from a d-quark Sivers function opposite in sign to the u-quark Sivers function: $f_{1T}^{\perp,u}(x) \approx -f_{1T}^{\perp,d}(x)$. This hypothesis is in agreement with previous extractions[13–15] of the Sivers function based on earlier HERMES data and with QCD predictions in the large-N_c limit.[16,17]

It is also supported by the vanishing COMPASS Sivers amplitudes measured on a deuteron target.[18]

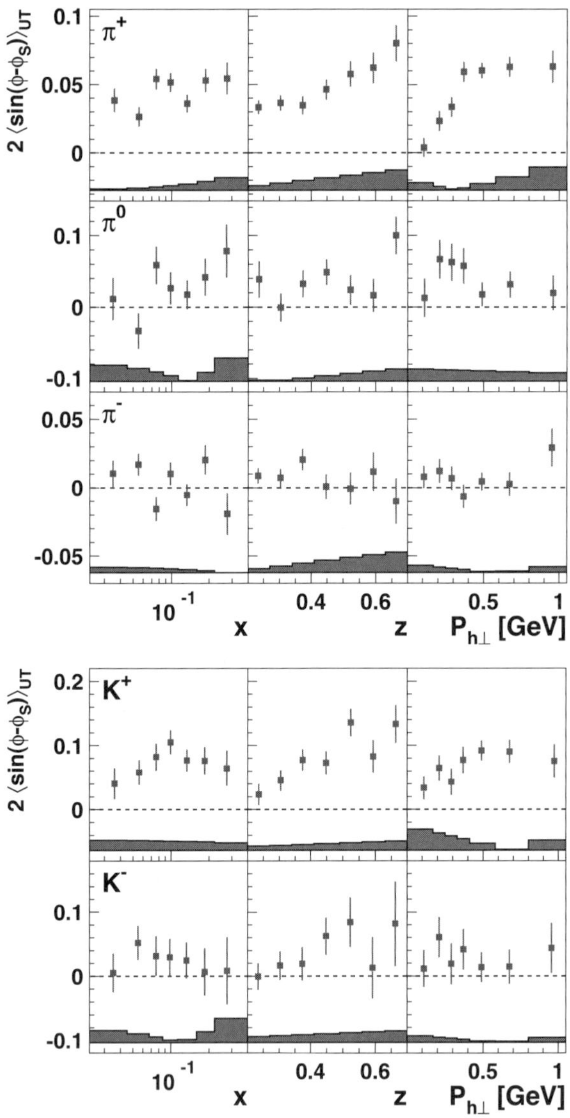

Fig. 2. Sivers amplitudes for pions (upper plot) and charged kaons (lower plot) as a function of x, z, or $P_{h\perp}$. A 7.3% scale uncertainty, not shown, arises from the accuracy in the measurement of the target polarization.

3. The other sine asymmetry amplitudes

The Collins and Sivers amplitudes were extracted together with other four sine azimuthal asymmetries, related to other terms of the SIDIS cross section. These Fourier amplitudes are all consistent with zero for all hadron

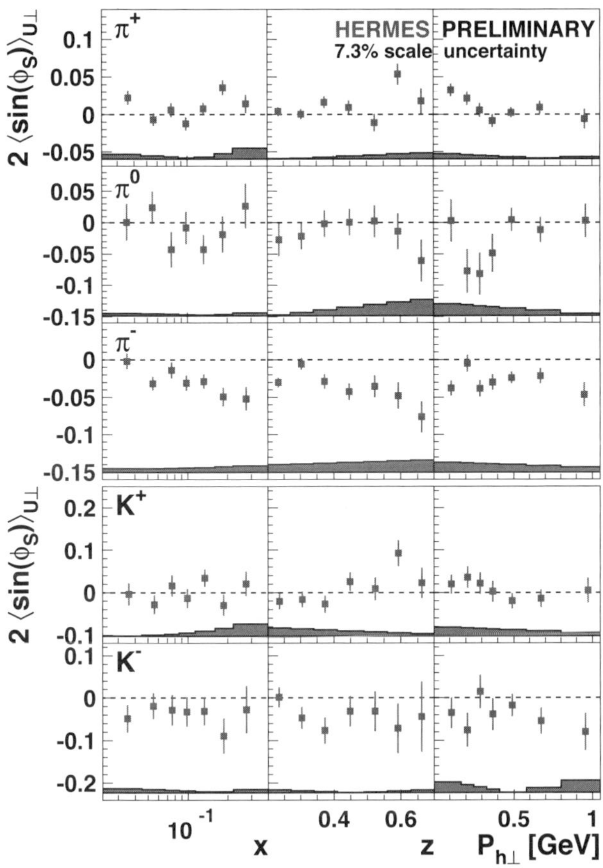

Fig. 3. $2\langle\sin(\phi_S)\rangle^h_{UT}$ amplitudes for pions and charged kaons as a function of x, z, or $P_{h\perp}$. A 7.3% scale uncertainty, not shown, arises from the accuracy in the measurement of the target polarization.

types, with the only exception of subleading-twist $2\langle\sin(\phi_S)\rangle^h_{UT}$ amplitude, which exhibits a large negative signal for π^-, similar to the corresponding Collins amplitude. Preliminary HERMES results for this Fourier component and for the leading twist $2\langle\sin(3\phi - \phi_S)\rangle^h_{UT}$ amplitude, are shown in Fig. 3

and Fig. 4, respectively. In particular, the latter is related to the chiral-odd $h_{1T}^{\perp}$ TMD distribution function, also known as *pretzelosity*, describing the correlation between the quark transverse momentum $\mathbf{p}_T$ and the transverse polarization of the quarks in a transversely polarized nucleon[19] . This function has recently been linked to a measure of the deviation of the nucleon shape from a sphere.[20]

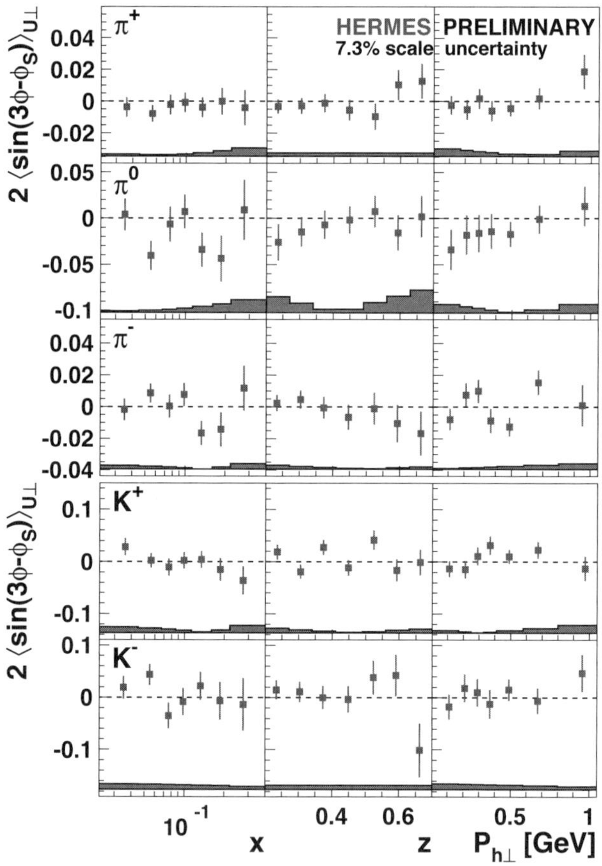

Fig. 4. $2\langle\sin(3\phi - \phi_S)\rangle^h_{UT}$ amplitudes for pions and charged kaons as a function of x, z, or $P_{h\perp}$. A 7.3% scale uncertainty, not shown, arises from the accuracy in the measurement of the target polarization.

4. Conclusions

The data collected by the HERMES experiment with a transversely polarized hydrogen target was analysed to extract several azimuthal asymmetry amplitudes. These amplitudes can be interpreted in the framework of the parton model in terms of parton distribution functions and fragmentation functions. Among them are the non-vanishing Collins and Sivers amplitudes, carrying precious information on two leading-twist parton distribution functions: the transversity distribution and the Sivers function, respectively. Significant positive (negative) Collins amplitudes were measured for π^+ (π^-), indicating that both the transversity and the Collins function are non-zero. Significantly positive Sivers amplitudes were observed for both positive pions and kaons. This result clearly indicates that the Sivers function is non-zero and, indirectly, that also the yet unmeasured quark orbital angular momentum is non-zero. In particular, the Sivers amplitude for K^+ was found to be significantly larger than that for π^+, suggesting, e.g., a significant Sivers function for the sea quarks. The additional azimuthal asymmetries, extracted together with the Collins and Sivers amplitudes, are all consistent with zero, with the only exception of $2\langle\sin(\phi_S)\rangle^h_{UT}$, which exhibits a large negative amplitude for π^-, similar to the Collins one. Future plans at HERMES include the extraction of the $2\langle\cos(\mathrm{m}\phi \pm \mathrm{n}\phi_\mathrm{S})\rangle^\mathrm{h}_\mathrm{LT}$ Fourier amplitudes, related to the *worm-gear* $g^\perp_{1T}$ TMD distribution function, and of several yet unmeasured dihadron structures of the SIDIS cross seciton.

References

1. Amsler C. *et al.* (Particle Data Group), *Phys. Lett. B* **667**, 1 (2008).
2. Barone V., Drago A. and Ratcliffe P.G., *Phys. Rep.* **359**,1 (2002).
3. Jaffe R.L. and Ji X., *Nucl. Phys. B* **375**, 527 (1992).
4. Mulders P.J. and Tangerman R.D., *Nucl. Phys. B* 461, 197 (1996), erratum-ibid. B **484**, 538 (1997).
5. Bacchetta A. *et al.*, *JHEP* **02**, 093 (2007).
6. Sivers D.W., *Phys. Rev. D* **41**, 83 (1990).
7. Brodsky S.J., Hwang D.S. and Schmidt I., *Phys. Lett. B* **530**, 99 (2002).
8. Airapetian A. *et al.* (HERMES), *Preprint: arXiv:1006.4221 [hep-ex]* (2010).
9. Abe A. *et al.* (BELLE), *Phys. Rev. Lett.* **96**, 232002 (2006).
10. Anselmino M. *et al.*, *Phys. Rev. D* **75**, 054032 (2007).
11. Anselmino M. *et al.*, *Preprint: arXiv:0812.4366v1 [hep-ph]* (2008).
12. Airapetian A. *et al.* (HERMES), *Phys. Rev. Lett.* **103**, 152002 (2007).
13. Vogelsang W. and Yuan F., *Phys. Rev. D* **72**, 054028 (2005).
14. Anselmino M. *et al.*, *Phys. Rev. D* **72**, 094007 (2005).
15. Collins J.C., *Phys. Rev. D* **73**, 014021 (2006).

16. Efremov A.V., Goeke K. and Schweitzer P., *Phys. Rev. D* **73**, 094025 (2006).
17. Pobylitsa P.V., *Preprint: arXiv:hep-ph/0301236* (2003).
18. Alekseev M. *et al.* (COMPASS), *Phys. Lett. B* **673**, 127 (2009).
19. Avakian H. *et al.*, *Phys. Rev. D* **78**, 114024 (2008).
20. Miller G.A., *Phys. Rev. C* **76**, 065209 (2007).

STUDIES OF TRANSVERSE MOMENTUM DEPENDENCE IN PP SCATTERING

L. C. BLAND

Brookhaven National Laboratory, Upton, NY

The Relativistic Heavy Ion Collider at Brookhaven National Laboratory is the first and only polarized collider in the world. Measurements with polarized proton collisions that are sensitive to gluon polarization find small values. Measurements of transverse single spin asymmetries find large values. Present understanding is that these spin effects are from spin-correlated transverse momentum dependence in distribution and fragmentation functions. A summary of existing measurements, and plans for the future are presented.

The nucleon is a complex object built from quarks, antiquarks and gluons. Describing nucleon structure in terms of collinear parton distribution functions (PDF), with their dependence only on longitudinal momentum fraction x and resolution scale, has been a highly successful endeavor through the years [1,2]. Experiments aimed at accounting for the nucleon spin were initially viewed as an important test as to whether our understanding of nucleon structure was complete. Arguably, recent results from nucleon spin structure studies provide the strongest hints that collinear PDF are inadequte descriptions of important aspects of nucleon structure. For example, the largest spin effects observed in particle production experiments at RHIC are found for transverse single spin asymmetries (SSA) [3–5]. Observables sensitive to gluon polarization are measured to be small at RHIC energies [6,7]. Consequently, an understanding of the spin structure of the proton is perhaps the strongest motivation to extend collinear descriptions to include transverse momentum dependence (TMD) in distribution functions [8,9]. Spin-correlated transverse momentum of the partons within the nucleon corresponds to parton orbital motion. This is a conceptually different view of the structure of the nucleon than provided by a collinear picture.

It is useful to have an intuitive understanding of why TMD distribution functions are important. The up and down quark masses are much smaller than the mass of the nucleon. Consequently, most of the mass of the nucleon is associated with the color electric and magnetic fields that confine the quarks within the nucleon. The quanta of these fields are the gluons. Gluons are radiated by an accelerated color charge, and so play a major role in the scale dependent evolution of PDF. Gluon radiation also generates transverse momentum, in addition to redistributing the longitudinal momentum. An arbitrary scale is introduced below which transverse momentum is viewed as intrinsic, and above which transverse momentum is viewed as being radiatively generated. TMD distributions naturally arise due to the small masses of the up and down quarks, and their strong couplings to the gluon.

This talk reviews findings of spin asymmetry measurements at RHIC, with a focus on transverse SSA results and plans. In the past nearly ten years, measurements of the two-spin helicity asymmetry observable (A_{LL}), sensitive to possible polarization of gluons in the proton, and transverse SSA (analyzing power, A_N), believed to be senstive to TMD distribution or fragmentation functions, have been completed for meson and jet production from spin-polarized proton collisions. An outlook to future measurements is also presented.

The Relativistic Heavy Ion Collider (RHIC) at Brookhaven National Laboratory is a unique collider. RHIC consists of two independent rings of dipole magnets [10]. At six symmetric locations around the rings, dipole insertions enable the counterrotating beams to move along the same line and collide. The two independent rings provides great flexibility in the types of positive ions that can be used for collisions. RHIC has collided beams of gold ions and copper ions to establish the existence of a (strongly coupled) quark-gluon plasma. In two separate runs, deuteron beam in one ring collided with Au ions in a separate ring, to study the nuclear modification of PDF, or more generally to study cold nuclear matter effects. Two helical dipole magnets in each ring, known as Siberian Snakes, overcome the influence of many depolarizing resonances thereby enabling the acceleration and collision of spin polarized proton beams [11].

The Siberian Snake magnets preserve polarization initially produced in an optically pumped source. A sequence of spin-up and spin-down beam bunches is injected into RHIC at the start of each fill. Acceleration ramps then follow RHIC filling, followed by other sequenced manipulations of the beams prior to the start of collisions. Polarization has been measured for colliding beams at $\sqrt{s}$=62, 200 and 500 GeV. The stable spin direction in the RHIC rings is transverse to the momentum. Spin rotator magnets

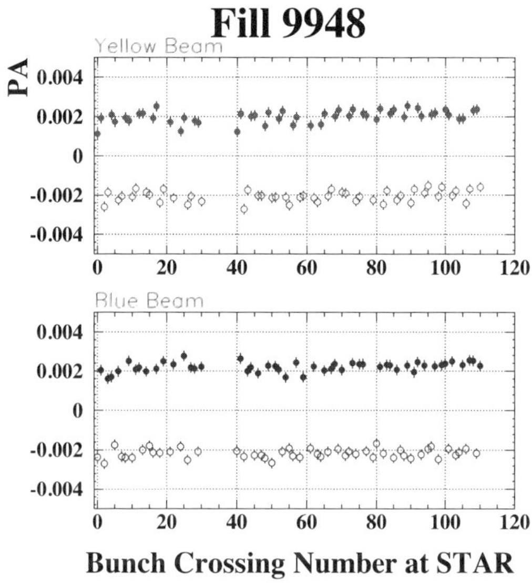

Fig. 1. Bunch-by-bunch left/right asymmetries from transversely polarized colliding beams at STAR, as measured by the beam-beam counters. Top (Bottom): asymmetries from the east (west) BBC which faces the yellow (blue) beam. Polarimetry results from RHIC yield Blue beam polarization of $0.488 \pm 0.050(\text{stat}) \pm 0.021(\text{sys})$ and Yellow beam polarization of $0.406 \pm 0.015(\text{stat}) \pm 0.020(\text{sys})$ for this fill.

precess the polarization prior to and after the interaction regions used by the STAR and PHENIX experiments, to enable the collisions of longitudinally polarized beams. Transversely polarized proton collisions can be studied at STAR and PHENIX by omitting the current ramp of the spin rotator magnets from the acceleration/storage sequence.

Polarization is measured for beams in collision by local polarimeters at STAR and PHENIX. Polarization of the individual beams is periodically monitored by the insertion of carbon fiber ribbons into the beams, to study p+C elastic scattering at small $|t|$ in the Coulomb-nuclear interference (CNI) region [12]. Absolute polarization of the beams is measured by p+p elastic scattering from a polarized gas jet target [13]. Longitudinal polarization of colliding beams corresponds to the absence of transverse SSA (from either vertical or radial polarization) from the local polarimeters when the spin rotators are properly tuned, and standard values for the CNI polarimeter, in a section of the rings where the stable spin direction is transverse. A typical asymmetry measurement from the local polarimeter at STAR is shown in Fig. 1 for collisions of transversely polarized proton

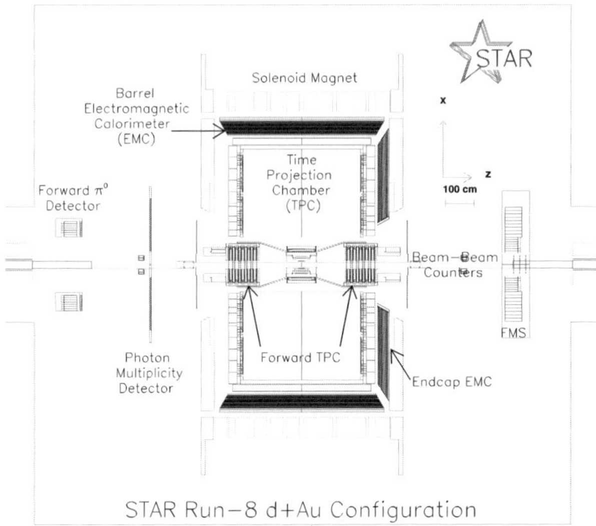

Fig. 2. Top view of STAR configuration for RHIC run 8. Scaling transverse to the beam is twice larger than scaling longitudinal to the beams. The TPC tracks charges through the 0.5T solenoid. The BEMC surrounds the TPC, to measure energetic photons and electrons and to provide event triggering. The EEMC is mounted on the west poletip of STAR. The FPD and FMS view particle production through the holes in the magnet poletips.

beams. The STAR local polarimeter [14] measures the spin-correlated topological distribution of multiplicity in the forward direction ($3.5 < |\eta| < 5.0$) with its azimuthally segmented beam-beam counters. Calibration of the effective analyzing power for this polarimeter is accomplished by concurrent measurements with the local polarimeter and single-beam measurements with the polarized gas jet target.

The STAR experiment is shown in Fig. 2. The focus in these proceedings is on results from the forward pion detector (FPD) and the forward meson spectrometer (FMS). These devices view particles produced at large rapidity ($|\eta| > 2.5$) through the holes in the STAR magnet poletips.

The FPD and FMS are primarily sensitive to energetic photons from meson decay. Both devices are matrices of lead glass detectors. Closely spaced photon pairs can be discriminated from single photons up to pair energy of 60 GeV, corresponding to $x_F = 0.6$ for $\sqrt{s}$=200 GeV collisions.

The cross section [15] for neutral pion production as a function of pion energy, and at three separate pion pseudorapidities, is shown in Fig. 3.

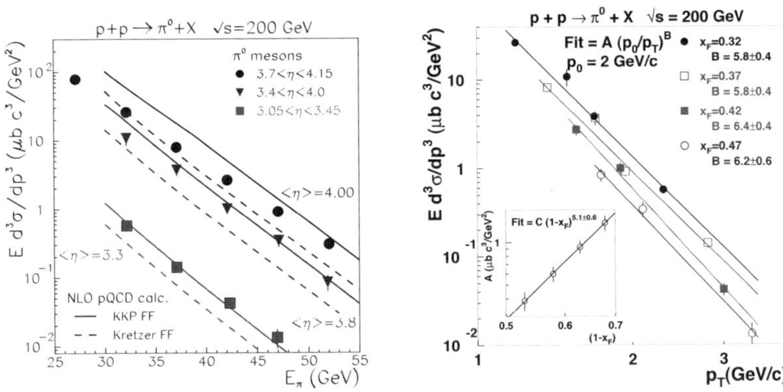

Fig. 3. Left: Measured cross sections [15] for neutral pion production at 200 GeV in comparison to next-to-leading order pQCD calculations. In general, leading twist collinear calculations describe the data, except at the largest η. Right: The same data fitted to determine dependence on x_F and p_T of the neutral pion.

These measurements were made with the FPD. A principal finding is that cross sections are well described by next-to-leading order pQCD calculations, for p_T >2 GeV/c. Fits to the x_F and p_T dependence show smooth behavior even to lower p_T, and the cross section is approximately given by $\sigma \propto (1 - x_F)^5/p_T^6$ throughout the measured range of p_T. The NLO pQCD calculations use collinear PDF and are done at leading twist. Measured cross sections at large x_F at lower $\sqrt{s}$ are much larger than NLO pQCD [16]. This suggests that hard scattering contributions are small in these kinematics at low $\sqrt{s}$ in comparison to soft processes, such as beam fragmentation. It has also been found that going beyond collinear calculations, by the introduction of transverse momentum dependence for the PDF, can describe the low energy cross sections in a hard scattering model [17], although there is no factorization theorem that supports use of TMD for the production of hadrons in p+p collisions.

In the NLO pQCD description of particle production one parton from each proton is involved in a hard scattering. At leading order, only elastic parton scattering (so-called $2 \to 2$ processes) enter. At NLO, inelastic parton scattering (so-called $2 \to 3$ processes) are included. The $2 \to 2$ partonic scatterings lead to the expectation that the transverse momentum of a produced jet of hadrons will be balanced by a second jet of hadrons, distributed in rapidity according to a convolution of the PDF with the hard scattering cross section. As shown in Fig. 4, this is what recent measurements of azimuthal correlations between pairs of neutral pions produced in the forward

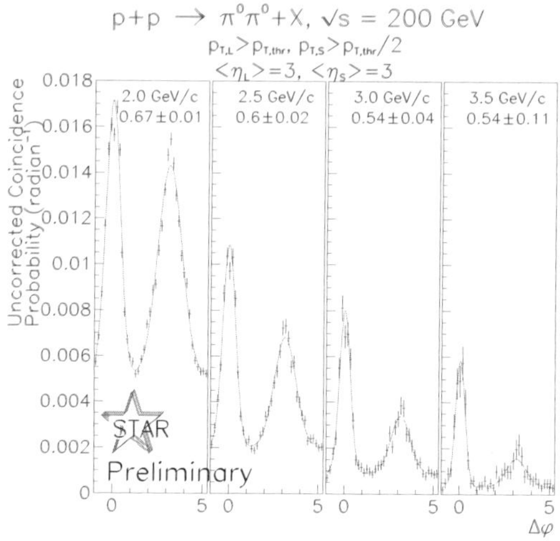

Fig. 4. Azimuthal correlations between pairs of neutral pions produced at large rapidity for increasing threshold p_T from 2 GeV/c (left) to 3.5 GeV/c [18] (right). Near-side and away-side jet-like peaks are observed. The data are well fit by two Gaussian peaks and a constant background. Peak widths decrease with increasing p_T as does the background level. Numeric values for the away side peak width are shown.

direction show. The correlation results are enabled by the large acceptance of the FMS. The systematic trend of decreasing background is consistent with a decreasing contribution from $2 \rightarrow 3$ processes as p_T increases.

Despite quantitative agreement of inclusive neutral pion yields with collinear pQCD at leading twist and qualitative agreement of two pion azimuthal correlations, there is a quandry. Specifically, neutral pion production in $\sqrt{s}$=200 GeV transversely polarized proton collisions has a measured large analyzing power that grows with increasing x_F (Fig. 5). This cannot be accomodated in collinear QCD at leading twist, due to the chiral nature of the theory [22]. Model calculations using spin-correlated TMD (Sivers functions [23]) that fit semi-inclusive deep inelastic scattering transverse SSA [24] can describe the data. The issue remains that the requisite gauge link [25] spoils factorization theorems. Spin-correlated TMD and the gauge link are approximated by a collinear, twist-3 calculation, for which factorization theorems for inclusive production have been proven [26]. Although there are relations between the qg correlators that appear in the collinear calculation with the spin-correlated TMD that is extracted from SIDIS, no detailed phenomenological global analysis has been completed to date. It is also the case that inclusive pion production can receive contribu-

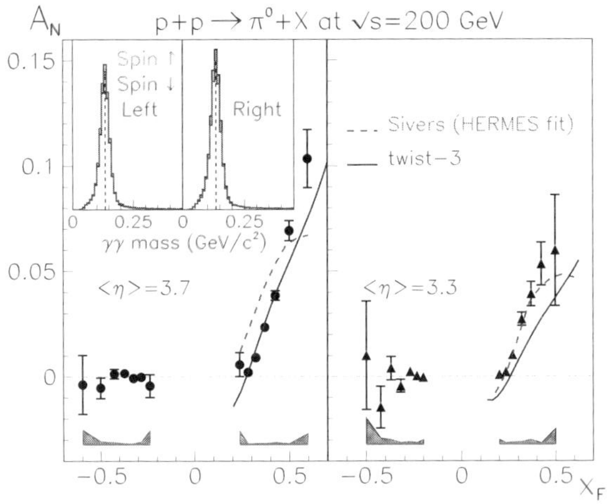

Fig. 5. Analyzing power for neutral pion production as a function of x_F at two different pseudorapidities [4]. These measurements are compared to model calculations [19] using spin-correlated TMD for quarks fit [20] to semi-inclusive deep inelastic scattering transverse single spin asymmetries. Twist-3 calculations are also shown in comparison to the data [21].

tions from spin-correlated TMD fragmentation through the Collins effect [27]. It is essential to go beyond inclusive production to separate Sivers and Collins contributions for pion production in polarized proton collisions.

Much has been learned in regard to the dynamics responsible for large-x_F particle production at RHIC energies, where large spin effects are observed. Attempts to robustly understand whether spin-correlated TMD are important in $p_\uparrow$+p collisions are hindered by complicated color charge interactions [28]. Based on present understanding, these color charge interactions are essential to enable a naively time-reveral odd operator to contribute to physical observables. Future transverse spin measurements at RHIC can focus on observables for which robust theoretical expectations exist.

There are robust theoretical expectations that the gauge link needed for naively T-odd TMD physically represents an attractive final state interaction in SIDIS and a repulsive initial state interaction in the Drell Yan production of a virtual photon (γ^*) in p+p collisions. This results in the transverse SSA having an opposite sign in these two processes [29]. Detailed factorization theorems have been proven for the use of spin-independent and spin-dependent TMD for the DY process [30]. This is a primary motivation for pursuing a measurement of A_N for DY production at RHIC.

Motivated by the pion production results from RHIC, and theoretical

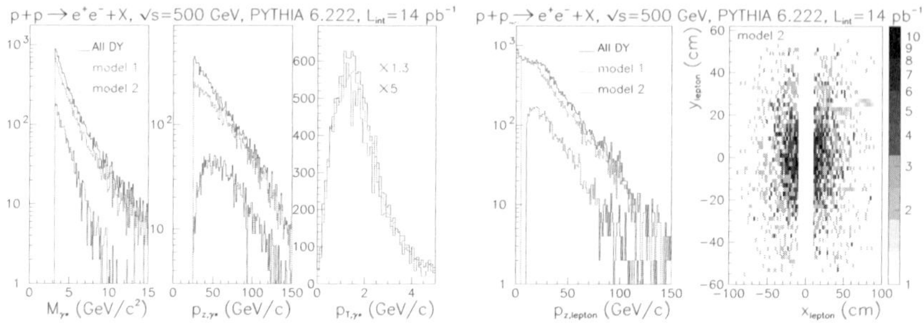

Fig. 6. Left: Distributions of virtual photon kinematic quantities. Right: Distributions of electron (positron) kinematic quantities from virtual photon production.

expectations for DY production, the expectation is that A_N will be largest for large rapidity γ^* production. It is expected that x_F for the virtual photon is the most efficient way of understanding the kinematic dependence.

Fig. 6 shows distributions of γ^* kinematic variables, and electron/positron kinematic variables, for DY production from p+p collisions at $\sqrt{s}$=500 GeV. In both sets of figures "all DY" refers to the yield absent any detector acceptance. "Model 1" refers to a future forward upgrade to STAR and/or PHENIX that aims to give good acceptance for di-electrons from DY production, and corresponds to a 4 m^2 electromagnetic and hadronic calorimeter, with a central 0.2 m side length hole for the beams, positioned 10m from the interaction point. "Model 2" represents a modular acceptance for a DY feasibility study we have proposed, to be conducted using primarily existing equipment.

The large rapidity DY feasibility experiment intends to make a statistically significant measurement of A_N for e^+e^- pairs with $M > 4$ GeV/c^2 produced with $x_F > 0.1$ (see Fig. 6) in $\sqrt{s}$=500 GeV p+p collisions. The energetic dielectrons will be detected by existing modular forward calorimetry, with additional apparatus to suppress backgrounds. A key question to address prior to embarking on major forward upgrades to STAR and/or PHENIX is whether charge sign identification is required to suppress backgrounds. We have a three phase plan to address this question. Staging of phase I will consist of forward hadronic calorimetery, and will address the impact of a third interaction region on RHIC operations. Phase II will add electromagnetic calorimetry and a preshower detector, with the goal of establishing dilepton backgrounds in the absence of magnetic analysis. Phase III will use a split dipole magnet, originally built and used by PHOBOS, with tracking as required for charge sign discrimination. Completion of a transverse spin DY measurement at RHIC is the most robust manner to test present understanding.

References

1. H.-L. Lai *et al.*, arXiv:1007.2241 [hep-ph].
2. A. D. Martin, R. G. Roberts, W. J. Stirling, R. S. Thorne, Eur. Phys. J. C**39** (2005) 155.
3. J. Adams *et al.* (STAR), Phys. Rev. Lett. **92** (2004) 171801.
4. B. I. Abelev *et al.* (STAR), Phys. Rev. Lett. **101** (2008) 222001.
5. I. Arsene *et al.* (BRAHMS), Phys. Rev. Lett. **101** (2008) 042001.
6. A. Adare *et al.* (PHENIX), Phys. Rev. Lett. **103** (2009) 012003.
7. B. I. Abelev *et al.* (STAR), Phys. Rev. Lett. **100** (2008) 232003.
8. D. E. Soper, Phys. Rev. D **15** (1977) 1141.
9. R. D. Tangerman, P. J. Mulders, Phys. Rev. D **51** (1995) 3357.
10. H. Hahn *et al.*, Nucl. Instr. Meth. A **499** (2003) 245.
11. I. Alekseev *et al.*, Nucl. Instr. Meth. A **499** (2003) 392.
12. J. Tojo *et al.*, Phys. Rev. Lett. **89** (2002) 052302.
13. H. Okada *et al.*, Phys. Lett. **B638** (2006) 450.
14. A. Gordon (for STAR), arXiv:1007.5329..
15. J. Adams *et al.* (STAR), Phys. Rev. Lett. **97** (2006) 152302.
16. C. Bourrely, J. Soffer, Eur. Phys. J. C 36 (2004) 371.
17. U. D'Alesio, F. Murgia, Phys. Rev. D **70** (2004) 074009.
18. A. Ogawa, contribution to CATHIE/TECHQM Workshop, BNL (2009).
19. M. Boglione, U. D'Alesio, F. Murgia, Phys. Rev. D **77** (2008) 051502.
20. M. Anselmino *et al.*, Phys. Rev. D **72** (2005) 094007.
21. C. Kouvaris, J. Qiu, W. Vogelsang, F. Yuan, Phys. Rev. D **74** (2006) 114013.
22. G. L. Kane, J. Pumplin,W. Repko, Phys. Rev. Lett. **41** (1978) 1689.
23. D. Sivers, Phys. Rev. D **41** (1990) 83; **43** (1991) 261.
24. A. Airapetian *et al.* (HERMES), Phys. Rev. Lett. **103** (2009) 152002.
25. S. J. Brodsky, D. S. Hwang, I. Schmidt, Phys. Lett. **B530** (2002) 99.
26. J. Qiu, G. Sterman, Phys. Rev. D **59** (1998) 014004.
27. J. Collins, Nucl. Phys. **B396** (1993) 161.
28. T. C. Rogers, P. J. Mulders, Phys. Rev. D **81** (2010) 094006.
29. J. Collins, Phys. Lett. **B536** (2002) 43.
30. G. Altarelli, R. K. Ellis, M. Greco, G. Martinelli, Nucl. Phys. **B246** (1984) 12; J. C. Collins, D. E. Soper, G. Sterman, Nucl. Phys. **B250** (1985) 199; X. Ji, J. P. Ma, F. Yuan, Phys. Lett. **B597** (2004) 299.

STUDIES OF TMDs WITH CLAS and CLAS12

P. ROSSI

Laboratori Nazionali di Frascati
Via Enrico Fermi 40, 00044 Frascati, Italy
**E-mail: rossi@lnf.infn.it*
www.lnf.infn.it

In recent years transverse momentum dependent parton distributions and fragmentation functions have been generalized to account for transverse degrees of freedom of quarks inside the nucleon. They offer novel views on QCD dynamics inside hadrons and help to increase our understanding of the intriguing transverse momentum and spin of hadrons. An overview of ongoing TMDs studies in Hall B at Jefferson Lab is given.

Keywords: Parton distributions, Transverse Momentum, Spin

1. Introduction

Modern developments in hadron physics emphasize the role of correlations between partonic transverse momentum and spin that are crucial for our understanding of the spin structure of the nucleon in terms of the quark and gluon degrees of freedom of QCD. Such spin-orbit correlations are described by transverse momentum dependent parton distribution and fragmentation functions (TMDs), which are currently under intense investigation both from the experimental and theoretical side. These functions give rise to interesting single-spin phenomena in semi-inclusive lepton- and hadron-induced processes. TMDs are probability densities for finding a polarized/unpolarized parton with a longitudinal momentum fraction x and transverse momentum $\vec{k}_T$ in a polarized/unpolarized nucleon, see Table 1, where U, L, and T stand for transitions of unpolarized, longitudinally polarized, and transversely polarized nucleons (rows), and correspondingly for quarks (columns). The diagonal elements of the table (f_1, g_1, h_1) are the partonic momentum, longitudinal spin and transverse spin distribution functions. Similar quantities arise in the hadronization process. Two fundamental mechanisms have been identified leading to single spin asym-

metries (SSAs) in hard processes: the Sivers mechanism,[1–5] which generates an asymmetry in the distribution of quarks due to orbital motion of partons and the Collins mechanism,[4,6] which generates an asymmetry during the hadronization of quarks.
Most of our information on TMDs is obtained from semi-inclusive deep inelastic scattering (SIDIS) experiments, $lN \rightarrow e'hX$ (with l = lepton, h = hadron). Here, one of the main players in the game is CLAS at JLab, where a broad experimental program is underway to access all TMDs listed in Table 1. Also new measurements of SSAs for both pions and kaons electroproduction have been approved by the JLab PAC for the 12 GeV beam using the CLAS12 detector.

Table 1. Leading-twist transverse momentum-dependent distribution functions.

N/q	U	L	T
U	$\mathbf{f_1}$		$h_1^\perp$
L		$\mathbf{g_1}$	$h_{1L}^\perp$
T	$f_{1T}^\perp$	g_{1T}	$\mathbf{h_1}$ $h_{1T}^\perp$

2. TMDs studies at JLab with CLAS

2.1. *Longitudinally Polarized Target*

For a longitudinally polarized target the only azimuthal asymmetry arising at leading order is the $\sin 2\phi$ moment,[7–9] involving the Collins fragmentation function $H_1^\perp$ and the Ralston-Soper-Mulders-Tangerman (RSMT) distribution function $h_{1L}^\perp$,[7,10] describing the transverse polarization of quarks in a longitudinally polarized proton.[7–9,11]
Measurements of SSAs in SIDIS kinematics have been done at Jefferson Lab using a 5.7 GeV electron beam and the CEBAF Large Acceptance Spectrometer (CLAS)[12] (EG1 experiment). Scattering of longitudinally polarized electrons off a polarized NH_3 and ND_3 targets was studied over a wide range of kinematics. The average beam polarization was $\sim 73\%$ and the average target polarization for NH_3 (ND_3) was $\sim 70\%$ ($\sim 30\%$). The scattered electrons and pions were detected in CLAS. In the DIS kinematics ($Q^2 > 1$ GeV^2, $W^2 > 4$ GeV^2), ~ 8 millions π^+ have been detected in the $y < 0.85$, $z > 0.4$ range.

The $\sin 2\phi$ moment $A_{UL}^{\sin 2\phi}$ as a function of x is plotted in Fig. 1 (full triangles).[13] Calculations[14,15] using $h_{1L}^{\perp}$ from the chiral quark soliton model[16] and the Collins function[17] extracted from HERMES[18] and Belle[19] data, are plotted as filled bands. In the plot $A_{UL}^{\sin 2\phi}$ measurements from HERMES[18] (open squares) are also plotted. The kinematic dependence of the SSA for π^+ from the CLAS data is roughly consistent with these predictions. The interpretation of the π^- data, which tend to have SSAs with a sign opposite to expectations, may require accounting for additional contributions. By the way, current statistical errors for π^-, and in particular for π^0, are large and do not allow strong conclusions from the measured SSAs. Data at 6 GeV with ten times higher statistics have been collected in 2009 (experiment E05-133). They will allow statistically significant measurement of the $\sin 2\phi$ moment. The projected sensitivity of results for this experiment are shown as full triangles in the upper part of Fig. 1.

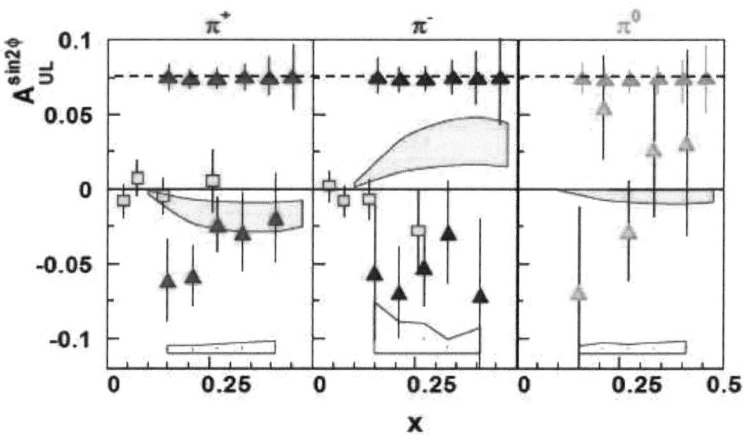

Fig. 1. The measured x-dependence of the longitudinal target SSA $A_{UL}^{\sin 2\phi}$ (full triangles). The squares show the existing $A_{UL}^{\sin 2\phi}$ measurement from HERMES.[18] The lower band shows the systematic uncertainty. The upper band shows the existing theory predictions with uncertainties due to the Collins function.[14,15] The full triangles in the upper part of the plots show the projected sensitivity of results for the experiment E05-133.

The data with longitudinally polarized beam and target provides important information input for the flavor decomposition through the measurement of double spin asymmetries, sensitive to the ratio of helicity and momentum distributions. The dependence of the double-spin asymmetry A_1 as a function of P_T and integrated over all x $(0.12 \div 0.48)$ for $Q^2 > 1$ GeV^2, $W^2 > 4$ GeV^2 and $y < 0.85$, is shown in Fig. 2.

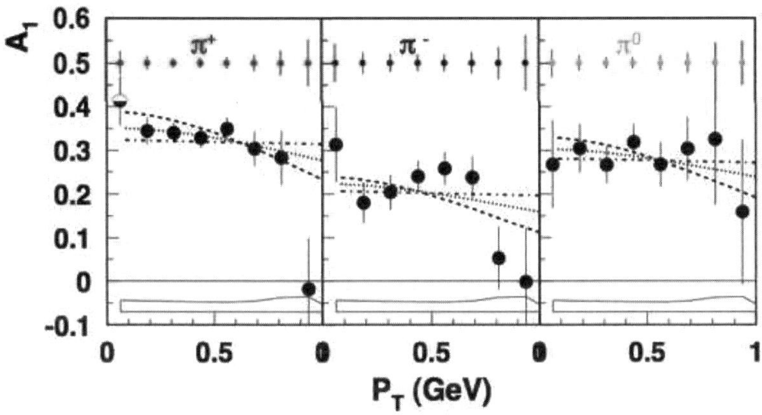

Fig. 2. The double spin asymmetry A_1 as a function of transverse momentum P_T, integrated over all kinematical variables. The open band corresponds to systematic uncertainties. The dashed, dotted and dash-dotted curves are calculations for different values of the ratio of the transverse momentum widths for g_1 and f_1 (0.40, 0.68, 1.0) for a fixed width for f_1 (0.25 GeV^2).[20]

Although these plots are consistent with flat distributions, $A_1(P_T)$ may decrease with P_T at moderately small P_T for π^+. The slope for π^- could be positive for moderate P_T (ignoring the first data point). A possible interpretation of the P_T-dependence of the double-spin asymmetry may involve different widths of the transverse momentum distributions of quarks with different flavor and polarizations[20] resulting from a different orbital structure of quarks polarized in the direction of the proton spin and opposite to it.[21,22] In Fig. 2 the measured A_1 is compared with calculations of the Torino group,[20] which uses different values of the ratio of widths in k_T for partonic helicity, g_1, and momentum, f_1, distributions, assuming Gaussian k_T distributions with no flavor dependence.

2.2. *Transversely Polarized Target*

In the last few years, first results on transverse SSAs have become available.[23,24] HERMES measurements for the first time directly indicated significant azimuthal moments generated both by Collins and Sivers effects. An experiment using a trasversely polarized target in a frozen-spin state, such as the HD-Ice target has been approved at JLab (CLAS experiment E08-015) and is planned to run late 2011. The advantages of an HD-Ice target compared to the solid state dynamically polarized NH_3 and ND_3 targets are manifold: it requires a small holding field and this potentially allows the target to be located even in the center of the detector, thus dramatically increasing the acceptance; in addition the HD-Ice target has almost no diluition; the polarization are expected to be very high: 75% for hydrogen and 40% for deuterium.

2.3. *Unpolarized Target*

Beam-spin asymmetries measurement in single neutral semi-inclusive pion electroproduction has been carried out using a 5.77 GeV electron beam and the CLAS detector. For this measurement scattering of longitudinally polarized electrons off a liquid-hydrogen target was studied over a wide kinematic region. The beam-spin asymmetries in single-pion production off the unpolarized target are higher-twist by their nature.[25,26] The higher twist observables are important for understanding the long-range quark-gluon dynamics. Different contributions to the beam SSAs, discussed so far, provide information on different leading and sub-leading parton distribution and fragmentation functions, related to both Collins and Sivers production mechanisms. At fixed and moderate values of the four momentum transfer Q^2 and at large values of x_B and z (where x_B is the fraction of the momentum of the nucleon carried by the struck quark and z is the fraction of the virtual photon energy carried by the detected π^0), the contribution of multiparton correlations or higher twist effects increases, eventually leading to a breakdown of the partonic description. Kinematical dependences of observables, will thus provide tests of the applicability of partonic descriptions.

In CLAS the kinematic dependences of the π^0 beam-spin asymmetry on x_B, z and P_T (where P_T is the transverse momentum of the π^0) have been measured. The asymmetry shows no significant x_B-dependence within the measured range. The $A_{UL}^{\sin\phi}$ moment as a function of P_T is shown in Fig. 3. It increases at low P_T and reaches a plateau at values of about 0.3 GeV, and there is an indication that its decrease at large P_T, expected from perturbative QCD, starts already at $P_T \sim 0.6$ GeV. In Fig. 3 the preliminary CLAS results are compared with HERMES data,[27] where, for the comparison a kinematic factor has been applied to take into account the different kinematical ranges covered by the two experiments. The observed large beam spin asymmetry for π^0 indicates that the major contribution in pion SSAs may be due to the Sivers mechanism. The obtained results provide a very significant improvement in statistical errors compared with published HERMES data, providing important input for studies of higher twist effects.

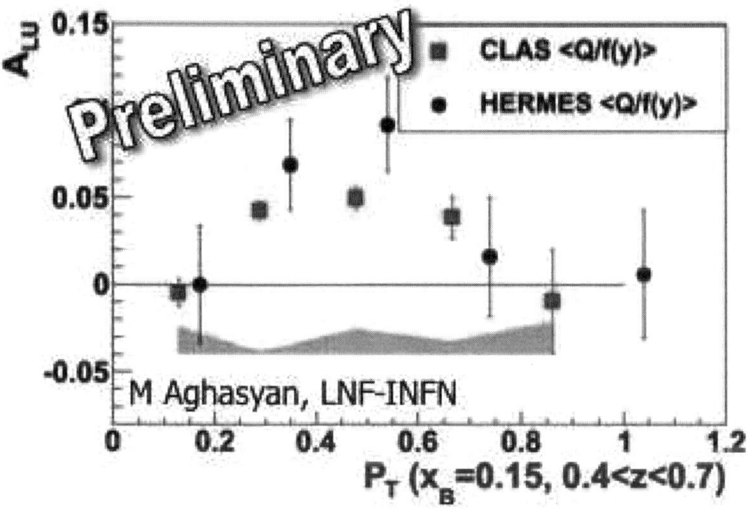

Fig. 3. The measured P_T-dependence of the $A_{UL}^{\sin\phi}$ for x_B=0.15 integrated over $0.4 < z < 0.7$. Error bars present the statistical and the bands the systematic uncertainties. The CLAS data are compared with the HERMES ones.[27]

3. Future TMDs studies at JLab with CLAS12

The study of the transverse structure of the nucleon represents a key challenge of the present research in hadron physics and is one of the main scientific goals for the 12 GeV upgrade of the Jefferson Laboratory,[28] which is expected to start to be operational in 2015. The complexity of the task of mapping out TMDs requires the interplay of different and complementary measurements of various observables over a wide kinematic range involving experiments with polarized beams and targets.

In Hall B the extended x and Q^2 range of the accessible kinematic plane, together with the expected luminosity of 10^{35} cm^{-2} s^{-1}, and a new large angle detector, CLAS12, will allow for the measurement of single-

and double-spin asymmetries with at least an order of magnitude better accuracy than the present, or soon to come, experimental data.

One of the key points of the new detector should be the possibility to identify semi-inclusive kaons with high rejection power from proton and pion backgrounds. In fact, kaon measurements are of fundamental importance in order to achieve flavor separation in the extraction of the distribution functions from the measured asymmetries. For this, the construction of a RICH detector, allowing in particular kaon to pion separation with a rejection factor better than 1:100 in the whole range of kaon momenta above 1 GeV/c, has been proposed. The Sivers asymmetry for kaons as expected with CLAS12, compared with existing data from HERMES[29] is shown in Fig. 4.

The CLAS12 experimental program on the TMDs includes measurements of all the eight leading twist parton distribution functions with the proper choice of beam and/or target polarizations. A full list of the relevant observables and the expected accuracy can be found in the several proposals already approved by the Jefferson Laboratory Program Advisory Committee.[30]

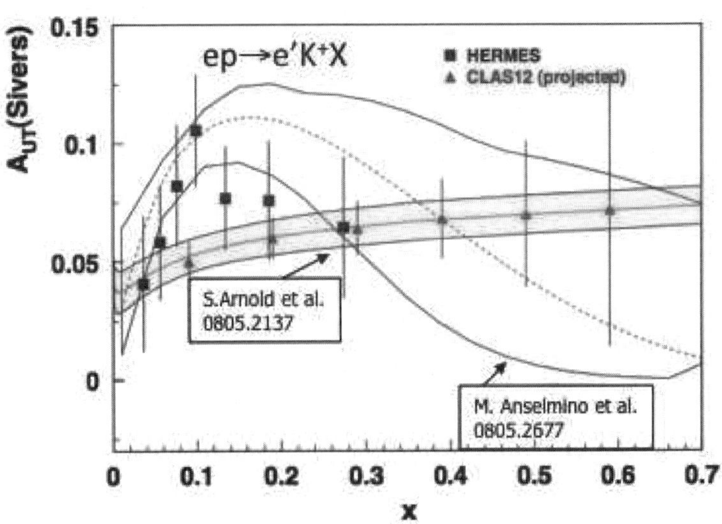

Fig. 4. Projected results for the kaon Sivers asymmetry for CLAS12 as a function of x compared with HERMES data[29] and theoretical predictions.

4. Conclusions and Outlook

Transverse momentum dependent distribution and fragmentation functions have been found to be one of the main tools to access spin-orbit correlations of the partons in the nucleon. TMDs can be accessed unambiguously in SIDIS with polarized/unpolarized leptons and nucleons by observing the azimuthal distributions of produced hadrons. CLAS in Hall B is playing a major role in these studies, which are also one of the main driving forces behind the upgrade of Jefferson Lab.

References

1. D. W. Sivers, *Phys. Rev. D* **43**, 261 (1991).
2. M. Anselmino *and* F. Murgia, *Phys. Lett. B* **442**, 470 (1998), hep-ph/9808426.
3. S.J. Brodsky, D.S. Hwang, *and* I. Schmidt, *Phys. Lett. B* **530**, 99 (2002), hep-ph/0201296.
4. J.C. Collins, *Phys. Lett. B* **536**, 43 (2002), hep-ph/0204004.
5. X.d. Ji *and* F. Yuan, *Phys. Lett. B* **543**, 66 (2002), hep-ph/0206057.
6. P.J. Mulders *and* R.D. Tangerman, *Nucl. Phys. B* **461**, 197 (1996), hep-ph/9510301.
7. P.J. Mulders *and* R.D. Tangerman, *Nucl. Phys. B* **461**, 197 (1996).
8. A. Kotzinian, *Nucl. Phys. B* **441**, 234 (1995).
9. A. Kotzinian *and* P.J. Mulders, *Phys. Rev. D* **54**, 1229 (1996).
10. J.P. Ralston *and* D.E. Soper, *Nucl. Phys. B* **152**, 109 (1979).
11. E. Di Salvo, *Int. J. Mod. Phys. A* **22**, 2145 (2007), hep-ph/0603005.
12. B. Mecking *et al.*, *Nucl. Instr. Meth. A* **503**, 513 (2003).
13. H. Avakian *et al.*, arXiv:1003.4549
14. H. Avakian *et al.*, *Phys. Rev. D* **77**, 014023 (2008), arXiv:0709.3253[hep-ph]
15. A. V. Efremov, K. Goeke, and P. Schweitzer, *Phys. Rev. D* **67**, 114014 (2003), hep-ph/0208124.
16. P. Schweitzer *et al.*, *Phys. Rev. D* **64**, 034013 (2001), arXiv:hep-ph/0101300.
17. A. V. Efremov, K. Goeke, *and* P. Schweitzer, *Phys. Rev. D* **73**, 094025 (2006), arXiv:hep-ph/0603054.
18. A. Airapetian *et al.*, (HERMES), *Phys. Rev. Lett. 94*, 012002 (2005), hep-ex/0408013.
19. K. Abe *et al.*, (Belle), *Phys. Rev. Lett. 96*, 232002 (2006), hep-ex/0507063.
20. M. Anselmino, A. Efremov, A. Kotzinian *and* B. Parsamyan, *Phys.Rev. D* **74**, 074015 (2006).
21. S.J. Brodsky *and* S.D. Drell, *Phys. Rev. D* **22**, 2236 (1980).
22. S.J. Brodsky, M. Burkardt *and* I. Schmidt, *Nucl. Phys. B* **441**, 197 (1995).
23. A. Airapetian *et al.*, *Phys. Rev. Lett.* **94**, 012002 (2005).
24. V.Y. Alexakhin *et al*, *Phys. Rev. Lett.* **94**, 202002 (2005).
25. A. Afanasev *and* C.E. Carlson, (2003), hep-ph/0308163.
26. F. Yuan, *Phys. Lett. B* **589**, 28 (2004), hep-ph/0310279.
27. A. Airapetian *et al.*, (HERMES) (2006), hep-ex/0612059.
28. *http : //www.jlab.org/12GeV/*

29. A. Airapetian *et al.* (HERMES), *Phys. Rev. Lett.* **103**, 152002 (2009).
30. H. Avakian *et al.*, CLAS proposal PR12-09-009; K. Hafidi *et al.*, CLAS proposal PR12-09-007, H. Avakian *et al.*, CLAS proposal PR12-09-008.

TRANSVERSE MOMENTUM DEPENDENT QUARK DENSITIES FROM LATTICE QCD

B. U. MUSCH

Theory Center, Jefferson Lab, 12000 Jefferson Avenue
Newport News, VA 23606, USA
E-mail: bmusch@jlab.org

PH. HÄGLER

Theoretische Physik T39, TU München, James-Franck-Straße 1,
85747 Garching, Germany
E-mail: phaegler@ph.tum.de

J. W. NEGELE

Center for Theoretical Physics, Massachusetts Institute of Technology,
77 Massachusetts Avenue, Bldg. 6-315, Cambridge, MA 02139, USA

A. SCHÄFER

Institut für Theoretische Physik, Universität Regensburg,
93040 Regensburg, Germany

We study transverse momentum dependent parton distribution functions (TMDs) with non-local operators in lattice QCD, using MILC/LHPC lattices. Results obtained with a simplified operator geometry show visible dipole deformations of spin-dependent quark momentum densities. We discuss the basic concepts of the method, including renormalization of the gauge link, and an extension to a more elaborate operator geometry that would allow us to analyze process-dependent TMDs such as the Sivers-function.

Keywords: transverse momentum; parton distribution functions; lattice; QCD

1. Introduction

Generalized parton distribution functions (GPDs) and transverse momentum dependent parton distribution functions (TMDs) provide us with a picture of the internal quark distributions in a nucleon at the instant of an interaction, see illustration Fig. 1 a). GPDs and TMDs have their natural interpretation at large nucleon momentum $\boldsymbol{P} = (0, 0, \boldsymbol{P}_z)$. The quark momentum k in terms of light cone coordinates $k^{\pm} \equiv (k^0 \pm k^3)/\sqrt{2}$, $\boldsymbol{k}_\perp = (\boldsymbol{k}_x, \boldsymbol{k}_y)$ scales like $k^+ : \boldsymbol{k}_\perp : k^- \sim P^+ : 1 : (P^+)^{-1}$ with the large momentum component P^+ of the nucleon. TMDs resolve the dependence on $x \equiv k^+/P^+$ and transverse momentum $\boldsymbol{k}_\perp$, but not on the suppressed component k^-. In spin-polarized channels at leading twist, TMDs encode dipole- or quadrupole-shaped deformations of the nucleon in the $\boldsymbol{k}_\perp$-plane. We have studied such deformations in first explorative lattice QCD calculations [1–3], see Fig. 1 and our discussion below. These studies have been motivated by a history of successful lattice computations of x-moments of GPDs, providing images of the nucleon in the impact parameter, $\boldsymbol{b}_\perp$-, plane, see [4] for a review. A remaining theoretical problem concerns the precise form of the correlator defining TMDs in the continuum, see [5,6] and references therein. In its basic form, it is given by [7]

$$
\begin{aligned}
&\Phi_q^{[\Gamma]}(x, \boldsymbol{k}_\perp; P, S; \mathcal{C}) \equiv \\
&\qquad \int dk^- \int \frac{d^4 l}{(2\pi)^4}\, e^{-ik\cdot l}\, \underbrace{\frac{1}{2} \langle P, S|\ \bar{q}(l)\,\Gamma\, \mathcal{U}[\mathcal{C}_l]\ q(0)\ |P, S\rangle}_{\widetilde{\Phi}_q^{[\Gamma]}(l, P, S; \mathcal{C})}\Big|_{k^+ = xP^+} \\
&\quad = \int \frac{dl^-}{2\pi}\, e^{-il^- k^+} \int \frac{d^2 \boldsymbol{l}_\perp}{(2\pi)^2}\, e^{i\boldsymbol{l}_\perp \cdot \boldsymbol{k}_\perp}\, \widetilde{\Phi}_q^{[\Gamma]}(l, P, S; \mathcal{C})\Big|_{l^+ = 0} \\
&\quad = \frac{1}{P^+} \underbrace{\int \frac{d(l\cdot P)}{2\pi}\, e^{-i(l\cdot P)x}}_{\not\int x}\ \underbrace{\int \frac{d^2 \boldsymbol{l}_\perp}{(2\pi)^2}\, e^{i\boldsymbol{l}_\perp \cdot \boldsymbol{k}_\perp}}_{\not\int l}\, \widetilde{\Phi}_q^{[\Gamma]}(l, P, S; \mathcal{C})\Big|_{l^+ = 0} \qquad (1)
\end{aligned}
$$

where Γ is a Dirac matrix. The Wilson line $\mathcal{U}[\mathcal{C}_l]$ running along a continuous path $\mathcal{C}_l$ from l to 0 ensures gauge invariance of the expression. For the SIDIS and Drell-Yan scattering process, the Wilson line extends to infinity along a direction v, see Fig. 4 a) below, such that the cross section factorizes into hard, perturbative parts and soft contributions. Based on its symmetry transformation properties, the above correlator can be parametrized in terms of TMDs [8–10], for example

$$
2\,\rho_{TL}^{(q)} \equiv \Phi_q^{[\gamma^+ + \lambda\gamma^+\gamma^5]} = f_{1,q} + \lambda \frac{\boldsymbol{k}_\perp \cdot \boldsymbol{S}_\perp}{m_N} g_{1T,q} + \left[\frac{\boldsymbol{S}_j \boldsymbol{\epsilon}_{ji} \boldsymbol{k}_i}{m_N}\, f_{1T,q}^\perp \right]_{\text{odd}}, \qquad (2)
$$

$$
2\,\rho_{LT}^{(q)} \equiv \Phi_q^{[\gamma^+ - s^j i\sigma^{+j}\gamma^5]} = f_{1,q} + \Lambda \frac{\boldsymbol{k}_\perp \cdot \boldsymbol{s}_\perp}{m_N} h_{1L,q}^\perp + \left[\frac{\boldsymbol{s}_j \boldsymbol{\epsilon}_{ji} \boldsymbol{k}_i}{m_N}\, h_{1,q}^\perp \right]_{\text{odd}}. \qquad (3)
$$

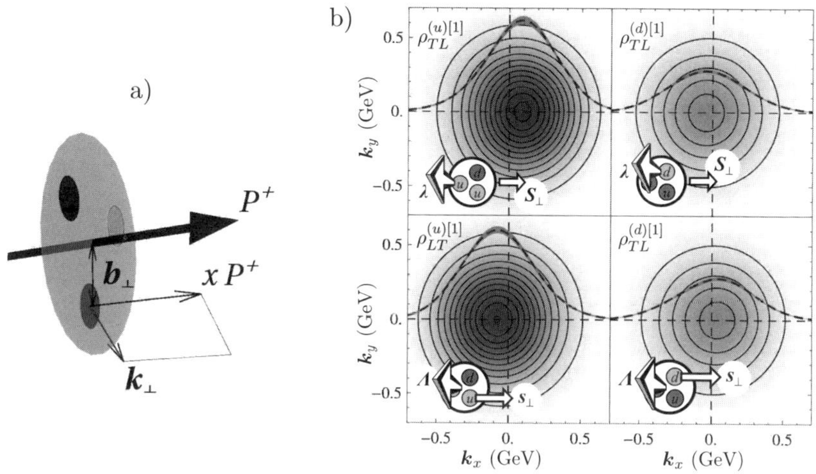

Fig. 1. a) Illustration of quark degrees of freedom in the nucleon at large momentum. b) Dipole-deformed x-integrated densities obtained with straight gauge links at a pion mass $m_\pi \approx 500$ MeV. The insets display the spin polarization s, λ of the quarks and S, Λ of the nucleon.

Here λ and $\boldsymbol{s}_\perp$ are longitudinal and transverse quark polarization, Λ and $\boldsymbol{S}_\perp$ longitudinal and transverse nucleon polarization, and the leading-twist TMDs $f_{1,q}$, $g_{1T,q}$, $f_{1T,q}^\perp$, $h_{1L,q}^\perp$ and $h_{1,q}^\perp$ are real-valued functions of x and $\boldsymbol{k}_\perp^2$. The "naively time-reversal odd" TMDs $f_{1T,q}^\perp$ and $h_{1,q}^\perp$ switch their sign when comparing the SIDIS- with the Drell-Yan process, because the direction v of the Wilson line changes from future- to past-pointing [11].

2. Straight link TMDs from the lattice

In light of the uncertainties about the precise form of the continuum correlator, and to develop our methods, our first lattice studies employ a simple operator geometry that does not relate to a specific scattering process: We connect the quark fields with a direct, straight Wilson line. For the resulting "process-independent" TMDs, the T-odd functions $f_{1T,q}^\perp$ and $h_{1,q}^\perp$ vanish exactly.

In our approach, we calculate matrix elements $\langle P, S|\, O \,|P, S\rangle$ from ratios of three- and two-point functions using the same techniques as GPD calculations by the LHP collaboration in Ref. [12]. We also use the same sequential propagators and quark propagators, calculated by LHPC with domain-wall valence fermions on top of asqtad-improved staggered MILC gauge configurations [13–15] with 2+1 quark flavors at a lattice spacing

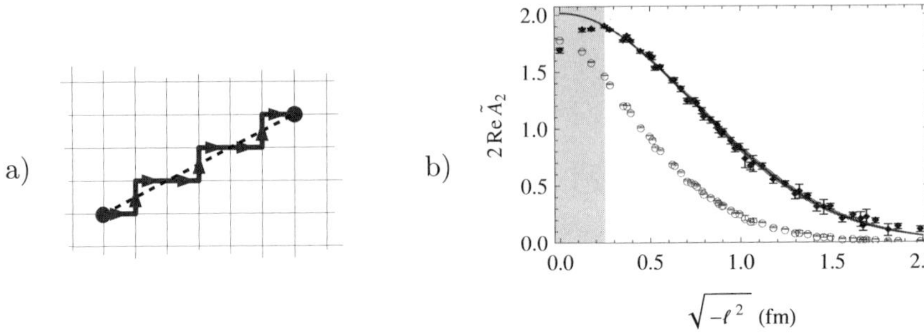

Fig. 2. a) Representation of a straight Wilson line (dashed line) as a step-like product of link variables. b) Amplitude $\widetilde{A}_2(l^2, 0)$ for up quarks at a pion mass $m_\pi \approx 500\,\mathrm{MeV}$, using straight gauge links.

$a \approx 0.12\,\mathrm{fm}$. The difference with respect to GPD calculations is that we directly insert the non-local operator $O \equiv \bar{q}(l)\Gamma\mathcal{U}[\mathcal{C}_l]q(0)$ in our three-point function. The Wilson line $\mathcal{U}[\mathcal{C}_l]$ is approximated as a step-like product of HYP-smeared link-variables as illustrated in Fig. 2 a). See also Ref. [2,3].

The connection between the matrix elements $\tilde{\Phi}^{[\Gamma]}$ and TMDs is established through a parametrization in terms of Lorentz-invariant amplitudes $\widetilde{A}_i(l^2, l{\cdot}P)$. For straight Wilson lines, we obtain in analogy to the parametrization in terms of amplitudes $A_i(k^2, k{\cdot}P)$ in Ref. [8] (here our sign conventions follow Ref. [10] with the substitution rule $k \to i m_N^2 l$):

$$\begin{aligned}
\tilde{\Phi}^{[\gamma^\mu]} &= 2\,P^\mu\,\widetilde{A}_2 + 2i\,m_N{}^2\,l^\mu\,\widetilde{A}_3\,,\\
\tilde{\Phi}^{[\gamma^\mu\gamma^5]} &= -2\,m_N\,S^\mu\,\widetilde{A}_6 - 2i\,m_N\,P^\mu(l\cdot S)\,\widetilde{A}_7 + 2\,m_N{}^3\,l^\mu(l\cdot S)\,\widetilde{A}_8\,,\\
\tilde{\Phi}^{[i\sigma^{\mu\nu}\gamma^5]} &= 2\,P^{[\mu}S^{\nu]}\,\widetilde{A}_9 + 2i\,m_N^2\,l^{[\mu}S^{\nu]}\,\widetilde{A}_{10} + 2\,m_N^2\,l^{[\mu}P^{\nu]}(l\cdot S)\widetilde{A}_{11}\,.
\end{aligned}$$

The TMDs are then obtained by

$$\begin{aligned}
f_1(x,\boldsymbol{k}_\perp^2) &= 2 \not{\!\!\int}\!\not{\!\!\int}\!\!\mathcal{M}\; \widetilde{A}_2(l^2, l{\cdot}P)\,, \quad h_1(x,\boldsymbol{k}_\perp^2) = -2 \not{\!\!\int}\!\not{\!\!\int}\!\!\mathcal{M}\; \widetilde{A}_{9m}(l^2, l{\cdot}P)\,,\\
h_{1L}^\perp(x,\boldsymbol{k}_\perp^2) &= 4m_N^2\partial_{\boldsymbol{k}_\perp^2}\left(\not{\!\!\int}\!\not{\!\!\int}\!\!\mathcal{M}\; \widetilde{A}_{10}(l^2, l{\cdot}P) + \partial_x \not{\!\!\int}\!\not{\!\!\int}\!\!\mathcal{M}\; \widetilde{A}_{11}(l^2, l{\cdot}P)\right),\\
g_{1T}(x,\boldsymbol{k}_\perp^2) &= 4m_N^2\partial_{\boldsymbol{k}_\perp^2} \not{\!\!\int}\!\not{\!\!\int}\!\!\mathcal{M}\; \widetilde{A}_7(l^2, l{\cdot}P)\,,
\end{aligned}$$

with $\widetilde{A}_{9m} \equiv \widetilde{A}_9 - \frac{1}{2}m_N^2 l^2 \widetilde{A}_{11}$. In the equations above, $\not{\!\!\int}$ only acts on $l{\cdot}P$, while $\not{\!\!\int}\!\!\mathcal{M}$ only acts on l^2. Thus $x \leftrightarrow l{\cdot}P$ and $\boldsymbol{k}_\perp^2 \leftrightarrow l^2$ are pairs of conjugate variables. Our Euclidean lattice approach is restricted to the determination of amplitudes $\widetilde{A}_i$ for $l^0 = -il_4 = 0$, i.e., to the region $l^2 < 0$, $|l{\cdot}P| \leq \sqrt{-l^2}|\boldsymbol{P}|$, where $\boldsymbol{P}$ is the selected three-momentum

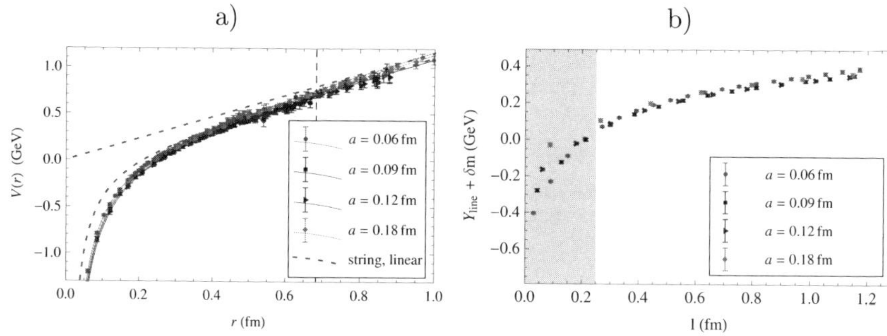

Fig. 3. a) Static quark potential from MILC lattices at several lattice spacings a, matched to the string potential at $r \approx 0.7\,\mathrm{fm}$. b) Test of the renormalization procedure with straight Wilson lines on a gauge fixed ensemble.

of the nucleon on the lattice. The limited range in $|l{\cdot}P|$ prohibits us from a direct evaluation of $\not\mathcal{X}$. However, first studies of x- and $\boldsymbol{k}_\perp$- correlations are possible [3,16]. Moreover, x-integrated TMDs and densities are directly accessible: Integrating Eq. (1) with respect to x removes $\not\mathcal{X}$ and sets $l{\cdot}P$ to zero. Correspondingly, the x-integral of, e.g., f_1 becomes $\int_{-1}^{1} dx\, f_1(x, \boldsymbol{k}_\perp^2) \equiv f_1^{[1]}(\boldsymbol{k}_\perp^2) = 2\not{M}\widetilde{A}_2(l^2, 0)$. In Fig. 2 b), open symbols correspond to unrenormalized lattice data for $\widetilde{A}_2(l^2, 0)$.

To obtain results independent of our lattice spacing a and our lattice action, we must renormalize our data. The Wilson line $\mathcal{U}[\mathcal{C}_l]$ introduces a length dependent renormalization factor $\exp(-\delta m\,\sqrt{-l^2})$ [17–19]. To fix δm, we follow the strategy of Refs. [20,21], and match the renormalized static quark potential $V^{\mathrm{ren}}(r) = V(r) + 2\delta m$ to the string potential $V_{\mathrm{string}} = \sigma r - \pi/(12r)$ [22] at a matching point $r = 1.5 r_0 \approx 0.7\,\mathrm{fm}$. In Fig. 3 a), we test the method for several lattice spacings a on four MILC lattices with similar pion masses $m_\pi \approx 500\,\mathrm{MeV}$. The renormalized lattice data agree very well with each other and are approximated well by the string potential (red dashed curve) near the matching point, indicated by a vertical dashed line. The procedure implements a gauge-invariant renormalization condition that we can formulate as the demand that the static quark potential asymptotically approach a straight line σr through the origin (shown as a red dashed line). In connection with TMDs, we lack at present an interpretation of this renormalization condition as a physical renormalization or factorization scale. In Figure 3 b), we check the applicability of the approach to Wilson lines by plotting $Y_{\mathrm{line}}^{\mathrm{ren}}(l) = -\ln(U_{l+a/2}/U_{l-a/2})/a + \delta m$, where U_l is the expectation value of the color trace of a straight Wilson line of length

l evaluated on a Landau gauge fixed ensemble, and where the length dependent renormalization has been carried out with the values δm obtained from the static quark potential. Only at short lengths, $l \lesssim 0.25\,\mathrm{fm}$, we find significant differences between lattice data from different lattice spacings, a sign of lattice cutoff effects. For our TMD calculations discussed below we exclude data obtained in this region from our fits. For $l \gtrsim 0.25\,\mathrm{fm}$, we assume that renormalization of the lattice operator can be carried out as in the continuum, $O^{\mathrm{ren}} = Z_{\Psi,z}^{-1} \exp(-\delta m \sqrt{-l^2})\, O$, where the renormalization constants $Z_{\Psi,z}^{-1}$ and δm are independent of the Dirac structure Γ [18].

Figure 2 b) shows the renormalized lattice data for $\widetilde{A}_2(l^2, 0)$ as solid data points. The curve and statistical error band correspond to a Gaussian fit to this data in the range $\sqrt{-l^2} \geq 0.25\,\mathrm{fm}$. Note that the renormalization constant $Z_{\Psi,z}^{-1}$ has been fixed (in the isovector, $u{-}d$-channel) such that the x-$\boldsymbol{k}_\perp$-integrated Gaussian density of unpolarized quarks yields the correct total number of valence quarks, $\int d^2\boldsymbol{k}_\perp f_{1,u-d}^{[1]} = 1$. Similar fits for the other amplitudes $\widetilde{A}_{10}$, and $\widetilde{A}_7$ enable us to calculate the "worm-gear" functions $g_{1T}^{[1]}$, $h_{1L}^{\perp[1]}$, and correspondingly, the dipole deformed x-integrated densities $\rho_{TL}^{(q)[1]}$ and $\rho_{LT}^{(q)[1]}$ defined in Eqns. (2), (3) and shown in Fig. 1 b). While the widths of our distributions depend strongly on our renormalization condition for δm, average transverse quark momenta can be expressed in terms of ratios of the Gaussian amplitudes at $l^2{=}0$:

$$\langle \boldsymbol{k}_x \rangle_{TL} \equiv \left. \frac{\int d\boldsymbol{k}_\perp\, \boldsymbol{k}_x\, \rho_{TL}^{[1]}}{\int d\boldsymbol{k}_\perp \rho_{TL}^{[1]}} \right|_{\substack{\lambda=1,\\ \boldsymbol{S}_\perp=(1,0)}} = -m_N \frac{\widetilde{A}_7(0,0)}{\widetilde{A}_2(0,0)} = \begin{cases} 67(5)\,\mathrm{MeV}\ (\mathrm{up}) \\ -30(5)\,\mathrm{MeV}\ (\mathrm{down}) \end{cases}$$

$$\langle \boldsymbol{k}_x \rangle_{LT} \equiv \left. \frac{\int d\boldsymbol{k}_\perp\, \boldsymbol{k}_x\, \rho_{LT}^{[1]}}{\int d\boldsymbol{k}_\perp \rho_{LT}^{[1]}} \right|_{\substack{\Lambda=1,\\ \boldsymbol{s}_\perp=(1,0)}} = -m_N \frac{\widetilde{A}_{10}(0,0)}{\widetilde{A}_2(0,0)} = \begin{cases} -60(5)\,\mathrm{MeV}\ (\mathrm{up}) \\ 16(5)\,\mathrm{MeV}\ (\mathrm{down}) \end{cases}$$

(errors statistical only). In these ratios, renormalization factors largely cancel. Reference [23] reveals a remarkable similarity of our results with a light-cone constituent quark model [24], despite the unphysically large quark masses employed in our lattice calculation: They find $\langle \boldsymbol{k}_x \rangle_{TL} = -\langle \boldsymbol{k}_x \rangle_{LT} = 55.8\,\mathrm{MeV}$ for up-, and $\langle \boldsymbol{k}_x \rangle_{TL} = -\langle \boldsymbol{k}_x \rangle_{LT} = -27.9\,\mathrm{MeV}$ for down-quarks.

3. Staple shaped gauge links

TMDs for SIDIS or the Drell-Yan process, in particular T-odd distributions like the Sivers function $f_{1T}^{\perp}$ in Eq. (2), require gauge links out to infinity in a direction v, see Fig. 4 a). To avoid rapidity divergences, the direction v can be chosen slightly off the light cone, see, e.g., [25,26]. The amplitudes $\widetilde{A}_i$

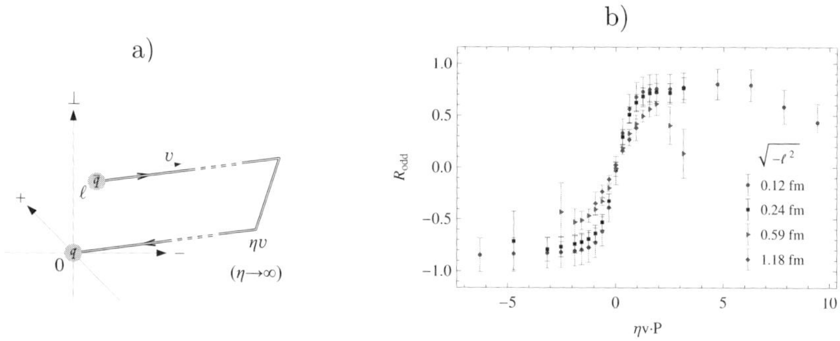

Fig. 4. a) Staple shaped Wilson line. b) Test calculation of a T-odd ratio of amplitudes using staple shaped links at $m_\pi \approx 800\,\text{MeV}$.

and the TMDs then also depend on Lorentz-invariant structures involving v, in particular on the parameter $\zeta \equiv (2v{\cdot}P)^2/v^2$, with respect to which the TMDs follow an evolution equation [25]. Our lattice operators are restricted to spacelike staple directions, $v^0 = 0$, i.e., our amplitudes are limited to $0 \leq -\zeta \leq |2\boldsymbol{P}|^2$. We can only realize staples of finite extent ηv, but we can look for a plateau at large η. Remaining challenges regarding the definition of TMDs mainly relate to the regularization of divergences attributed to the Wilson line and possible soft-factor modifications of Eq. (1) needed to ensure QCD factorization. First lattice studies are planned for ratios of amplitudes, in which renormalization factors and potential soft factors cancel. Figure 4 b) shows results from a test calculation [2] of a ratio of T-odd over T-even amplitudes $R_\text{odd} \equiv (\widetilde{A}_{12} - (m_N/|\boldsymbol{P}|)^2 \widetilde{B}_8)/\widetilde{A}_2$, evaluated at $l{\cdot}P = 0$, $|\boldsymbol{P}| \approx 0.5\,\text{GeV}$, for selected values of l^2 and a range of staple extents η. Note that $\widetilde{A}_{12}$ would correspond to the Sivers function $f_{1T}^{\perp}$ for lightlike v, while the amplitudes $\widetilde{B}_i$ parametrize explicit v-dependence [10]. As expected, the result is an odd function of $\eta v{\cdot}P$. Moreover, we see the onset of a plateau at $|\eta v{\cdot}P| \gtrsim 2$. This is a promising indication that lattice estimates could be feasible for, e.g., the transverse momentum shift $\langle \boldsymbol{k}_y \rangle_{TU}$ due to the Sivers function.

4. Conclusions

We have performed first lattice studies of TMDs using non-local operators with a simplified, straight gauge link. Resulting average momentum shifts $\langle \boldsymbol{k}_x \rangle_{TL}$ and $\langle \boldsymbol{k}_x \rangle_{LT}$ corroborate model results. Staple-shaped gauge links can potentially address TMDs specific to SIDIS or the Drell-Yan process, including T-odd functions responsible for single-spin asymmetries.

Acknowledgments

We are very grateful to the LHP and MILC collaborations, for providing us gauge configurations and propagators. We thank Vladimir Braun, Meinulf Göckeler, Gunnar Bali, Markus Diehl, Alexei Bazavov, and Dru Renner for very helpful discussions. Our software uses the `Chroma`-library [27], and we use USQCD computing resources at Jefferson Lab. We acknowledge support by the Emmy-Noether program and the cluster of excellence "Origin and Structure of the Universe" of the DFG (Ph.H. and B.M.), SFB/TRR-55 (A.S.) and the US Department of Energy grant DE-FG02-94ER40818 (J.N.). Authored by Jefferson Science Associates, LLC under U.S. DOE Contract No. DE-AC05-06OR23177. The U.S. Government retains a non-exclusive, paid-up, irrevocable, world-wide license to publish or reproduce this manuscript for U.S. Government purposes.

References

1. P. Hägler *et al.*, *Europhys. Lett.* **88**, p. 61001 (2009).
2. B. U. Musch, phd thesis, TU München (2009). arXiv:0907.2381.
3. B. U. Musch, P. Hägler, A. Schäfer, and J. W. Negele, tbp.
4. P. Hägler, *Phys. Rept.* **490**, 49 (2010).
5. J. Collins, *PoS* **LC2008**, p. 028 (2008).
6. I. Cherednikov, A. Karanikas and N. Stefanis (2010), arXiv:1004.3697.
7. J. C. Collins and D. E. Soper, *Nucl. Phys.* **B194**, p. 445 (1982).
8. P. J. Mulders and R. D. Tangerman, *Nucl. Phys.* **B461**, 197 (1996).
9. D. Boer and P. J. Mulders, *Phys. Rev.* **D57**, 5780 (1998).
10. K. Goeke, A. Metz and M. Schlegel, *Phys. Lett.* **B618**, 90 (2005).
11. J. C. Collins, *Phys. Lett.* **B536**, 43 (2002).
12. P. Hägler *et al.*, *Phys. Rev.* **D77**, p. 094502 (2008).
13. C. W. Bernard *et al.*, *Phys. Rev.* **D64**, p. 054506 (2001).
14. C. Aubin *et al.*, *Phys. Rev.* **D70**, p. 094505 (2004).
15. A. Bazavov *et al.* (2009), arXiv:0903.3598.
16. B. U. Musch *et al.*, *PoS* **LC2008**, p. 053 (2008).
17. N. S. Craigie and H. Dorn, *Nucl. Phys.* **B185**, p. 204 (1981).
18. H. Dorn, *Fortsch. Phys.* **34**, 11 (1986).
19. P. Boucaud, C. L. Lin and O. Pene, *Phys. Rev.* **D40**, p. 1529 (1989).
20. M. Cheng *et al.*, *Phys. Rev.* **D77**, p. 014511 (2008).
21. A. Bazavov *et al.*, *Phys. Rev.* **D80**, p. 014504 (2009).
22. M. Lüscher, K. Symanzik and P. Weisz, *Nucl. Phys.* **B173**, p. 365 (1980).
23. B. Pasquini, S. Boffi, A. Efremov and P. Schweitzer (2009), arXiv:0912.1761.
24. B. Pasquini, S. Cazzaniga and S. Boffi, *Phys. Rev.* **D78**, p. 034025 (2008).
25. J. C. Collins and D. E. Soper, *Nucl. Phys.* **B193**, p. 381 (1981).
26. X.-d. Ji, J.-p. Ma and F. Yuan, *Phys. Rev.* **D71**, p. 034005 (2005).
27. R. G. Edwards and B. Joo, *Nucl. Phys. Proc. Suppl.* **140**, p. 832 (2005).

MODELING THE TRANSVERSE MOMENTUM DEPENDENT PARTON DISTRIBUTIONS

B. PASQUINI

Dipartimento di Fisica Nucleare e Teorica, Università degli Studi di Pavia, and INFN, Sezione di Pavia, I-27100 Pavia, Italy
E-mail: Barbara.Pasquini@pv.infn.it

C. LORCÉ

Institut für Kernphysik, Johannes Gutenberg-Universitaet, Mainz, D55099, Germany
E-mail: lorce@kph.uni-mainz.de

We review quark model calculations of the transverse momentum dependent parton distributions (TMDs). For the T-even TMDs, we discuss the physical origin of model relations which hold in a large class of quark models. For the T-odd TMDs we review results in a light-cone constituent quark model (CQM) with the final state interaction effects generated via single-gluon exchange mechanism. As phenomenological application, we show the good agreement between results in the light-cone CQM and available experimental data for the Collins asymmetry.

Keywords: transverse momentum dependent parton distributions, quark models, semi-inclusive deep inelastic scattering

1. Introduction

Parton distributions entering many hard and exclusive processes play a key role to describe the nonperturbative structure of hadrons. The most prominent examples are the ordinary parton distributions which describe the probability to find in a fast-moving hadron a parton with a certain fraction of the longitudinal momentum of the parent hadron. However, in addition to the information on the longitudinal momentum distribution, a complete three-dimensional picture of the nucleon also requires knowledge of the transverse motion of partons. A full account of the orbital motion, which is also an important issue to understand the spin structure of the nucleon, can be given in terms of transverse-momentum dependent parton distributions (TMDs). There are eight leading-twist TMDs. Two of them, the Boer-Mulders and Sivers functions are T-odd, i.e. they change sign under naïve time reversal, which is defined as usual time reversal, but without interchange of initial and final states. The other six leading-twist TMDs are T-even. The latter will be reviewed in Sect. 2, discussing in particular the origin of model relations which were found in a large class of quark models.[1–7] Understanding the physical origin of these relations provides useful insights for building up realistic quark models. Indeed, from phenomenological applications to observables one can verify to which extent the assumptions at the basis of the model relations are also satisfied in "nature". The T-odd TMDs will be the subject of Sect. 3, where we will show predictions in a light-cone constituent quark model (CQM)[8] which was successfully applied to describe many nonperturbative properties of the nucleon, like generalized parton distributions,[9] form factors[10] and distribution amplitudes.[11] The same light-cone CQM was also applied to describe single-spin asymmetries in semi-inclusive deep inelastic (SIDIS) which allow one to extract information on TMDs.[12] In Sect. 4 we will show as an example the results for the Collins asymmetry.

2. T-even TMDs and quark model relations

In the light-cone gauge $A^+ = 0$, the physical meaning of the correlations encoded in the T-even TMDs becomes especially transparent when using the canonical expansion in terms of light-cone Fock operators for the quark fields. Restricting ourselves to the quark contribution, one finds

$$f_1^q(x, \boldsymbol{k}_\perp^2) = \langle P\Lambda | V^q(\tilde{k}) | P\Lambda \rangle, \tag{1}$$

$$\Lambda\, g_{1L}^q(x, \boldsymbol{k}_\perp^2) = \langle P\Lambda | A^q(\tilde{k}) | P\Lambda \rangle, \tag{2}$$

$$\frac{\boldsymbol{k}_\perp \cdot \boldsymbol{S}_\perp}{M}\, g_{1T}^q(x, \boldsymbol{k}_\perp^2) = \langle PS_\perp | A^q(\tilde{k}) | PS_\perp \rangle, \tag{3}$$

$$\Lambda\, \frac{\boldsymbol{k}_\perp}{M}\, h_{1L}^{\perp q}(x, \boldsymbol{k}_\perp^2) = \langle P\Lambda | \boldsymbol{T}_\perp^q(\tilde{k}) | P\Lambda \rangle, \tag{4}$$

$$\boldsymbol{S}_\perp\, h_{1T}^q(x, \boldsymbol{k}_\perp^2) + \frac{\boldsymbol{k}_\perp \cdot \boldsymbol{S}_\perp}{M} \frac{\boldsymbol{k}_\perp}{M}\, h_{1T}^{\perp q}(x, \boldsymbol{k}_\perp^2) = \langle PS_\perp | \boldsymbol{T}_\perp^q(\tilde{k}) | PS_\perp \rangle. \tag{5}$$

where Λ ($S_\perp$) is the longitudinal (transverse) light-cone polarization of the nucleon, and P is the nucleon four-momentum. In Eqs. (1)-(5), the quark operators V^q, A^q, and $\boldsymbol{T}_\perp^q$ can be written in terms of creation and annihilation operators of quarks with flavor q, light-cone helicity λ and light-cone momentum $\tilde{k} = (xP^+, \boldsymbol{k}_\perp)$ as

$$V^q(\tilde{k}) = \sum_\lambda q_\lambda^\dagger(\tilde{k}) q_\lambda(\tilde{k}), \quad A^q(\tilde{k}) = \sum_\lambda \text{sgn}(\lambda)\, q_\lambda^\dagger(\tilde{k}) q_\lambda(\tilde{k}),$$
$$T_R^q(\tilde{k}) = \left[T_L^q(\tilde{k})\right]^\dagger = 2q_+^\dagger(\tilde{k}) q_-(\tilde{k}), \tag{6}$$

where for a generic two-component vector $\boldsymbol{a}_\perp$ we have defined $a_{R,L} = a^1 \pm ia^2$. The quark operators V^q and A^q have a probabilistic interpretation since they are written just in terms of number operators $N_\lambda^q = q_\lambda^\dagger(\tilde{k}) q_\lambda(\tilde{k})$. The operator $\boldsymbol{T}_\perp^q$ has also a probabilistic interpretation but only when written in terms of transversely polarized operators

$$\boldsymbol{S}_\perp^q \cdot \boldsymbol{T}_\perp^q = \sum_{s_\perp} \text{sgn}(s_\perp) q_{s_\perp}^\dagger(\tilde{k}) q_{s_\perp}(\tilde{k}). \tag{7}$$

Choosing the frame system such that $\hat{e}_z = \hat{P}$ and $\hat{e}_y = \hat{k}_\perp$, the T-even TMDs can be classified in terms of the quark and nucleon light-cone polarizations as summarized in Table 1. In QCD all TMDs are independent

Table 1. T-even TMDs according to quark (columns) and nucleon (rows) light-cone polarizations.

nucleon pol → quark pol ↓	U	L	T$_y$	T$_x$
U	f_1			
L		g_{1L}	$\frac{k_\perp}{M} h_{1L}^\perp$	
T$_y$		$\frac{k_\perp}{M} g_{1T}$	h_{1T}^+	
T$_x$				h_{1T}^-

Note: $h_{1T}^{\pm q} = h_1 \pm \frac{k_\perp^2}{2M} h_{1T}^\perp$.

functions. However, in a large class of quark models[1–7] there appear relations among different TMDs. In particular, for the T-even TMDs one finds *i*) three linear relations

$$C^q f_1^q + g_{1L}^q - 2h_1^q = 0, \tag{8}$$

$$g_{1T}^q + h_{1L}^{\perp q} = 0, \tag{9}$$

$$g_{1L}^q - \left[h_1^q + \frac{k_\perp^2}{2M^2} h_{1T}^{\perp q}\right] = 0, \tag{10}$$

with C^q in Eq. (8) a real number, and *ii*) a quadratic relation

$$2h_1 h_{1T}^{\perp q} = -(g_{1T}^q)^2. \tag{11}$$

Although such model relations hold in quark models based on different descriptions of the quark dynamics, their origin can be traced back to a few common assumptions for treating the quark and nucleon spin degrees of freedom. Indeed, Eqs. (8)-(11) connect the reciprocal polarizations of quark and nucleon, and we expect an underlying symmetry under rotation of the polarization degrees of freedom, common to all models. In order to describe the behavior of the polarizations under rotations, it is convenient to work in the basis of canonical (instant form) spin instead of light-cone helicity. This is because rotations in light-cone quantization depend non trivially on the interaction, while they become kinematical operators within instant-form quantization. A first common assumption to all the quark models satisfying the relations is that there is no mutual interaction among the quarks at the interaction vertex. In this case, light-cone helicity and canonical spin operators are just related by a Wigner rotation in the polarization space with axis orthogonal to both $\hat{e}_z$ and $\boldsymbol{k}_\perp$ directions

$$q_\lambda(\tilde{k}) = D^{(1/2)*}_{\lambda s}\, q_s(\tilde{k}), \qquad D^{(1/2)*}_{\lambda s} = \begin{pmatrix} \cos\frac{\theta}{2} & \hat{k}_L \sin\frac{\theta}{2} \\ -\hat{k}_R \sin\frac{\theta}{2} & \cos\frac{\theta}{2} \end{pmatrix}, \tag{12}$$

where λ and s refer to the light-cone helicity and canonical spin, respectively, while the explicit expression of the angle θ in the different models can be found in Ref. [13]. Applying the rotation in Eq. (12) to the polarization operators in Eq. (6) and the nucleon states, one finds the expressions of the T-even TMDs in the basis of canonical spin given in Table 2. If we

Table 2. T-even TMDs according to quark (columns) and nucleon (rows) canonical polarizations.

nucleon pol → quark pol ↓	U	L	T_y	T_x
U	f_1			
L		$\cos\theta\, g_{1L} - \sin\theta\, \frac{k_\perp}{M} h_{1L}^\perp$	$\cos\theta\, g_{1L} + \sin\theta\, \frac{k_\perp}{M} h_{1L}^\perp$	
T_y		$\cos\theta\, \frac{k_\perp}{M} g_{1T} - \sin\theta\, h_{1T}^+$	$\sin\theta\, \frac{k_\perp}{M} g_{1T} + \cos\theta\, h_{1T}^+$	
T_x				h_{1T}^-

assume rotational symmetry around one of the coordinate axis, we find:

- for the $\hat{e}_x$ axis: the probability to find a quark with spin parallel to the longitudinal direction in a longitudinally polarized nucleon is equal to the probability to find both quark and nucleon transversely polarized in the $\hat{e}_y$ direction. As a result, one can derive from Table 2 the two linear relations of Eqs. (9) and (10);

- for $\hat{e}_z$ axis: the momentum distribution is axially symmetric when the canonical spins of the active quark and nucleon are aligned in the same transverse direction. Using the expressions in Table 2, one finds the quadratic relation of Eq. (11).

The cylindrical symmetry around the $\hat{e}_y$ leads to relations among TMDs which are not independent from the previous ones. The last linear relation in Eq. (8), relating unpolarized and polarized spin amplitudes, holds within more restrictive conditions. It requires spherical symmetry and $SU(6)$ symmetry for the spin-isospin dependence. We note that all the models discussed in the literature which satisfy the relations (9) and (10) among polarized TMDs and the quadratic relation (11) satisfy spherical symmetry in the instant-form, which is equivalent to assuming cylindrical symmetry around all the three spatial axis. Note that this condition is sufficient but not necessary for the validity of the relations.

3. T-odd TMDs

The gauge-link operator entering the quark correlation function which defines the TMDs is crucial to obtain non-zero T-odd quark distributions. In the light-cone CQM[8], we consider the expansion of the gauge-link operator at leading order, taking into account the one-gluon exchange mechanism between the struck quark and the nucleon spectators described by (real) light-cone wave functions (LCWFs). This approach is complementary to a recent work[8] where the rescattering effects are incorporated in augmented LCWFs containing an imaginary phase. The approximation of truncating the expansion of the gauge-link operator at leading order is common to most of the quark-model calculations in literature.[15,16] However, studies beyond the one-gluon exchange approximation where one resums all order contributions have recently been presented in Ref. [17]. Following Ref. [8], one can represent the T-odd TMDs in terms of overlap of light-cone amplitudes corresponding to different orbital angular momentum. Both T-odd TMDs are obtained from the interference of wave components which differ by one unit of orbital angular momentum. In particular, we found that the Sivers function for both up and down quarks is dominated by the interference of S and P wave components, while the $P-D$ wave interference terms contribute at most by 20%. On the other side, the relative weight of the $P-D$ wave interference terms increases in the case of the Boer-Mulders function, in particular for the down-quark component. In Fig. 1 we show the results for the first transverse-momentum moments of the Sivers and

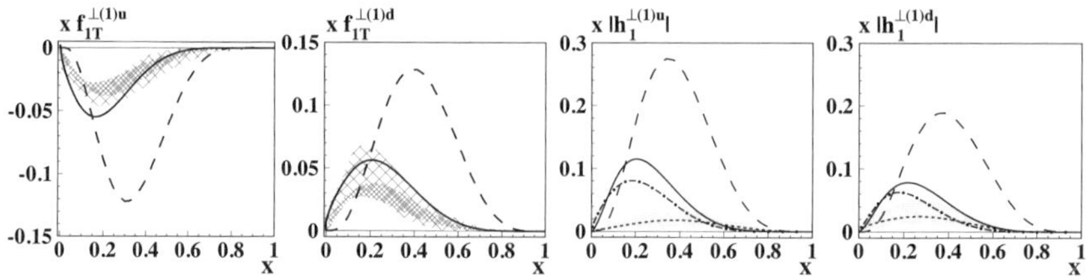

Fig. 1. Results for the first transverse-momentum moment of the Sivers and Boer Mulders functions for up and down quarks, as function of x. See text for the explanation of the different curves.

Boer-Mulders functions. The dashed curves correspond to the results at the hadronic scale of the model. The solid curves are the results evolved to $Q^2 = 2.5$ GeV^2. Since the exact evolution equations for the T-odd quark distributions are still under study, we used those evolution equations which seem most promising to simulate the correct evolution. We evolved the first transverse-momentum moment of the Sivers function by means of the evolution pattern of the unpolarized parton distribution, while for the first transverse-momentum moment of the Boer-Mulders we used the evolution pattern of the transversity. After evolution, the model results are consistent with the available parametrizations. For the Sivers function, the lighter and darker shaded areas are the uncertainty bands due to the statistical error of the parametrizations of Ref. [18] and Ref. [19], respectively, which refer to an average scale of $Q^2 = 2.5$ GeV^2. For the Boer-Mulders function, the dashed-dotted curves are the results of the phenomenological parametrization of Refs. [20,21] at the average scale of $Q^2 = 2.4$ GeV^2, and the short-dashed curves show the results of Refs. [22,23] valid at $Q^2 \approx 1$ GeV^2, with the shaded area describing the variation ranges allowed by positivity bounds. We also note that a non trivial constraint in model calculations of the Sivers function is given by the Burkardt sum rule,[24] which corresponds to require that the net (summed over all partons) transverse momentum due to final-state interaction is zero. Our model calculation of the Sivers function reproduces exactly this sum rule.

4. Results for the Collins asymmetry

In Ref. [12] the present results for the T-even TMDs were applied to estimate azimuthal asymmetries in SIDIS, discussing the range of applicability of the model, especially with regard to the scale dependence of the ob-

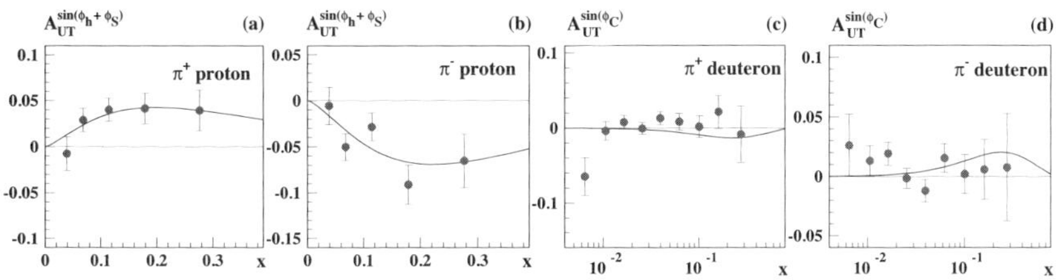

Fig. 2. The single-spin asymmetry $A_{UT}^{\sin(\phi_h+\phi_S)} \equiv -A_{UT}^{\sin\phi_C}$ in DIS production of charged pions off proton and deuterium targets, as function of x. The solid curves are obtained using the light-cone CQM predictions for $h_1(x, Q^2)$ from Refs. [3,30]. The proton target data are from HERMES[28], the deuterium target data are from COMPASS[29] .

servables and the transverse-momentum dependence of the distributions. Here we review the results for the Collins asymmetry $A_{UT}^{\sin(\phi+\phi_S)}$, due to the Collins fragmentation function and to the chirally-odd TMDs h_1. For the Collins function we use the results extracted in Ref. [25]. In the denominator of the asymmetry we take f_1 from Ref. [26] and the unpolarized fragmentation function from Ref. [27], both at the scale $Q^2 = 2.5$ GeV^2.

In Fig. 2 the results for the Collins asymmetry in DIS production of charged pions off proton and deuterium targets are shown as function of x. The model results for h_1 evolved from the low hadronic scale of the model to $Q^2 = 2.5$ GeV^2 ideally describe the HERMES data[28] for a proton target (panels (a) and (b)). This is in line with the good agreement between our model predictions[30] and the phenomenological extraction of the transversity and the tensor charges.[31] Our results are also compatible with the COMPASS data[29] for a deuterium target (panels (c) and (d)) which extend down to much lower values of x.

Acknowledgments

B.P. acknowledges a fruitful collaboration with S. Boffi, A. Efremov, P. Schweitzer, and F. Yuan with whom some of the results presented in this paper have been obtained. This work was supported in part by the Research Infrastructure Integrating Activity "Study of Strongly Interacting Matter" (acronym HadronPhysics2, Grant Agreement n. 227431) under the Seventh Framework Programme of the European Community, by the Italian MIUR through the PRIN 2008EKLACK "Structure of the nucleon: transverse momentum, transverse spin and orbital angular momentum".

References

1. H. Avakian, et al., Phys. Rev. D **81**, 074035 (2010).
2. H. Avakian, et al., Phys. Rev. D **78**, 114024 (2008).
3. B. Pasquini, S. Cazzaniga and S. Boffi, Phys. Rev. D **78**, 034025 (2008).
4. J. She, J. Zhu and B. Q. Ma, Phys. Rev. D **79**, 054008 (2009).
5. A. V. Efremov, et al., Phys. Rev. D **80**, 014021 (2009).
6. R. Jakob, P. J. Mulders and J. Rodrigues, Nucl. Phys. A **626**, 937 (1997).
7. C. Lorce, Phys. Rev. D **79**, 074027 (2009).
8. B. Pasquini and F. Yuan, Phys. Rev. D **81**, 114013 (2010).
9. S. Boffi and B. Pasquini, Riv. Nuovo Cim. **30**, 387 (2007).
10. B. Pasquini and S. Boffi, Phys. Rev. D **76**, 074011 (2007).
11. B. Pasquini, M. Pincetti and S. Boffi, Phys. Rev. D **80**, 014017 (2009); S. Boffi and B. Pasquini, Mod. Phys. Lett. A **24**, 2882 (2009).
12. S. Boffi, A. V. Efremov, B. Pasquini and P. Schweitzer, Phys. Rev. D **79** (2009) 094012; B. Pasquini, S. Boffi and P. Schweitzer, Mod. Phys. Lett. A **24**, 2903 (2009).
13. C. Lorce, B. Pasquini, in preparation, (2010).
14. S. J. Brodsky, B. Pasquini, B. W. Xiao and F. Yuan, Phys. Lett. B **687**, 327 (2010).
15. A. Bacchetta, F. Conti and M. Radici, Phys. Rev. D **78**, 074010 (2008).
16. A. Courtoy, S. Scopetta and V. Vento, Phys. Rev. D **79**, 074001 (2009); Phys. Rev. D **80**, 074032 (2009).
17. L. Gamberg and M. Schlegel, Phys. Lett. B **685**, 95 (2010).
18. M. Anselmino *et al.*, Eur. Phys. J. A **39** (2009) 89; Phys. Rev. D **71**, 074006 (2005); Phys. Rev. D **72**, 094007 (2005) [Erratum-ibid. D **72**, 099903 (2005)];
19. J. C. Collins, et al., Phys. Rev. D **73**, 014021 (2006);
A. V. Efremov, et al., Phys. Lett. B **612**, 233 (2005).
20. V. Barone, A. Prokudin and B. Q. Ma, Phys. Rev. D **78** (2008) 045022.
21. V. Barone, S. Melis and A. Prokudin, Phys. Rev. D **81**, 114026 (2010).
22. B. Zhang, Z. Lu, B. Q. Ma and I. Schmidt, Phys. Rev. D **78**, 034035 (2008); Phys. Rev. D **77** (2008) 054011.
23. Z. Lu and I. Schmidt, Phys. Rev. D **81**, 034023 (2010).
24. M. Burkardt, Phys. Rev. D **69** (2004) 091501(R); Phys. Rev. D **69** (2004) 057501 .
25. A. V. Efremov, K. Goeke and P. Schweitzer, Phys. Rev. D **73**, 094025 (2006).
26. M. Glück, E. Reya and A. Vogt, Eur. Phys. J. C **5**, 461 (1998).
27. S. Kretzer, Phys. Rev. D **62**, 054001 (2000).
28. M. Diefenthaler, AIP Conf. Proc. **792**, 933 (2005); L. Pappalardo, Eur. Phys. J. A **38**, 145 (2008).
29. M. Alekseev *et al.* [COMPASS Coll.], Phys. Lett. B **673**, 127 (2009).
30. B. Pasquini, M. Pincetti and S. Boffi, Phys. Rev. D **72**, 094029 (2005); Phys. Rev. D **76**, 034020 (2007).
31. M. Anselmino, et al., and C. Turk, Phys. Rev. D **75**, 054032 (2007); Nucl. Phys. Proc. Suppl. **191**, 98 (2009).

QUARK ORBITAL ANGULAR MOMENTUM: CAN WE LEARN ABOUT IT FROM GPDs AND TMDs?

H. AVAKIAN[1], A. V. EFREMOV[2],
P. SCHWEITZER[3], O. V. TERYAEV[2], P. ZAVADA[4]

[1] *Jefferson National Accelerator Facility, Newport News, VA 23606, U.S.A.*
[2] *Bogoliubov Laboratory of Theoretical Physics, JINR, 141980 Dubna, Russia*
[3] *Department of Physics, University of Connecticut, Storrs, CT 06269, U.S.A.*
[4] *Institute of Physics, Academy of Sciences of the Czech Republic, Na Slovance 2, CZ-182 21 Prague 8, Czech Republic*

It is known how to access information on quark orbital angular momentum from generalized parton distribution functions, in a certain specified framework. It is intuitively expected, that such information can be accessed also through transverse momentum dependent distribution functions, but not known how. Now quark models provide promising hints. Recent results are reviewed.

Keywords: nucleon spin structure, quark models, orbital angular momentum

1. Introduction

Transverse parton momentum dependent distribution functions (TMDs) and generalized parton distribution functions (GPDs) describe complementary aspects of the transverse nucleon structure. TMDs[1–3] describe the *momentum distribution* of partons in the transverse plane. GPDs[4–7] describe their *spatial distribution* in the transverse plane[8] (and much more[9]).

It is known how to learn from GPDs about orbital angular momentum of partons in the nucleon, namely[10] (using impact parameter presentation[8])

$$L^q = \frac{1}{2}\int \mathrm{d}x \int \mathrm{d}^2b \left(xH^q(x,b) + xE^q(x,b) - \widetilde{H}^q(x,b)\right) \tag{1}$$

where $\int \mathrm{d}^2b\, \widetilde{H}^q(x,b)$ is the helicity distribution and b the impact parameter. This decomposition has the advantage that all spin contributions are measurable quantities. Other decomposition schemes exist[11] and give, in gauge theories, in general different results.[12]

In this way, one obtains from the spatial distribution of partons in the transverse plane information about orbital angular momentum. TMDs con-

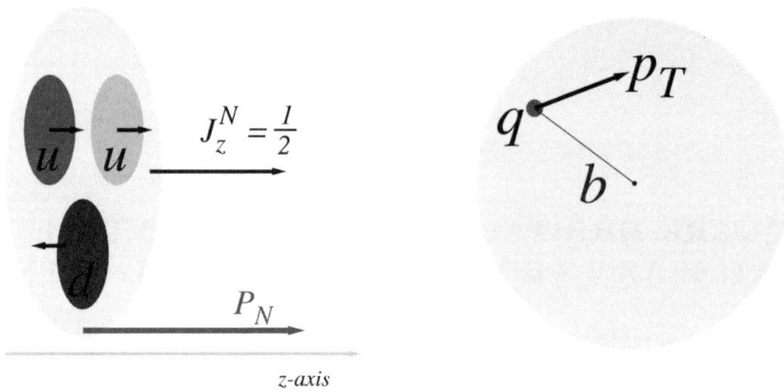

Fig. 1. Left: Naive picture of the spin decomposition of a nucleon moving along z-axis with the large momentum $P_N \to \infty$. Right: Now the nucleon moves towards us. The impact parameter b-distribution of the quark q is described by GPDs. The complimentary information on its transverse momentum p_T-distribution is described by TMDs.

tain information on the parton momenta p_T in the transverse plane. This is in some sense complimentary to GPDs, see Fig. 1. Intuitively, one would therefore expect TMDs to contain also information about orbital angular momentum. However, so far no rigorous connection of orbital angular momentum and TMDs could be established.

Recent results from quark models could indicate a possible connection, and the key to that is "pretzelosity." We review the recent developments.

2. The key TMD: pretzelosity

The light-front correlator (with[2,3] a process-dependent gauge-link $\mathcal{W}$)

$$\phi(x,\vec{p}_T)_{ij} = \int \frac{\mathrm{d}z^- \mathrm{d}^2\vec{z}_T}{(2\pi)^3}\, e^{ipz}\, \langle N(P,S)|\bar{\psi}_j(0)\,\mathcal{W}\,\psi_i(z)|N(P,S)\rangle \Big|_{\substack{z^+=0 \\ p^+=xP^+}} \tag{2}$$

allows us to define the twist-2 chiral-odd TMDs of the nucleon as

$$\frac{1}{2}\,\mathrm{tr}\left[i\sigma^{+j}\gamma_5\,\phi(x,\vec{p}_T)\right] = S_T^j\,h_1 + S_L\,\frac{p_T^j}{M_N}\,h_{1L}^\perp + \frac{\varepsilon^{jk}p_T^k}{M_N}\,h_1^\perp + \frac{(p_T^j p_T^k - \frac{1}{2}\,\vec{p}_T^{\,2}\delta^{jk})S_T^k}{M_N^2}\,h_{1T}^\perp\,. \tag{3}$$

All TMDs in (3) depend on x and $p_T = |\vec{p}_T|$, and can be accessed in semi-inclusive deeply inelastic scattering (SIDIS)[1] in combination with the Collins fragmentation function[13] $H_1^\perp$ by measuring various azimuthal (single spin) asymmetries. In particular

$$A_{UT}^{\sin(3\phi-\phi_S)} \propto \frac{\sum_q e_q^2\, h_{1T}^{\perp q} \otimes H_1^{\perp q}}{\sum_q e_q^2\, f_1^q \otimes D_1^q} \,. \tag{4}$$

Positivity bounds[14] constrain $|h_{1T}^{\perp(1)q}(x,p_T)| \leq \frac{1}{2}(f_1^q(x,p_T) - g_1^q(x,p_T))$ with the (1)-moment defined as $h_{1T}^{\perp(1)q}(x,p_T) \equiv p_T^2/(2M^2)\, h_{1T}^{\perp q}(x,p_T)$.

In the limit of a large number N_c of colors in QCD[15] pretzelosity behaves as $(h_{1T}^{\perp u} + h_{1T}^{\perp d})/(h_{1T}^{\perp u} - h_{1T}^{\perp d}) \sim 1/N_c$. Interesting aspects are[16,17] that it describes the "non-sphericity" of the spin distribution of quarks in a transversely polarized nucleon, and requires[16,17] the presence of nucleon wave-function components differing by two units of orbital angular momentum. At large x it is predicted[17,18] to behave as $h_{1T}^{\perp q} \sim (1-x)^5$. That is all that is known about this function model-independently.

What raised much interest about this TMD are results from quark models. The following relation was found in the bag model in Ref. [19]

$$h_{1T}^{\perp(1)q}(x,p_T) = g_1^q(x,p_T) - h_1^q(x,p_T)\,. \tag{5}$$

This relation is supported also in other[20–27] though not all[27] quark models, and is broken when gauge field degrees of freedom are present,[21] see [28] for a review. Notice that $h_{1T}^{\perp(1)q}(x,p_T) \equiv p_T^2/(2M^2)\, h_{1T}^{\perp q}(x,p_T)$, and if we recall that the difference of g_1 and h_1 vanishes in the non-relativistic limit,[29] we see that this (1)-moment of pretzelosity is a 'measure of relativistic effects' in the nucleon. This is not surprizing because in this limit[23]

$$\lim_{\text{non-rel}} h_{1T}^{\perp q}(x,p_T) = -\frac{N_c^2}{2}\, P_q\, \delta\!\left(x - \frac{1}{N_c}\right)\, \delta^{(2)}(\vec{p}_T) \tag{6}$$

where $P_u = \frac{4}{3}$, $P_d = -\frac{1}{3}$, and similarly for other TMDs, i.e. in the non-relativistic limit the (1)-moments of all TMDs vanish. Another interesting model relation is the remarkable *non-linear relation* first observed in the covariant parton model[23] connecting all T-even, chiral-odd, twist-2 TMDs:

$$\frac{1}{2}\left[h_{1L}^{\perp q}(x,p_T)\right]^2 \overset{\text{model}}{=} -\, h_1^q(x,p_T)\, h_{1T}^{\perp q}(x,p_T)\,. \tag{7}$$

This relation implies an attractive prediction. The transversity distribution gives rise to the single spin asymmetry $A_{UT}^{\sin(\phi+\phi_S)} \propto \sum_q e_q^2\, h_1^q \otimes H_1^{\perp q}$ whose sign is known experimentally.[30] From (7) it immediately follows that the pretzelosity asymmetry in (4) must have opposite sign,[28] i.e.

$$\text{sign}\,[\,A_{UT}^{\sin(3\phi-\phi_S)}\,] = (-1)\cdot \text{sign}\,[\,A_{UT}^{\sin(\phi+\phi_S)}\,]\,, \tag{8}$$

which is expected[28] to hold in the valence-x region, where this prediction is confirmed by COMPASS for negative hadrons from a deuteron target.[31]

3. Pretzelosity and Orbital Angular Momentum

Many more quark model relations among TMDs were found, see [25] for the derivation of a complete set of relations in the bag model, and it is understood why they are widely supported in a large class of quark models.[32]

The pretzelosity-relation (5) plays a particularly important role in what follows for the following reason. Namely, it was shown in the light-cone SU(6) quark-diquark model[33] that the contribution to the nucleon spin from the orbital angular momentum of quarks is related to the difference of transversity and helicity distributions, i.e. to the right-hand-side of the pretzelosity-relation (5) found in.[19] It was subsequently shown that the pretzelosity-relation (5) is valid also in the light-cone SU(6) quark-diquark model.[24] In other words, the (1)-moment of pretzelosity is, in this quark model, a measure for the contribution of quark orbital angular momentum to the nucleon spin. This exciting finding was subsequently confirmed in the bag model[25] and the covariant parton model.[26]

More precisely, three different quark models[24–26] support the relation

$$L_z^q = -\int \mathrm{d}x\, h_{1T}^{\perp(1)q}(x) \tag{9}$$

(in principle, there is a fourth model where it holds: in the non-relativistic limit one has the consistent result[23] $L_z^q = -\int \mathrm{d}x\, h_{1T}^{\perp(1)q}(x) = 0$).

An interesting question in this context: how can chiral-odd (pretzelosity) and chiral-even (orbital angular momentum) quantities be related?

The answer in the bag model is as follows. Here the quark wave-function has an upper-(s-wave-)component and a lower-(p-wave-)component. The expectation value of the orbital angular momentum operator in the s-wave is zero, i.e. only the p-wave contributes. Next, we know[16,17] pretzelosity requires $\Delta L = 2$ which in the bag model is possible only through interference of the $L_z = \pm 1$ components of the p-wave, i.e. again only the p-wave contributes. Finally, knowing that only the p-wave (i.e. the lower component of the Dirac-spinor) matters, we can "replace" in the operator of pretzelosity $\gamma^0 = \mathrm{diag}(1, -1)$ (in Bjorken-Drell notation) by $(-1)\times$(unit matrix). This changes the number of gamma-matrices by one unit, and "transforms" a chiral-odd operator into a chiral-even one.[25]

From this exercise we learn: the relation between pretzelosity and orbital angular momentum is at best at the level of matrix elements. In other words, there is no operator-identity between these quantities — not even in quark models. It is interesting to stress that in quark models the result for L_z^q does not depend on which orbital angular momentum definition[10,11] is used.[12]

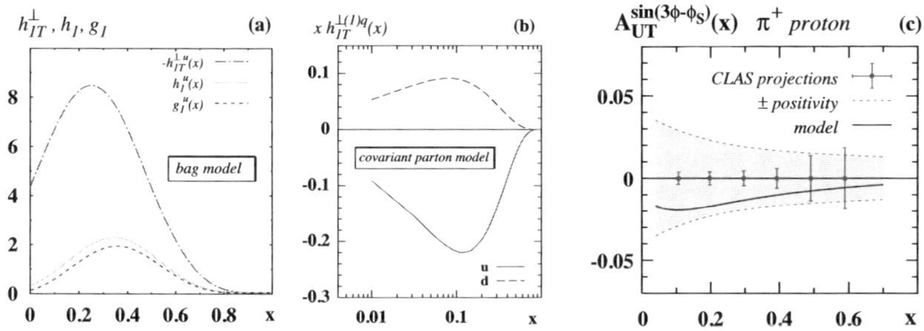

Fig. 2. (a) Bag model predictions at the low scale for $(-1)h_{1T}^{\perp u}(x)$, $h_1^u(x)$, $g_1^u(x)$ vs. x. (b) $h_{1T}^{\perp(1)q}(x)$ from the covariant parton model[23] at the high scale $Q^2 = 4\,\text{GeV}^2$ vs. x. (c) The azimuthal single spin asymmetry $A_{UT}^{\sin(3\phi-\phi_s)}$, Eq. (4) for π^+ production from a proton target in the kinematics of the CLAS experiment with 12 GeV upgrade vs. x. The shaded areas show[19] what is allowed by positivity (quark models predict negative sign). The solid curve is the prediction[23] from the covariant parton model shown in Fig. 2b. The error projections are for a 2000 hours run time.[36]

4. How does pretzelosity look like, and how to access it?

Having discussed that in quark models pretzelosity is related to orbital angular momentum, it is interesting to ask what quark models actually predict. Fig. 2a shows the bag model predictions for $h_{1T}^{\perp u}(x)$ in comparison to $h_1^u(x)$ and $g_1^u(x)$. The negative sign and the large magnitude of $h_{1T}^{\perp u}(x)$ can be understood from the non-relativistic limit,[23] where pretzelosity is enhanced by the factor $\frac{1}{2}N_c^2$ compared to $h_1^u(x)$ or $g_1^u(x)$, see Sec. 2. Relativistic models (like the bag model) preserve this enhancement. Concerning the p_T-dependence of pretzelosity (and other TMDs), we remark that in the bag[25] and covariant parton[23] model it is approximately Gaussian which is supported by phenomenology.[35]

Note that $h_{1T}^{\perp q}(x)$ is not constrained by positivity, but $h_{1T}^{\perp(1)q}(x)$ is, see Sec. 2. Moreover, $h_{1T}^{\perp q}$ enters cross sections with a prefactor of $\mathcal{O}(p_T^2/M^2)$ from the correlator (2). So in observables effectively the (1)-moment of pretzelosity is relevant, and the latter is not large in the bag[19] or light-cone constituent[22] model. In fact, the latter predicts[34] a small asymmetry (4).

The covariant parton model[23] predicts large $h_{1T}^{\perp(1)q}(x)$, see Fig. 2b, and hence a sizable pretzelosity asymmetry (4) shown in Fig. 2c for the kinematics of the CLAS experiment with 12 GeV upgrade.[19,23] The predictions are consistent with preliminary SIDIS data[37] showing a zero within error bars effect. Recent COMPASS data[31] show a small but non-vanishing effect and confirm the signs predicted for $h_{1T}^{\perp q}$, as was discussed in Sec. 2.

One has to bear in mind that so far the exciting relation (9) between TMDs and orbital angular momentum is established only in quark models. An important question is: to what extent can we trust such models? One way to address this question consists in reproducing (SI)DIS spin observables within a given quark model. On the basis of the comparison of model results and data it was found[34] that (light-cone constituent) quark models work in the valence x-region $0.2 \lesssim x \lesssim 0.6$ with an accuracy of (10–30) %.

In this context we recall that the absence of gauge degrees of freedom implies in quark models certain relations among TMDs (called "LIRs"), which hold approximately in QCD upon the neglect of quark-gluon-quark correlators and current quark mass terms.[38] For the collinear twist-3 parton distribution function $g_T^q(x)$ such an approximation works reasonably well[39] but it needs to be tested for other TMDs.[40] If LIRs were confirmed to be reasonably good approximations, this would be a necessary (not sufficient) condition for quark model predictions of the type (5, 9) to work similarly.

5. Conclusions

The concept of quark orbital angular momentum is difficult to address rigorously in gauge field theories. GPDs and TMDs, which describe complementary aspects of the nucleon structure, give rise to a dual (in quark models equivalent) picture of quark orbital angular momentum as follows

$$L_z^q \overset{\text{QCD+models}}{=} \int \mathrm{d}x \int \mathrm{d}^2 b \ \mathrm{GPDs}(x, b) \quad \overset{\textbf{!?}}{=} \tag{10}$$

$$L_z^q \overset{\text{quark models}}{=} \int \mathrm{d}x \int \mathrm{d}^2 p_T \mathrm{TMDs}(x, p_T) \tag{11}$$

where GPDs $= (xH^q + xE^q - \widetilde{H}^q)/2$ and TMDs $= -h_{1T}^{\perp(1)q}$. The first relation (10) holds exactly in QCD and in consistent models.[10] The second relation (11) holds in a large class of relativistic quark models.[24–26] Quark models catch important features of QCD, and could provide useful insights also in the context of GPDs, TMDs and orbital angular momentum, provided one uses them responsibly within their range of applicability. Recent advances[41] allow us to test these relations in lattice QCD, where due the presently often practioned omission of disconnected diagrams valence quarks are probed.

Note added

In the discussion following the talk it was pointed out that in [42] the predictions from quark models were shown to be compatible with phenomenology and lattice data, resolving the "spin crisis" from quark model point of view.

Acknowledgments

P. S. is grateful to the organizers of the "4th Workshop on Exclusive Reactions at High Momentum Transfer," 18-21 May 2010, Jefferson Lab, where this work was reported. A. E. and O. T. are supported by the Grants RFBR 09-02-01149 and 07-02-91557, RF MSE RNP.2.2.2.2.6546 (MIREA) and by the Heisenberg-Landau Program of JINR. P. Z. is supported by the project AV0Z10100502 of the Academy of Sciences of the Czech Republic. The work was supported in part by DOE contract DE-AC05-06OR23177, under which Jefferson Science Associates, LLC, operates the Jefferson Lab.

References

1. D. Boer and P. J. Mulders, Phys. Rev. D **57**, 5780 (1998). A. Bacchetta, M. Diehl, K. Goeke, A. Metz, P. J. Mulders and M. Schlegel, JHEP **0702**, 093 (2007).
2. J. C. Collins, Phys. Lett. B **536**, 43 (2002). S. J. Brodsky, D. S. Hwang and I. Schmidt, Phys. Lett. B **530**, 99 (2002); Nucl. Phys. B **642**, 344 (2002).
3. A. V. Belitsky, X. Ji and F. Yuan, Nucl. Phys. B **656**, 165 (2003). X. D. Ji and F. Yuan, Phys. Lett. B **543**, 66 (2002). D. Boer, P. J. Mulders and F. Pijlman, Nucl. Phys. B **667**, 201 (2003). I. O. Cherednikov and N. G. Stefanis, Phys. Rev. D **77**, 094001 (2008); Nucl. Phys. B **802**, 146 (2008); Phys. Rev. D **80**, 054008 (2009).
4. D. Müller, D. Robaschik, B. Geyer, F. M. Dittes and J. Horejsi, Fortsch. Phys. **42**, 101 (1994).
5. X. D. Ji, Phys. Rev. D **55**, 7114 (1997).
6. A. V. Radyushkin, Phys. Rev. D **56**, 5524 (1997).
7. K. Goeke, M. V. Polyakov and M. Vanderhaeghen, Prog. Part. Nucl. Phys. **47**, 401 (2001). M. Diehl, Phys. Rept. **388**, 41 (2003). A. V. Belitsky and A. V. Radyushkin, Phys. Rept. **418**, 1 (2005).
8. M. Burkardt, Int. J. Mod. Phys. A **18**, 173 (2003); Phys. Rev. D **62**, 071503 (2000); D **66**, 119903E (2002)
9. M. V. Polyakov, Phys. Lett. B **555**, 57 (2003).
   K. Goeke *et al*, Phys. Rev. D **75**, 094021 (2007).
10. X. D. Ji, Phys. Rev. Lett. **78**, 610 (1997).
    P. Hoodbhoy, X. D. Ji and W. Lu, Phys. Rev. D **59**, 014013 (1999).
11. R. L. Jaffe and A. Manohar, Nucl. Phys. B **337**, 509 (1990).
12. M. Burkardt and B. C. Hikmat, Phys. Rev. D **79**, 071501 (2009).
13. J. C. Collins, Nucl. Phys. B **396**, 161 (1993).
14. A. Bacchetta *et al.*, Phys. Rev. Lett. **85**, 712 (2000).
15. P. V. Pobylitsa, arXiv:hep-ph/0301236.
16. G. A. Miller, Phys. Rev. C **76**, 065209 (2007).
17. M. Burkardt, arXiv:0709.2966 [hep-ph].

18. H. Avakian, S. J. Brodsky, A. Deur and F. Yuan, Phys. Rev. Lett. **99**, 082001 (2007).
19. H. Avakian, A. V. Efremov, P. Schweitzer and F. Yuan, Phys. Rev. D **78**, 114024 (2008); arXiv:0808.3982 [hep-ph].
20. R. Jakob, P. J. Mulders and J. Rodrigues, Nucl. Phys. A **626**, 937 (1997).
21. S. Meissner, A. Metz and K. Goeke, Phys. Rev. D **76**, 034002 (2007).
22. B. Pasquini, S. Cazzaniga and S. Boffi, Phys. Rev. D **78**, 034025 (2008).
23. A. V. Efremov, P. Schweitzer, O. V. Teryaev and P. Zavada, Phys. Rev. D **80**, 014021 (2009); AIP Conf. Proc. **1149**, 547 (2009); arXiv:0912.3380 [hep-ph].
24. J. She, J. Zhu and B. Q. Ma, Phys. Rev. D **79**, 054008 (2009).
25. H. Avakian, A. V. Efremov, P. Schweitzer and F. Yuan, Phys. Rev. D **81**, 074035 (2010).
26. A. V. Efremov, P. Schweitzer, O. V. Teryaev and P. Zavada, forthcoming.
27. A. Bacchetta, F. Conti and M. Radici, Phys. Rev. D **78**, 074010 (2008).
28. H. Avakian, A. V. Efremov, P. Schweitzer, O. V. Teryaev, F. Yuan and P. Zavada, Mod. Phys. Lett. A **24**, 2995 (2009).
29. R. L. Jaffe, X. D. Ji, Nucl. Phys. B **375**, 527 (1992); Phys. Rev. Lett. **67**, 552 (1991).
30. A. Airapetian *et al.* [HERMES Coll.], Phys. Rev. Lett. **94**, 012002 (2005).
V. Y. Alexakhin *et al.* [COMPASS Coll.], Phys. Rev. Lett. **94**, 202002 (2005).
E. S. Ageev *et al.* [COMPASS Coll.], Nucl. Phys. B **765**, 31 (2007).
M. G. Alekseev *et al.* [COMPASS Coll.], arXiv:1005.5609 [hep-ex].
A. Airapetian *et al.* [HERMES Coll.], arXiv:1006.4221 [hep-ex].
31. M. G. Alekseev *et al.* [COMPASS Coll.], arXiv:1007.1562 [hep-ex].
32. B. Pasquini and C. Lorcé, arXiv:1008.0945 [hep-ph].
33. B. Q. Ma and I. Schmidt, Phys. Rev. D **58**, 096008 (1998). B. Q. Ma, I. Schmidt and J. Soffer, Phys. Lett. B **441**, 461 (1998).
34. S. Boffi, A. V. Efremov, B. Pasquini and P. Schweitzer, Phys. Rev. D **79**, 094012 (2009); AIP Conf. Proc. **1149**, 471 (2009); arXiv:0910.1677 [hep-ph].
35. P. Schweitzer, T. Teckentrup and A. Metz, Phys. Rev. D **81**, 094019 (2010).
36. H. Avakian, *et al.*, JLab LOI 12-06-108 (2008).
37. A. Kotzinian [COMPASS Coll.], arXiv:0705.2402 [hep-ex].
Naomi Makins [HERMES Coll.], at the "Workshop on Partonic Transverse Momentum in Hadrons," March 12-13, 2010 Duke University.
38. A. Metz, P. Schweitzer and T. Teckentrup, Phys. Lett. B **680**, 141 (2009); Mod. Phys. Lett. A **24**, 2950 (2009).
39. A. Accardi, A. Bacchetta, W. Melnitchouk and M. Schlegel, JHEP **0911**, 093 (2009).
40. A. Kotzinian, B. Parsamyan and A. Prokudin, Phys. Rev. D **73**, 114017 (2006). H. Avakian, A. V. Efremov, K. Goeke, A. Metz, P. Schweitzer and T. Teckentrup, Phys. Rev. D **77**, 014023 (2008).
41. Ph. Hägler, Phys. Rept. **490**, 49 (2010). Ph. Hägler, B. U. Musch, J. W. Negele and A. Schäfer, Europhys. Lett. **88**, 61001 (2009).
42. A. W. Thomas, Phys. Rev. Lett. **101**, 102003 (2008).
M. Wakamatsu, Eur. Phys. J. A **44**, 297 (2010).

TRANSVERSE-MOMENTUM PARTON DENSITIES: GAUGE LINKS, DIVERGENCES AND SOFT FACTOR

I. O. CHEREDNIKOV[†*]

INFN Cosenza, Università della Calabria
I-87036 Rende (CS), Italy
and
Bogoliubov Laboratory of Theoretical Physics
JINR RU-141980 Dubna, Russia
[†]*E-mail: igor.cherednikov@jinr.ru*

N. G. STEFANIS[‡]

Institut für Theoretische Physik II,
Ruhr-Universität Bochum, D-44780 Bochum, Germany
[‡]*E-mail: stefanis@tp2.ruhr-uni-bochum.de*

We discuss the state-of-the-art of the theory of transverse-momentum dependent parton densities (TMDs), paying special attention to their renormalization properties, the structure of the gauge links in the operator definition, and the role of the soft factor in the factorization formula within the TMD approach to the semi-inclusive processes. We argue that the use of the lightcone axial gauge offers certain advantages for a consistent definition of TMDs as compared to the off-the-light-cone gauges, or covariant gauges with off-the-lightcone gauge links.

Keywords: Parton distribution functions, Wilson lines, renormalization.

1. Introduction

The distribution functions of partons (in what follows we consider only quark distributions), depending on the longitudinal components x, as well as on the transverse components $\boldsymbol{k}_\perp$, of their momenta (TMDs), accumulate useful information about the intrinsic motion of the hadron's constituents and enter as a nonperturbative input in the QCD approach to the semi-inclusive hadronic processes (see, e.g., Refs. [1–4]). The QCD factorization formula for a semi-inclusive structure function is expected to have the fol-

[*]Also at: ITPM, Moscow State University, Russia

lowing symbolic form [5–7]

$$F(x_B, z_h, \boldsymbol{P}_{h\perp}, Q^2) = \sum_i e_i^2 \cdot H \otimes \mathcal{F}_D \otimes \mathcal{F}_F \otimes S \ , \tag{1}$$

where z_h and $\boldsymbol{P}_{h\perp}$ are the longitudinal and transverse fractions of the momentum of the produced hadron, respectively. This expression contains the hard (perturbatively calculable) part H, the (nonperturbative) distribution and fragmentation functions $\mathcal{F}_D$ and $\mathcal{F}_F$, and the soft part S. The latter is absent in the collinear (fully inclusive) picture, and will be discussed below. However, several problems arise, in attempting to formulate the TMD approach in terms of the quantum field operators and their matrix elements: (*i*) Extra (rapidity) divergences appear already at the one-loop level, which invalidate the standard renormalization procedure [3,8–10]. (*ii*) A much more complicated (compared to the collinear case) structure of the gauge links leads to the non-universality of distribution or fragmentation functions (see, e.g., Refs. [11–13]). (*iii*) Several counter-examples have been given showing that the straightforward factorization formula (1) may fail, at least in some specific situations [14,15]. (*iv*) The role and explicit expression of the soft factor S can be different in different schemes. In what follows, we basically concentrate on the first and the last problem.

2. Divergences and renormalization properties of TMDs

The operator definition of the quark TMD (without the soft term) reads [5,6,9,10,16–21]:

$$\tilde{\mathcal{F}}_{i/h}\left(x, \boldsymbol{k}_\perp\right) = \frac{1}{2}\int \frac{d\xi^- d^2\boldsymbol{\xi}_\perp}{2\pi(2\pi)^2} \mathrm{e}^{-ik^+\xi^- + i\boldsymbol{k}_\perp\cdot\boldsymbol{\xi}_\perp} \ .$$

$$\times \Big\langle h|\bar{\psi}_i(\xi^-, \boldsymbol{\xi}_\perp)[\xi^-, \boldsymbol{\xi}_\perp; \infty^-, \boldsymbol{\xi}_\perp]^\dagger_{[n]}[\infty^-, \boldsymbol{\xi}_\perp; \infty^-, \boldsymbol{\infty}_\perp]^\dagger_{[\boldsymbol{l}]} \cdot$$

$$\times \gamma^+[\infty^-, \boldsymbol{\infty}_\perp; \infty^-, \boldsymbol{0}_\perp]_{[\boldsymbol{l}]}[\infty^-, \boldsymbol{0}_\perp; 0^-, \boldsymbol{0}_\perp]_{[n]}\psi_i(0^-, \boldsymbol{0}_\perp)|h\Big\rangle \ . \tag{2}$$

This expression may be given a physical meaning, because it is formally gauge invariant. The gauge invariance is ensured by means of the path-ordered gauge links

$$\begin{aligned} [\infty^-, \boldsymbol{\xi}_\perp; \xi^-, \boldsymbol{\xi}_\perp]_{[n]} &\equiv \mathcal{P}\exp\left[ig\int_0^\infty d\tau \ n^-_\mu \ A^\mu_a t^a(\xi + n^-\tau)\right] \ , \\ [\infty^-, \boldsymbol{\infty}_\perp; \infty^-, \boldsymbol{\xi}_\perp]_{[\boldsymbol{l}]} &\equiv \mathcal{P}\exp\left[ig\int_0^\infty d\tau \ \boldsymbol{l}\cdot\boldsymbol{A}_a t^a(\boldsymbol{\xi}_\perp + \boldsymbol{l}\tau)\right] \ , \end{aligned} \tag{3}$$

where we distinguish between longitudinal (lightlike, $n^2 = 0$) $[...]_{[n]}$ and transverse $[...]_{[\boldsymbol{l}]}$ links. (A generalized definition, which includes into the

Wilson lines the spin-dependent Pauli term $F^{\mu\nu}[\gamma_\mu, \gamma_\nu]$, is worked out in Ref. [22]).

Of course, the above expressions have to be quantized, using, for instance, functional-derivative techniques. This means that the gluon potential in the gauge link has to be Wick contracted with corresponding terms in the interaction Lagrangian, accompanying the Heisenberg fermion (quark) field operators.

At the tree-level, the "distribution of a quark in a quark" (here we consider only UV and rapidity divergences, which are independent of the particular hadronic state) is normalized as

$$\tilde{\mathcal{F}}^{(0)}_{q/q}(x, \boldsymbol{k}_\perp) = \frac{1}{2} \int \frac{d\xi^- d^2\xi_\perp}{2\pi(2\pi)^2} \mathrm{e}^{-ik^+\xi^- + ik_\perp \cdot \xi_\perp} \cdot$$
$$\times \langle p|\bar{\psi}(\xi^-, \xi_\perp)\gamma^+\psi(0^-, 0_\perp)|p\rangle = \delta(1-x)\delta^{(2)}(\boldsymbol{k}_\perp) \ , \tag{4}$$

Eq (4) and, formally, the integration over $\boldsymbol{k}_\perp$ yields the usual collinear (integrated) PDF:

$$\int dk^{(2)}_\perp \tilde{\mathcal{F}}^{(0)}_{q/q}(x, \boldsymbol{k}_\perp) = F^{(0)}_{q/q}(x) = \delta(1-x) \ . \tag{5}$$

However, already in the calculation of the one-gluon contributions, one encounters—besides the normal UV divergences—certain pathological singularities. Namely, one has at the one-loop level the following singular terms:

(1) Standard UV poles $\sim \frac{1}{\varepsilon}$ in the dimensional regularization: they can be removed by the usual $R-$operation and are controlled by renormalization-group evolution equations (analoguous to the DGLAP equation in the integrated case).
(2) Pure *rapidity divergences*: they give rise to logarithmic and double-logarithmic terms of the form $\sim \ln \eta \ , \ \ln^2 \eta$. These terms, although they depend on the additional rapidity parameter η [1,5,6,8], do not affect the UV renormalization properties, and can be safely resummed, e.g., by means of the Collins-Soper equation.
(3) Pathological *overlapping divergences*: they contain the UV and rapidity poles simultaneously $\sim \frac{1}{\varepsilon} \ln \eta$ and are considered to be *highly problematic*. The reason is that they prevent the removal of *all* UV-singularities by the standard $R-$procedure. Therefore, a special generalized renormalization procedure is needed in order to take care of those terms and enable the construction of well-defined renormalizable TMD.

It is interesting to note that working in the lightcone gauge with the Mandelstam-Leibbrandt prescription [23,24], one doesn't get any overlapping divergences, at least in the leading loop order [25]. The renormalization properties of operators and matrix elements containing Wilson lines and loops with or without obstructions have been extensively studied in various situations—see, e.g., Refs. [26–30]. The specifics of the TMD consist in the fact that, while the fermion fields are separated by a spacelike distance, the gauge links lay on pure lightlike rays, or on the $2D-$transverse plane at lightcone infinity.

The analysis of the one-loop anomalous dimension of the TMD (2) shows that the contribution of the overlapping singularity is nothing else, but the *cusp anomalous dimension* [29]. Therefore, in order to renormalize expression (2), one can apply, apart from the standard $R-$operation, an additional renormalization factor, which depends on the cusp angle and can be written as a vacuum matrix element of the Wilson lines evaluated along a special contour with an obstruction (cusp):

$$Z_{\chi}^{-1} = \left\langle 0 \left| \mathcal{P} \exp \left[ig \int_{\chi} d\zeta^{\mu}\ t^{a} A_{\mu}^{a}(\zeta) \right] \right| 0 \right\rangle . \tag{6}$$

The UV singularity of this factor cancels the cusp anomalous dimension from the overlapping divergence, thus rendering the re-defined TMD (2) renormalizable [9,10]. Therefore, the generalized renormalization procedure for the TMD can be formulated as

$$\tilde{\mathcal{F}}_{\rm ren}(x, \boldsymbol{k}_{\perp}, \chi, ...) = Z_{\rm R} \cdot Z_{\chi} \cdot \tilde{\mathcal{F}}(x, \boldsymbol{k}_{\perp}, \chi, ...) \ , \tag{7}$$

where $Z_{\rm R}$ is the usual renormalization constant, while Z_{χ} can be included in the definition of the TMD itself. In that case, it is treated as a "soft factor":

$$Z_{\chi} \equiv [\text{Soft Factor}] \ , \tag{8}$$

which is defined as

$$[\text{Soft Factor}] = \\ \langle 0 | \ \mathcal{P} e^{ig \int_{\mathcal{C}_{\chi}} d\zeta^{\mu}\ t^{a} A_{\mu}^{a}(\zeta)} \cdot \mathcal{P}^{-1} e^{-ig \int_{\mathcal{C'}_{\chi}} d\zeta^{\mu}\ t^{a} A_{\mu}^{a}(\xi+\zeta)} |0\rangle \ , \tag{9}$$

where the contours $\mathcal{C}$ and $\mathcal{C}'$ are explicitly given in Ref. [10]. Therefore, the generalized definition of the TMD reads [9,10,25]

$$\mathcal{F}_{i/h}\left(x, \boldsymbol{k}_{\perp}\right) = \frac{1}{2} \int \frac{d\xi^{-} d^{2}\boldsymbol{\xi}_{\perp}}{2\pi(2\pi)^{2}} \mathrm{e}^{-ik^{+}\xi^{-} + i\boldsymbol{k}_{\perp}\cdot\boldsymbol{\xi}_{\perp}}$$
$$\times \left\langle h \middle| \bar{\psi}_{i}(\xi^{-}, \boldsymbol{\xi}_{\perp})[\xi^{-}, \boldsymbol{\xi}_{\perp}; \infty^{-}, \boldsymbol{\xi}_{\perp}]^{\dagger}_{[n]}[\infty^{-}, \boldsymbol{\xi}_{\perp}; \infty^{-}, \boldsymbol{\infty}_{\perp}]^{\dagger}_{[\boldsymbol{l}]}\right.$$
$$\left. \times \gamma^{+}[\infty^{-}, \boldsymbol{\infty}_{\perp}; \infty^{-}, \boldsymbol{0}_{\perp}]_{[\boldsymbol{l}]}[\infty^{-}, \boldsymbol{0}_{\perp}; 0^{-}, \boldsymbol{0}_{\perp}]_{[n]}\psi_{i}(0^{-}, \boldsymbol{0}_{\perp}) \middle| h \right\rangle$$
$$\times \text{[Soft Factor]} . \tag{10}$$

This function is free (at least, at the one-loop order) of pathological divergences and is a well-defined renormalizable quantity.

3. Factorization and role of the soft factor

The soft factor, introduced above, naturally enters in the factorization formula (1): [Soft Factor] $= S$. However, its interpretation is twofold.

On the one hand, it formally looks similar to the "intrinsic" Coulomb phase found by Jakob and Stefanis [31] in QED for Mandelstam charged fields involving a gauge contour which is a timelike straight line. The name "intrinsic" derives from the fact that this phase is different from zero even in the absence of external charge distributions. Its origin was ascribed in [31] to the long-range interaction of the charged particle with its oppositely charged counterpart that was removed "behind the moon" after their primordial separation. Note that the existence of a balancing charge "behind the moon" was postulated before by several authors—see [31] for related references—in an attempt to restore the Lorentz covariance of the charged sector of QED. This phase is acquired during the parallel transport of the charged field along a timelike straight line from infinity to the point of interaction with the photon field and is absent in the local approach, i.e., for local charged fields joined by a connector. It is different from zero only for Mandelstam fields with their own gauge contour attached to them and keeps track of its full history since its primordial creation. Keep in mind that the connector is introduced ad hoc in order to restore gauge invariance and is not part of the QCD Lagrangian. In contrast, when one associates a distinct contour with each quark field, one, actually, implies that these Mandelstam field variables should also enter the QCD Lagrangian (see [31] for more details). However, a consistent formulation of such a theory for QCD is still lacking and not without complications of its own.

The analogy to our case is the following. First, formally adopting a direct contour for the gauge-invariant formulation of the TMD in the light-cone gauge, the connector gauge link does not contribute any anomalous dimension—except at the endpoints; this anomalous dimension being, however, irrelevant for the issue at stake. Hence, there is no intrinsic Coulomb phase in that case. Second, splitting the contour and associating each branch to a quark field, transforms it into a Mandelstam field and, as a result, adding together all gluon radiative corrections at the one-loop order, a η-dependent term survives that gives rise to an additional anomalous dimension. We have shown that this extra anomalous dimension can be viewed as originating from a contour with a discontinuity in the four-velocity $\dot{x}(\sigma)$ at light-cone infinity—a cusp obstruction.

Classically, it is irrelevant how the two distinct contours are joined, i.e., smoothly or by a sharp bend. But switching on gluon quantum corrections, the renormalization effect on the junction point reveals that the contours are not smoothly connected, but go instead through a cusp [10]. Here, we have a second analogy to the QED case discussed above. Similarly to the "particle behind the moon", this cusp-like junction point is "hidden" and manifests itself only through the path-dependent phase.

In that case, the soft factor looks like an intrinsic property of the gauge-invariant operators containing the fermion fields and must be taken into account in order to construct consistently the gauge-invariant renormalizable two-particle matrix elements.

On the other hand, the soft factor appears as the result of the separation the "soft" contributions from the one-loop graphs [7]. In this situation, it is needed in order to avoid the double-counting in the factorization formula (1). In general, these two soft factors might be different. In particular, it has been shown in Ref. [32] that the anomalous dimensions of the TMD within different subtraction schemes of the soft factors, are different. The relationship between these frameworks will be studied elsewhere.

4. Conclusions

The above mentioned results have been obtained by using the lightcone axial gauge with a proper regularization of the gluon propagator [10,25]. The extra rapidity divergences can be treated by different methods, e.g., one may shift the gauge links off the lightcone, or use, alternatively, the off-the-lightcone axial gauge [5–7]. In these cases, the additional rapidity variable parameterizes the deviation of the gauge links from the light rays in terms of the axial-gauge fixing vector. An advantage of the "pure lightcone" frameworks is the more straightforward physical interpretation of the factorization and the role of the collinear Wilson lines in the definition of TMDs, as well as the direct relationship between the (unintegrated) TMDs and the (integrated) collinear PDFs: one can get the collinear PDF, satisfying the DGLAP evolution equation, by simple $\boldsymbol{k}_\perp$-integration. In contrast, the "off-the-lightcone" frameworks don't allow us to perform such a procedure, so that more sophisticated methods must be invented. The complete proof of the QCD factorization, within the TMD approach (in particular within the "pure lightcone" scheme), as well as the clarification of the role played by the soft factors in different approaches, are still lacking.

References

1. D. E. Soper, *Phys. Rev. D* **15**, 1141 (1977); *Phys. Rev. Lett.* **43**, 1847 (1979).
2. J. C. Collins, *Phys. Rev. Lett.* **42**, 291 (1979).
3. J.C. Collins, *Acta Phys. Pol. B* **34**, 3103 (2003).
4. A.V. Belitsky, A.V. Radyushkin, *Phys. Rept.* **418**, 1 (2005).
5. J. C. Collins and D. E. Soper, *Nucl. Phys. B* **193**, 381 (1981); *Nucl. Phys. B* **213**, 545 (1983) Erratum.
6. J. C. Collins and D. E. Soper, *Nucl. Phys. B* **194**, 445 (1982).
7. X. Ji, J. Ma, and F. Yuan, *Phys. Rev. D* **71**, 034005 (2005).
8. J. Collins, *Proc. Sci.* **LC2008**, 028 (2008) [arXiv:0808.2665 [hep-ph]].
9. I. O. Cherednikov and N. G. Stefanis, *Phys. Rev. D* **77**, 094001 (2008).
10. I. O. Cherednikov and N. G. Stefanis, *Nucl. Phys. B* **802**, 146 (2008).
11. C. J. Bomhof and P. J. Mulders, *Nucl. Phys. B* **795**, 409 (2008).
12. J. C. Collins, T. C. Rogers, and A. M. Stasto, *Phys. Rev. D* **77**, 085009 (2008).
13. A. Bacchetta, D. Boer, M. Diehl, and P. J. Mulders, *JHEP* **0808**, 023 (2008).
14. J. Collins and J. W. Qiu, *Phys. Rev. D* **75**, 114014 (2007).
15. T. C. Rogers and P. J. Mulders, *Phys. Rev. D* **81**, 094006 (2010).
16. J. C. Collins and F. Hautmann, *Phys. Lett. B* **472**, 129 (2000).
17. X. Ji and F. Yuan, *Phys. Lett. B* **543**, 66 (2002).
18. A. V. Belitsky, X. Ji, and F. Yuan, *Nucl. Phys. B* **656**, 165 (2003).
19. D. Boer, P. J. Mulders, and F. Pijlman, *Nucl. Phys. B* **667**, 201 (2003).
20. J. C. Collins and A. Metz, *Phys. Rev. Lett.* **93**, 25001 (2004).
21. F. Hautmann, *Phys. Lett. B* **655**, 26 (2007).
22. I. O. Cherednikov, A. I. Karanikas and N. G. Stefanis, *Nucl. Phys. B* (2010), in print; *arXiv:1004.3697 [hep-ph]*.
23. S. Mandelstam, *Nucl. Phys. B* **213**, 149 (1983).
24. G. Leibbrandt, *Phys. Rev. D* **29**, 1699 (1984).
25. I. O. Cherednikov and N. G. Stefanis, *Phys. Rev. D* **80**, 054008 (2009).
26. A. M. Polyakov, *Nucl. Phys. B* **164**, 171 (1979).
27. N. S. Craigie and H. Dorn, *Nucl. Phys. B* **185**, 204 (1981).
28. N. G. Stefanis, *Nuovo Cim. A* **83**, 205 (1984).
29. G. P. Korchemsky and A. V. Radyushkin, *Nucl. Phys. B* **283**, 342 (1987).
30. I. I. Balitsky and V. M. Braun, *Nucl. Phys. B* **311**, 541 (1989).
31. R. Jakob and N. G. Stefanis, *Annals Phys.* **210** (1991) 112.
32. I. O. Cherednikov and N. G. Stefanis, *arXiv:0911.1031 [hep-ph]*.

PHOTOPRODUCTION OF A $\pi\rho$ PAIR AND TRANSVERSITY GPDs

M. EL BEIYAD[*,**], B. PIRE[*], L. SZYMANOWSKI[$], S. WALLON[**,$$]

[*] *CPhT, École Polytechnique, CNRS Palaiseau, 91128, France*
[**] *LPT, Université Paris XI, CNRS, 91404 Orsay, France*
[$] *Soltan Inst. for Nuclear Studies, Warsaw, 00-681, Poland*
[$$] *UPMC Univ. Paris 06, Faculté de Physique, 75252 Paris, France*

We demonstrate that the chiral-odd transversity generalized parton distributions (GPDs) of the nucleon can be accessed through the exclusive photoproduction process $\gamma + N \to \pi + \rho + N'$, in the kinematics where the meson pair has a large invariant mass and the final nucleon has a small transverse momentum, provided the vector meson is produced in a transversely polarized state. We calculate perturbatively the scattering amplitude at leading order in α_s. We build a simple model for the dominant transversity GPD $H_T(x, \xi, t)$ based on the concept of double distribution. Counting rates estimates show that the experiment looks feasible with the real photon beam characteristics expected at JLab@12 GeV, in low Q^2 leptoproduction at Jlab@12 GeV and in the COMPASS experiment.

Keywords: Generalised Parton Distributions, Exclusive Processes

1. Chiral-odd GPDs and factorization

Transversity quark distributions in the nucleon remain among the most unknown leading twist hadronic observables, mainly due to their chiral-odd character which enforces their decoupling in most hard amplitudes. After the pioneering studies of Ref. 1, much work[2] has been devoted to the study of many channels but experimental difficulties have challenged the most promising ones.

On the other hand, tremendous progress has been recently witnessed on the QCD description of hard exclusive processes, in terms of generalized parton distributions (GPDs) describing the 3-dimensional content of hadrons. Access to the chiral-odd transversity GPDs,[3] noted H_T, E_T, $\tilde{H}_T$, $\tilde{E}_T$, has however turned out to be even more challenging[4] than the usual transversity distributions: one photon or one meson electroproduction leading twist amplitudes are insensitive to transversity GPDs. The strategy which we follow here, as initiated in Ref. 5, is to study the leading twist contribution to processes where more mesons are present in the final state. A similar strategy has also been advocated recently in Ref. 6 for chiral-even GPDs. We advocate that the hard scale which allows to probe the short distance structure of the nucleon is $s = M^2_{\pi\rho} \sim |t'|$ in the fixed angle regime. In the example developed previously,[5] the process under study was the high energy photo (or electro) diffractive production of two vector mesons, the hard probe being the virtual "Pomeron" exchange (and the hard scale was the virtuality of this pomeron), in analogy with the virtual photon exchange occuring in the deep electroproduction of a meson.

We study here[7] a process involving a transversely polarized ρ meson in a 3-body final state:

$$\gamma + N \rightarrow \pi + \rho_T + N' \,. \tag{1}$$

It is a priori sensitive to chiral-odd GPDs due to the chiral-odd character of the leading twist distribution amplitude (DA) of ρ_T. The estimated rate depends of course much on the magnitude of the chiral-odd GPDs. Not much is known about them, but model calculations have been developed in Refs. 5,8–10 and a few moments have been computed on the lattice.[11] To factorize the amplitude of this process we use the now classical proof of the factorization of exclusive scattering at fixed angle and large energy.[12] The amplitude for the process $\gamma + \pi \rightarrow \pi + \rho$ is written as the convolution of mesonic DAs and a hard scattering subprocess amplitude $\gamma + (q + \bar{q}) \rightarrow (q + \bar{q}) + (q + \bar{q})$ with the meson states replaced by collinear quark-antiquark pairs. We then extract from the factorization procedure of the deeply virtual Compton scattering amplitude near the forward region the right to replace one entering meson DA by a $N \rightarrow N'$ GPD, and thus get Fig. 1. The needed skewness parameter ξ is written in terms of the meson pair squared invariant mass $M^2_{\pi\rho}$ as

$$\xi = \frac{\tau}{2-\tau} \quad , \quad \tau = \frac{M^2_{\pi\rho}}{S_{\gamma N} - M^2} \,. \tag{2}$$

Indeed the same collinear factorization property bases the validity of the leading twist approximation which either replaces the meson wave function by its DA or the $N \rightarrow N'$ transition by nucleon GPDs. A slight difference is that light cone fractions $(z, 1-z)$ leaving the DA are positive, while the corresponding fractions $(x+\xi, \xi-x)$ may be positive or negative in the case of the GPD. Our Born order calculation shows that this difference does not ruin the factorization property.

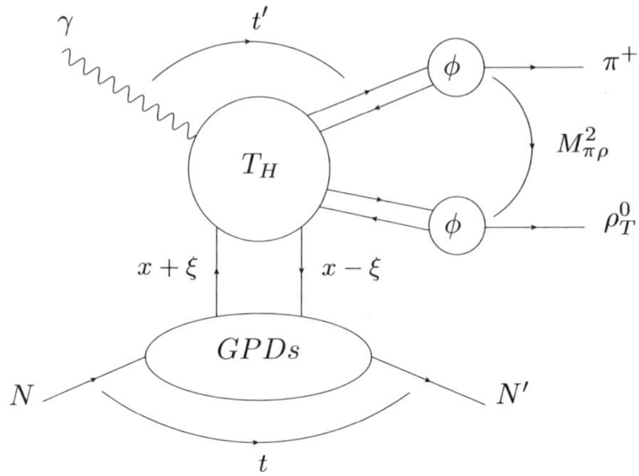

Fig. 1. Factorization of the amplitude for $\gamma + N \to \pi + \rho + N'$ at large $M^2_{\pi\rho}$.

In order for the factorization of a partonic amplitude to be valid, and the leading twist calculation to be sufficient, one should avoid the dangerous kinematical regions where a small momentum transfer is exchanged in the upper blob, namely small $t' = (p_\pi - p_\gamma)^2$ or small $u' = (p_\rho - p_\gamma)^2$, and the resonance regions for each of the invariant squared masses $(p_\pi + p_{N'})^2$, $(p_\rho + p_{N'})^2$, $(p_\pi + p_\rho)^2$.

Let us finally stress that our discussion applies as well to the case of electroproduction where a moderate virtuality of the initial photon may help to access the perturbative domain with a lower value of the hard scale $M_{\pi\rho}$.

2. The scattering amplitude

The scattering amplitude of the process (1) is written in the factorized form:

$$\mathcal{A}(t, M^2_{\pi\rho}, u') = \int_{-1}^{1} dx \int_{0}^{1} dv \int_{0}^{1} dz\; T^q(x,v,z)\, H^q_T(x,\xi,t)\Phi_\pi(z)\Phi_\perp(v)\,, \tag{3}$$

where T^q is the hard part of the amplitude and the transversity GPD of a parton q in the nucleon target which dominates at small momentum transfer is defined[3] by

$$\begin{aligned}&\langle N'(p_2),\lambda'|\bar{q}\left(-\tfrac{y}{2}\right)\sigma^{+j}\gamma^5 q\left(\tfrac{y}{2}\right)|N(p_1),\lambda\rangle\\ &= \bar{u}(p',\lambda')\sigma^{+j}\gamma^5 u(p,\lambda)\int_{-1}^{1} dx\; e^{-\frac{i}{2}x(p_1^+ + p_2^+)y^-}\, H^q_T\,,\end{aligned} \tag{4}$$

where λ and λ' are the light-cone helicities of the nucleon N and N'. The chiral-odd DA for the transversely polarized meson vector ρ_T, is defined, in leading twist 2, by the matrix element

$$\langle 0|\bar{u}(0)\sigma^{\mu\nu}u(x)|\rho_T^0(p,\epsilon_\pm)\rangle = \frac{i}{\sqrt{2}}(\epsilon_\pm^\mu(p)p^\nu - \epsilon_\pm^\nu(p)p^\mu)f_\rho^\perp \int_0^1 du\ e^{-iup\cdot x}\ \phi_\perp(u) \quad (5)$$

where $\epsilon_\pm^\mu(p_\rho)$ is the ρ-meson transverse polarization and with $f_\rho^\perp = 160$ MeV. Two classes of Feynman diagrams without (Fig. 2) and with (Fig. 3) a 3-gluon vertex describe this process.

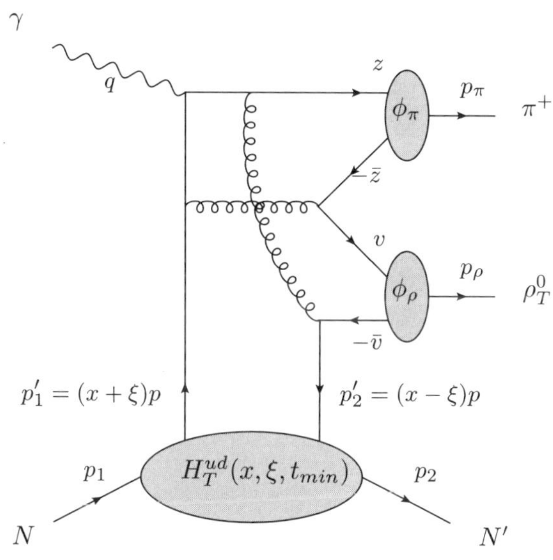

Fig. 2. Sample diagram without a 3-gluon vertex.

The scattering amplitude gets both a real and an imaginary part. Integrations over v and z have been done analytically whereas numerical methods are used for the integration over x.

3. Results

Various observables can be calculated with this amplitude. We stress that even the unpolarized differential cross-section $\frac{d\sigma}{dt\,du'\,dM_{\pi\rho}^2}$ is sensitive to the transversity GPD. To estimate the rates, we modelize the dominant

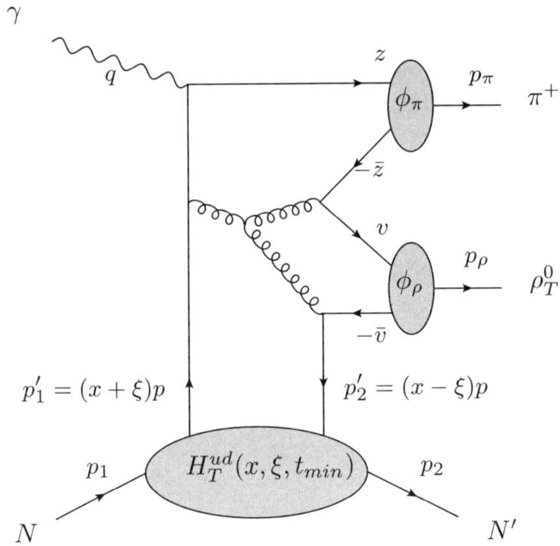

Fig. 3. Sample diagram with a 3 gluon vertex.

transversity GPD $H_T^q(x,\xi,t)$ $(q = u,\ d)$ in terms of double distributions

$$H_T^q(x,\xi,t=0) = \int_\Omega d\beta\, d\alpha\ \delta(\beta+\xi\alpha-x) f_T^q(\beta,\alpha,t=0)\,, \tag{6}$$

with $\Omega = \{|\alpha|^2 + |\beta|^2 \leq 0\}$ and where f_T^q is the quark transversity double distribution written as

$$f_T^q(\beta,\alpha,t=0) = \Pi(\beta,\alpha)\,\delta\, q(\beta)\Theta(\beta) - \Pi(-\beta,\alpha)\,\delta\bar{q}(-\beta)\,\Theta(-\beta)\,, \tag{7}$$

where $\Pi(\beta,\alpha) = \frac{3}{4}\frac{(1-\beta)^2-\alpha^2}{(1-\beta)^3}$ is a profile function and δq, $\delta\bar{q}$ are the quark and antiquark transversity parton distribution functions of Ref. 13. The t-dependence of these chiral-odd GPDs - and its Fourier transform in terms of the transverse localization of quarks in the proton[14] - is very interesting but completely unknown. We describe it in a simplistic way using a dipole form factor:

$$H_T^q(x,\xi,t) = H_T^q(x,\xi,t=0) \times \frac{C^2}{(t-C)^2} \qquad (C = 0.71\ \text{GeV}^2)\,. \tag{8}$$

In Fig. 4, we show the $M^2_{\pi\rho}$ dependence of the differential cross section $d\sigma/dM^2_{\pi\rho}$. for JLab kinematics ($s_{\gamma N} = 20\,\text{GeV}^2$). On the same plot, we show the similar cross section for the case where the ρ meson is longitudinally polarized. Note that in this case (which will be discussed in detail in a forthcoming publication), the GPDs which contribute are the usual chiral even ones, parametrized here through the usual double distribution ansatz.

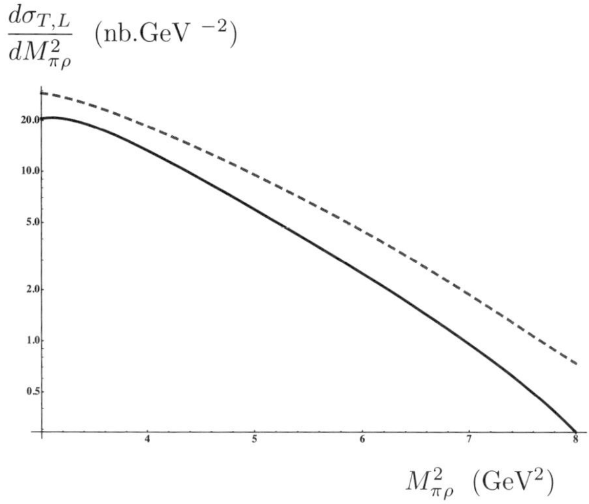

Fig. 4. Differential cross section $\frac{d\sigma}{dM^2_{\pi\rho}}$ at $s_{\gamma N} = 20$ GeV^2 for the process $\gamma p \to \pi^+ \rho^0 n$ for a transversely polarized (full curve) and a longitudinally polarized (dashed curve) ρ meson. The cuts are discussed in detail in Ref. 7.

Rate estimates[7] show that expected luminosities at JLab (after the 12 GeV energy increase) and at the Compass experiment at CERN are sufficient to gather a sample of interesting events allowing this physics to be experimentally tested.

Note that in the case of leptoproduction, the virtuality Q^2 of the exchanged photon plays no crucial role in our process, and the virtual photoproduction cross section is almost Q^2-independent if we choose to select events in a sufficiently narrow Q^2−window ($.02 < Q^2 < 1$ GeV^2), which is legitimate since the effective photon flux is strongly peaked at very low values of Q^2. This procedure applied to muoproduction at Compass yields a rate sufficient to get an estimate of the transversity GPDs in the region of small ξ (~ 0.01).

Target transverse spin asymetries need to consider nucleon helicity flip amplitudes, i.e. the effect of $E_T(x, \xi, t)$ and/or $\tilde{E}_T(x, \xi, t)$ GPDs. This will be studied later on.

In conclusion, we expect the process discussed here to be observable in two quite different energy ranges, which should give complementary information on the chiral-odd transversity GPDs: the large ξ region may be scrutinized at JLab and the smaller ξ region may be studied at COMPASS. Let us stress that our study is built on known leading twist factorization theorems, i.e. the scattering amplitude involves only leading twist non-perturbative components. This stays in contrast with other attempts to access transversity GPDs.[15]

Acknowledgments

This work is partly supported by the ANR-06-JCJC-0084 and the Polish Grant N202 249235.

References

1. J. P. Ralston and D. E. Soper, Nucl. Phys. B **152**, 109 (1979); X. Artru and M. Mekhfi, Z. Phys. C **45**, 669 (1990); J. L. Cortes *et al.*, Z. Phys. C **55**, 409 (1992); R. L. Jaffe and X. D. Ji, Phys. Rev. Lett. **67**, 552 (1991).
2. V. Barone, A. Drago and P. G. Ratcliffe, Phys. Rept. **359**, 1 (2002); M. Anselmino, arXiv:hep-ph/0512140; B. Pire and L. Szymanowski, Phys. Rev. Lett. **103**, 072002 (2009).
3. M. Diehl, Eur. Phys. J. C **19**, 485 (2001).
4. M. Diehl *et al.*, Phys. Rev. D **59**, 034023 (1999); J. C. Collins and M. Diehl, Phys. Rev. D **61**, 114015 (2000).
5. D. Yu. Ivanov *et al.*, Phys. Lett. B **550**, 65 (2002); R. Enberg *et al.*, Eur. Phys. J. C **47**, 87 (2006).
6. S. Kumano *et al.*, Phys. Rev. D **80**, 074003 (2009).
7. M. E. Beiyad *et al.* Phys. Lett. B **688**, 154 (2010) and arXiv:0911.2611 [hep-ph] and 1006.0740 [hep-ph].
8. S. Scopetta, Phys. Rev. D **72**, 117502 (2005).
9. M. Pincetti *et al.*, Phys. Rev. D **72**, 094029 (2005) and Czech. J. Phys. **56**, F229 (2006).
10. M. Wakamatsu, Phys. Rev. D **79**, 014033 (2009); D. Chakrabarti *et al.*, Phys. Rev. D **79**, 034006 (2009).
11. M. Gockeler *et al.*, Phys. Rev. Lett. **98**, 222001 (2007) and Phys. Lett. B **627**, 113 (2005).
12. A. V. Efremov and A. V. Radyushkin, Theor. Math. Phys. **44** (1980) 573 [Teor. Mat. Fiz. **44** (1980) 17]; *ibid* Theor. Math. Phys. **44** (1981) 664 [Teor. Mat. Fiz. **44** (1980) 157]; G. P. Lepage and S. J. Brodsky, Phys. Rev. D **22** (1980) 2157.
13. M. Anselmino *et al.*, Phys. Rev. D **75** (2007) 054032.
14. M. Burkardt, Phys. Rev. D **62** (2000) 071503; J. P. Ralston and B. Pire, Phys. Rev. D **66** (2002) 111501; M. Burkardt, Phys. Rev. D **72**, 094020 (2005); M. Diehl and Ph. Hagler, Eur. Phys. J. C **44**, 87 (2005); A. Mukherjee, D. Chakrabarti and R. Manohar, AIP Conf. Proc. **1149**, 533 (2009).
15. S. Ahmad, G. R. Goldstein and S. Liuti, Phys. Rev. D **79**, 054014 (2009); S. V. Goloskokov and P. Kroll, Eur. Phys. J. C **65**, 137 (2010).

LIGHT-FRONT DENSITIES FOR TRANSVERSELY POLARIZED HADRONS

C. LORCÉ* and M. VANDERHAEGHEN
Institut fuer Kernphysik, Johannes Gutenberg-Universitaet, Mainz, D55099, Germany
**E-mail: lorce@kph.uni-mainz.de*

B. PASQUINI
Dipartimento di Fisica Nucleare e Teorica, Università degli Studi di Pavia, and INFN, Sezione di Pavia, I-27100 Pavia, Italy
E-mail: Barbara.Pasquini@pv.infn.it

We discuss the recent interpretation of quark-distribution functions in the plane transverse to the light-cone direction. Such a mapping is model independent and allows one to build up multidimensional pictures of the hadron and to develop a semi-classical interpretation of the quark dynamics. We comment briefly the results obtained from the form factors of the nucleon. A generalization to a target with arbitrary spin led to a set of preferred values for the electromagnetic coupling characterizing structureless particles. Generalized polarizabilities can also be interpreted in that frame as the distortion of the charge densities due to a quasi-static external electromagnetic field. Finally, we present results for the generalized transverse-momentum dependent distributions which encode the most complete information about quark distributions.

Keywords: charge densities, generalized polarizabilities, Wigner distributions

1. Introduction

Hadrons are composite objects. Their interaction with external probes like *e.g.* photons is parameterized in terms of Lorentz-invariant functions. The latter encode information about the distribution of partons in momentum and/or position space. In order to extract this information, one has to go to a frame where the hadron moves with almost the speed of light allowing for a (quasi-)probabilistic/charge-density interpretation. This interpretation is in principle model independent. A brief discussion can be found in Section 2.

We first focus on form factors (FFs) which, by Fourier transform, give information on the charge distribution in the transverse plane, or impact parameter space, see Section 3. While for a longitudinally polarized target the pattern is axially symmetric, two-dimensional multipoles appear for a transversely polarized target. Such multipoles turn out to be a manifestation of the target composite nature and to be intimately connected to quark orbital angular momentum. Since a structureless particle does not contain any intrinsic orbital angular momentum, it should not show any multipole pattern. This constraint then leads to a set of preferred values for the electromagnetic (EM) coupling ("natural" EM moments) of particles with arbitrary spin.

Interpretation of generalized polarizabilities (GPs) is then discussed in Section 4 in the infinite momentum frame. They describe how the charge distribution is affected by a quasi-static external electromagnetic field. Similarly to FFs, interesting multipole patterns also show up. However, working in the infinite momentum frame we can arrive to a spatial representation for only three combinations of the six independent GPs.

Finally, we present in Section 5 the recently introduced generalized transverse-momentum dependent distributions (GTMDs) and their connection with the generalized parton distributions (GPDs), transverse-momentum dependent distributions (TMDs) and Wigner distributions. The last ones encode all the possible correlations between quark transverse position and momentum. As it is not possible to directly access Wigner distributions from experiments, we use relativistic quark models to learn about the quark dynamics encoded in these quantities.

2. Impact parameter space

According to the standard interpretation [1,2] the charge density can be identified with the three-dimensional Fourier transform of the electric Sachs form factor G_E, i.e.

$$\rho(\vec{r}) = \int \frac{\mathrm{d}^3 q}{(2\pi)^3}\, e^{-i\vec{q}\cdot\vec{r}}\, G_E(\vec{q}), \tag{1}$$

where G_E is defined as the matrix element of the charge current between the initial and final nucleon with momentum $\vec{P}$ and $\vec{P}' = \vec{P}+\vec{q}$, respectively. The identification in Eq. (1) is valid only in the nonrelativistic approximation. To work out the Fourier transform, one has to know the FFs for every Q^2. In the Breit frame the latter is identified with the three-momentum of the virtual photon, i.e. $Q^2 = \vec{q}^{\,2}$. This means that for every value of Q^2 we have to move to a different frame and the charge density undergoes naturally a different Lorentz contraction. Moreover, in order to have a probabilistic/charge-density interpretation, the number of particles should be conserved. However, in the Breit frame nothing prevents the photon to create or annihilate a quark-antiquark pair.

All these problems are cured in the infinite momentum frame (IMF) with $q^+ = q^0 + q^z = 0$ (the so-called Drell-Yan-West frame). In such a frame the photon is kinematically not allowed to change the number of quarks since the light-cone momentum of a massive particle is strictly positive $p^+ > 0$. Moreover, in the limit $p^+ \to \infty$ the hadron undergoes an extreme Lorentz contraction. Only a two-dimensional charge density [3–5] is then meaningful and can be identified with the two-dimensional Fourier transform of the matrix element of J^+

$$\rho_{\vec{s}}(\vec{b}) = \int \frac{\mathrm{d}^2 q_\perp}{(2\pi)^2} \frac{e^{-i\vec{q}_\perp \cdot \vec{b}}}{2p^+} \langle p^+, \tfrac{\vec{q}_\perp}{2}, \vec{s} | J^+(0) | p^+, -\tfrac{\vec{q}_\perp}{2}, \vec{s} \rangle, \tag{2}$$

with the photon virtuality $Q^2 = \vec{q}_\perp^{\,2}$, and $\vec{s}$ denoting the hadron polarization.

3. Form factors

For a spin-1/2 hadron with definite light-cone helicity λ, the charge density is simply given by the Fourier transform of the Dirac FF

$$\rho_\lambda(\vec{b}) = \int_0^\infty \frac{\mathrm{d}Q}{2\pi} Q \, J_0(bQ) \, F_1(Q^2), \tag{3}$$

where J_0 is the cylindrical Bessel function. Since no preferred direction appears in the transverse plane, it comes without any surprise that the corresponding charge density is axially symmetric. Using phenomenological parameterizations of the experimental nucleon FFs, one observes a counterintuitive negative core in the neutron charge distribution [6]. For a transverse polarization, the charge density receives a dipole contribution [7] from the Fourier transform of the Pauli FF

$$\rho_{s_\perp}(\vec{b}) = \rho_\lambda(\vec{b}) + \sin\phi_b \int_0^\infty \frac{\mathrm{d}Q}{2\pi} \frac{Q^2}{2M} J_1(bQ) \, F_2(Q^2), \tag{4}$$

where ϕ_b is the angle between quark transverse position $\vec{b}$ and transverse hadron polarization $\vec{s}_\perp$. This transverse polarization gives a preferred direction in the transverse plane allowing for multipole patterns. It also involves matrix elements with helicity flip, where quark orbital angular momentum is necessary. The dipole moment in natural units $(e/2M)$ is given by $F_2(0)$, *i.e.* the anomalous magnetic moment. Higher multipole moments can be observed for hadrons with higher spin [8–11]. There is therefore a connection between anomalous EM moments, quark orbital angular momentum and distortions of the charge densities. In the case of a particle *without* internal structure, there can not be any intrinsic orbital angular momentum and the charge densities remain axially symmetric. This provides us with a sufficient condition to derive the natural values of the EM moments [11] characterizing the interaction of a structureless particle of any spin with the electromagnetic field. The general results agree with the known lower spin cases (Standard Model and Supergravity) and support the prediction of a $g = 2$ gyromagnetic ratio for any spin.

Table 1. Natural EM moments of a spin-j particle with electric charge $Z = +1$ are organized according to a pseudo Pascal triangle, when expressed in terms of natural units of e/M^l and $e/2M^l$ for $G_{El}(0)$ and $G_{Ml}(0)$, respectively.

j	$G_{E0}(0)$	$G_{M1}(0)$	$G_{E2}(0)$	$G_{M3}(0)$	$G_{E4}(0)$	$G_{M5}(0)$	$G_{E6}(0)$
0	1	0	0	0	0	0	0
1/2	1	1	0	0	0	0	0
1	1	2	−1	0	0	0	0
3/2	1	3	−3	−1	0	0	0
2	1	4	−6	−4	1	0	0
5/2	1	5	−10	−10	5	1	0
3	1	6	−15	−20	15	6	−1

4. Generalized Polarizabilities

In the presence of an external electric field, any charge density is deformed and develops an induced polarization. The "ease" by which such a charge density is deformed is described by the electric polarizability. Virtual Compton scattering (VCS) allows us to study [12] the distortion of transverse charge densities in an external electric field. The nucleon VCS tensor is given by

$$H^{\mu\nu} = -i \int d^4x \, e^{-iq\cdot x} \, \langle p', \vec{s}_N | T\{J^\mu(x) J^\nu(0)\} | p, \vec{s}_N \rangle. \tag{5}$$

In the limit of low energy of the final real photon, it is described in terms of six Generalized Polarizabilities (GPs). To obtain a spatial representation, we consider the VCS process in a light-front frame. In the low-energy limit $\nu = q \cdot P/M \to 0$ ($P = \frac{p'+p}{2}$) the VCS amplitude $H = \epsilon'^*_\nu \, H^{+\nu}$ describes the linear response to an external quasi-static electric dipole field, allowing us to define an induced polarization $\vec{P}(\vec{q})$

$$i\vec{\epsilon}'^*_\perp \cdot \vec{P}(\vec{q}) \equiv \frac{1+\tau}{2P^+} \frac{\partial H}{\partial \nu}\bigg|_{\nu=0}. \tag{6}$$

By Fourier transforming $\vec{P}(\vec{q})$ we obtain the spatial distortion of the charge density induced by the external electric field ($\hat{a} = \vec{a}/|\vec{a}|$), i.e.

$$\begin{aligned}\vec{P}(\vec{b}) &= \int \frac{d^2 q_\perp}{(2\pi)^2}\, e^{-i\vec{q}_\perp\cdot\vec{b}}\, \vec{P}(\vec{q}) \\ &= \hat{b}\int_0^\infty \frac{dQ}{2\pi}\, Q\, J_1(bQ)\, A(Q^2) \\ &\quad - \hat{b}\,(\vec{s}_\perp \times \hat{e}_z)\cdot\hat{b}\int_0^\infty \frac{dQ}{2\pi}\, Q\, J_2(bQ)\, B(Q^2) \\ &\quad + (\vec{s}_\perp \times \hat{e}_z)\int_0^\infty \frac{dQ}{2\pi}\, Q\left[J_0(bQ)\, C(Q^2) + \frac{J_1(bQ)}{bQ}\, B(Q^2)\right].\end{aligned} \tag{7}$$

Only three multipoles are at our disposal to interpret six GPs. The first term corresponds just to the standard dipole distortion of the charge density, irrespective of the nucleon polarization. The other two terms appear only when the nucleon is transversely polarized. In absence of external electric field and for a transversely polarized nucleon, the charge distribution already contains a dipole structure (see Eq. (4)). Applying an external electric field on this dipole charge density then leads to a monopole and a quadrupole distortions.

5. Generalized Transverse-Momentum dependent Distributions

FFs provide information about the quark transverse position while parton distribution functions (PDFs) give their longitudinal-momentum distribution. The full correlation between quark transverse position and longitudinal momentum is contained in GPDs leading then to a three-dimensional picture of the nucleon. A dual three-dimensional picture of the nucleon is provided by TMDs which give the full quark three-momentum distribution. The full correlation between quark three-momentum and transverse position is captured by the Wigner distributions defined as Fourier transform in the transverse plane of the so-called GTMDs. All FFs, PDFs, GPDs, and TMDs appear just as particular limits and/or projections of GTMDs. There are sixteen GTMDs [13] which parameterize the general parton correlator

$$W^{[\Gamma]}_{\lambda'\lambda} = \frac{1}{2}\int \frac{d^4 z}{(2\pi)^3}\, \delta(z^+)\, e^{iq\cdot z}\langle p', \lambda'|\overline{\psi}(-\tfrac{1}{2}z)\Gamma\mathcal{W}(-\tfrac{1}{2}z, \tfrac{1}{2}z|n)\psi(\tfrac{1}{2}z)|p, \lambda\rangle, \tag{8}$$

where Γ is a matrix in Dirac space, ψ is the quark field, and $\mathcal{W}$ a Wilson line which preserves gauge invariance. These GTMDs $X(x, \xi, \vec{k}_\perp^2, \vec{k}_\perp\cdot\vec{\Delta}_\perp, \vec{\Delta}_\perp^2; \eta)$

are functions of quark mean momentum $(x, \vec{k}_\perp)$ and momentum transfer $(\xi, \vec{\Delta}_\perp)$. The parameter η indicates whether the Wilson line is past- or future-pointing.

Wigner Distributions in the context of quantum field theory have already been discussed to some extent in the Breit frame [14,15]. For the reason mentioned before, it is preferable to work in the IMF. The distribution of an unpolarized quark in an unpolarized hadron is given by the GTMD F^e_{11}, and is related to the GPD H and the TMD f_1 as follows

$$\begin{aligned} H(x, 0, \vec{\Delta}^2_\perp) &= \int \mathrm{d}^2 k_\perp \, F^e_{11}(x, 0, \vec{k}^2_\perp, \vec{k}_\perp \cdot \vec{\Delta}_\perp, \vec{\Delta}^2_\perp), \\ f_1(x, \vec{k}^2_\perp) &= F^e_{11}(x, 0, \vec{k}^2_\perp, 0, 0). \end{aligned} \tag{9}$$

By Fourier transforming F^e_{11} with respect to $\vec{\Delta}_\perp$ and integrating over x, we obtain a (transverse) phase-space distribution $\rho(\vec{k}^2_\perp, \vec{k}_\perp \cdot \vec{b}, \vec{b}^2)$. The only two available transverse vectors are $\vec{k}_\perp$ and $\vec{b}$. This means that for fixed $\hat{k}_\perp \cdot \hat{b}$ and $|\vec{k}_\perp|$ the distribution is axially symmetric, which is physically meaningful since a global rotation of the distribution around $\hat{e}_z$ should not have any effect.

If we are now interested in the amplitude to find a fixed transverse momentum $\vec{k}_\perp$ in the transverse plane, the distribution is not axially symmetric anymore due to the $\vec{k}_\perp \cdot \vec{b}$ dependence. This amplitude can not be directly accessed from experiments. We therefore use successful phenomenological constituent quark models (namely the Light-Cone Constituent Quark Model [16] and the Chiral Quark-Soliton Model [17]) restricted to the valence sector [18] to compute F^e_{11}. Both models give the same qualitative picture with a larger distribution amplitude when $\vec{k}_\perp \perp \vec{b}$ and smaller when $\vec{k}_\perp \parallel \vec{b}$, see Fig. 1. This can be understood with naive semi-classical arguments. The radial momentum $(\vec{k}_\perp \cdot \hat{b})\,\hat{b}$ of a quark has to decrease rapidly in the periphery because of confinement. The polar momentum $\vec{k}_\perp - (\vec{k}_\perp \cdot \hat{b})\,\hat{b}$ receives a contribution from the orbital motion of the quark which can still be significant in the periphery (in an orbital motion, one does not need to reduce the momentum to avoid a quark escape). This naive picture also tells us that this phenomenon should become more pronounced as we go to peripheral regions ($|\vec{b}| \gg$) and to high quark momenta ($|\vec{k}_\perp| \gg$). This tendency is supported by both models.

It is also interesting to compare up and down quark distributions. Let us fix once again $\hat{k}_\perp \cdot \hat{b}$ and $|\vec{k}_\perp|$. The distribution being axially symmetric in the transverse plane, we focus on the radial distribution, see Fig. 2. The up quark distribution has been divided by two for comparison with the down

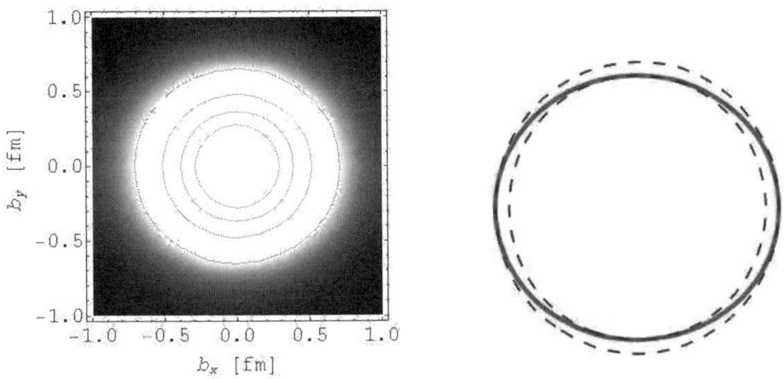

Fig. 1. Distribution (left) and equi-amplitude curve (right) in the transverse plane with $\hat{k}_\perp = \hat{e}_y$ and $|\vec{k}_\perp| = 0.45$ GeV for an unpolarized quark in an unpolarized nucleon.

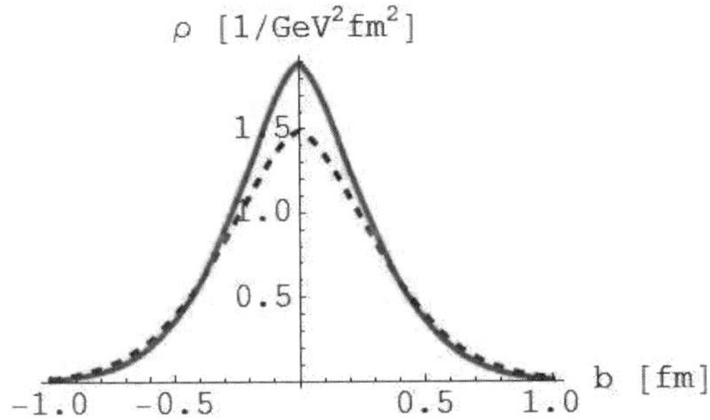

Fig. 2. Radial distributions in the transverse plane with $\hat{k}_\perp \cdot \hat{b} = 1$ and $|\vec{k}_\perp| = 0.55$ GeV for an unpolarized quark in an unpolarized proton. Solid curve corresponds to half of the up quark distribution. Dashed curve corresponds to the down quark distribution.

quark distribution. Up quarks appear to be more concentrated around the center than down quarks. For a neutron, we just have to exchange up and down quarks. We can therefore see that the center of the neutron is negative, in agreement with the conclusion obtained using the phenomenological neutron FFs.

6. Summary

The standard interpretation of FFs in the Breit frame is not fully consistent with the principles of quantum mechanics and relativity. However, a probabilistic/charge-density interpretation is available in a frame where the target is moving with almost the speed of light. It then turns out that FFs describe the distribution of quark in the transverse plane. While this distribution is axially symmetric for a longitudinally polarized target, multipoles appear when the target is transversely polarized. This distortion comes from the intrinsic orbital angular momentum and hence the composite nature of the target. This observation allowed us to derive the natural EM moments characterizing the interaction of any structureless particle with the electromagnetic field. We also tried to interpret the Generalized Polarizabilities in a light-front frame based on the VCS process in the low-energy limit. It appeared that only three combinations out of six GPs can be mapped out. Finally, we discussed briefly phase-space (Wigner) distribution using constituent quark models. Non-trivial pattern have been obtained and interpreted semi-classically as related to orbital angular momentum.

Acknowledgments

This work was supported in part by the Research Infrastructure Integrating Activity "Study of Strongly Interacting Matter" (acronym HadronPhysics2, Grant Agreement n. 227431) under the Seventh Framework Programme of the European Community, by the Italian MIUR through the PRIN 2008EKLACK "Structure of the nucleon: transverse momentum, transverse spin and orbital angular momentum".

References

1. F. J. Ernst, R. G. Sachs and K. C. Wali, *Phys. Rev.* **119**, 1105 (1960).
2. R. G. Sachs, *Phys. Rev.* **126**, 2256 (1962).
3. D. E. Soper, *Phys. Rev. D* **15**, 1141 (1977).
4. M. Burkardt, *Phys. Rev. D* **62**, 071503 (2000).
5. M. Burkardt, *Int. J. Mod. Phys. A* **18**, 173 (2003).
6. G. A. Miller, *Phys. Rev. Lett.* **99**, 112001 (2007).
7. C. E. Carlson and M. Vanderhaeghen, *Phys. Rev. Lett.* **100**, 032004 (2008).
8. C. E. Carlson and M. Vanderhaeghen, *Eur. Phys. J. A* **41**, 1 (2009).
9. C. Alexandrou *et al.*, *Phys. Rev. D* **79**, 014507 (2009).
10. C. Alexandrou *et al.*, *Nucl. Phys. A* **825**, 115 (2009).
11. C. Lorce, *Phys. Rev. D* **79**, 113011 (2009).
12. M. Gorchtein, C. Lorce, B. Pasquini and M. Vanderhaeghen, *Phys. Rev. Lett.* **104**, 112001 (2010).
13. S. Meissner, A. Metz and M. Schlegel, *JHEP* **0809**, p. 056 (2009).
14. X. D. Ji, *Phys. Rev. Lett.* **91**, 062001 (2003).
15. A. V. Belitsky, X. D. Ji and F. Yuan, *Phys. Rev. D* **69**, 074014 (2004).
16. B. Pasquini and S. Boffi, *Phys. Lett. B* **653**, 23 (2007).
17. C. Lorce, *Phys. Rev. D* **79**, 074027 (2009).
18. C. Lorce and B. Pasquini, Wigner distributions on the light cone, in preparation, (2010).

INTRODUCTION TO EIC/DETECTOR CONCEPT

T. HORN* et al.

Physics Department, The Catholic University of America, Washington, DC 20064, USA

The main goal of a future electron-ion collider is to allow us to explore the sea quark and gluon structure of the nucleon. The key physics measurements require a high figure of merit for both polarized and unpolarized cases. Such measurements could ideally be performed with a high-luminosity collider of moderate center of mass energies. The concept of this setup is described and examples of the main physics categories are discussed. The detector design, which is largely driven by exclusive DIS measurements, is introduced as well.

Building an Electron-Ion Collider (EIC) has several advantages compared to upgrading current fixed target facilities. Perhaps the most obvious one is that it is significantly easier to reach higher center of mass energies, $E_{CM}^2 = s$. Indeed, even a small deviation from fixed target configurations gives a large gain in s. For instance, while the Jefferson Lab at 12 GeV [a] will reach s=$2E_eM_p$=20 GeV^2, and COMPASS an s of 375 GeV^2 using a 200 GeV beam, already for collisions of 9 GeV electrons on 60 GeV protons is s=$4E_eE_p$=2027 GeV^2. Another great advantage of a collider is that it allows access to spin physics with a high Figure Of Merit (FOM). In the unpolarized case, it is given by $FOM = Rate = Luminosity \times CrossSection \times Acceptance$. In a comparison, the latter two are important to consider as many cross sections rise with energy, and beam energies can be adjusted to optimize the collider kinematics to improve the acceptance for the reaction products. The polarized figure of merit is in turn given by $FOM = Rate \times (TargetPolarization)^2 \times (TargetDilution)^2$, where the contribution of the product of the target polarization and dilution can be large. In the case of a collider, one can reach both high longitudinal and transverse ion polarizations without limitations on current (luminosity)

[a]with 11 GeV electrons

or acceptance from transverse holding fields.

A recently decommissioned collider (HERA), which operated between 1992-2007, collided 27 GeV electrons on 920 GeV protons at luminosities of $L{=}5 \times 10^{31}$. Proposals for future $e-p$ and $e-A$ colliders are the Large Hadron electron Collider (LHeC) at CERN[2] and the EIC in the US, which would be built at either the Jefferson Laboratory (JLab) in Newport News, VA, or the Brookhaven National Laboratory (BNL) in Upton, NY.[3]

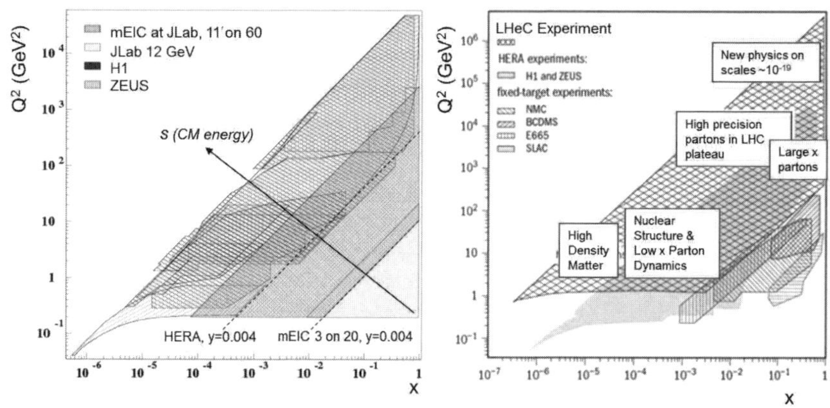

Fig. 1. The Q^2/x_B kinematic coverage of the EIC in comparison to JLab 12 GeV, HERA, the LHeC, and other existing facilities. The diagonal represents the center of mass energy. (We thank P. Newman for the code used to generate one of these figures).

Figure 1 shows the kinematic coverage in the Q^2/x_B plane for the medium-energy EIC and other facilities. The large-x_B, low-Q^2 region is largely covered by JLab 12 GeV, while the low-x_B and high-Q^2 regions will be covered by the LHeC, if it is built. As shown on the right-hand side in Figure 1, the LHeC will cover most of the kinematics of the former HERA collider at DESY. A medium-energy EIC would cover the range between JLab 12 GeV and the LHeC, providing measurements with high luminosity and excellent polarization with common systematics. It would thus be complementary to both JLab 12 GeV and the LHeC, and supersede all low-luminosity fixed-target experiments (including COMPASS) which cover small portions of this region. A high-energy EIC, which could be a future upgrade, would increase the overlap with HERA and the LHeC, although it could by itself not connect with JLab 12 GeV. It would retain the heavy ion capability and good polarization for light ions of the medium-energy

EIC - two things that HERA lacked. If the LHeC is not built, this may be the only machine that can see gluon saturation in $e-A$ collisions.

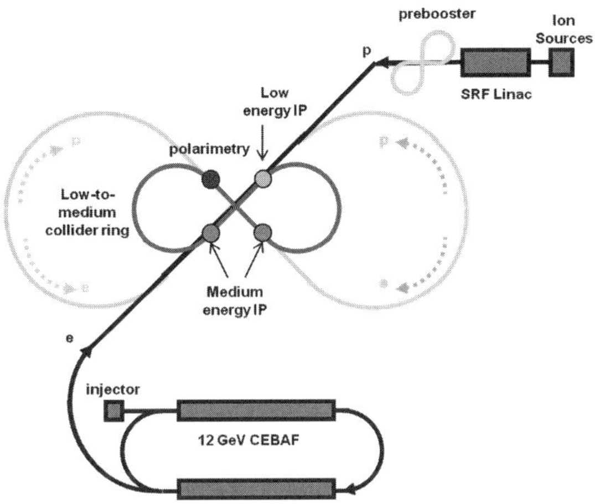

Fig. 2. The layout of the EIC@JLAB. This proposal takes advantage of the existing CEBAF electron beam after the 12 GeV upgrade. For the medium-energy EIC, intersecting figure-8 rings with 1 km circumference would be added to provide electron-ion collisions. The figure-8 design is advantageous as is minimizes the impact of depolarizing resonances and supports multiple interaction regions. The gray figure-8 represents a possible future high-energy upgrade option.

The layout of the high-luminosity EIC at JLab is shown in Figure 2. The medium-energy collider figure-8 storage ring has a circumference of 1 km, which is comparable to the 1.4 km of the CEBAF accelerator. The ring would store 3-11 GeV electrons or positrons injected from CEBAF, which would collide with proton and ion beams. With arc dipole magnets of conservative strength (6.7 T), it could operate with 20-70 GeV protons. With LHC-style magnets, the maximum energy would increase to 80-100 GeV. It will thus cover a range in s from 250 to 3000-4000 GeV^2 (an order of magnitude higher than COMPASS). The design provides a large dynamic range where the high luminosity is not limited to the highest energies, allowing for a meaningful connection to the results from the 12 GeV program. This is, for instance, essential for exclusive reactions. The JLab EIC design can also be operated in parallel with the fixed-target program.

The physics opportunities for such a medium-energy machine may be best visualized in the context of the x_B/Q^2 landscape pictured in Figure 3. The luminosity profile along the diagonal (s-axis), shown in Figure 4, helps

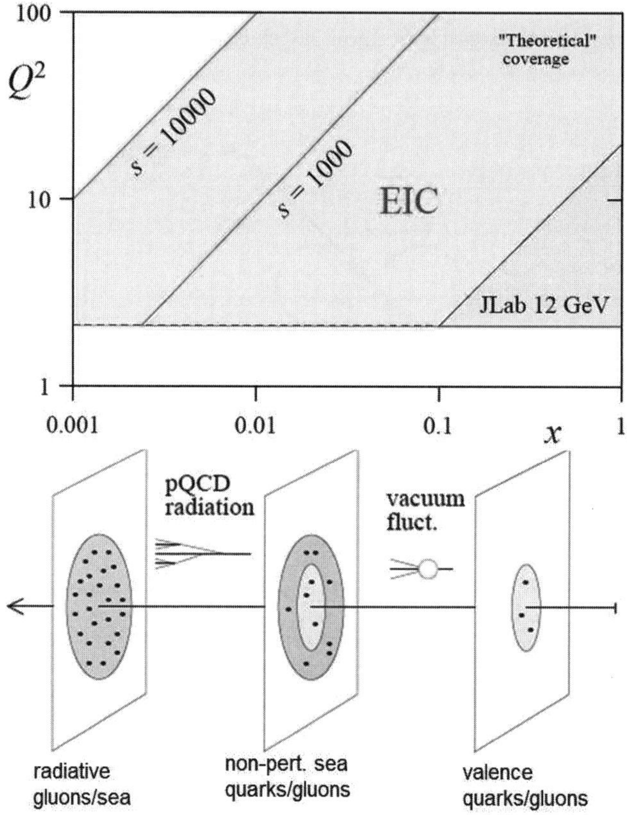

Fig. 3. The nuclear physics landscape.[1] Depending on x_B one probes different components of the hadron wave function. For instance, JLab 12 GeV covers the valence quark/gluon region, while the non-perturbative sea quarks and gluons could be accessed by a medium-EIC. Radiative gluons are the dominant degrees of freedom at even lower values of x_B.

to understand the requirements of various parts of the physics program. In this report, the focus is on nucleon structure studies that require a high luminosity in the intermediate energy range, for which the medium-energy EIC at JLab has been optimized.

The main goals of the EIC at JLab will include the following categories: (transverse) spatial imaging of gluon and sea quarks in the nucleon, the nucleon spin (orbital angular momentum of guarks and gluons as well as ΔG vs. lnQ^2), nuclei in QCD (gluons in nuclei, quark/gluon energy loss), QCD vacuum and hadron structure/creation, and electroweak physics.

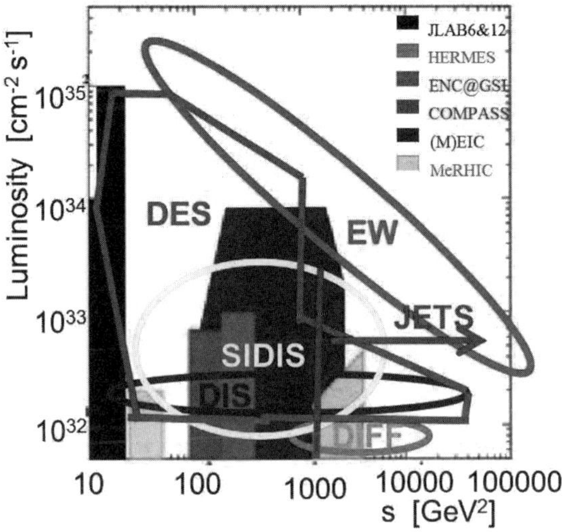

Fig. 4. The luminosity profile showing the luminosity/energy requirements for certain classes of reactions.

Consider as an example the interplay of the first two categories (inclusive and exclusive physics). There are at least two pictures for the decomposition of the proton spin (intrinsic spin and angular momenta of the components).[4] The traditional one (Jaffe)[5] is more intuitive and connects to well known PDFs. However, since it is not gauge invariant, it cannot be directly related to, for instance, quantities calculated on the lattice. The other one (Ji)[6] is based on the language of Generalized Parton Distributions (GPDs), the moments of which are calculable on the lattice. The former can be probed with inclusive, the latter with exclusive reactions. An EIC offers the opportunity to do both, although they pose somewhat contradictory requirements, and one thus needs to ensure that one strikes the right balance.

The spin carried by the quarks is common to both decompositions. In the Jaffe framework, one can in addition experimentally probe the gluon polarization (Δg) by measuring the g_1 structure function over a sufficiently wide range of x_B, while the orbital angular momentum of the quarks and gluons, although it formally appears in the sum rule, is difficult to access. In terms of x_B, one would clearly like to go beyond RHIC spin, which measured $\Delta g \sim 0$, but since Δg is expected to be small at small x_B, one needs to consider what range is sufficient to make the error on the integral satisfactory. For these inclusive measurements, the range in x_B drives the maximum energy, while the luminosity requirements are more modest despite the need for range in Q^2.

Recent measurements from RHIC suggest that the net contribution of the gluon polarization to the proton spin is zero. Secondly, what is the uncertainty in Δg? Since all values of x_B contribute to the final uncertainty, this will be large without data at small x_B. Looking at the Q^2 dependence, one can see that the EIC coverage will help extending measurements of $g_1(x, Q^2)$. In addition, there are opportunities to use charm production to probe $\Delta g/g$ directly.

For the GPD and TMD studies related to the angular momenta in the Ji decomposition the situation is different. Here the most important requirement is luminosity and resolution at intermediate energies. Recent lattice results from the LHPC collaborations suggest that as expected the contribution from the quark spin is about 30%. However, the orbital angular momentum for quarks cancel in this prediction, which may imply that the gluon angular momentum (J_g) is large in the Ji decomposition.[6]

A simpler interpretation in the GPD framework is transverse spatial imaging at small values of skewness. At high enough virtuality of the photon, Q^2, exclusive processes can be understood in terms of the virtual photon interacting with partonic degrees of freedom. In that regime, a QCD factorization theorem allows one to calculate the electroproduction amplitude in terms of GPDs. Depending on the skewness, one can access the transverse spatial distribution of partons with longitudinal momentum x_B or the longitudinal correlations in the nucleon. The moments of GPDs are related to fundamental static properties like the electromagnetic tensor. While the small size regime may have been reached by some exlusive processes at JLab energies, for exclusive meson production, which allows for probing the spin-flavor structure, this still remains to be shown. However, in EIC kinematics factorization should be less of an issue.

An important goal of the imaging studies is to map the transverse spatial structure of the non-perturbative sea. One of the best classes of reactions to do this is exclusive meson production, where the t-dependence is related to the transverse spatial image of the nucleon through a Fourier transform. Figure 5 shows the t slopes for π^+ and K^+ production for three bins in Q^2 around x_B=0.1. Comparing strange and non-strange pseudo-scalar production processes also allows for probing the spatial strange sea quark distributions. A priori it is not clear that these areas should be the same over the full x_B range as the wavefunction in pion production is dominated by πN contributions, while for kaon production the $K\Lambda$ components are dominant.

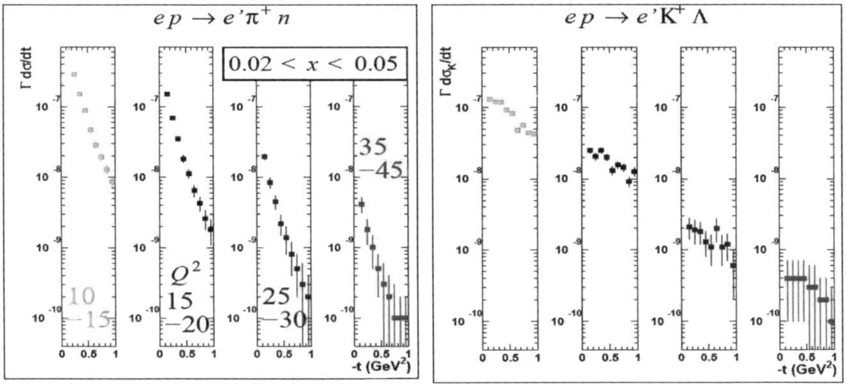

Fig. 5. Simulation of the t slopes of the exclusive π^+ and K^+ production cross section. The simulation assumes 5 GeV electrons, 50 GeV protons, and 100 days run time at 10^{34} luminosity. It uses a cross section model based on Regge theory and an empirical parameterization.[1]

Exclusive processes can provide similar information about the spatial distribution of quarks and gluons. Pseudo-scalar meson measurements, *e.g.*, pion and kaon production, over the full t range are needed to enable the Fourier transformation. The gluon distribution can be probed directly with J/Ψ production. While J/Ψ production probes only gluons, Deeply Virtual Compton Scattering (DVCS) is also sensitive to singlet quarks. By comparing these two processes one can address the question if singlet quarks and gluons have the same spatial distribution.

The EIC detector design is largely driven by exclusive reactions. Their detection requires hermeticity, particle identification, momentum resolution, forward detection of recoil baryons, muon detection (J/Ψ), and photon detection. Some of these requirements will also be important for other classes of experiments such as Deep Inelastic Scattering (DIS) and Semi-Inclusive DIS (SIDIS). The detector design also includes very forward detection capabilities for spectator tagging, diffractive, and coherent nuclear processes, as well as good vertex resolution for charm, and hadronic calorimetry for jet reconstruction.

Before looking into the details of the detector, it is beneficial to first consider the orientation and momentum of the produced particles. Figure 6

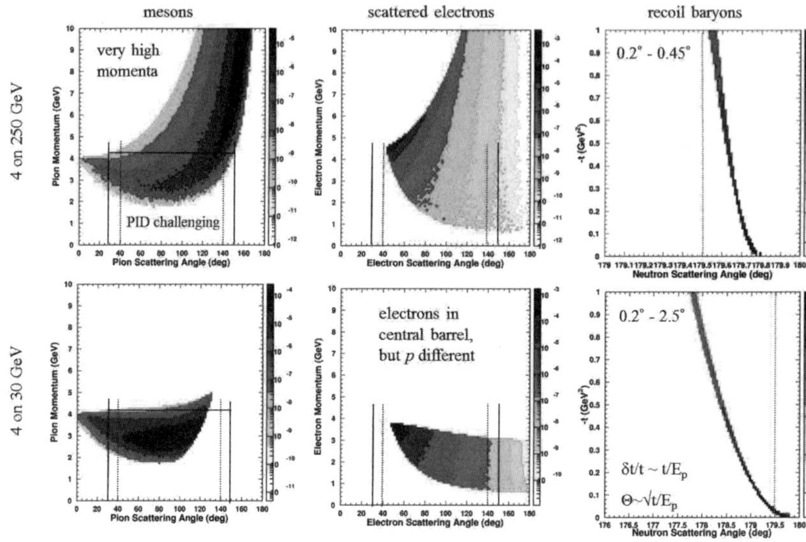

Fig. 6. Kinematic distributions of the scattered meson, the scattered electron, and the recoiling baryon in exclusive light meson production. A cut of $Q^2 > 10$ GeV^2 is applied in all panels. The vertical lines at 30° and 40° denote the transition from the central barrel to the endcaps on either side of the solenoid. The horizontal lines in the scattered meson panels illustrates the 4 GeV limit of a DIRC for π/K separation. The vertical lines in the recoil baryon panels denote a scattering angle of 0.5°.

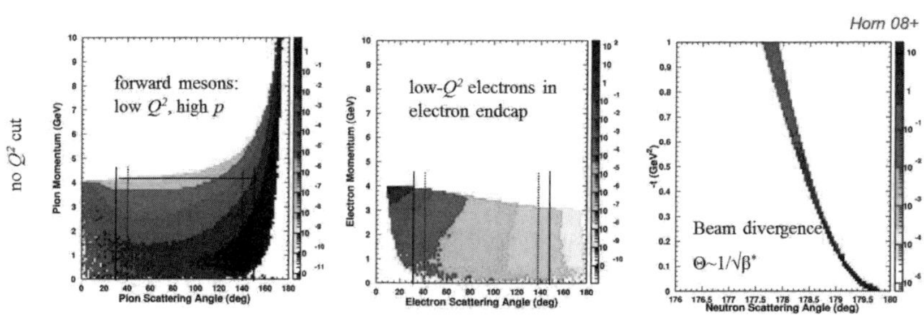

Fig. 7. Kinematic distributions for low-Q^2 meson production.

illustrates the kinematic distributions for exclusive light meson production. In these reactions, one needs to detect three particles, the scattered meson, the scattered electron, and the recoiling baryon. Due to the requirement of $Q^2 > 10$ GeV^2, one can see that the meson is predominantly scattered into the central detector/barrel region. Comparing a high and a low energy configuration, one can see that a DIRC will work for π/K separation at 30 GeV ion energies, but that PID becomes more challenging at higher energies. A DIRC detector would then no longer be sufficient and would have to complemented by a gas Cherenkov or replaced by a dual-radiator RICH to double its momentum reach. The scattered high-Q^2 electrons all go into the central/barrel region, and thus e/π PID is needed there. The scattering angle of the recoiling baryon depends on the ion energy. At higher beam energies the cone becomes narrower, which translates into a poor t-resolution. Figure 7 compares the kinematic distributions of low-Q^2 production with those of high Q^2 light meson production. In contrast to high-Q^2 production, photoproduction creates very high momentum particles at small angles - but not small enough to be detected downstream of the machine quadrupoles. For these particles, as well as those from target fragmentation in non-exclusive processes, the detector utilizes a 3° crossing angle for the ion beam, and a dipole between the endcap and the quadrupoles.

For measurements involving coherent nuclear processes or spectator particles such as the proton in a deuteron, there will be about 0.5° acceptance (smaller if one needs to go to higher ion energies) around zero degrees where they can go through the quadrupoles and then pass through a large ($\sim$ 20 Tm) analyzing dipole downstream. On the opposite size of the central detector, a similar arrangement allows for tagging of low-Q^2 electrons.

The central detector layout is shown in Figure 8. It consists of a central barrel with endcaps on either side built around a solenoidal magnet. The return yoke of the solenoid also serves as an integrated hadron calorimeter and muon detector. At low momenta time-of-flight can be used for PID. At higher momenta different PID methods will be applied. For instance, a DIRC can provide p/K separation up to 7 GeV/c, and together with a low-threshold Cherenkov (LTCC) also do π/K separation up to 9 GeV/c. The LTCC and an electromagnetic calorimeter will be used for e/π separation. The central tracker will contain a vertex detector, a small time projection chamber, and outer GEM layers. On either side there will be vertical drift chambers for tracking of particles going into the endcaps.

In summary, a medium-energy EIC is well suited for taking JLab beyond 12 GeV, providing an excellent tool for accessing nucleon/nuclear structure. A collider is particularly appealing for measurements requiring transverse target polarization and/or good resolution and particle identification, for instance, TMD (SIDIS) and GPD (exclusive) measurements. These processes benefit from high luminosity, excellent polarization, and more symmetric collision kinematics. The hermetic detector concept allows

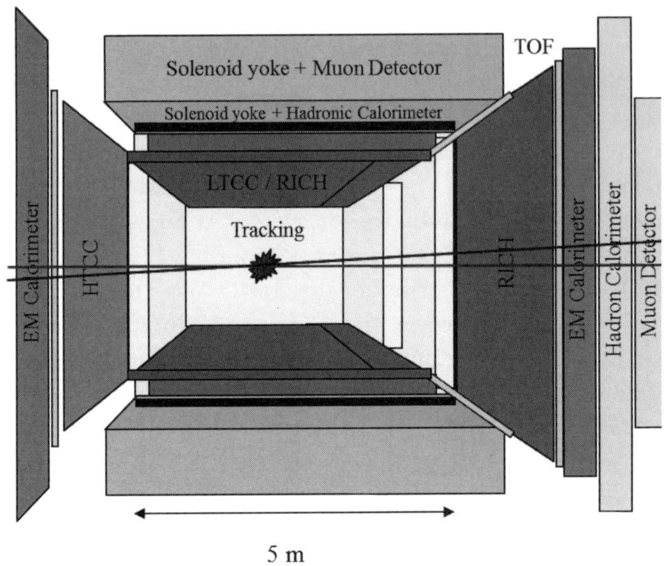

Fig. 8. Layout of the EIC at JLAB central detector.

excellent coverage of all kinematics in exclusive reactions and SIDIS.

References

1. Slides: http://conferences.jlab.org/exclusive2010/talks/Wednesday/horn_exclusive_workshop_eic_final.pdf
2. M. Klein, P. Newman, CERN Courier 1 April, (2009); J. Dainton et al., JINST 1, 10001 (2006); http://www.lhec.org.uk.
3. A. Deshpande, R. Ent, and R. Milner, CERN Courier 30 October, (2009); http://web.mit.edu/eicc.
4. M. Stratmann, INT09-43W, (2009); http://www.int.washington.edu/talks/WorkShops/int_09_43W/People/Stratmann_M/Stratmann.pdf.
5. S. Bashinsky, R.L. Jaffe, Nucl. Phys. **B536**, 303 (1998); R.L. Jaffe, A. Manohar, Nucl. Phys. **B337**, 509 (1990).
6. X.-D. Ji, Phys. Rev. Lett. **78**, 610 (1997).

SEMI/EXCLUSIVE MESON PRODUCTION AT EIC ENERGIES

MARK STRIKMAN

Physics Department, Penn State University ,
Univeristy Park, PA 16802, U.S.A.
**E-mail: strikman@phys.psu.edu*

We discuss interplay of the leading twist and higher twist effects in vector meson production in DIS, kinematics which is necessary to study nucleon GPDs. It is explained that the DGLAP evolution of dipole - parton elastic amplitude is suppressed in large t rapidity gap processes with leading J/psi allowing to explain the the recent HERA observations. This effect may mimic effect of softening of t-dependence of the GPDs with decrease of x. A pattern of the energy dependence of the t-slope for the hard amplitude with non-vacuum exchange in t-channel is suggested. It could be studied in a novel class of $2 \to 3$ hard processes which can also probe dynamics of the large angle $2 \to 2$ processes as well as various hadrons and photon.

Keywords: Perturbative QCD, exclusive hard processes

1. Introduction

The QCD factorization theorems for the exclusive vector meson production at small x [1] for the general case of meson production[2] and for exclusive photon production[3–7] were proven more than a decade ago. They provided a way to use the exclusive processes to probe nucleon and meson structure as well as the QCD dynamics.

The dynamics of these processes depends strongly on the interplay between Q^2, s and t. There are several interesting directions in this 3D landscape sketched in Fig. 1:

$$x = \text{const}, Q^2 \to \infty,\ t = \text{const};\ -t \sim Q^2, s \to \infty;\ -t > Q^2, s \to \infty, \quad (1)$$

which probe different aspects of the QCD dynamics. In this talk I will review the current status of some of these studies as well as discuss novel $2 \to 3$ processes which may provide a new direction for study of generalized parton distributions (GPDs) different hadrons as well as the QCD dynamics.

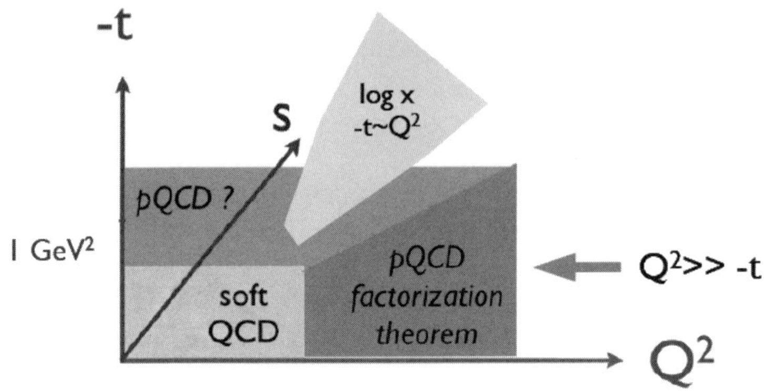

Fig. 1. 3D landscape of the hard exclusive processes

2. Exclusive meson production at small t

2.1. *At what Q the leading twist regime sets in*

The proof of the factorization theorem for the meson production in the process $\gamma_L^* + T \to M + T'$ heavily uses ideas of squeezing, suppression of the interaction strength for small dipoles - color transparency (CT), and freezing - ability to neglect space- time evolution of small wave packets during the time of the collision. The squeezing is ensured by localization of the $q\bar{q}$ wave function of the longitudinal photon, which in turn ensures suppression of interactions between a small size $q\bar{q}$ system which at a later stage would form the meson system "M" and the residual target system – color transparency (CT).

Factorization and the link to the CT are best seen in the Breit frame. Before the interaction photon is static: $q = (0, \vec{q})$. After the interaction, the meson system consists of fast left movers, and baryon system of fast right movers. No soft interactions between left and right movers is possible provided the meson system has a small size which is insured by the choice of γ_L^*. Note here that large Q^2 is not enough – large enough W is necessary to insure that expansion of the wave packet happens far away from the target. For γ_T^* nonperturbative contribution appears to be suppressed only by $\ln Q^2$ terms. Hence in the pQCD regime one expects two signature differences between the vector meson production for γ_T^* and γ_L^* channels – a larger t-slope for γ_T^*, and increase of σ_L/σ_T ratio with increase of W at fixed Q^2. The experimental studies at HERA did not reach a definite conclusion: H1 data do indicate a steeper decrease of σ_T with t, while ZEUS does not see an evidence for such an effect. There is no evidence for the dependence of σ_L/σ_T on W for fixed Q^2. Hence the theoretical understanding of the $\gamma_T p \to \rho p$ reaction remains a challenge.

Theoretical analysis of Ref. [8] extended the LT approximation for the vector meson production to include HT effects due to finite size of the $q\bar{q}$ configuration at intermediate Q^2. It concluded that a significant squeezing should occur in the case of the light vector mesons already at relatively moderate $Q^2 \geq 2 \div 3$ GeV^2, while the HT effects for the absolute cross section remain large up to much larger Q^2. In the case of the J/ψ squeezing is effective already for $Q^2 = 0$, leading to a prediction that the t-slope, B should be practically Q^2 independent. The HERA data confirm both the weak dependence of $B_{J/\psi}$ on Q^2 and graduate decrease of $B_{\rho,\phi}$ with Q^2. The pattern of the decrease of $B(Q^2)$ is consistent with predictions of Ref. [8] as well as convergence of B_ρ to $B_{J/\psi}$.

We can estimate the effective size of the $q\bar{q}$ dipole as

$$\frac{B(Q^2) - B_{2g}}{B(Q^2 = 0) - B_{2g}} \sim \frac{R^2(\text{dipole})}{R_\rho^2}, \tag{2}$$

where B_{2g} is the slope of the square of the two gluon form factor. Based on the HERA data[9] we conclude that

$$R^2(\text{dipole})(Q^2 \geq 3\,\text{GeV}^2)/R_\rho^2 \leq 1/2 \div 1/3 \tag{3}$$

for the collider energies.

At the same time the HERMES data on exclusive production of ρ-mesons off hydrogen[10] at lower energies $W \sim 5$ GeV do not observe a change of the t-slope between $Q^2 = 0$ and $Q^2 = 2\,\text{GeV}^2$, indicating that squeezing is still small for $Q^2 = 2$ GeV^2 though there is an indication of the onset of squeezing for $Q^2 = 3$ GeV^2. Consequently, the extraction of information on the GPDs from the HERMES data on exclusive meson production, which is dominated by the data at $Q^2 \leq 2 \div 2.5$ GeV^2 appears to be problematic. Note here that Q^2 at which transition from the soft to the hard regime occurs is likely to decrease with the increase of W as the hard dipole - nucleon cross section grows faster with increase of W than the soft one. A seeming tension between the collider data and HERMES data could be due to this effect.

To establish more reliably the Q^2 range where squeezing sets in it is critical to perform CT studies with nuclei both in the incoherent and coherent channels. High precision data[11] on the exclusive production of pions off nuclei at Jlab energies are consistent with the CT predictions of

Ref.[12] indicating that in the case of pion production sets in already at $Q^2 \sim 3\,\mathrm{GeV}^2$.

2.2. *GPDs via meson production*

One of the prime aims of the EIC ep program is to determine 3D structure of the nucleon on a single parton level both for gluons and for quarks.

For small x there are channels which are essentially purely gluonic with no contribution from the quark GPDs - like production of J/ψ and Υ. In the case of GDS linked to the sea quarks situation is more complicated. In the case of the DVCS amplitude in the NLO, quark and gluon contributions enter with opposite signs and are of a comparable magnitude.[13] As a result a relatively small difference of the transverse sizes of sea quarks and gluons expected due to pionic effects is amplified in the t-slopes of DVCS.[14] Hence it is critical to measure at the EIC the gluon GPDs both for its own sake and for precision determination of quark GPDs.

The HT effects modify the t-distribution of the light mesons up to $Q^2 \sim 15\,\mathrm{GeV}^2$. Therefore it appears that the only practical chance to perform a precision measurement of the gluon GPDs is production of onium states.

The effects of non-diagonality in the GPD appear to be small. For J/ψ case it is due to large transverse momenta in the wave function which lead to the comparable light-cone fractions of the gluons attached to $c\bar{c}$: $x_1/x_2 \sim 2, x_1 - x_2 = x = (Q^2 + m^2_{J/\psi})/W^2$. In the Υ case non-diagonality is washed away due to the DGLAP evolution. As a result one can write

$$\frac{d\sigma^{\gamma+p\to J/\psi+p}}{dt} \propto F_g^2(x,t)\exp(\Delta B\, t), \tag{4}$$

where $F_g(x,t)$ is the two-gluon form factor of the nucleon and the second factor takes into account a small but finite correction due to the finite size of J/ψ which was estimated in Ref.[8] to be $\Delta B \approx 0.3\,\mathrm{GeV}^2$. The t–dependence of the measured differential cross sections of exclusive processes at $|t| < 1\,\mathrm{GeV}^2$ is commonly described either by an exponential, or by a dipole form inspired by analogy with the nucleon elastic form factors. The data are not precise enough to distinguish between two forms as they mostly differ at small t, where the resolution of the measurements is moderate, and at large $-t \geq 0.8\,\mathrm{GeV}^2$ where the measurements rather strongly depend on the procedure of subtraction of the inelastic background contribution.

The results of the current measurements of the t-dependence of the J/ψ

photoproduction cross section analyzed using exponential fits to the data are summarized in Fig. 2 adopted from Ref. [15] which provided an update of the analysis of Ref. [16].

.

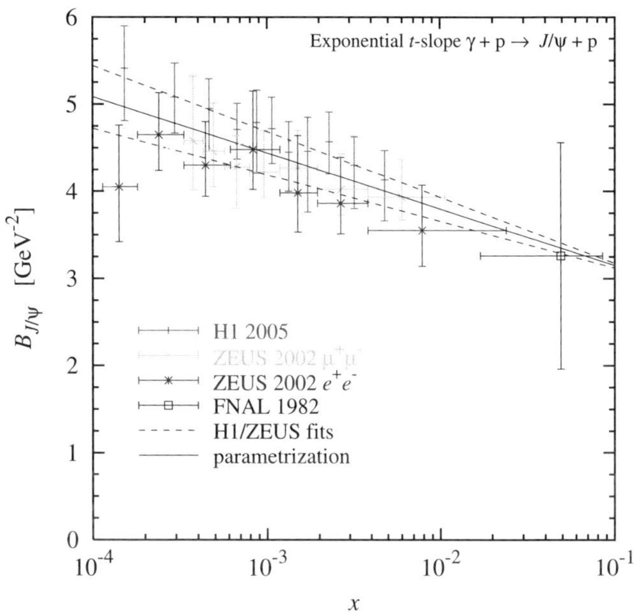

Fig. 2. The exponential t–slope, $B_{J/\psi}$, of the differential cross section of exclusive J/ψ photoproduction measured in the FNAL E401/E458,[17] HERA H1,[18] and ZEUS[19] experiments, as a function of $x = M^2_{J/\psi}/W^2$. (In the H1 and ZEUS results the quoted statistical and systematic uncertainties were added linearly.) The dashed lines represent the published two–dimensional fits to the H1 and ZEUS data.[18,19] The parameter B_g in the exponential two–gluon form factor is related to the measured J/ψ slope by Eq. (4). Our parametrization Eqs. (5)–(8) is shown by the solid line.

The data can be fitted as

$$B_g(x) = B_{g0} \; + \; 2\alpha'_g \; \ln(x_0/x), \tag{5}$$

$$x_0 = 0.0012, \tag{6}$$

$$B_{g0} = 4.1 \; (^{+0.3}_{-0.5}) \; \text{GeV}^{-2}, \tag{7}$$

$$\alpha'_g = 0.140 \; (^{+0.08}_{-0.08}) \; \text{GeV}^{-2}. \tag{8}$$

Fits of similar quality are produced with a dipole with

$$F_g(x, t|Q^2) = (1 - t/m_g^2)^{-2}, B_g = 3.2/m_g^2. \tag{9}$$

The spatial distributions of gluons in the transverse plane for two fits are given by

$$F_g(x,\rho|Q^2) = \begin{cases} (2\pi B_g)^{-1}\exp[-\rho^2/(2B_g)], \\ [m_g^2/(2\pi)]\,(m_g\rho/2)\,K_1(m_g\rho), \end{cases} \tag{10}$$

These transverse distributions are similar for average ρ, leading, for example, to nearly identical distributions over the impact parameter for production of the dijets in *pp* collisions.[15] At the same time, dipole fits gives a significantly larger $F_g(x,\rho|Q^2)$ at small b. As a result, analyses of the proximity to the black disc regime for the interaction of the small dipole with the nucleon for small b are sensitive to the choice of the model for $F_g(x,\rho|Q^2)$. A related effect is a factor of 1.6 difference of $\langle b_g^2\rangle$ for two fits. The conclusion[20] that gluons are localized in a smaller transverse area than given by the e.m. form factor is based on the use of the same shape for the e.m. and two gluon form factors.

It is critical that the future experiments produce much more accurate data for the t-dependence. This would allow to address of the shape of t-dependence as well as the energy dependence which is somewhat puzzling. Indeed the central value for α'_g in Eq.(8) is very close to that for the ρ-meson photoproduction obtained from the global analysis of the world data[21] : $\alpha'_{I\!P} = 0.126 \pm 0.013(stat.) \pm 0.012(syst.)\,\mathrm{GeV}^{-2}$, though one expects a reduction of α' with virtuality due to the suppression of the Gribov diffusion [1] [a]. Note also, that the data on the J/ψ electroproduction seem to suggest a smaller α'_g, though these data have larger errors than the photoproduction data.

Two effects maybe relevant for the explanation of the the discussed effects - one is is fluctuations in the transverse size of the J/ψ dipole. For photoproduction there is a significant amplitude for a dipole to fluctuate to a large size configuration for which diffusion is not suppressed.[22] This effect should quickly dye out with increase of Q^2. Another effect is related to a transition to the limit when $-t$ becomes comparable to the intrinsic hard scale of the exclusive process which we will discuss next.

---

[a]The data[21] may indicate that the effective Pomeron trajectory is nonlinear with larger $\alpha'_{I\!P}$ for small t. However average t for J/ψ photoproduction is larger than for the ρ-meson case. So it is not clear how this effect could increase the difference between α''s for two cases.

3. Break up processes with gaps - from small to large t

3.1. $t \sim 0$ *probing fluctuations of gluon field*

In a high energy process $\gamma^* + p \to M + X$ at $t \approx 0$ a two gluon ladder can couple only to one parton in the target. If the strength of the couple to all configurations containing partons with given x is the same it is impossible to produce an inelastic final system X. As a result, similar to the case of the inelastic diffraction of hadrons off hadrons the discussed process measures the degree to what the strength of the gluon field is different in different configurations.

The gluon density is average over the gluon fields in different configurations in the nucleon. An important characteristics of the nucleon wave function is a degree to what the strength of the gluon field is different in different configurations. The variance of the gluon field is expressed through the ratio diffraction dissociation and elastic cross section for the vector meson production at $t = 0$ as follows:[23]

$$\omega_{\text{hard}} = \frac{d\sigma(x, Q^2)/dt(\gamma^* + p \to VM + M_X)}{d\sigma(x, Q^2)/dt(\gamma^* + p \to VM + p)} = \frac{\langle G^4 \rangle - \langle G^2 \rangle^2}{\langle G^2 \rangle^2} . \tag{11}$$

Our estimates of the strength of the fluctuations due to fluctuations of the overall size of nucleon as seen in the soft inelastic diffraction find the value of the ratio ω_{hard} of the order 0.15 at small x which is consistent with the current data which however have large systematic uncertainties. Experiments at EIC will be in a good position to study this phenomenon provided a good acceptance in the target fragmentation region is ensured.

3.2. *Onset of new regime at large t*

In order to determine spacial distribution of gluons at $b \sim 0$ one needs to study exclusive processes at large t since in this case the integral over $-t$ in the Fourier transform converges very slowly. For example in the case of the dipole fit a half of the contribution to $F_g(x, b = 0)$ originates from $-t \geq m_g^2$ and one quarter from $-t \geq 3m_g^2$. This implies that one needs a detector which will be able to separate exclusive processes from diffractive dissociation in a very broad range of t where the dissociation dominates by far.

Further complication is that the QCD factorization theorem[2] was derived in the limit t =const, x =const, $Q^2 \to \infty$. Our recent studies[24] indicate that the pattern of the QCD evolution changes in the limit when $-t$ becomes comparable to the intrinsic hardness scale of the process

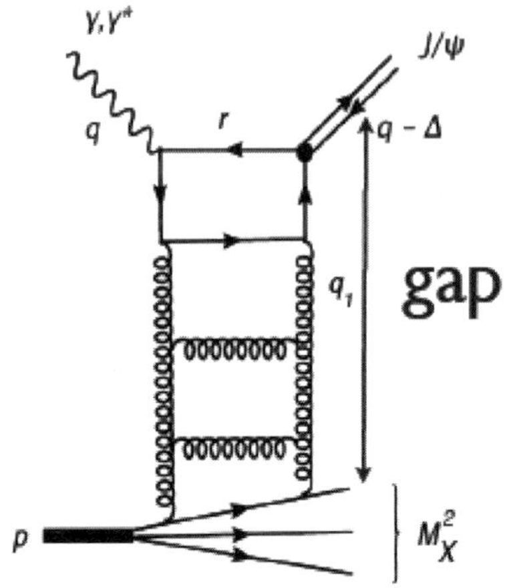

Fig. 3. Typical QCD diagram for the rapidity gap process (12).

($Q^2_{\rm eff} \sim 3\,GeV^2$ for the J/ψ photoproduction). The DGLAP equations require modification in this case. Let us for example consider the process

$$\gamma^* + p \to J/\psi + \text{rapidity gap} + M_X, \tag{12}$$

for large $-t$ which are still smaller than $Q^2 + M^2_{J/\psi}$. The typical leading QCD diagrams (Fig. 3) correspond to attachment of the two gluon ladder to a parton with $x_J = -t/(-t + M_X^2 - m_N^2)$. The cross section can be written in the factorized form similar to:[26]

$$\frac{d\sigma_{\gamma+p\to V+X}}{dt dx_J} = \frac{d\sigma_{\gamma+quark\to V+quark}}{dt}\left[\frac{81}{16} g_p(x_J,t) + \sum_i (q_p^i(x_J,t) + \bar{q}_p^i(x_J,t))\right]. \tag{13}$$

The energy dependence for fixed x_j is determined by the evolution with x/x_J of the dipole - parton elastic scattering amplitude. In difference from the small t limit the DGLAP evolution is strongly suppressed and completely disappears for $-t$ close to the intrinsic scale. The HERA data on the energy dependence of the process (12) are consistent with our results.

Consequently, the effective $\alpha_{I\!P}(t)$ in this limit stays close to one until very small x/x_J (not available at HERA) where Pomeron type behavior may reveal itself.

The same mechanism maybe responsible for a part of the drop of $\alpha_{I\!P}(t)$ with $-t$ increase observed for the elastic J/ψ production. This may indicate that precision measurements of the t-dependence of the two gluon form factor at high energies would require using electroproduction rather than

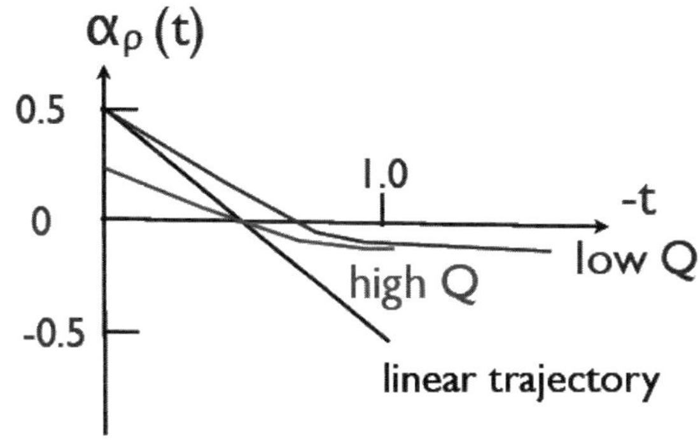

Fig. 4. Sketch of the Q^2 and t dependence of the effective reggeon trajectory.

photoproduction. The discussed phenomenon maybe relevant also for the explanation of the pattern observed in photoproduction of ρ - mesons where $\alpha_{I\!P}(t)$ appears to flatten out around $\alpha_{I\!P}(t) = 1$ for large t.

It maybe easier to observe a variation of the energy dependence as a function of t and Q^2 in the non-vacuum channel since α'_R is much larger than $\alpha_{I\!P}$. One can consider for example a process with production of a leading pion with rapidity gap $\gamma^* + N \to \pi + \text{gap} + M_X$. In this case in the soft regime the energy dependence is given by the Reggeon exchange with intercept $\alpha_R(t=0) \approx 1/2$. One can expect that with increase of Q^2 or/and $-t$ the energy dependence would approach the one expected in the perturbative regime for an exchange of $q\bar{q}$: $\alpha_R^{\rm pert} = 0$ with a small downward correction due to the quark reggeization. This pattern sketched in Fig. 4 seems to be consistent with the limited data on the energy dependence of the two body hadronic processes with exchange of meson and baryon Reggeons in t-channel for $-t \geq 0.7\,\mathrm{GeV}^2$ [25]. The collider kinematics is well suited for such measurements, while ability to study this reaction both with proton and nuclear targets has another advantage – the A-dependence of the process allows to determine at what Q^2 and $-t$ the squeezing of the meson and photon wave functions becomes significant.

4. Studies of QCD dynamics in 2 → 3 Exclusive processes

Recently it was realized that there exists a new class of high processes where one of the incoming particles scatters of a small size white cluster in the second particle. An extensive study of these processes was reported in Ref.[27]. Schematically these processes are described by Fig. 5. If the upper block ($2 \to 2$ process) is hard, "b", "d", "c" are in small size configurations

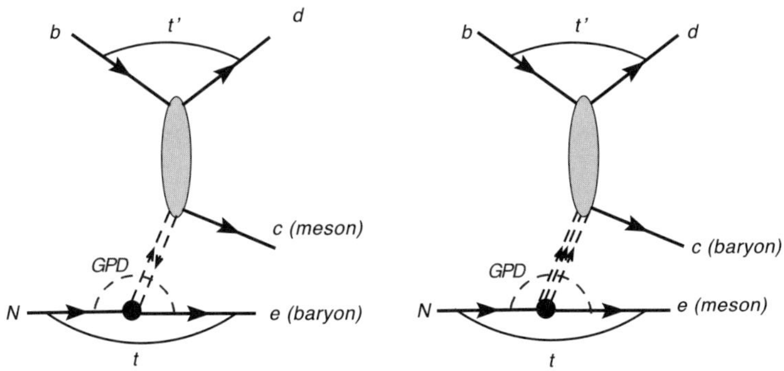

Fig. 5. The block structure of the 2 → 3 process in the hard regime.

as well as exchange system $(q\bar{q}, qqq)$. One can use the same line of logic (based on CT) as in the proof of the QCD factorization for the meson exclusive production in DIS.[2]

In case of the ep scattering one such process is $\gamma^* + p \to \pi^+\pi^+ n$ where pions carry practically all energy of the photon and have balancing transverse momenta $p_t^{(1)} \approx -p_t^{(2)} \geq 1 \div 2\,\mathrm{GeV}/c$. These processes allow to measure both nucleon and photon GPDs as well as to study two - body reactions of photon - $q\bar{q}$ scattering at large momentum transfer - large scattering angles in the c.m. of the $\pi\pi$ system. The dependence of the squeezing of hadrons on $p_t^{(i)}$ in these processes can be tested using electron - nucleus scattering at the EIC energies. Since the coherence length in this kinematics is much larger than the nucleus size, the initial and final state can be treated as frozen leading to much stronger sensitivity to the CT than at the intermediate energies.[28] The studies of the 2 → 3 process will allow to discover the pattern of interplay of large and small transverse distance effects (soft and hard physics) in wide range of the processes including elastic scattering, large angle two body processes, measure a variety of GPDs including GPDs of photon, compare wave function of different mesons, map the space-time evolution of small wave packets at distances.

5. Conclusions

HERA left plenty of open questions related to the dynamics of exclusive vector meson production and characteristics of GPDs – especially the gluon GPD which dominates at small x. Rapidity gap processes provide tests of elastic hard scattering in QCD at large t and also serve as a new tool for studying $N \to N^*$ form factors involving gluons. Many novel processes in QCD which will reveal QCD high energy dynamics and hadron structure on multiparton level are not explored yet. A key for a successful experimental research in this field is a sufficiently hermetic detector with a good acceptance in the nucleon/nucleus fragmentation region.

Acknowledgements

The work reported in this talk was done in collaboration with a number of people. I thank them for the critical role they played in these studies. The research was supported by DOE grant No. DE-FG02-93ER40771.

References

1. S. J. Brodsky, L. Frankfurt, J. F. Gunion, A. H. Mueller and M. Strikman, *Phys. Rev. D* **50**, 3134 (1994) [arXiv:hep-ph/9402283].
2. J. C. Collins, L. Frankfurt and M. Strikman, *Phys. Rev. D* **56**, 2982 (1997) [arXiv:hep-ph/9611433].
3. D. Mueller, D. Robaschik, B. Geyer, F. M. Dittes and J. Horejsi, *Fortsch. Phys.* **42**, 101 (1994) [arXiv:hep-ph/9812448].
4. H. Abramowicz, L. Frankfurt and M. Strikman, eConf **C940808**, 033 (1994) [Surveys High Energ. Phys. **11**, 51 (1997)] [arXiv:hep-ph/9503437].
5. X. D. Ji, *Phys. Rev. D* **55**, 7114 (1997) [arXiv:hep-ph/9609381].
6. A. V. Radyushkin, *Phys. Rev. D* **56**, 5524 (1997) [arXiv:hep-ph/9704207].
7. J. C. Collins and A. Freund, *Phys. Rev. D* **59**, 074009 (1999) [arXiv:hep-ph/9801262].
8. L. Frankfurt, W. Koepf and M. Strikman, Phys. Rev. D **54**, 3194 (1996); ibid **57**, 512 (1998).
9. S. Chekanov *et al.* [ZEUS Collaboration], PMC Phys. A **1**, 6 (2007) [arXiv:0708.1478 [hep-ex]].
10. M. Tytgat, Ph.D. Thesis, Gent University, DESY- THESIS-2001-018.
11. B. Clasie *et al.*, Phys. Rev. Lett. **99**, 242502 (2007) [arXiv:0707.1481 [nucl-ex]].
12. A. Larson, G. A. Miller and M. Strikman, Phys. Rev. C **74**, 018201 (2006) [arXiv:nucl-th/0604022].
13. A. Freund and M. F. McDermott, Phys. Rev. D **65**, 091901 (2002) [arXiv:hep-ph/0106124].
14. M. Strikman and C. Weiss, Phys. Rev. D **80**, 114029 (2009) [arXiv:0906.3267 [hep-ph]].

15. L. Frankfurt, M. Strikman and C. Weiss, arXiv:1009.2559 [hep-ph].
16. L. Frankfurt, M. Strikman and C. Weiss, Phys. Rev. D **69**, 114010 (2004) [arXiv:hep-ph/0311231].
17. M. Binkley *et al.*, Phys. Rev. Lett. **48**, 73 (1982).
18. A. Aktas *et al.* [H1 Collaboration], Eur. Phys. J. C **46**, 585 (2006).
19. S. Chekanov *et al.* [ZEUS Collaboration], Eur. Phys. J. C **24**, 345 (2002).
20. L. Frankfurt and M. Strikman, Phys. Rev. D **66**, 031502 (2002) [arXiv:hep-ph/0205223].
21. B. List [H1 Collaboration], arXiv:0906.4945 [hep-ex].
22. L. Frankfurt, M. McDermott and M. Strikman, JHEP **0103**, 045 (2001) [arXiv:hep-ph/0009086].
23. L. Frankfurt, M. Strikman, D. Treleani and C. Weiss, Phys. Rev. Lett. **101**, 202003 (2008) [arXiv:0808.0182 [hep-ph]].
24. B. Blok, L. Frankfurt and M. Strikman, Eur. Phys. J. C **67**, 99 (2010) [arXiv:1001.2469 [hep-ph]]; Phys. Lett. B **690**, 159 (2010) [arXiv:1002.3048 [hep-ph]].
25. M. Strikman, Nucl. Phys. A **805**, 369 (2008) [arXiv:0711.1634 [hep-ph]].
26. L. Frankfurt and M. Strikman, Phys. Rev. Lett. **63** (1989) 1914 [Erratum-ibid. **64** (1990) 815].
27. S. Kumano, M. Strikman and K. Sudoh, Phys. Rev. D **80**, 074003 (2009) [arXiv:0905.1453 [hep-ph]].
28. S. Kumano and M. Strikman, Phys. Lett. B **683**, 259 (2010) [arXiv:0909.1299 [hep-ph]].

NUCLEAR GENERALIZED PARTON DISTRIBUTIONS AND COHERENT NUCLEAR PROCESSES

V. GUZEY*

Thomas Jefferson National Accelerator Facility
12000 Jefferson Avenue, Newport News, VA 23606, U.S.A.
**E-mail: vguzey@jlab.org*

We discuss nuclear generalized parton distributions that can be accessed in coherent hard exclusive processes with nuclei. We consider two examples of kinematics and relevant nuclear effects: medium values of Bjorken x_B, $x_B > 0.05$, and the interplay of the coherent and incoherent nuclear deeply virtual Compton scattering (DVCS), and small values of x_B, $x_B < 0.05$, where the leading twist nuclear shadowing significantly suppresses nuclear GPDs, which leads to clear experimental signals.

Keywords: Nuclear generalized parton distributions, coherent nuclear processes

1. Nuclear generalized parton distributions

Generalized parton distributions (GPDs) parameterize non-perturbative structure of hadrons probed in various hard exclusive processes: deeply virtual Compton scattering (DVCS), exclusive vector meson production, Timelike Compton scattering, etc. These distributions are universal, related to the familiar elastic form factors and usual parton distribution functions, and contain information on the three-dimensional and orbital parton structure of hadrons.

Hard exclusive processes with nuclei and GPDs of nuclei are additionally interesting in their own right.

(i) Nuclear targets enable us to study different spin and flavor combinations of GPDs of the nucleon. Inelastic nuclear DVCS gives an access to DVCS on quasi-free protons and neutrons.
(ii) Traditional nuclear effects (EMC effect, nuclear shadowing) are predicted to be enhanced in exclusive (off-diagonal) kinematics.
(iii) It is argued that several not well-established nuclear effects such as the medium modifications of bound nucleon GPDs and non-nucleonic

degrees of freedom are more prominent in exclusive reactions with nuclei than in traditional inclusive deep inelastic scattering (DIS).

In this contribution, I will consider nuclear DVCS and GPDs of nuclei in two distinct kinematics regimes: at medium $x_B > 0.05$, where the details of the nuclear structure and the interplay between the coherent and incoherent DVCS are important, and at small $x_B < 0.05$, where the effect of nuclear shadowing dominates.

2. Nuclear GPDs at medium x_B

Nuclear DVCS is more complex and versatile than DVCS on a proton because many more final states can be excited. In particular, the nuclear target can recoil intact (coherent nuclear process), can break-up into nucleons (incoherent nuclear process), can break-up into nucleons which in turn also break up, can produce an excited nuclear state, etc.

When the final nucleus is not detected and one measures the sum of all products of the nuclear decay, one can use the completness of the final states to show that the nuclear DVCS amplitude squared can be expressed as a sum of two terms:[1]

$$\sum_{A'} |\mathcal{T}_{\rm DVCS}^{A \to A'}|^2 = A(A-1) F_A^2(t') |\mathcal{T}_{\rm DVCS}^A|^2 + A |\tilde{\mathcal{T}}_{\rm DVCS}^N|^2 \,. \tag{1}$$

In Eq. (1), the first term is the coherent contribution dominating at small values of the momentum transfer t; $F_A(t)$ is the nuclear form factor; $t' = A/(A-1)t$; $\mathcal{T}_{\rm DVCS}^A$ is the nuclear DVCS amplitude with the factored-out t dependence. The second term is the incoherent contribution which dominates at large values of t; $|\tilde{\mathcal{T}}_{\rm DVCS}^N|^2$ is the DVCS amplitude squared of the bound nucleon. Equation (1) is an interpolation formula connecting the small t and large t regimes. Similar equations can be derived for the interference between the DVCS and Bethe-Heitler (BH) amplitudes and for the BH amplitude squared.

To model $\mathcal{T}_{\rm DVCS}^A$, one usually uses the two-step convolution approximation for nuclear GPDs:[1–3]

$$H_A^q(x,\xi,t) = \sum_N \int_x^1 \frac{dz}{z} h_A^N(z,\xi,t) H_N^q\left(\frac{x}{z}, \frac{\xi}{z}, t\right) , \tag{2}$$

where H_A^q is the quark flavor q GPD H of a nucleus; H_N^q is the GPD of the nucleon; $\sum_N$ is the sum over nucleons; x, ξ, and t are the usual variables that GPDs depend on; z is the light-cone fraction of the active nucleon. More involved versions of the convolution approximation have also been considered in the literature, where the convolution also involves the transverse momentum of the active nucleon and the possible medium modifications of its GPDs.[2]

To model $|\tilde{\mathcal{T}}^N_{\rm DVCS}|^2$, as a first approximation one can ignore the medium modifications of the bound nucleon GPDs and use any reliable model for the free nucleon GPDs.

Having specified the model for nuclear GPDs, one can use the standard formalism to make predictions for various DVCS observables.[4] We focus on the DVCS beam-spin asymmetry measured with a linearly polarized lepton beam and unpolarized target. Figure 1 presents the ratio of the nuclear to free proton beam-spin asymmetries, $A^A_{\rm LU}/A^p_{\rm LU}$, as a function of the momentum transfer t in average Hermes kinematics, $x_B = 0.065$ and $Q^2 = 1.7$ GeV^2. As one can see from the figure, at small t, we predict the enhancement of $A^A_{\rm LU}/A^p_{\rm LU}$ above unity, which is model-independent and combinatoric in its nature. Indeed, the DVCS beam-spin asymmetry is proportional to the ratio of the interference between the DVCS and BH amplitudes and the BH cross section. While the former involves both protons and neutrons of the nuclear target, the BH cross section at small t predominantly involves only the protons, which leads to the combinatoric enhancement of $A^A_{\rm LU}/A^p_{\rm LU}$ above unity by approximately the factor of A/Z.

At large t, $A^A_{\rm LU}$ is dominated by the incoherent contribution [second term in Eq. (1)] and we predict that $A^A_{\rm LU}/A^p_{\rm LU}$ is somewhat suppressed compared to unity due to the negative neutron contribution to $A^A_{\rm LU}$.

Our predictions for $A^A_{\rm LU}/A^p_{\rm LU}$ at small t do not seem to be corroborated by the final result of the Hermes analysis of nuclear DVCS that reported $A^A_{\rm LU}/A^p_{\rm LU} = 0.91 \pm 0.19$ at $\langle -t \rangle = 0.018$ GeV^2 (coherent) and $A^A_{\rm LU}/A^p_{\rm LU} = 0.93 \pm 0.23$ at $\langle -t \rangle = 0.2$ GeV^2 (incoherent).

This discrepancy will clarified by the analysis of the new JLab data on DVCS on ^{4}He.[6] The advantages of the JLab experiment compared to the Hermes one include the detection of the recoiled ^{4}He, which guarantees the exclusivity of the process, and higher luminosity.

The JLab DVCS on ^{4}He experiment will also extract the incoherent DVCS on a bound proton which will probe possible nuclear modifications of the bound nucleon GPDs.[2,7]

3. Nuclear shadowing and nuclear GPDs at small x_B

Nuclear shadowing is a small Bjorken x nuclear effect: for $x < 0.05$, the nuclear structure function $F_{2A}(x, Q^2)$ measured in inclusive DIS with nuclei is smaller than the sum of the free nucleon structure functions $F_{2N}(x, Q^2)$,

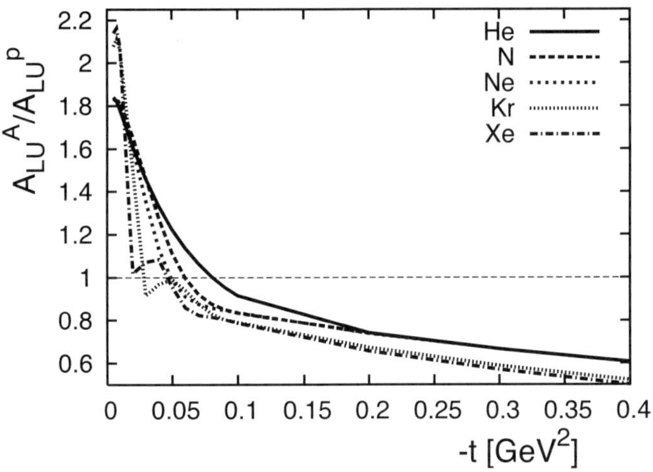

Fig. 1. The ratio of the nuclear to free proton DVCS beam-spin asymmetries, $A^A_{\rm LU}/A^p_{\rm LU}$, as a function of the momentum transfer t in average Hermes kinematics, $x_B = 0.065$ and $Q^2 = 1.7$ GeV^2.

i.e., $F_{2A}(x, Q^2) < AF_{2N}(x, Q^2)$.

Using the QCD factorization theorem, one extracts nuclear parton distributions (PDFs) from the available data on $F_{2A}(x, Q^2)$ and other observables and finds that nuclear shadowing is also present in nuclear PDFs, i.e., $f_{j/A}(x, Q^2) < Af_{j/N}(x, Q^2)$, where $f_{j/A}(x, Q^2)$ is the nuclear PDF of flavor j and $f_{j/N}(x, Q^2)$ is the free nucleon PDF. However, the amount of nuclear shadowing is rather uncertain, especially in the gluon channel, since the available data covers the shadowing region with sufficiently large Q^2 insignificantly and because the gluon PDF is found indirectly through the scaling violations. One should note that the determination of the nuclear gluon distribution is of fundamental importance for theory and phenomenology and this will be studied at the LHC and a future Electron-Ion Collider (EIC).

An alternative to the extraction of nuclear PDFs using global fits is to use dynamical models of shadowing. We use the leading twist theory of nuclear shadowing[8] which enables us to calculate nuclear PDFs as functions of x and the impact parameter b at a certain initial scale Q_0^2. The subsequent Q^2 dependence of nuclear PDFs is given by the standard DGLAP evolution.

The leading twist theory of nuclear shadowing is based on the generalization of the Gribov theory of nuclear shadowing to DIS and arbitrary

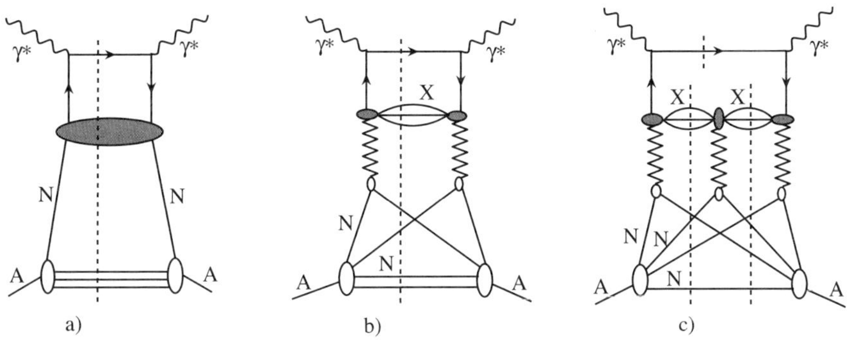

Fig. 2. A graphical representation of the multiple scattering series for nuclear quark parton distributions.

nuclei, the collinear factorization theorems for inclusive and hard diffractive DIS, and QCD fits to HERA measurement of diffraction in ep DIS.

The Gribov theory of nuclear shadowing is based on the space-time picture of high-energy scattering in the target rest frame, when the projectile interacts coherently with all nucleons of the target. Since the nucleons partially overlap or shadow each other, the projectile-nucleus scattering cross section is smaller than the sum of individual projectile-nucleon cross sections, which is the essence of nuclear shadowing. A graphical representation of the multiple scattering series for nuclear quark PDFs is given in Fig. 2. Note that the graphs corresponding to the interaction with $N > 3$ nucleons of the nucleus are not shown, but are assumed.

The corresponding analytic expression for nuclear PDFs reads:

$$\begin{aligned} & xf_{j/A}(x,Q^2) = Axf_{j/N}(x,Q^2) \\ & - xf_{j/N}(x,Q^2) 8\pi A(A-1)\,\Re e\frac{(1-i\eta)^2}{1+\eta^2} B_{\text{diff}} \int_x^{0.1} dx_{I\!P}\,\beta f_j^{D(3)}(\beta,Q^2,x_{I\!P}) \\ & \times \int d^2b \int_{-\infty}^{\infty} dz_1 \int_{z_1}^{\infty} dz_2 \rho_A(\vec{b},z_1)\rho_A(\vec{b},z_2) e^{i(z_1-z_2)x_{I\!P}m_N} \\ & \times e^{-\frac{A}{2}(1-i\eta)\sigma_3^j(x,Q^2)\int_{z_1}^{z_2} dz'\rho_A(\vec{b},z')} . \end{aligned} \quad (3)$$

In this equation, $f_j^{D(3)}$ is the diffractive parton distribution of the nucleon that depends on the light-cone fractions β and $x_{I\!P}$ and Q^2; $B_{\text{diff}} = 6\ \text{GeV}^{-2}$ is the slope of the t dependence of the ep diffractive cross section; $\rho_A(b,z)$ is the nuclear density; b is the transverse position of the nucleons (impact parameter) and z_1, z_2, and z' are the longitudinal positions of the nucleons; $\sigma_3^j(x,Q^2)$ is the effective rescattering cross section that controls the strength of the interaction with $N \geq 3$ nucleons.

Removing the integration over the impact parameter in Eq. (3), we obtain the impact parameter dependent nuclear PDFs:

$$\begin{aligned}
& xf_{j/A}(x,Q^2,b) = AT_A(b)xf_{j/N}(x,Q^2) \\
- & xf_{j/N}(x,Q^2)8\pi A(A-1)\,\Re e\frac{(1-i\eta)^2}{1+\eta^2}B_{\text{diff}}\int_x^{0.1} dx_{I\!P}\beta f_j^{D(3)}(\beta,Q^2,x_{I\!P}) \\
\times & \int_{-\infty}^{\infty} dz_1 \int_{z_1}^{\infty} dz_2 \rho_A(\vec{b},z_1)\rho_A(\vec{b},z_2)e^{i(z_1-z_2)x_{I\!P}m_N} \\
\times & \, e^{-\frac{A}{2}(1-i\eta)\sigma_3^j(x,Q^2)\int_{z_1}^{z_2} dz'\rho_A(\vec{b},z')}\,, \qquad (4)
\end{aligned}$$

where $T_A(b) = \int dz \rho_A(b,z)$ is the nuclear optical density. One can show that the impact parameter dependent nuclear PDFs are nothing else but the nuclear generalized parton distributions (GPDs) in the $\xi = 0$ limit and in the impact parameter representation:[9]

$$f_{j/A}(x,Q^2,b) = H_A^j(x,\xi=0,b,Q^2)\,. \qquad (5)$$

We quantify our predictions for $H_A^j(x,\xi=0,b,Q^2)$ by considering the ratio $R^j(x,b,Q^2)$,

$$R^j(x,b,Q^2) = \frac{f_{j/A}(x,Q^2,b)}{AT_A(b)f_{j/N}(x,Q^2)} = \frac{H_A^j(x,\xi=0,b,Q^2)}{AT_A(b)f_{j/N}(x,Q^2)}\,. \qquad (6)$$

Figure 3 presents the ratio $R^j(x,b,Q^2)$ for $\bar{u}$ quarks in ^{40}Ca as a function of x and b at $Q^2 = 4$ GeV^2. One can see from the figure that nuclear shadowing is larger at smaller x and smaller impact parameters b, where the density of nucleons is larger. Hence, nuclear shadowing introduces non-trivial correlations between x and b. Figure 3 essentially presents the spatial picture of nuclear shadowing.

The impact parameter dependent nuclear shadowing leads to an increase of the transverse size of partons in nuclei at small x. This increase has experimentally testable consequences such as the shift of the position of the minima of the DVCS cross section towards smaller t and dramatic oscillations of DVCS asymmetries as functions of t.[9]

4. Summary

Nuclear generalized parton distributions and coherent nuclear processes contain and may give an access to unique information on such nuclear effect as the EMC effect, nuclear shadowing, and medium modifications of the bound nucleons. In addition, nuclear GPDs give complimentary information

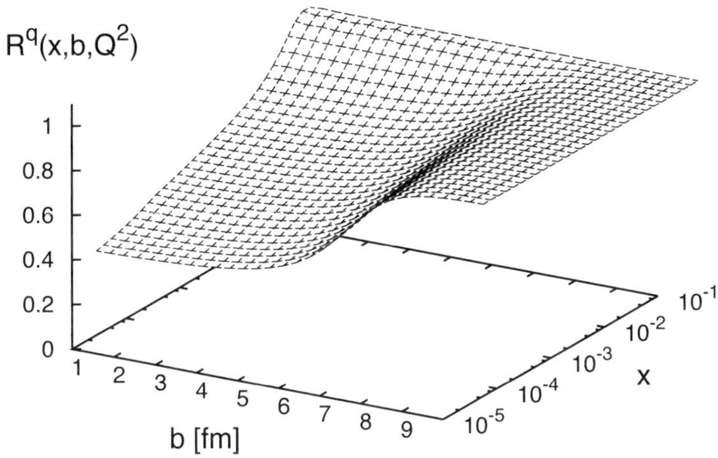

Fig. 3. Nuclear shadowing in nuclear GPDs. The ratio $R^j(x, b, Q^2)$ of Eq. (6) for $\bar{u}$ quarks in ^{40}Ca as a function of the light-cone fraction x and impact parameter b.

on the nucleon GPDs compared to that extracted from exclusive processes with a hydrogen target.

At medium values of Bjorken x_B, the important effect that we discussed is the interplay between the coherent and incoherent nuclear processes. As an example, we considered the ratio of the nuclear to free proton DVCS beam-spin asymmetries, $A^A_{\rm LU}/A^p_{\rm LU}$, and predicted that $A^A_{\rm LU}/A^p_{\rm LU} \approx A/Z > 1$ for small momentum transfers t and $A^A_{\rm LU}/A^p_{\rm LU} < 1$ for large values of t. The enhancement of $A^A_{\rm LU}/A^p_{\rm LU}$ above unity is essentially model-independent and combinatoric and seems to be in contradiction with the Hermes result. The situation will be clarified by the analysis of the JLab data on DVCS on ^{4}He.

At small value of Bjorken x_B, nuclear GPDs are significantly suppressed compared to the sum of the GPDs of the free nucleon due to nuclear shadowing. This leads to an increase of the transverse size of partons in nuclei which has measurable consequences: a shift of the positions of the minima of the nuclear DVCS cross section towards smaller t and dramatic oscillations of DVCS cross section asymmetries.

References

1. V. Guzey and M. Strikman, *Phys. Rev. C* **68**, 015204 (2003);
   V. Guzey, *Phys. Rev. C* **77**, 025211 (2008).
2. S. Liuti and S.K. Taneja, *Phys. Rev. C* **72**, 032201(R) and 034902 (2005).
3. S. Scopetta, *Phys. Rev. C* **79**, 025207 (2009).
4. A.V. Belitsky, D. Mueller and A. Kirchner, *Nucl. Phys. B* **629**, 323 (2002).
5. A. Airapetian *et al.* [Hermes Collaboration], *Phys. Rev. C* **81**, 035202 (2010).
6. K. Hafidi, F.-X. Girod, E. Voutier, H. Egiyan *et al.*, *Deeply Virtual Compton Scattering off* ^4He, Jefferson Lab Experiment E08-024 (2008).
7. V. Guzey, A.W. Thomas and K. Tsushima, *Phys. Lett. B* **673**, 9 (2009).
8. L.Frankfurt and M. Strikman, *Eur. Phys. J. A* **5**, 293 (1999);
   L. Frankfurt, V. Guzey and M. Strikman, *Phys. Rev. D* **71**, 054011 (2005).
9. K. Goeke, V. Guzey and M. Siddikov, *Phys. Rev. C* **79**, 035210 (2009).

DEEPLY VIRTUAL EXCLUSIVE PROCESSES WITH CHARM

S. LIUTI[1]

University of Virginia - Physics Department
382, McCormick Rd., Charlottesville, Virginia 22904 - USA
[1] *E-mail: sl4y@virginia.edu*

G.R. GOLDSTEIN[2]

Department of Physics and Astronomy
Tufts University, Medford, MA 02155 - USA.
[2] *E-mail: gary.goldstein@tufts.edu*

We propose to investigate a largely unexplored sector that is unique to the formulation of hard exclusive processes in terms of GPDs, namely the electroproduction of strange and charmed mesons in the kinematical ranges of Jefferson Lab's 12 GeV upgrade, and of the proposed Electron Ion Collider (EIC). In this contribution we focus on charmed meson production that is unique to the EIC. Exclusive strange and charmed meson production provides new insights in the connection of the quark/gluon degrees of freedom with the meson-baryon description, both in the unpolarized and polarized sectors. However, as particularly evident in polarized scattering, the underlying mechanisms are still far from being fully understood. We present an approach in terms of generalized parton distributions. As an application, we show that through exclusive electroproduction of pseudoscalar charmed mesons one can uniquely single out the non-perturbative charmed component in the nucleon structure function.

1. Introduction

Many experiments have been conducted to determine the charm content of the proton. Different types of hadronic reactions allow us in principle to access the charm quark distributions in a varied set of kinematical ranges, from hadron scattering at the Tevatron to inclusive and semi-inclusive leptoproduction at colliders (HERA) and in various fixed target experiments.

Two main competing interpretations can be given on the way charm gets excited. Either $c-\bar{c}$ pairs are produced through a Perturbative QCD (PQCD) mechanism such as gluon-gluon or photon-gluon fusion, when the scale of the probe, $Q^2 \geq 4m_c^2$, m_c being the charm quark mass, or such pairs exist in the nucleon as a so-called "Intrinsic Charm" (IC) component, with a non-negligible probability density at $Q^2 = m_c^2$, due to some underlying non-perturbative mechanism. Predictions from the two mechanisms at $Q^2 > m_c^2$, could differ in an observable way. The problem of disentangling them lies, however, unsolved to date. Figure 1 illustrates the rather large impact of non-perturbative charm contributions on the state of the art PDF parametrizations (dot-dashed curves) compared to the radiatively generated ones (shaded area) (Ref.[1]).

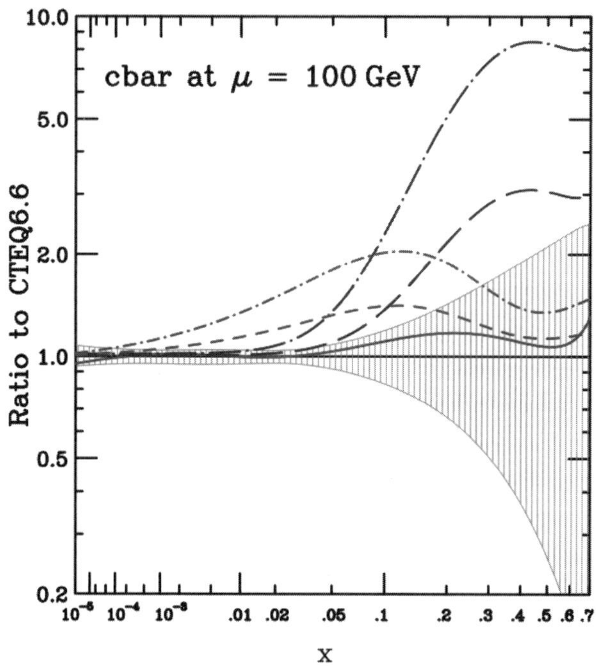

Fig. 1. Illustration of the status of present studies of the charmed quarks parton distribution. The arrow, and the visibly smaller error band, indicate the region in Bjorken x were most experiments are concentrated. The dashed and dot-dashed curves were obtained including different models of IC at the scale $\mu = 100$ GeV (adapted from [1]).

It has now become particularly pressing to study the heavy quark components of the nucleon because of the advent of the LHC. The heavy quark components will be key in the study of QCD matrix elements in the unprecedented multi-TeV CM energy regimes accessible at the LHC. At the same time, as the LHC opens new horizons for studies of physics beyond the Standard Model, many "candidate theories" will provide similar signatures of a departure from SM predictions. For these types of precision measurements it will be necessary to provide accurately determined QCD inputs.

The analyses in Ref.[2] have shown how the inclusion of IC could modify the outcome of global PDF analyses. However, the situation is not clear-cut. PDF analyses do not provide direct evidence that IC exists, but can be considered as indications in that direction.

In this contribution we make the point that in order to substantiate these findings it is important to identify new observables that would allow us to extend the studies to parton distribution size observables such as in Deeply Virtual Meson Production (DVMP), and to spin correlations. We present preliminary results involving the following electroproduction exclusive processes (Fig.2):

(1) $\gamma^* p \to J/\psi\, p'$,
(2) $\gamma^* p \to D\,\overline{D}\, p'$,
(3) $\gamma^* p \to \overline{D}\,\Lambda_c$,
(4) $\gamma^* p \to \eta_C\, p'$.

As we explain on what follows, these processes necessitate: *i)* high luminosity because they are exclusive; *ii)* high enough Q^2, in order to produce the various charmed mesons, and *iii)* a wide kinematical range in Bjorken x (see discussion relative to Fig.1). Their ideal realization would be at an Electron Ion Collider (EIC) that could guarantee high luminosity conditions also at relatively large x.

2. Windows into Heavy Flavor Production at the EIC

QCD factorization for exclusive meson electroproduction was proven initially in Ref.[3]. Factorization arguments can be extended to the production of charmed mesons, thus allowing us to write the various contributions to the cross section, originating from the amplitudes for different beam and target spin configurations, as a convolution over the longitudinal momentum fraction X of the hard scattering process with the quark-proton amplitudes. Using the approach in Ref.[4] (see Section 3), we can describe the exclusive production of pseudoscalar charmed mesons mainly in terms of *chiral-odd* Generalized Parton Distributions (GPDs) for the u, d and c

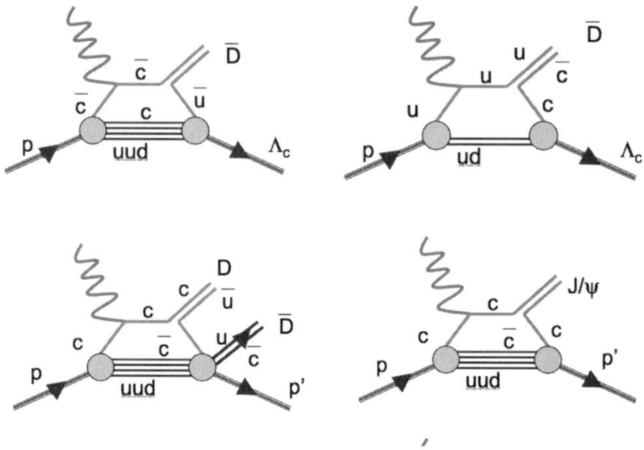

Fig. 2. Deeply Virtual Charmed Meson Production (DVCMP). All produced mesons are unique probes of the IC charm content.

quarks, $F_T^q \equiv \{H_T\, E_T, \widetilde{H}_T, \widetilde{E}_T\}$, with $q = u, d, c$. [a] These by definition do not couple to gluons, or they evolve perturbatively similarly to the flavor non singlet distributions, Refs.[5–7]. All GPDs involved in pseudoscalar charmed mesons electroproduction are therefore of the "intrinsic" kind since they cannot evolve from gluons. By finding observables that can isolate the charmed GPDs, one can therefore uniquely gauge the size of the IC contribution.

A somewhat similar perspective was presented in Ref.[8] where it was shown that the IC content of the proton can be studied by measuring asymmetries in inclusive heavy quark jet production, $ep \to e'QX(\overline{Q})$. In fact, IC does not contribute at all to azimuthal asymmetries of the $cos(2\phi)$ form $\frac{\sigma_{TT}}{\sigma_T + \epsilon\sigma_L}$. Hence the reduction at high x_{Bj} of large asymmetry from their perturbative gluon model signals the intrinsic charm contribution. This, however is dependent on their model and requires measuring an asymmetry that is reduced from the gluon contribution. While these studies present yet another opportunity for studying the multifaceted and still elusive connections between semi-inclusive and exclusive processes, our approach is a more stringent test for IC since the suggested exclusive production processes occur solely via the IC channel.

The hard scattering amplitudes for the four processes above are re-

[a] A small contribution from the chiral even GPD is in principle also present, that we disregard here.

spectively (1) $\gamma^* c(\bar{c}) \rightarrow J/\psi\, c(\bar{c})$, (2) $\gamma^*\, c \rightarrow D\, u$, (3) $\gamma^* \bar{c} \rightarrow \overline{D}\, (\bar{u})$ and $\gamma^*\; u \rightarrow \overline{D}\, c$, (4) $\gamma^* c(\bar{c}) \rightarrow \eta_C\, c(\bar{c})$ (Fig.2). All of these involve the charm mass scale in some way, along with a heavy meson distribution amplitude. For D and $\bar{D}$ creation in the beam direction, the distribution or wavefunction will be of the same order for each. The J/Ψ will be more difficult to produce in the hard process, aside from its sizable diffractive production, which occurs through the virtual photon dissociation into $c\bar{c}$. The pair scatter diffractively and recombine as the J/Ψ. The diffractive scattering is implemented through two gluon exchange, so that the overall process would involve gluon-nucleon GPDs rather than intrinsic charm, at least in the diffractive region.

Amplitudes (1) and (4) could be smaller than (2) and (3) for several reasons. From the GPD perspective, the soft contribution is given only by the charm GPD, F_c that, based on the constraint from inclusive scattering – giving the forward limit of GPDs– we expect to be a few per cent of the light quarks' GPDs. Moreover process (4) requires the detection of an η_c meson, which seems to be a forbidding task even in a future EIC. Process (2) does not seem the best way to access GPDs because of the exotic $\overline{D}\, p$ final hadronic configuration. Finally, process (3) can occur in the two configurations described in Fig.1c and Fig.1d. By assuming $SU(4)$ symmetry, these are proportional to linear combinations of the non-singlet GPDs H_c, H_u and H_d. So the mechanism involving IC is dominated by scattering off the light flavor quarks, at least before symmetry breaking is implemented. However, by detecting different baryons in the final state, that is considering the reactions:

$$\gamma^* p \rightarrow \overline{D}^0\, \Lambda_c^+ \Rightarrow \left(2F_T^u - F_T^d + F_T^c\right)/\sqrt{6},$$
$$\gamma^* p \rightarrow \overline{D}^0\, \Sigma_c^+ \Rightarrow \left(F_T^d - F_T^c\right)/\sqrt{2},$$
$$\gamma^* n \rightarrow D^-\, \Sigma_c^{++} \Rightarrow F_T^d - F_T^c,$$

one can extract the charmed GPDs, F_T^c.

We related the GPDs that involve charmed quarks or charmed hadrons to their uncharmed counterparts through flavor SU(4). In the GPDs involving open charm production, process (3), there are the off-diagonal amplitudes for $N \rightarrow u\,(d) : c \rightarrow \Lambda_C^+$, $\Sigma_C^+\,(\Sigma_C^{++})$. The spin $\frac{1}{2}$ hadrons belong to the 20-plet. To simplify the determination of the SU(4) Clebsch-Gordon coefficients, notice that the $n\,\&\,p$ are in the same relation to these 3 charmed baryons as the strange set Λ^0, Σ^0, and Σ^+, which corresponds to replacing c by s and vice versa. This result can be seen as a rotation in the SU(4) space of the usual baryon octet with no s, one s, two s (the non-charm subset of the 20) to the non-strange no c, one c and two c octet (the non-strange subset of the 20). The same correspondence applies to the meson 15-plet. The 4×4 matrix can have s & c entries reversed. Now there is strong breaking of the SU(4) symmetry. There are many ansätze for such symmetry breaking, e.g. Ref. [9]. Obviously, symmetry breaking must suppress the production of charm. Phase space does enter, but other effects may be significant. This is being studied.

3. Pseudoscalar Charmed Mesons Electroproduction and Chiral Odd GPDs

The next step is to provide a model for charmed GPDs that could guide experimenters for future simulations and possible extractions of F_T^c from data. Our approach is similar to Refs.[10,11], where a Regge improved quark-diquark model was devised to provide a parametrization of the chiral even, unpolarized GPDs H^q and E^q, $q = u, d$ that is quantitatively constrained by the nucleon form factors data, and from deep inelastic scattering structure functions data through

$$\int_{-1+\zeta}^{1} dX H^q(X,\zeta,t) = F_1^q(t) \qquad \int_{-1+\zeta}^{1} dX E^q(X,\zeta,t) = F_2^q(t)$$

$$F^q(X,0,0;Q^2) \equiv q(X,Q^2)$$

where we introduced the kinematical variables $X = k^+/P^+$, $\zeta = \Delta^+/P^+$, and $t = \Delta^2$, P being the proton momentum, k the struck quark momentum, and Δ the momentum transfer between the initial and final proton. $F^q_{1(2)}$ are the quark contributions to the nucleon Dirac and Pauli form factors, and $q(X)$ is the quark parton distribution at a given scale, Q^2 (for a review of GPDs see Ref.[12]).

In order to constrain charmed GPD, $H_T^{c(\bar{c})}$, we used the same set of non-perturbative models of IC that were used in Ref.[2]. These were found to be consistent with present DIS data, posing an upper limit on the intrinsic charm content measured by the momentum fraction $\langle x \rangle_{c+\bar{c}} \leq 2\%$. The constraint from the form factor is less obvious since no measurements or *ab initio* evaluations are available (see however, Ref.[13]). We therefore took as an upper limit the recent experimental extraction of the strangeness axial form factor $G_A^s(t)$ [14]. This predicts negative values of G_A^s in the range $0.47 \leq -t \leq 1$ GeV^2, that are consistent with recent findings of Δs. Initial results based on these estimates are shown in Fig.3. We have chiral odd

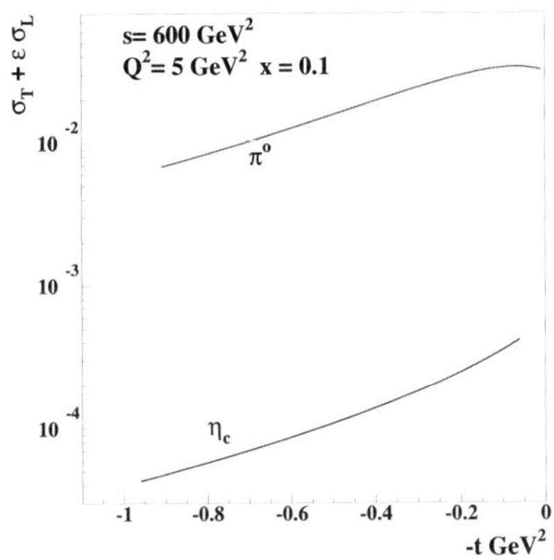

Fig. 3. Comparison of π^o and η_c cross sections. Although η_c is hard to detect, the range between the two lines gives an estimate of where the cross sections for the other processes will lie.

GPDs for these pseudoscalar processes, just as we introduced for the π^0 &η, Ref.[15]. The logic followed the observation that transverse virtual photons were significant, while the chiral even contributions, severely limited by the C-parity odd $t-$channel, were small, in principle. Once the dominant behaviour is established as due to transversity $(H_T(X, \zeta, t))$, the admixture of gluons at leading twist is forbidden. So focusing on charmed pseudoscalars provides access to purely intrinsic charm rather than photon-gluon fusion.

4. Conclusions

The emergence of the EIC as a top priority of the nuclear physics community presents a unique opportunity to explore the non-perturbative sea and gluonic structure of the nucleon and nuclei. Here we presented a connection between intrinsic and open charm and transverse spatial distributions of the nucleon's sea structure through electroproduction and GPDs.

In particular the issue of whether perturbative gluons produce $c\,\bar{c}$ pairs, or intrinsic, non-perturbative charm mechanisms are present in the nucleon, still remains unsettled. One possibility that we studied is that various asymmetries for exclusive charm production providing *e.g.* the azimuthal dependence for the D or D^* meson, the $\cos(2\phi)$ term in the unpolarized differential cross section, may provide that discrimination.

The proposed EIC presents a unique opportunity to test the distinction between intrinsic charm and competing models by exploiting its kinematic reach in x and the final state variables (z, and P_T) in semi-inclusive DIS, and t in exclusive proccesses. The various design scenarios range from $E_e/E_P = 10/100-250$ GeV, $E_e/E_P = 3-7/30-150$GeV with luminosity projections ranging from few $\times\ 10^{33}\,\mathrm{cm}^{-2}\mathrm{s}^{-1}$ to few $\times\ 10^{35}\,\mathrm{cm}^{-2}\mathrm{s}^{-1}$, and to a recently proposed medium energy collider $E_e/E_P = 3-7/30-60$ GeV (Ref.[16]).

Another feature of exclusive charm production is that the produced charmed hyperon can carry significant polarization through a mechanism similar to strange hyperons. The Λ_c hyperon's polarization is determined by its weak decay products and is known to provide favorable polarization analysis (Ref.[17]). Utilizing polarized electron and proton beams in the proposed kinematic regime would allow the disentangling of the multiple mechanisms as well. We are studying these multiple spin correlation observables to see what their impact can be in both LHC and EIC kinematical regimes (Ref.[18]).

Acknowledgements

The authors appreciate the work of the organizers of Exclusives 2010. We also thank our colleague L. Gamberg for initial participation in this project.This work is partially supported by the U.S. Department of Energy grants DE-FG02-01ER4120 (S.L.), and DE-FG02-92ER40702 (G.R.G.).

References

1. P. M. Nadolsky *et al.*, Phys. Rev. D **78**, 013004 (2008)
2. J. Pumplin, H. L. Lai and W. K. Tung, Phys. Rev. D **75**, 054029 (2007)
3. J.C. Collins, L. Frankfurt and M. Strikman, Phys. Rev. D **56**, 2982 (1997)
4. S. Ahmad, G. R. Goldstein and S. Liuti, Phys. Rev.D**79**, 054014 (2009)
5. V. Barone, T. Calarco and A. Drago, Phys. Lett. B **390**, 287 (1997)
6. W. Vogelsang, Phys. Rev. D **57**, 1886 (1998)
7. M. Hirai, S. Kumano and M. Miyama, Comput. Phys. Commun. **111**, 150 (1998); S. Kumano and M. Miyama; Phys. Rev. D **56**, 2504 (1997)
8. L. N. Ananikyan and N. Y. Ivanov, Nucl. Phys. B **762**, 256 (2007)
9. X. Song, arXiv:hep-ph/9910515; *ibid* Int. J. Mod. Phys. A **18**, 1501 (2003).
10. S. Ahmad, H. Honkanen, S. Liuti and S.K.Taneja, Phys. Rev.D**75**, 094003 (2007)
11. S. Ahmad, H. Honkanen, S. Liuti and S.K.Taneja, Eur. Phys. J. C **63**, 407 (2009)

12. A. V. Belitsky and A. V. Radyushkin, Phys. Rept. **418**, 1 (2005)
13. G. Bali, S. Collins and A. Schafer, PoS **LATTICE2008**, 161 (2008)
14. S. F. Pate, D. W. McKee and V. Papavassiliou, Phys. Rev. C **78**, 015207 (2008)
15. G.R. Goldstein, and S. Liuti, *these proceedings.*
16. T. Horn, P. Nadel-Turonski and C. Weiss, arXiv:0908.1999 [hep-ph].
17. W. G. D. Dharmaratna and G. R. Goldstein, Phys. Rev. D **53**, 1073 (1996); G. R. Goldstein, arXiv:9907573 [hep-ph].
18. G.R. Goldstein, and S. Liuti, *in preparation.*

COLOR TRANSPARENCY IN EXCLUSIVE MESON AND TWO-PROTON PHOTOPRODUCTION

W. COSYN*† and J. RYCKEBUSCH

Department of Physics and Astronomy, Ghent University,
Proeftuinstraat 86, Gent, B9000, Begium
* *E-mail: wim.cosyn@ugent.be*
http://fens.ugent.be/
† Present address: *Florida International University,*
Miami, Florida, 33199, USA

We present transparency calculations obtained in a relativistic multiple-scattering Glauber approximation model. For the pion photo- and electroproduction reaction results are compared with Jefferson Lab data and predictions are shown for energies available after the 12 GeV upgrade. For the $A(\gamma, pp)$ reaction transparencies we consider two possible reaction mechanisms. Both the one-step knockout of a correlated pair and a two-step hard rescattering reaction picture yield low transparencies.

Keywords: Color Transparency; Relativistic scattering theory

1. Introduction

Color transparency is an effect originating from QCD first described by Mueller and Brodsky.[1,2] It deals with the expansion of a small-sized configuration of quarks into a regular hadron state. To the nuclear medium, this small-sized configuration (also called Point Like Configuration or PLC) behaves like a colorless object. As a consequence the strongly interacting hadron propagates with reduced final-state interactions (FSI) along a certain formation length before it evolves to its normal hadronic state.

In search for CT at intermediate energies, the measured observable is the nuclear transparency. Detection of a CT signals the onset of partonic degrees of freedom. The nuclear transparency is defined as the ratio of the cross section per target nucleon for a process on a nucleus to the cross section of the process on a free nucleon

$$T = \frac{\sigma^{\text{nucleus}}}{A\sigma^{\text{free nucleon}}} . \tag{1}$$

Accordingly, the nuclear transparency provides a measure of the combined effect of nuclear attenuation on the hadrons produced in some reaction. One can study the hard-scale dependence of the transparency for a certain target nucleus A, or the A dependence at a fixed value of the hard-scale parameter. If CT effects were to appear at a certain energy, the nuclear transparency would be observed to overshoot the predictions from traditional nuclear physics models. The measurement of the onset and magnitude of the CT effect allows one to constrain models describing the evolution of a PLC into a hadron. The CT effect also plays a role in the access to generalized parton distributions (GPDs) in deeply virtual Compton scattering and meson production processes.[3,4] Indeed, it is a necessary condition for the assumption of factorization which states that from a certain virtuality onwards the soft and hard physics can be separated.

In this paper we present transparency calculations obtained in a model that uses a relativistic multiple-scattering Glauber approximation[5] (RMSGA) to describe the FSI of the particles in the final state. We compare to data from several Jefferson Lab (JLab) experiments and show some predictions for the kinematic region accessible after the 12 GeV upgrade.

2. Relativistic multiple-scattering Glauber approximation

Originally developed in optics, Glauber theory can be applied in situations where the wavelength of the fast scattering particle is small in comparison with the typical interaction range between the fast particle and the spectators. In Glauber theory, the eikonal approximation is used and each spectator adds a phase to the wave function of the outgoing scatterer. The RMSGA model[5] provides a relativistic extension of the Glauber theory and can be applied to a variety of reactions.[6–11]

In the RMSGA model, the eikonal phase that enters in the amplitude for a reaction with a leptonic probe and multiple hadrons in the final state takes the following form:

$$\mathcal{G}(\vec{b}, z) = \prod_{\alpha_{occ} \neq \alpha} \int d\vec{r}\,' \left| \phi_{\alpha_{occ}} (\vec{r}\,') \right|^2 \prod_i \left[1 - \theta(z_i' - z_i) \Gamma_{iN} \left(\vec{b}_i' - \vec{b}_i \right) \right] . \quad (2)$$

Here, α stands for the quantum numbers of the nucleon that interacts with the initial probe, $\vec{r} = (\vec{b}_i, z_i)$ is the coordinate of the hard interaction, with z_i along the momentum of the hadron subject to the attenuations. The product α_{occ} runs over all residual nucleons, with $\phi_{\alpha_{occ}} (\vec{r}\,')$ the relativistic single particle bound-state wave functions obtained from the Serot-Walecka model.[12] The product i runs over all particles subject to FSI. The profile functions Γ_{iN} depend on three energy-dependent parameters wich are fitted to nucleon-nucleon or meson-nucleon scattering data.

The effects of CT are included in the model through the quantum diffusion model.[13] For the ΔM^2 parameter that enters in the formation length in this model, we take the values $\Delta M^2 = 1$ GeV^2 for the proton and $\Delta M^2 = 0.7$ GeV^2 for the pion.

The interaction of the target nucleus with the impinging probe at some position requires the presence of a nucleon. The finite size of the target nucleon will induce local fluctuations in the density of the target nucleus due to short-range correlations (SRC). We correct our treatment of the FSI for this medium effect. This is achieved by substituting the single-particle densities $|\phi_{\alpha_{occ}}(\vec{r}')|^2$ in Eq. (2) with an effective correlated density in the following manner:[14]

$$\begin{aligned} |\phi_\alpha(\vec{r_2})|^2 &\rightarrow \frac{\rho_A^{[1]}(\vec{r_2})}{A} \\ &\rightarrow \frac{1}{A-1}\frac{\rho_{A,\mathrm{corr.}}^{[2]}(\vec{r_2},\vec{r})}{\rho_A^{[1]}(\vec{r})} = \frac{1}{A}\gamma(\vec{r})\rho_A^{[1]}(\vec{r})\rho_A^{[1]}(\vec{r_2})\gamma(\vec{r_2})g(|\vec{r}-\vec{r_2}|)\,. \end{aligned} \tag{3}$$

Here, $\rho_A^{[1]}(\vec{r})$ is the one-body density, $\rho_{A,\mathrm{corr.}}^{[2]}(\vec{r_2},\vec{r})$ the correlated two-body density, and $g(|\vec{r}-\vec{r_2}|)$ the Jastrow correlation function. The function $\gamma(\vec{r})$ imposes the correct normalization of the two-body density and can be obtained as the solution to an integral equation.

3. Results

In Fig. 1, we show the results of transparency calculations[9] for $^4\mathrm{He}(\gamma, p\pi^-)$ together with the experimental data and the predictions of a semiclassical model.[15] The calculations extend to the energy region that will become accessible after the JLab 12 GeV upgrade. For $\theta_{\mathrm{c.m.}} = 90°$, this implies that the treshold for charm resonance production can be reached. This treshold has been linked[16] to the drop in the measured $^{12}\mathrm{C}(p, 2p)$ transparencies at higher energies.[17–19] As can be appreciated from Fig. 1, the predicted role of CT on the transparencies is comparable in the semiclassical and RMSGA model. We wish to stress that the semiclassical results including CT are normalized to the ones without CT at the data point with the lowest $|t|$. No renormalization has been applied to the RMSGA results of Fig. 1. Our calculations including CT systematically overestimate the measured transparencies by 10-15%. The hard-scale dependence of the data, on the other hand, is better reproduced after including CT. As can be seen in Fig. 1, our model predicts a rise in the transparency for $|t| \leq 1.2$ GeV^2. This phenomenon can be attributed to the behavior of the proton-proton and (to a lesser extent) proton-neutron cross section for proton momenta of about 1 GeV.[20] The proton-proton cross section drops from ~ 40 mb to about 20-30 mb as one approaches 1 GeV from higher momenta. One observes that SRC mechanisms induce an overall increase in the nuclear transparency of about 2%. This reflects the fact that the medium effectively reduces the free pion-nucleon and nucleon-nucleon cross sections. The predicted increase is almost independent of $|t|$ which points towards no direct dependence on the hard scale. The CT phenomenon, on the other hand, leads to a linear rise in the transparencies from almost 0 to over 20% at the largest values of the hard-scale parameter $|t|$. At high $|t|$ the short-range correlations have a modest impact on the magnitude of the CT effects. At those energies, the CT phenomenon leads to almost vanishing interactions close to the initial interaction point thereby mitigating the impact of the SRC on the computed magnitude of the FSI.

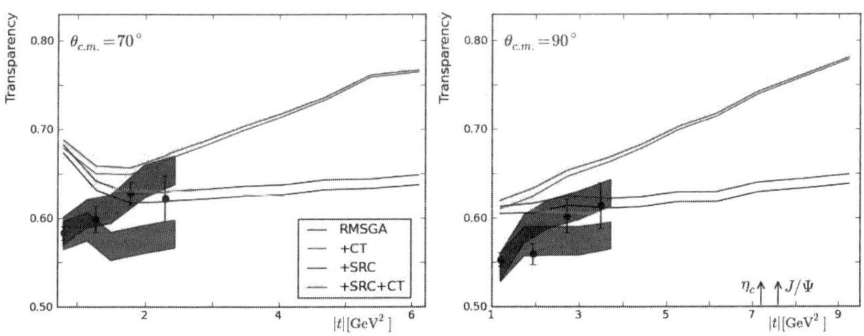

Fig. 1. (Color online) The nuclear transparency extracted from ^{4}He$(\gamma, p\pi^-)$ versus the squared momentum transfer $|t|$ at $\theta_{\mathrm{c.m.}} = 70°$ and $\theta_{\mathrm{c.m.}} = 90°$. The semiclassical model[15] results are presented by the shaded areas: the red (blue) area is a calculation without (with) CT. Data from Ref. [21].

Figure 2 presents the results from our transparency calculations for the pion electroproduction $(A(e, e'\pi^+))$ reaction on three different nuclei up to the highest energies available in JLab after the 12 GeV upgrade, together with the experimental data[22] and results from a semiclassical model.[23] The RMSGA and semiclassical calculations excluding CT display a modest increase over the Q^2 range. This behaviour finds a simple explanation in the p_π dependence of the $\sigma^{\mathrm{tot}}_{\pi^+ p}$. The results contained in Fig. 2 cover a range in pion momenta given by $2.8 \leq p_\pi \leq 4.4$ GeV. In this range, $\sigma^{tot}_{\pi^+ p}$ displays a soft decrease, which reflects itself in a soft increase of the nuclear transparency. The SRC induce a Q^2 independent 2% increase of the transparencies. The calculations including CT enhance the transparency up to 50% at the highest energies and are in very good agreement with the experimental data which extend up to about $\mid t \mid \approx 5$ GeV^2. Upon comparing the RMSGA+CT results to the semiclassical calculations that include CT (denoted as *Larson + CT* in the figure), we see that the slopes of both calculations are in excellent agreement. This indicates the robustness of the role of CT. Indeed, the semiclassical and RMSGA model are quite different with regard to the modeling of nuclear attentuation. Using an identical parametrization for the quantum diffusion model, however, both models attribute a similar role to CT mechanism.

The RMSGA transparency predictions are somewhat larger than the semiclassical ones for the ^{12}C target, and evolve to similar values for the ^{63}Cu target. A more recent model developed by Kaskulov et al.[24] also finds excellent agreement between the calculations including CT and the data.

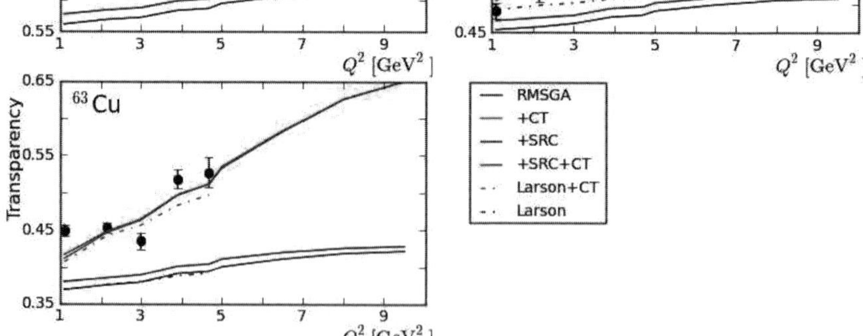

Fig. 2. (Color online) The Q^2 dependence of the nuclear transparency for the $A(e, e'\pi^+)$ process in ^{12}C, ^{27}Al and ^{63}Cu. The dot-dashed curves are the results of the semiclassical model by Larson, Miller and Strikman.[23] JLab data are taken from Ref. [22].

The Hall A experiment E03-101[25] at Jefferson Lab has measured the transparency for the ^{3}He(γ, pp) and ^{12}C(γ, pp) reactions, and data analysis is underway. Data are collected for proton c.m. angles $\theta_{c.m.} = 90^\circ$ in coplanar and symmetric kinematics. In a direct knockout picture for the $A(\gamma, pp)$ reaction the photon interacts directly with a correlated pair. The two nucleons constitute a high density region in the nucleus, have a high relative momentum and are separated by a very small distance. By interacting with one of the nucleons, the photon breaks up the pair. Due to the high relative momentum, the second nucleon can also be ejected from the nucleus and detected if the kinematics is carefully tuned. We computed the transparencies for ^{3}He(γ, pp) and ^{12}C(γ, pp) in such a reaction picture. The results are contained in Fig. 3 and are plotted as a function of momentum transfer squared $|t| = (q^\mu - p^\mu_{N_1})^2$. As for the pion photoproduction channel, the transparencies rise for $|t| \leq 2$ (GeV/c^2) due to the local minimum of the total nucleon-nucleon cross section for momenta around 1 GeV/c. The inclusion of CT enhances the transparency up to over 20% for carbon at the highest energies. The most striking feature, however, is the low values of the ^{12}C(γ, pp) transparencies. Naively, one would expect that they are approximately equal to the squared transparency observed in the single-nucleon knockout reaction ^{12}C$(e, e'p)$ at these energies. In line with the measurements, the RMSGA model predicts a transparency of about 0.6 for the ^{12}C$(e, e'p)$ reaction at high energies.[6] Upon squaring one finds 0.36 which is significantly higher than the ~ 0.23 from the ^{12}C(γ, pp) calculations. This is a consequence of the fact that on average the $A(\gamma, pp)$ reaction probes higher density regions of the nucleus than the $A(e, e'p)$ reaction.[26] Higher densities imply stronger nuclear attenuation, and hence a lower transparency.

Figure 4 shows predictions for the ^{3}He(γ, pp) transparencies whereby the reaction is described as a hard rescattering process. In this two-step reaction picture a single nucleon obtains a high momentum by absorbing an energetic photon. Upon propagating to the surface of the target nucleus the nucleon is not only subject to soft collisions with the residual nucleons but experiences a hard rescattering with one of them. As a result, two nucleons are ejected. In our calculations, we account for the soft collisions of the initial and secondary proton. We have also performed calculations under the assumption that the initial nucleon moves freely between the photon-nucleon interaction point and the nucleon-nucleon rescattering point. As can be appreciated from Fig. 4, the FSI in between these points lower the transparency by about 5%. Upon comparing the ^{3}He results of Fig.

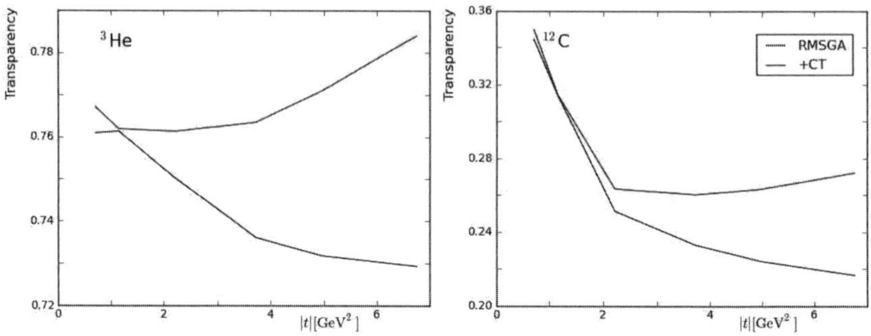

Fig. 3. (Color online) The nuclear transparency extracted from $^{12}C(\gamma, pp)$ (right) and $^3He(\gamma, pp)$ (left) versus the squared momentum transfer $|t|$. These calculations consider $A(\gamma, pp)$ as the breakup of a correlated pair.

4 with those of Fig. 3 one can conclude that the direct reaction picture and the rescattering reaction picture result in a similar $|t|$ dependence. The calculations including all FSI in the hard rescattering picture yield lower transparencies than those in Fig. 3, indicating that the two-step two-nucleon knockout probes even higher densities than the correlated pair knockout. A detailed investigation of the $A(\gamma, NN)$ transparencies requires calculations with the two competing mechanisms summed at amplitude level. In this way, the size of the interference between the one- and two-step processes can also be determined.

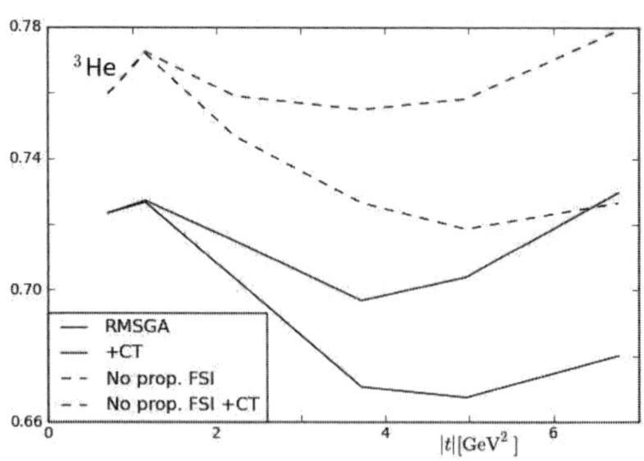

Fig. 4. (Color online) The nuclear transparency extracted from $^3He(\gamma, pp)$ versus the squared momentum transfer $|t|$. Solid curves include all FSI, dashed curves do not include the FSI of the intermediate propagator. These calculations consider $^3He(\gamma, pp)$ as a hard-rescattering mechanism.

4. Conclusion

As illustrated in this paper, the RMSGA model is a flexible and robust framework for computing nuclear transparencies at intermediate energies. The hard-scale dependence of the RMSGA transparencies for the pion production reaction show a fair (in the case of photoproduction) and very nice (in the case of electroproduction) agreement with the JLab data when including the effects of CT. Both SRC and CT enhance the transparencies, but they can be clearly separated due to their different behavior upon changing the magnitude of the hard-scale parameter. For the $A(\gamma, pp)$ reaction, transparency calculations for both the knockout of a correlated pair and the hard rescattering process yield unexpectedly low values. They are a result of the fact that for both reaction mechanisms the process is highly sensitive to the high density regions in the deep interior of the target nucleus.

This work was supported by the Research Foundation Flanders and the Research Board of Ghent University.

References

1. A. Mueller, *Proceedings of the Seventeenth Recontre de Moriond, Moriond* , p. 13 (1982).
2. S. J. Brodsky, G. T. Bodwin and G. P. Lepage, *Proceedings of the 13th Int. Symp. on Multiparticle Dynamics, Volendam* , p. 963 (1982).
3. S. Liuti and S. K. Taneja, *Phys. Rev.* **D70**, p. 074019 (2004).
4. M. Burkardt and G. A. Miller, *Phys. Rev.* **D74**, p. 034015 (2006).
5. J. Ryckebusch *et al.*, *Nucl. Phys.* **A728**, 226 (2003).
6. P. Lava, M. C. Martinez, J. Ryckebusch, J. A. Caballero and J. M. Udias, *Phys. Lett.* **B595**, 177 (2004).
7. B. Van Overmeire, W. Cosyn, P. Lava and J. Ryckebusch, *Phys. Rev.* **C73**, p. 064603 (2006).
8. B. Van Overmeire and J. Ryckebusch, *Phys. Lett.* **B644**, 304 (2007).
9. W. Cosyn, M. C. Martinez, J. Ryckebusch and B. Van Overmeire, *Phys. Rev.* **C74**, p. 062201 (2006).
10. W. Cosyn, M. C. Martinez and J. Ryckebusch, *Phys. Rev.* **C77**, p. 034602 (2008).
11. M. C. Martinez *et al.*, *Phys. Rev.* **C73**, p. 024607 (2006).
12. R. J. Furnstahl, B. D. Serot and H.-B. Tang, *Nucl. Phys.* **A615**, 441 (1997).
13. L. L. Frankfurt and M. I. Strikman, *Phys. Rept.* **160**, 235 (1988).
14. S. Frankel, W. Frati and N. Walet, *Nucl. Phys.* **A580**, 595 (1994).
15. H. Gao, R. J. Holt and V. R. Pandharipande, *Phys. Rev.* **C54**, 2779 (1996).
16. S. J. Brodsky and G. F. de Teramond, *Phys. Rev. Lett.* **60**, p. 1924 (1988).
17. A. S. Carroll *et al.*, *Phys. Rev. Lett.* **61**, 1698 (1988).
18. I. Mardor *et al.*, *Phys. Rev. Lett.* **81**, p. 5085 (1998).

19. A. Leksanov *et al.*, *Phys. Rev. Lett.* **87**, p. 212301 (2001).
20. W. M. Yao *et al.*, *J. Phys.* **G33**, 1 (2006).
21. D. Dutta *et al.*, *Phys. Rev.* **C68**, p. 021001 (2003).
22. B. Clasie *et al.*, *Phys. Rev. Lett.* **99**, p. 242502 (2007).
23. A. Larson, G. A. Miller and M. Strikman, *Phys. Rev.* **C74**, p. 018201 (2006).
24. M. M. Kaskulov, K. Gallmeister and U. Mosel, *Phys. Rev.* **C79**, p. 015207 (2009).
25. D. Margaziotis *et al.* (2003), JLAB-PR-03-101,JLAB-PR-03-005.
26. W. Cosyn and J. Ryckebusch, *Phys. Rev.* **C80**, p. 011602 (2009).

HOW TO MEASURE THE CHARM DENSITY IN THE PROTON AT EIC

N.Ya. IVANOV

Yerevan Physics Institute, Alikhanian Bros. Str. 2, 0036 Yerevan, Armenia
E-mail: nikiv@yerphi.am
www.yerphi.am

We study two experimental ways to measure the heavy-quark content of the proton: using the Callan-Gross ratio $R(x,Q^2) = F_L/F_T$ and/or the azimuthal $\cos(2\varphi)$ asymmetry in deep inelastic lepton-nucleon scattering. Our approach is based on the following observations. First, unlike the production cross sections, the ratio $R(x,Q^2) = F_L/F_T$ and the azimuthal $\cos(2\varphi)$ asymmetry in heavy-quark leptoproduction are sufficiently stable, both parametricallly and perturbatively, in a wide region of variables x and Q^2 within the fixed-flavor-number scheme of QCD. Second, both these quantities, $R(x,Q^2) = F_L/F_T$ and $\cos(2\varphi)$ asymmetry, are sensitive to resummation of the mass logarithms of the type $\alpha_s \ln\left(Q^2/m^2\right)$ within the variable-flavor-number schemes. These two facts together imply that the heavy-quark densities in the nucleon can, in principle, be determined from high-Q^2 data on the Callan-Gross ratio and/or the azimuthal asymmetry in heavy-quark leptoproduction. In particular, the charm content of the proton can be measured in future studies at the proposed Large Hadron-Electron (LHeC) and Electron-Ion (EIC) Colliders.

Keywords: Perturbative QCD, Heavy-Quark Leptoproduction, Mass Logarithms Resummation, Callan-Gross Ratio, Azimuthal Asymmetry

1. Introduction

The notion of the intrinsic charm (IC) content of the proton has been introduced about 30 years ago in Ref. [1]. It was shown that, in the light-cone Fock space picture,[2] it is natural to expect a five-quark state contribution, $|uudc\bar{c}\rangle$, to the proton wave function. This component can be generated by $gg \to c\bar{c}$ fluctuations inside the proton where the gluons are coupled to different valence quarks. The original concept of the charm density in the proton[1] has nonperturbative nature since a five-quark contribution $|uudc\bar{c}\rangle$ scales as $1/m^2$ where m is the c-quark mass.[3]

In the middle of nineties, another point of view on the charm content of the proton has been proposed in the framework of the variable-flavor-number scheme (VFNS).[4,5] The VFNS is an approach alternative to the traditional fixed-flavor-number scheme (FFNS) where only light degrees of freedom (u, d, s and g) are considered as active. Within the VFNS, the mass logarithms of the type $\alpha_s \ln\left(Q^2/m^2\right)$ are resummed through the all orders into a heavy quark density which evolves with Q^2 according to the standard DGLAP[6] evolution equation. Hence this approach introduces the parton distribution functions (PDFs) for the heavy quarks and changes the number of active flavors by one unit when a heavy quark threshold is crossed. Note also that the charm density arises within the VFNS perturbatively via the $g \rightarrow c\bar{c}$ evolution. Some recent developments concerning the VFNS are presented in Refs. [7–9]. So, the VFNS was introduced to resum the mass logarithms and to improve thus the convergence of original pQCD series.

Presently, both nonperturbative IC and perturbative charm density are widely used for a phenomenological description of available data. (A recent review of the theory and experimental constraints on the charm quark distribution may be found in Ref. [10]). In particular, practically all the recent versions of the CTEQ[11–13] and MRST[14] sets of PDFs are based on the VFN schemes and contain a charm density. At the same time, the key question remains open: How to measure the charm content of the proton? The basic theoretical problem is that radiative corrections to the heavy-flavor production cross sections are large: they increase the leading order (LO) results by approximately a factor of two. Moreover, soft-gluon resummation of the threshold Sudakov logarithms indicates that higher-order contributions can also be substantial (for reviews, see Refs. [15,16]). On the other hand, perturbative instability leads to a high sensitivity of the theoretical calculations to standard uncertainties in the input QCD parameters: the heavy-quark mass, m, the factorization and renormalization scales, μ_F and μ_R, $\Lambda_{\rm QCD}$ and the PDFs. For this reason, one can only estimate the order of magnitude of the pQCD predictions for charm production cross sections in the entire energy range from the fixed-target experiments[17] to the RHIC collider.[18]

Since production cross sections are not perturbatively stable, they cannot be a good probe of the charm density in the proton. For this reason, it is of special interest to study those observables that are well-defined in pQCD. Nontrivial examples of such observables were proposed in Refs. [19–24], where the azimuthal $\cos(2\varphi)$ asymmetry and Callan-Gross ratio $R(x,Q^2) = F_L/F_T$ in heavy quark leptoproduction were analyzed.[a] It was shown that, contrary to the production cross sections, the azimuthal asymmetry[20,22] and Callan-Gross ratio[24] in heavy flavor leptoproduction are stable within the FFNS, both parametrically and perturbatively.

In the present talk, we discuss resummation of the mass logarithms of the type $\alpha_s \ln\left(Q^2/m^2\right)$ in heavy quark leptoproduction:[23,26]

$$l(\ell) + N(p) \to l(\ell - q) + Q(p_Q) + X[\overline{Q}](p_X). \tag{1}$$

The cross section of the reaction (1) may be written as

$$\frac{\mathrm{d}^3\sigma_{lN}}{\mathrm{d}x\mathrm{d}Q^2\mathrm{d}\varphi} = \frac{2\alpha_{em}^2}{Q^4}\frac{y^2}{1-\varepsilon}\Big[F_T(x,Q^2) + \varepsilon F_L(x,Q^2) + \varepsilon F_A(x,Q^2)\cos(2\varphi) + 2\sqrt{\varepsilon(1+\varepsilon)}F_I(x,Q^2)\cos\varphi\Big], \tag{2}$$

where $F_2(x,Q^2) = 2x(F_T + F_L)$ while the quantity ε measures the degree of the longitudinal polarization of the virtual photon in the Breit frame:[27] $\varepsilon = \frac{2(1-y)}{1+(1-y)^2}$. The quantities x, y, and Q^2 are the usual Bjorken kinematic variables while the azimuth φ is defined in Fig. 1.

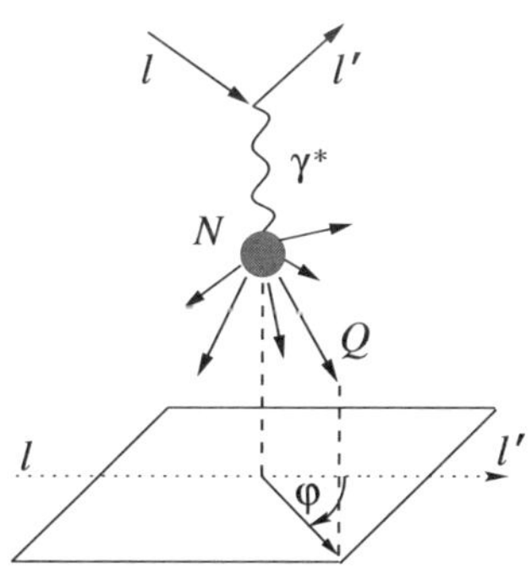

Fig. 1. Definition of the azimuthal angle φ in the nucleon rest frame.

In next sections, we will consider resummation of the mass logarithms for the quantities $R(x,Q^2)$ and $A(x,Q^2)$ defined as

$$R(x,Q^2) = \frac{F_L}{F_T}(x,Q^2), \qquad A(x,Q^2) = 2x\frac{F_A}{F_2}(x,Q^2). \tag{3}$$

[a]Note also the recent paper [25], where the perturbative stability of the QCD predictions for the charge asymmetry in top-quark hadroproduction has been observed.

2. Resummation for F_2 and Callan-Gross Ratio

To estimate the charm-initiated contributions, we use the ACOT(χ) VFNS proposed in Ref. [7].[b] In Figs. 2 and 3, we present numerical analysis of the NLO corrections[28] and charm-initiated contributions to the structure function $F_2(x, Q^2)$ and Callan-Gross ratio $R(x, Q^2) = F_L/F_T$ in charm leptoproduction. In our calculations, we use the CTEQ6M parametrization of the gluon and charm PDFs together with the value $m_c = 1.3$ GeV [13].[c] The default value of the factorization and renormalization scales is $\mu = \sqrt{4m_c^2 + Q^2}$.

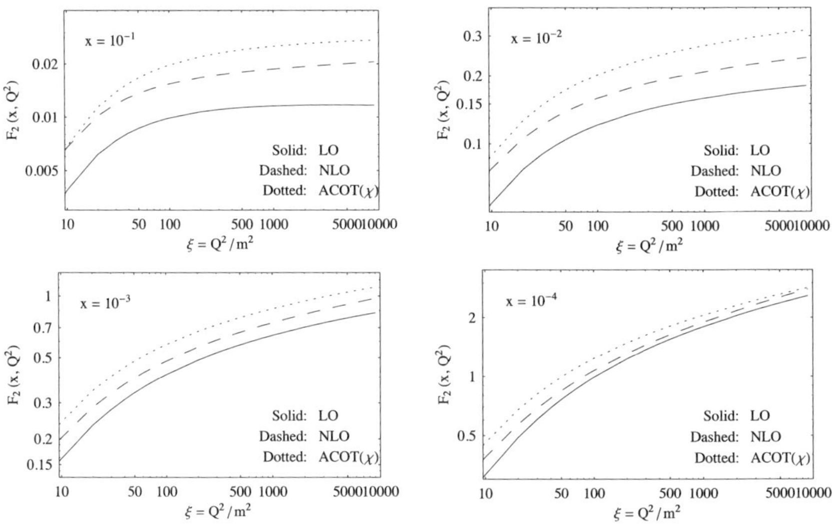

Fig. 2. Q^2 dependence of the structure function $F_2(x, Q^2)$ in charm leptoproduction at $x = 10^{-1}$, 10^{-2}, 10^{-3} and 10^{-4}. Plotted are the LO (solid lines) and NLO (dashed lines) FFNS predictions, as well as the ACOT(χ) VFNS (dotted curves) results.

One can see from Fig. 2 that, at $x \sim 10^{-1}$, both the radiative corrections and charm-initiated contributions to $F_2(x, Q^2)$ are large: they increase the LO FFNS results by approximately a factor of two for all Q^2. At the same

[b] For more details, see Refs. [23,26].

[c] Note that we convolve the NLO CTEQ6M distribution functions with both the LO and NLO partonic cross sections that makes it possible to estimate directly the degree of stability of the FFNS predictions under radiative corrections.

time, the relative difference between the dashed and dotted lines does not exceed 25% for $\xi = Q^2/m^2 < 10^3$.

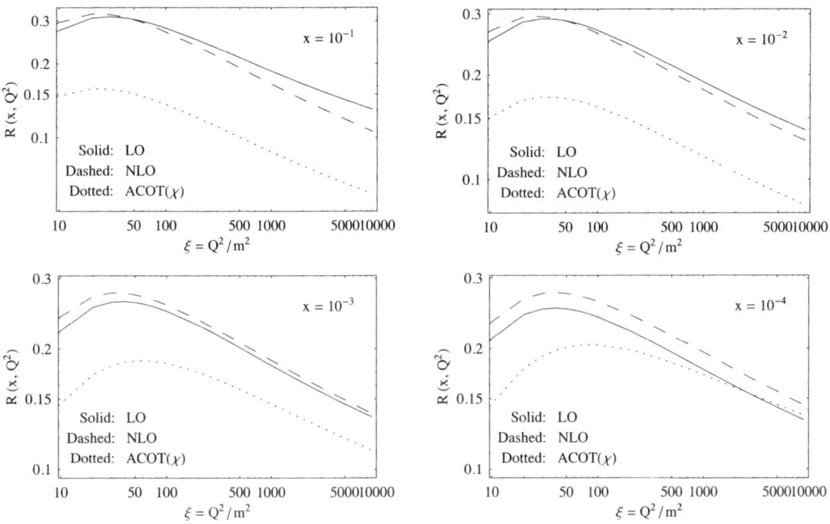

Fig. 3. Q^2 dependence of the Callan-Gross ratio, $R(x, Q^2) = F_L/F_T$, in charm leptoproduction at $x = 10^{-1}$, 10^{-2}, 10^{-3} and 10^{-4}. Plotted are the LO (solid lines) and NLO (dashed lines) FFNS predictions, as well as the ACOT(χ) VFNS (dotted curves) results.

Considering the corresponding predictions for the ratio $R(x, Q^2)$ presented in Fig. 3, we see that, in this case, the NLO and charm-initiated contributions are strongly different. The NLO corrections to $R(x, Q^2)$ are small, less than 15%, for $x \sim 10^{-3}$–10^{-1} and $\xi < 10^4$. On the other hand, the corresponding charm-initiated contributions are large: they decrease the LO FFNS predictions by about 50% practically for all values of $\xi > 10$. This is due to the fact that resummation of the mass logarithms has different effects on the structure functions $F_T(x, Q^2)$ and $F_L(x, Q^2)$ because they have different dependences on the quantities $\alpha_s^n \ln^k(Q^2/m^2)$. In particular, contrary to the transverse structure function, $F_T(x, Q^2)$, the longitudinal one, $F_L(x, Q^2)$, does not contain potentially large mass logarithms at both LO and NLO.[29,30] We conclude that, contrary to the the production cross sections, the Callan-Gross ratio $R(x, Q^2) = F_L/F_T$ could be good probe of the charm density in the proton at $x \sim 10^{-3}$–10^{-1}.

Note that this observation depends weakly on the PDFs we use. We have verified that all the recent CTEQ versions[11–13] of the PDFs lead to a sizeable reduction of the LO FFNS predictions for the ratio $R(x, Q^2)$.

As to the low $x \to 0$ behavior of the Callan-Gross ratio, this problem requires resummation of the BFKL[31] terms of the type $\ln(1/x)$ and will be considered in a forthcoming publication.

3. Resummation for Azimuthal Asymmetry

Fig. 4 shows the ACOT(χ) predictions for the asymmetry parameter $A(x, Q^2) = 2xF_A/F_2$ at several values of variable x: $x = 10^{-1}$, 10^{-2}, 10^{-3} and 10^{-4}. For comparison, we plot also the LO FFNS predictions (solid curves). Again, we use the CTEQ6M parametrization of PDFs, $m_c = 1.3$ GeV, and $\mu = \sqrt{4m_c^2 + Q^2}$.

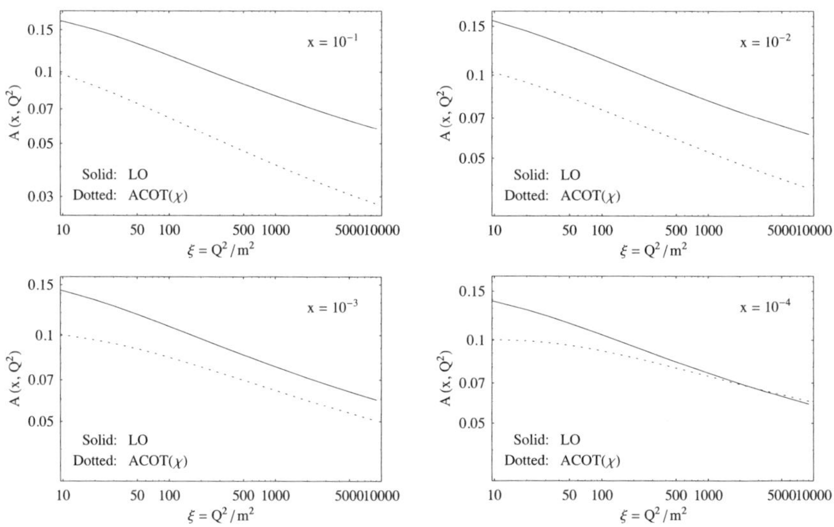

Fig. 4. Q^2 dependence of the azimuthal asymmetry, $A(x, Q^2) = 2xF_A/F_2$, in charm leptoproduction at $x = 10^{-1}$, 10^{-2}, 10^{-3} and 10^{-4}. Plotted are the LO FFNS (solid lines) and ACOT(χ) VFNS (dotted curves) results.

One can see from Fig. 4 the following properties of the azimuthal asymmetry. The mass logarithms resummation leads to a sizeable decreasing of the LO FFNS predictions for the $\cos(2\varphi)$-asymmetry. In the ACOT(χ) scheme, the charm-initiated contribution reduces the FFNS results for $A(x, Q^2)$ by about (30–40)% at $x \sim 10^{-2}$–10^{-1}. The origin of this reduction is the same as in the case of $R(x, Q^2)$: contrary to F_2, the azimuth dependent structure function F_A is safe in the limit $m^2 \to 0$ at least at LO.

Presently, the exact NLO predictions for the azimuth dependent structure function F_A are not available. However, in Ref. [23] the NLO corrections to the $\cos(2\varphi)$-asymmetry have been estimated within the so-called soft-gluon approximation at $Q^2 \lesssim m^2$.[d] It was demonstrated that large soft-gluon corrections to both F_A and F_2 cancel each other in their ratio $A = 2xF_A/F_2$ with a good accuracy. For this reason, one can exepct that the $\cos(2\varphi)$-asymmetry is also stable, both parametrically and perturbatively, in a wide kinematic range of variables x and Q^2 within the FFNS.

We have also analyzed how the VFNS predictions depend on the choice of subtraction prescription. In particular, the schemes proposed in Refs. [8,32] have been considered. We have found that, sufficiently above the production threshold, these subtraction prescriptions also reduce the LO FFNS results for the asymmetry by approximately (30–50)%.

One can conclude that impact of the mass logarithms resummation on the $\cos(2\varphi)$ asymmetry is essential at $x \sim 10^{-2}$–10^{-1} and therefore can be tested experimentally.

4. Conclusion

In the present talk, we compare the structure function F_2, Callan-Gross ratio $R = F_L/F_T$ and azimuthal asymmetry $A = 2xF_A/F_2$ in charm leptoproduction as probes of the charm content of the proton. To estimate the charm-initiated contributions, we used the ACOT(χ) VFNS[7] and recent CTEQ sets[11–13] of PDFs. Our analysis of the radiative and charm-initiated corrections indicates that, in a wide kinematic range, both contributions to the structure function $F_2(x, Q^2)$ have similar x and Q^2 behaviors. For this reason, it is difficult to estimate the charm content of the proton using only data on $F_2(x, Q^2)$.

The situation with the Callan-Gross ratio and azimuthal asymmetry seems to be more optimistic. Our analysis shows that resummation of the mass logarithms leads to reduction of the FFNS predictions for $R(x, Q^2)$ and $A(x, Q^2)$ by (30–50)% at $x \sim 10^{-2}$–10^{-1} and $Q^2 \gg m^2$. Taking into account the perturbative stability of the Callan-Gross ratio and azimuthal asymmetry within the FFNS,[23,26] this fact implies that the charm density in the proton can, in principle, be determined from high-Q^2 data on $R = F_L/F_T$ and $A = 2xF_A/F_2$.

[d]The soft-gluon approximation is unreliable for high $Q^2 \gg m^2$.

Concerning the experimental aspects, the quantities $R(x, Q^2)$ and $A(x, Q^2)$ in charm leptoproduction can be measured in future studies at the proposed EIC[33] and LHeC[34] colliders at BNL/JLab and CERN, correspondingly.

Acknowledgments

We thank S. I. Alekhin and J. Blümlein for providing us with fast code[28] for numerical calculations of the NLO DIS cross sections. The author is grateful to H. Avakian, S.J. Brodsky and C. Weiss for useful discussions. This work was supported in part by the ANSEF 2010 grant PS-2033.

References

1. S. J. Brodsky, P. Hoyer, C. Peterson, and N. Sakai, Phys. Lett. B **93**, 451 (1980); S. J. Brodsky, C. Peterson, and N. Sakai, Phys. Rev. D **23**, 2745 (1981).
2. S. J. Brodsky, *"Light-front QCD"* [hep-ph/0412101]; S. J. Brodsky, Few Body Syst. **36**, 35 (2005).
3. M. Franz, V. Polyakov, and K. Goeke, Phys. Rev. D **62**, 074024 (2000).
4. M. A. G. Aivazis, J. C. Collins, F. I. Olness, and W. -K. Tung, Phys. Rev. D **50**, 3102 (1994).
5. J. C. Collins, Phys. Rev. D **58**, 094002 (1998).
6. V. N. Gribov and L. N. Lipatov, Sov. J. Nucl. Phys. **15**, 438 (1972); Y. L. Dokshitzer, Sov. Phys. JETP **46**, 641 (1977); G. Altarelli and G. Parisi, Nucl. Phys. B **126**, 298 (1977).
7. W. -K. Tung, S. Kretzer, and C. Schmidt, J. Phys. G **28**, 983 (2002).
8. M. Kramer, F. I. Olness, and D. E. Soper, Phys. Rev. D **62**, 096007 (2000).
9. R. S. Thorne, Phys. Rev. D **73**, 054019 (2006); C. D. White and R. S. Thorne, Phys. Rev. D **74**, 014002 (2006); W. K. Tung, H .L. Lai, A. Belyaev, J. Pumplin, D. Stump, and C. -P. Yuan, JHEP **0702**, 053 (2007); S. Kretzer, H. L. Lai, F. I. Olness and W. -K. Tung, Phys. Rev. D **69**, 114005 (2004); R. S. Thorne and W. K. Tung, arXiv:0809.0714 [hep-ph]; P. M. Nadolsky, arXiv:0809.0945 [hep-ph].
10. J. Pumplin, Phys. Rev. D **73**, 114015 (2006); S. J. Brodsky, B. Kopeliovich, I. Schmidt, and J. Soffer, Phys. Rev. D **73**, 113005 (2006); J. Pumplin, H. L. Lai, and W. K. Tung, Phys. Rev. D **75**, 054029 (2007).
11. H.L. Lai, J. Huston, S. Kuhlmann, F. Olness, J. Owens, D. Soper, W.K. Tung and H. Weerts, Phys. Rev. D **55**, 1280 (1997).
12. H.L. Lai, J. Huston, S. Kuhlmann, J. Morfin, F. Olness, J.F. Owens, J. Pumplin and W.K. Tung, Eur. Phys. J. C **12**, 375 (2000).
13. J. Pumplin, D. R. Stump, J. Huston, H. L. Lai, P. Nadolsky, and W. K. Tung, JHEP **0207**, 012 (2002).
14. A. D. Martin, R. G. Roberts, W. J. Stirling, and R. S. Thorne, Phys. Lett. B **604**, 61 (2004).

15. E. Laenen and S. -O. Moch, Phys. Rev. D **59**, 034027 (1999).
16. N. Kidonakis, Phys. Rev. D **64**, 014009 (2001); N. Kidonakis, Phys. Rev. D **73**, 034001 (2006).
17. M. L. Mangano, P. Nason, and G. Ridolfi, Nucl. Phys. B **373**, 295 (1992); S. Frixione, M. L. Mangano, P. Nason, and G. Ridolfi, Nucl. Phys. B **412**, 225 (1994).
18. R. Vogt, Eur. Phys. J. ST **155**, 213 (2008).
19. N. Ya. Ivanov, A. Capella, and A. B. Kaidalov, Nucl. Phys. B **586**, 382 (2000).
20. N. Ya. Ivanov, Nucl. Phys. B **615**, 266 (2001).
21. N. Ya. Ivanov, P. E. Bosted, K. Griffioen, and S. E. Rock, Nucl. Phys. B **650**, 271 (2003).
22. N. Ya. Ivanov, Nucl. Phys. B **666**, 88 (2003).
23. L. N. Ananikyan and N. Ya. Ivanov, Phys. Rev. D **75**, 014010 (2007); L. N. Ananikyan and N. Ya. Ivanov, Nucl. Phys. B **762**, 256 (2007).
24. N. Ya. Ivanov and B. A. Kniehl, Eur. Phys. J. C **59**, 647 (2009).
25. L. G. Almeida, G. Sterman, and W. Vogelsang, Phys. Rev. D **78**, 014008 (2008).
26. N. Ya. Ivanov, Nucl. Phys. B **814**, 142 (2009).
27. N. Dombey, Rev. Mod. Phys. **41**, 236 (1969).
28. S. I. Alekhin and J. Bluemlein, Phys. Lett. B **594**, 299 (2004).
29. E. Laenen, S. Riemersma, J. Smith, and W. L. van Neerven, Nucl. Phys. B **392**, 162 (1993).
30. M. Buza, Y. Matiounine, J. Smith, R. Migneron, and W. L. van Neerven, Nucl. Phys. B **472**, 611 (1996).
31. E. A. Kuraev, L. N. Lipatov, and V. S. Fadin, Sov. Phys. JETP **44**, 443 (1976); I. I. Balitzki and L. N. Lipatov, Sov. J. Nucl. Phys. **28**, 822 (1978); L. N. Lipatov, Sov. Phys. JETP **63**, 904 (1986).
32. S. Kretzer and I. Schienbein, Phys. Rev. D **58**, 094035 (1998).
33. A. Deshpande, R. Milner, R. Venugopalan, and W. Vogelsang, Ann. Rev. Nucl. Part. Sci. **55**, 165 (2005).
34. J. B. Dainton, M. Klein, P. Newman, E. Perez, and F. Willeke, J. Inst. **1**, P10001 (2006).

SOME FUNDAMENTAL ISSUES IN EXCLUSIVE AND SEMI-INCLUSIVE REACTIONS

A. W. THOMAS

CSSM, School of Chemistry and Physics, University of Adelaide, Adelaide, SA 5005, Australia

We describe some of the recent developments in testing the Standard Model, beginning with the tremendous progress that has been made concerning our theoretical understanding of strangeness in the proton (the analogue of the Lamb shift in QED). We also summarise the contribution of parity violation, from electron-proton to neutrino-nucleus scattering, in testing the evolution of the weak charge of the proton predicted by the Standard Model. Finally, we describe recent progress in the calculation of fragmentation functions, which may prove important in the exploration of new observables such as transversity.

Keywords: Weak charge of the proton, Standard Model tests, parity violation, electron scattering, neutrino-nucleus interaction, charge symmetry violation, fragmentation functions

1. Introduction

It is almost universally believed that QCD is the correct theory for the strong interaction. This belief is being strengthened by every success of lattice QCD, with (for example) the masses of the octet baryons now being reproduced within a few percent.[1–4] Nevertheless, some observables hold a more distinguished place in the pantheon and amongst them the determination of the strange quark contributions to nucleon form factors is especially important, given that it involves entirely disconnected terms analogous to the Lamb shift that confirmed QED as the correct theory of electromagnetism. We briefly review the successful comparison of lattice predictions with experimental results before discussing another aspect of strangeness which recently yielded a surprise.

The study of parity violating electron scattering, which was used as a tool to determine the strange form factors of the proton, can also be used to test the electro-weak (EW) sector of the Standard Model itself. We shall review the progress in that area which included an increase by a factor of two in the lower mass limit above which new physics which can lead to parity violation is allowed.[5]

Another famous test of the EW sector involves neutrino-nucleus scattering. Some years ago the NuTeV collaboration reported a 3 σ deviation discrepancy[6] in the value of $sin^2\theta_W$ derived from neutrino nucleus scattering. This discrepancy is now understood[7] in terms of some physics which is fascinating in its own right - including a further hint that the quark substructure of the nucleon plays a critical role in nuclear structure. Another piece of the resolution of this anomaly involves charge symmetry violation in parton distribution functions, a topic on which we have strong theoretical constraints but little in the way of direct experimental interest. We stress the potential importance of an electron-ion collider in providing such evidence.

Finally, as the experimental capacity to study semi-inclusive reactions at high energy and momentum transfer improves, for example to put experimental constraints on quark orbital angular momentum,[8] it will be important to develop a first principles understanding of the fragmentation functions which are required. We briefly outline some promising steps made recently towards this end.

2. Strangeness in the nucleon

The correct description of pair creation in the vacuum is an essential part of any quantum field theory and the quantitative understanding of the effects of $e^+ e^-$ creation on the hyperfine structure of hydrogen, seen through the Lamb shift, was critical to the acceptance of QED. For QCD, obtaining a quantitative understanding of the effects of $q-\bar{q}$ loops on proton properties has a similar level of importance. Since there are no valence strange quarks in the proton, the *only* way that they can contribute to, say, the electromagnetic form factors of the proton is through such vacuum polarization effects – in the language of lattice QCD, "disconnected terms". (That is, the strange quark loop is disconnected from the valence quarks except through the exchange of infinitely many gluons.)

Over the past two decades there has been a tremendous experimental effort to measure the contribution of strange quarks to the electric and magnetic form factors of the proton using parity violating electron scattering (PVES).[9–11] Assuming charge symmetry, by measuring three form factors (e.g., $G^{\gamma p}_{E,M}(q^2), G^{\gamma n}_{E,M}(Q^2), G^{Z^0 p}_{E,M}(Q^2)$) one can solve for the individual (electric, magnetic) contributions of the three light quark flavors. While there was a some hint of a relatively large contribution from strange quarks initially, after a careful global analysis it is now clear that they contribute less than 3% of the mean-square charge radius and less than 4% of the magnetic moment of the proton (at 68% confidence level).[12]

The initial progress in lattice QCD used charge symmetry and chiral extrapolation of the valence quark magnetic moments (and charge radii), together with octet data to indirectly the strange quark form factors.[13] The results were both remarkably accurate, with $G_M^s(0.1\text{GeV}^2) = -0.046 \pm 0.019\mu_N$[14] and $G_E^s(0.1\text{GeV}^2) = 0.001 \pm 0.004 \pm 0.004$[15] and in excellent agreement with the global analysis once that was possible. Later measurements at higher Q^2 in both PVES[16] and lattice QCD[17] are consistent with these results. Consistent results have also been obtained in the first direct lattice QCD evaluation of these disconnected terms.[18]

In conclusion, not only are the results of lattice QCD in excellent agreement with the measurements of $G_{E,M}^s$ but this is a unique case in hadronic physics where the theoretical calculations are an order of magnitude more precise than the state of the art experiments – a remarkable triumph for QCD.

3. PVES as a test of the Standard Model

The global analysis of the data for PVES on the proton was so successful that it suggested that it might be possible to turn the problem around and actually use the PVES data to constrain the weak charge of the proton. At Jefferson Lab a precision experiment of this kind (Q_{weak}) had been planned for a long time[19] and is now underway but it was totally unexpected that the data taken to determine strangeness might actually yield a meaningful constraint. In fact, it turned out that the world data set was indeed able to provide a very important constraint on physics beyond the Standard Model (specifically, new physics that could lead to parity violation). Fig. 1 shows the constraint from the global analysis of data on the proton, compared with the best previous constraint from atomic parity violation in Cs. Because atomic Cs is almost isoscalar, its constraint is almost orthogonal to that from the proton and by combining them we obtain a *much* stronger result. Indeed, the analysis of Young *et al.*[5] decreased the allowed range of the low energy parameters $C_{1u,1d}$ by a factor of 5. In turn, this raised the lower mass limit above which new physics might be allowed from 0.4 TeV to over 0.9 TeV!

The success of this analysis also strengthens the case for Q_{weak}, which aims to measure the weak charge of the proton at low energy to about 4%. This would further raise the lower mass limit on new physics, to 2

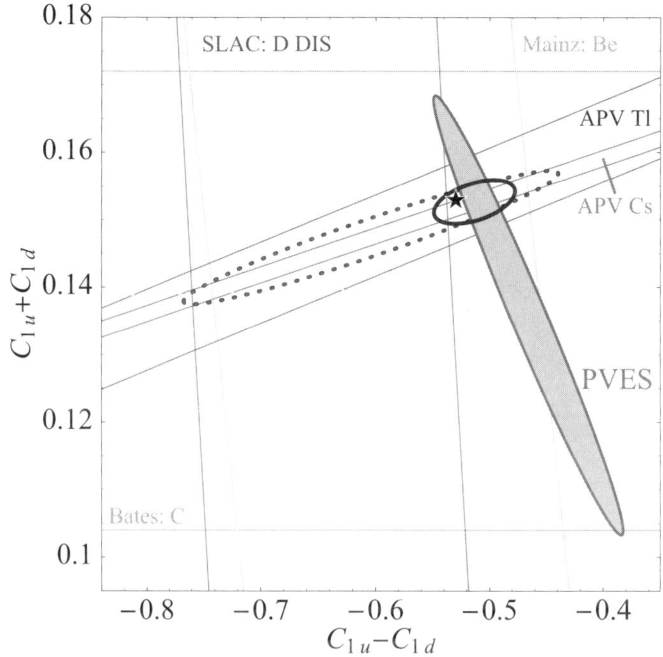

Fig. 1. Comparison of the constraints on the low energy parity violation coefficients $C_{1u,1d}$ arising from atomic parity violation (predominantly in Cs) – more or less horizontal ellipse – and PVES on hydrogen – more or less vertical ellipse. The constraint from combining the data into a global analysis is given by the smallest ellipse, which represents a factor 5 reduction in the allowed range – from Ref. [5]

TeV, which is both competitive with and complementary to the LHC. In particular, if the LHC were to find, for example, a new Z' in the relevant mass range before Q_{weak}, the latter would be able to pin down the relative couplings to the u and d quarks much more effectively. This, in turn, would help to distinguish between popular extensions of the Standard Model containing such a Z'.

3.1. *Radiative corrections revisited*

In order to test the Standard Model using atomic parity violation or PVES, in the way we just outlined, it is essential that the radiative corrections be accurately calculated. This is especially true for an experiment like Q_{weak} which aims for a major step up in precision. In this regard, it was recently realized[20] that a correction which can be ignored in atomic parity violation because it is down by order m_e/M_p,[21] must be included at energies of order a GeV (because $E_e/M_p \sim 1$). Their first estimate of this suggested that the errors might be large enough to put success of the Q_{weak} experiment in some doubt.

This issue was sufficiently important that it was urgently re-examined by Sibirtsev *et al.*[22] They followed Ref. [20] in using a dispersion relation to evaluate the correction. After correcting some minor problems in that work they did in fact find a 50% bigger result. Morover, by taking advantage of the tremendous amount of high quality data from CLAS and HERA, it was possible to constrain the uncertainties in the result quite adequately for the purposes of Q_{weak}.

4. The NuTeV anomaly resolved

About 8 years ago the NuTeV Collaboration reported a 3 σ discrepancy with the Standard Model for the Weinberg angle, $sin^2\theta_W$, determined from neutrino nucleus scattering. Their work was motivated by the Paschos-Wolfenstein relation, which (assuming the validity of charge symmetry) gives the ratio of the difference of neutral current cross sections for neutrinos and anti-neutrinos on an isoscalar target to the corresponding difference of charged current cross sections as $0.5 - sin^2\theta_W$.

4.1. *Charge symmetry violation*

Almost a decade before the result was published, theoretical studies of charge symmetry violation (CSV) in parton distribution functions (PDFs)[23,24] had shown that the $u - d$ mass difference led to sizeable CSV differences, $\delta d \equiv d^{p-n}$ and $\delta u \equiv u^p - d^n$, which were roughly equal in magnitude and of opposite sign. This effect, which was overlooked in the NuTeV analysis, was of a sign and magnitude to reduce the discrepancy with the Standard Model by more than 1 σ. The major discovery concerning this contribution since the original calculations is that for the moment relevant for NuTeV, namely $\langle x \rangle_{\delta d - \delta u}$, the correction is independent of the model to a very good approximation.[25]

More recently, another, independent source of CSV has been suggested.[26,27] The origin is physically very transparent, namely that the larger charge of the u quark means that it radiates photons (in a process analogous to the familiar DGLAP evolution within QCD) faster than a d quark. This depletes the momentum of u quarks faster than d quarks and hence leads to CSV. The only ambiguity is the scale at which this effect is set to zero – the starting scale. The groups who have worked on this usually set this CSV effect of QED radiation to zero at a very low scale and, of course, it will then increase logarithmically above that. At 20 GeV^2, the relevant scale for NuTeV, it is perhaps 2/3 of the size of the effect described above coming from the $u-d$ mass difference – and of the same sign. Altogether, allowing for the NuTeV functional which accounts for the details of the experimental acceptance, the total correction for to the value of $sin^2\theta_W$ is about half of the discrepancy.

4.2. *The iso-vector EMC effect*

The discovery by the European Muon Collaboration that the nuclear structure function differs dramatically from the free nucleon structure function in the valence region[28] suggests that the immersion in the nuclear medium changes the structure of the bound "nucleon".[29] Microscopic models of the atomic nucleus have been constructed, starting at the quark level with either the MIT bag[30] or the NJL model[31] including simulated confinement, which reproduce the existing data very well. These models also make important but as yet untested predictions that the change in the spin structure function in the nuclear medium should be even larger.

Only recently though has it been realized that the effect of the isovector potential, generated by the excess of neutrons over protons in the Fe target used by NuTeV, will fundamentally alter the structure functions of all of the nucleons in the target – shifting momentum from the u to the d quarks (because of the repulsion felt by the latter). This effect, to which we refer as the "iso-vector EMC effect",[32] *cannot* be removed by simply subtracting the contribution to the structure function of Fe from the excess neutrons – its effect remains for *all* of the nucleons in the nucleus!

The importance of this very simple piece of physics was only realized very recently and applied to the NuTeV data, again using the NuTeV functional. As it happens, the sign and magnitude are the same as the CSV contribution arising from the $u-d$ mass difference. The effect depends primarily on the difference in the densities of protons and neutrons in Fe and on the symmetry energy for nuclear matter, both of which are very well known. This effect is therefore more or less independent of model. Together with the CSV corrections described above, this correction totally eliminates the NuTeV anomaly. Indeed, the recent analysis of Bentz *et al.*[7] leads to $sin^2\theta_W = 0.2232 \pm 0.0013 \pm 0.0024$, versus $sin^2\theta_W = 0.2227 \pm 0.0004$ within the Standard Model. The result is compared with the standard evolution of the Weinberg angle[33] and other determinations of it in Fig. 2.

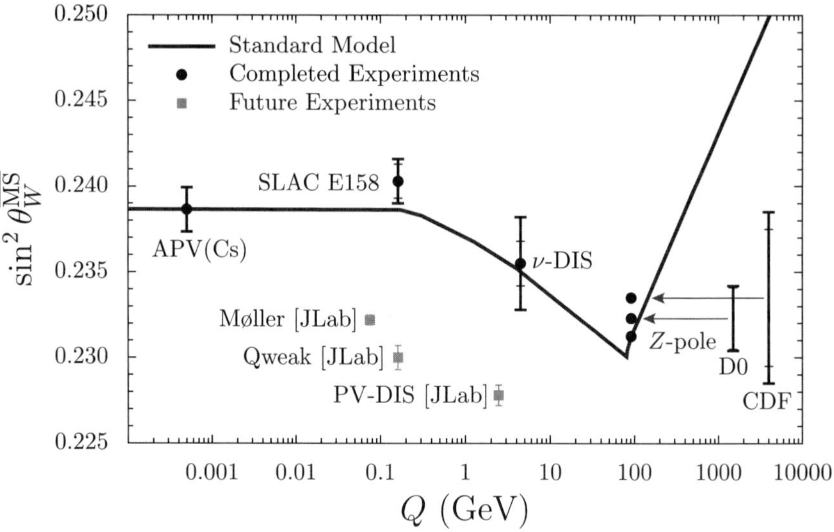

Fig. 2. Evolution of the Weinberg angle with scale, from the Z-pole probed so accurately at LEP to the lower energy regions associated with NuTeV, PVES and atomic parity violation. The neutrino-nucleus point shows the recent re-analysis of the NuTeV data by Bentz *et al.*[7]

The remaining problem in reducing the errors on this determination arise from the possible asymmetry between s and $\bar{s}$ quarks in the proton. The best data comes from NuTeV but the relevant integral, $\int dx\, x\,(s(x) - \bar{s}(x))$, is highly dependent on the assumed position of the zero in $s - \bar{s}$. The analysis reported above is based on a zero above $x = 0.1$, as strongly preferred on theoretical grounds. It really does need to be a high priority to improve the experimental determination of $s - \bar{s}$, as well as to pin down the extent of CSV in the nucleon's PDFs.[34] For the latter the study of charge current scattering at a future electron-ion collider[35] could be definitive. Finally, a study of the dependence of nuclear structure functions on $N \neq Z$ is not only essential with regard to the analysis of the NuTeV data but to our understanding of the role of quark and gluon degrees of freedom in nuclear structure.

5. Semi-inclusive scattering and the fragmentation functions

Semi-inclusive deep-inelastic scattering has already proven extremely valuable in defining some key aspects of the flavor structure of the nucleon. For example, the data from Hermes concerning the $\bar{d}-\bar{u}$ asymmetry provided important confirmation of the Drell-Yan data for this quanitity which is crucial to understanding the chiral structure of the nucleon. When it comes to fundamental problems such as the distribution of spin and orbital angular momentum on quarks in the nucleon and particularly the exploitation of generalized parton distributions and transversity, it will be vital to understand the unfavoured as well as the favoured fragmentation functions as well as their dependence on transverse momentum. Then it will no longer be enough to rely on phenomenological recipes but one will really need guidance from soundly based theoretical calculations.

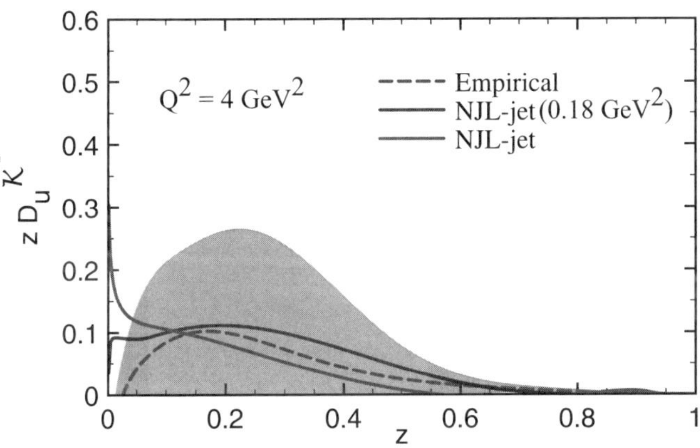

Fig. 3. Unfavoured kaon fragmentation function calculated within the NJL model - from Ref. [36].

A significant step in this direction was taken recently using the NJL model.[37] For the first time it was possible to calculate the absolute value of the fragmentation functions for $u \to \pi^+$ and π^-, without the need for a phenomenological rescaling. The essential ingredient in this success was the summation over the chain of processes in which an arbitrary number of pions was omitted but only one detected. Since that first paper, it as been possible to include the strange quark as well, with the decay chain including emission of arbitrary numbers of pions and kaons.[36] Figures 3 and 4 show, respectively, the favoured and unfavoured fragmentation functions for a u quark to a kaon, where we stress that the normalization has not been chosen to match data but comes from the calculation.

There are many improvements to be made in this calculation of the

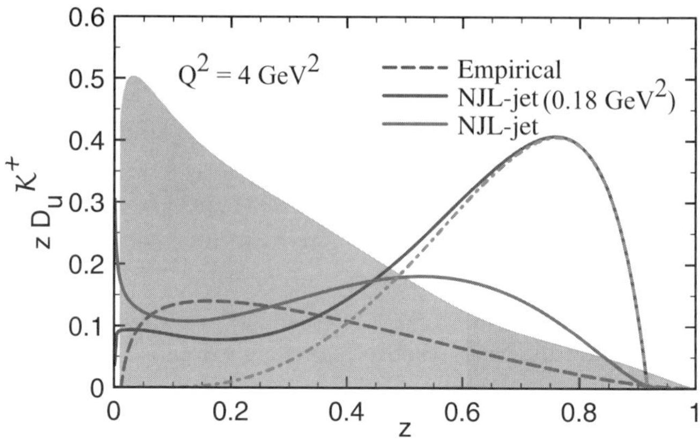

Fig. 4. Favoured kaon fragmentation function calculated within the NJL model - from Ref. [36].

fragmentation functions – such as including vector meson production, fragmentation to baryons etc. Nevertheless, the level of agreement with the phenomenological distributions is already very promising. We also stress that while the current calculations correspond to the Bjorken limit, the application of Monte Carlo techniques to the same formalism should enable us to investigate changes in the kinematics of real experiments. Since some of the most interesting future studies will need accurate estimates of the unfavoured fragmentation functions as well as the transverse momentum and spin dependence, the development of the techniques developed in Refs. [36,37] may be expected to be very helpful if not essential.

6. Conclusion

We have reviewed some remarkable examples of recent tests of the Standard Model, from the strangeness in the nucleon (analogous to the Lamb shift in QED) to the application of parity violating electron scattering in a more accurate measurement of the Weinberg angle. We saw that the existence of accurate modern data on photoproduction, especially from Jefferson Lab, as well as the deep-inelastic data from HERA, has helped enormously in the computation of the radiative corrections for parity violation. With regard to the outstanding deviation from the Standard Model, the NuTeV anomaly, we saw that charge symmetry violation plus a new discovery concerning nuclear structure functions, namely the iso-vector EMC effect completely removes any significant discrepancy.

Finally, we noted the necessity for many future studies of fundamental importance for understanding the strong interaction of having reliable fragmentation functions – including their transverse momentum dependence. We saw that there has recently been a major development in the microscopic derivation of these functions, with the potential for future very close collaboration with the relevant experimental groups.

Acknowledgments

This work was supported by the University of Adelaide and by the Australian Research Council through an Australian Laureate Fellowship.

References

1. R. D. Young and A. W. Thomas, Phys. Rev. D **81**, 014503 (2010) [arXiv:0901.3310 [hep-lat]].
2. S. Durr *et al.*, Science **322**, 1224 (2008) [arXiv:0906.3599 [hep-lat]].
3. S. Aoki *et al.* [PACS-CS Collaboration], Phys. Rev. D **79**, 034503 (2009) [arXiv:0807.1661 [hep-lat]].
4. H. W. Lin *et al.* [Hadron Spectrum Collaboration], Phys. Rev. D **79**, 034502 (2009) [arXiv:0810.3588 [hep-lat]].
5. R. D. Young, R. D. Carlini, A. W. Thomas and J. Roche, Phys. Rev. Lett. **99**, 122003 (2007) [arXiv:0704.2618 [hep-ph]].
6. G. P. Zeller *et al.* [NuTeV Collaboration], Phys. Rev. Lett. **88**, 091802 (2002). [hep-ex/0110059].
7. W. Bentz *et al.*, [arXiv:0908.3198 [nucl-th]].
8. A. W. Thomas, Phys. Rev. Lett. **101**, 102003 (2008). [arXiv:0803.2775 [hep-ph]].
9. F. E. Maas, K. Aulenbacher, S. Baunack *et al.*, Phys. Rev. Lett. **94**, 152001 (2005). [nucl-ex/0412030].
10. K. A. Aniol *et al.* [HAPPEX Collaboration], Phys. Rev. Lett. **96**, 022003 (2006). [nucl-ex/0506010].
11. D. S. Armstrong *et al.* [G0 Collaboration], Phys. Rev. Lett. **95**, 092001 (2005). [nucl-ex/0506021].
12. R. D. Young, J. Roche, R. D. Carlini and A. W. Thomas, Phys. Rev. Lett. **97**, 102002 (2006) [arXiv:nucl-ex/0604010].
13. D. B. Leinweber, A. W. Thomas, Phys. Rev. **D62**, 074505 (2000). [hep-lat/9912052].
14. D. B. Leinweber, S. Boinepalli, I. C. Cloet *et al.*, Phys. Rev. Lett. **94**, 212001 (2005). [hep-lat/0406002].
15. D. B. Leinweber, S. Boinepalli, A. W. Thomas *et al.*, Phys. Rev. Lett. **97**, 022001 (2006). [hep-lat/0601025].
16. S. Baunack, K. Aulenbacher, D. Balaguer Rios *et al.*, Phys. Rev. Lett. **102**, 151803 (2009). [arXiv:0903.2733 [nucl-ex]].

17. P. Wang, D. B. Leinweber, A. W. Thomas *et al.*, Phys. Rev. **C79**, 065202 (2009). [arXiv:0807.0944 [hep-ph]].
18. T. Doi, M. Deka, S. -J. Dong *et al.*, Phys. Rev. **D80**, 094503 (2009). [arXiv:0903.3232 [hep-ph]].
19. W. T. H. van Oers [Qweak Collaboration], Nucl. Phys. **A805**, 329-337 (2008). [arXiv:0708.1972 [nucl-ex]].
20. M. Gorchtein, C. J. Horowitz, Phys. Rev. Lett. **102**, 091806 (2009). [arXiv:0811.0614 [hep-ph]].
21. W. J. Marciano, A. Sirlin, Phys. Rev. **D27**, 552 (1983).
22. A. Sibirtsev *et al.*, [arXiv:1002.0740 [hep-ph]].
23. E. N. Rodionov, A. W. Thomas, J. T. Londergan, Mod. Phys. Lett. **A9**, 1799-1806 (1994).
24. E. Sather, Phys. Lett. **B274**, 433-438 (1992).
25. J. T. Londergan, A. W. Thomas, Phys. Rev. **D67**, 111901 (2003). [hep-ph/0303155].
26. A. D. Martin, R. G. Roberts, W. J. Stirling *et al.*, Eur. Phys. J. **C39**, 155-161 (2005). [hep-ph/0411040].
27. M. Gluck, P. Jimenez-Delgado, E. Reya, Phys. Rev. Lett. **95**, 022002 (2005). [hep-ph/0503103].
28. J. J. Aubert *et al.* [European Muon Collaboration], Phys. Lett. **B123**, 275 (1983).
29. D. F. Geesaman, K. Saito, A. W. Thomas, Ann. Rev. Nucl. Part. Sci. **45**, 337-390 (1995).
30. K. Saito, A. Michels, A. W. Thomas, Phys. Rev. **C46**, 2149-2152 (1992).
31. I. C. Cloet, W. Bentz, A. W. Thomas, Phys. Lett. **B642**, 210-217 (2006). [nucl-th/0605061].
32. I. C. Cloet, W. Bentz, A. W. Thomas, Phys. Rev. Lett. **102**, 252301 (2009). [arXiv:0901.3559 [nucl-th]].
33. J. Erler, M. J. Ramsey-Musolf, Phys. Rev. **D72**, 073003 (2005). [hep-ph/0409169].
34. J. T. Londergan, J. C. Peng, A. W. Thomas, [arXiv:0907.2352 [hep-ph]].
35. A. W. Thomas, [arXiv:0907.4785 [hep-ex]].
36. H. H. Matevosyan, A. W. Thomas, W. Bentz, [arXiv:1004.3075 [nucl-th]].
37. T. Ito, W. Bentz, I. C. Cloet *et al.*, Phys. Rev. **D80**, 074008 (2009). [arXiv:0906.5362 [nucl-th]].

Participants list

Afanasev, Andrei
Hampton University /JLab
afanas@jlab.org

Avakian, Harut
Jefferson Lab
avakian@jlab.org

Bedir, Islam
ODU
ibedir@gmail.com

Bland, Les
BNL
bland@bnl.xxx

Braun, Vladimir
Regensburg University
vladimir.braun@physik.uni-regensburg.de

Brodsky, Stanley
SLAC
sjbth@slac.stanford.edu

Burkardt, Matthias
NMSU
burkardt@nmsu.edu

Burkert, Volker
Jefferson Lab
burkert@jlab.org

Camsonne, Alexandre
Jefferson Laboratory
camsonne@jlab.org

Canan, Mustafa
Old Dominion University
mcanan@odu.edu

Carlson, Carl
William & Mary
carlson@physics.wm.edu

Chen, Jian-ping
Jefferson Lab
jpchen@jlab.org

Cherednikov, Igor
INFN Cosenza
igor.cherednikov@jinr.ru

Cosyn, Wim
FIU/UGent
wim.cosyn@gmail.com

Dalton, Mark
University of the Witwatersrand
dalton@jlab.org
de Jager , Cornelis
Jefferson Lab
kees@jlab.org

D'Hose, Nicole
CEA- Scalay
nicole.dhose@cea.fr

Elouadrhiri, Latifa
Jefferson Lab
latifa@jlab.org

Ent, Rolf
Jefferson Lab
eljones@jlab.org

Fradi, Ahmed
IPN-ORSAY
fradi@ipno.in2p3.fr

Fuchey, Eric
Universite Blaise Pascal
efuchey@jlab.org

Gamberg, Leonard
Penn State Berks
lpg10@psu.edu

Gilfoyle, Gerard
University of Richmond
gilfoyle@jlab.org

Girod, Francois-Xavier
Jefferson Lab
fxgirod@jlab.org

Goldstein, Gary
Tufts University
gary.goldstein@tufts.edu

Granados, Carlos
Florida International University
cgran005@fiu.edu

Guzey, Vadim
Jefferson Lab
vguzey@jlab.org

Haegler, Philipp
TU Munich
phaegler@ph.tum.de

Higinbotham, Douglas
Jefferson Lab
doug@jlab.org

Horn, Tanja
Catholic University of America
hornt@jlab.org

Huber, Garth
University of Regina
huberg@uregina.ca

Ivanov, Nikolay
Yerevan Physics Institute
nivanov@jlab.org

Jones, Mark
Jefferson Lab
jones@jlab.org

Joo, Kyungseon
University of Connecticut
kjoo@phys.uconn.edu

Kageya, Tsuneo
Jefferson Lab
kageya@jlab.org

Khandaker, Mahbub
Norfolk State U / JLab
mahbub@jlab.org
Khetarpal, Puneet
Rensselaer Polytechnic Institute
puneetk@jlab.org

Kroll, Peter
Universitaet Wuppertal
kroll@physik.uni-wuppertal.de

Kubarovskiy, Alexey
RPI
alexeyk@jlab.org

Kubarovsky, Valery
Jefferson Lab
vpk@jlab.org

Kunkel, Michael
ODU
mkunkel@jlab.org

Laget, Jean-Marc
JLab
laget@jlab.org

Lansberg, Jean-Philippe
Ecole Polytechnique
Jean-Philippe.Lansberg@
cpht.polytechnique.fr

LeRose, John
JLab
lerose@jlab.org

Lin, Huey-Wen
University of Washington
hwlin@phys.washington.edu

Liuti, Simonetta
University of Virginia
sl4y@virginia.edu

Lorce, Cedric
Mainz University
lorce@kph.uni-mainz.de

Luo, Wei
Lanzhou University
weiluo@jlab.org

Meziane, Mehdi
The College of William and Mary
mezianem@jlab.org

Mkrtchyan, Arthur
YerPhi
mkrtchya@jlab.org

Mkrtchyan, Hamlet
JLab/YerPhI
hamlet@jlab.org

Mokeev, Victor
JLAB
mokeev@jlab.org

Montgomery, Hugh
Jefferson Lab
mont@jlab.org

Moreno, Brahim
CEA Saclay
brahim.moreno@cea.fr

Moutarde, Hervé
CEA Saclay
herve.moutarde@cea.fr

Mueller, Dieter
Ruhr-University Bochum
dieter.mueller@physik.uni-
regensburg.de

Musch, Bernhard
Jefferson Lab
bmusch@jlab.org

Nadel-Turonski, Pawel
Jefferson Lab
turonski@jlab.org

Nepali, Chandra
Old Dominion University
cnepali@odu.edu

Pappalardo, Luciano
INFN - Ferrara
pappalardo@fe.infn.it

Park, Kijun
Jefferson Lab
parkkj@jlab.org

Pasquini, Barbara
University of Pavia & INFN, Pavia
pasquini@pv.infn.it

Perdrisat, Charles
William and Mary
perdrisa@jlab.org

Prokudin, Alexei
JLAB
prokudin@jlab.org

Puckett, Andrew
Los Alamos National Laboratory
puckett@jlab.org

Radyushkin, Anatoly
ODU/JLab
radyush@jlab.org

Ramalho, Gilberto
Instituto Superior Tecnico, Lisbon
gilberto.ramalho@cftp.ist.utl.pt

Richards, David
Jefferson Lab
dgr@jlab.org

Roberts, Craig
Argonne National Laboratory
cdroberts@anl.gov

Roche, Julie
Ohio University
rochej@ohio.edu

Rossi, Patrizia
INFN-LNF
rossi@lnf.inf.it

Saha, Arun
Jefferson Lab
saha@jlab.org

Saylor, Nicholas
Rensselaer Polytechnic Institute
saylon@rpi.xxx

Schulte, Elaine
Temple University
schulte@temple.edu

Schweitzer, Peter
UConn
peter.schweitzer@phys.uconn.xxx

Sterman, George
Stony Brook University
george.sterman@stonybrook.edu

Stoler, Paul
Rensselaer Polytechnic Institute
stoler@rpi.edu

Strikman, Mark
Penn State University
strikman@phys.psu.edu

Szymanowski, Lech
Soltan Institute for Nuclear Studies,
Warsaw
lechszym@fuw.edu.pl

Tandogan, Asli
ODU
atand002@0du.edu

Thomas, Anthony
CSSM, School of Chemistry
and Physics,
University of Adelaide
awthomas@jlab.org

Vanderhaeghen, Marc
University Mainz
marcvdh@kph.uni-mainz.de

Weiss, Christian
Jefferson Lab
weiss@jlab.org

Wojtsekhowski, Bogdan
TJNAF
bogdanw@jlab.org

Wollny, Heiner
University of Freiburg
heiner.wollny@cern.ch

Zhang, Jixie
Jlab
jixie@jlab.org

Zhao, Zhiwen
Univ. of South Carolina
zwzhao@jlab.org

WILL DETMOLD
CARL CARLSON
A205
Exclusive Reactions

HIGH MOMENTUM TRANSFER
May 18 - 21, 2010
Tuesday, May 18, 2010
Latifa
Elouadrhiri
Paul
Stoler

Jefferson Lab
Site Wide Information System
Experimental Hall B

AT A GLANCE
Bogdan
Wojtsekhowski
Jefferson Lab
Charles
Perdrisat

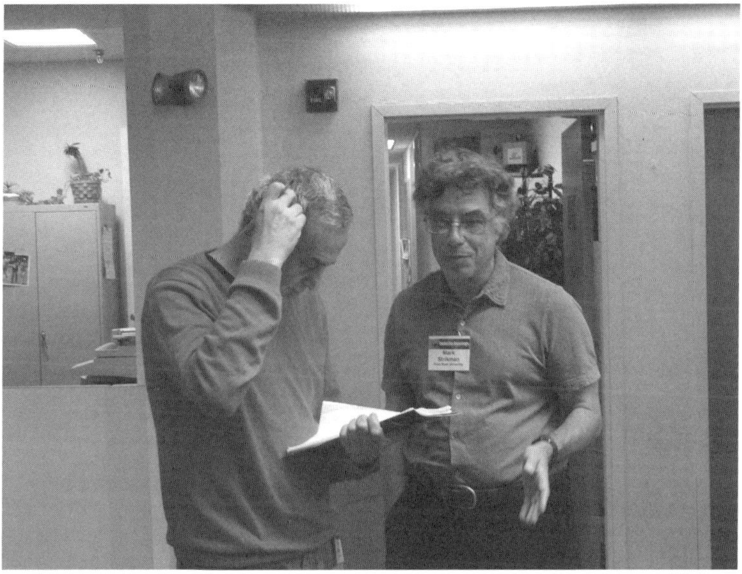

GILDAN
Matthias
Burkardt

GLANCE

AT A GLANCE

EXIT

AT A GLANCE

Hall A
Hall B
Hall C

George
Sterman